Learning and Analytics in Intelligent Systems

Volume 20

The main aim of the series is to make available a publication of books in hard copy form and soft copy form on all aspects of learning, analytics and advanced intelligent systems and related technologies. The mentioned disciplines are strongly related and complement one another significantly. Thus, the series encourages cross-fertilization highlighting research and knowledge of common interest. The series allows a unified/integrated approach to themes and topics in these scientific disciplines which will result in significant cross-fertilization and research dissemination. To maximize dissemination of research results and knowledge in these disciplines, the series publishes edited books, monographs, handbooks, textbooks and conference proceedings.

More information about this series at http://www.springer.com/series/16172

C. Kiran Mai · B. V. Kiranmayee ·
Margarita N. Favorskaya ·
Suresh Chandra Satapathy · K. Srujan Raju
Editors

Proceedings of International Conference on Advances in Computer Engineering and Communication Systems

ICACECS 2020

Springer

Editors
C. Kiran Mai
Department of Computer Science
and Engineering
VNR Vignana Jyothi Institute
of Engineering and Technology
Hyderabad, Telangana, India

B. V. Kiranmayee
Department of Computer Science
and Engineering
VNR Vignana Jyothi Institute
of Engineering and Technology
Hyderabad, Telangana, India

Margarita N. Favorskaya
Department of Informatics and Computer
Techniques
Siberian State Aerospace University
of Science and Technology
Krasnoyarsk, Krasnoyarsk Territory, Russia

Suresh Chandra Satapathy
School of Computer Engineering
KIIT Deemed to be University
Bhubaneswar, Odisha, India

K. Srujan Raju
CMR Technical Campus
Hyderabad, Telangana, India

ISSN 2662-3447 ISSN 2662-3455 (electronic)
Learning and Analytics in Intelligent Systems
ISBN 978-981-15-9295-9 ISBN 978-981-15-9293-5 (eBook)
https://doi.org/10.1007/978-981-15-9293-5

This Springer imprint is published by the registered company Springer Nature Singapore Pte Ltd.
The registered company address is: 152 Beach Road, #21-01/04 Gateway East, Singapore 189721, Singapore

ICACECS20 Organizing Committee

Patrons

Dr. D. N. Rao, President, Vignana Jyothi Society.
Sri K. Harishchandra Prasad, General Secretary, Vignana Jyothi Society.

Chief Patrons

Dr. C. D. Naidu, Principal, VNR VJIET.
Dr. B. Chennakesava Rao, Director-Advancement, VNR VJIET.
Dr. A. Subhananda Rao, Director-Research & Development, VNR VJIET.

Programme Chairs

Dr. C. Kiran Mai, VNR VJIET, Hyderabad, India.
Dr. B. V. Kiranmayee, VNR VJIET, Hyderabad, India.

Programme Co-chairs

Dr. A. Brahmananda Reddy, VNR VJIET, Hyderabad, India.
Dr. P. Subhash, VNR VJIET, Hyderabad, India.
Dr. Deepak Sukheja, VNR VJIET, Hyderabad, India.

General Chair

Dr. Raghava Rao Mukkamala, CBDA, CBS, Denmark.

Honorary Chairs

Dr. Lakhmi C. Jain, University of Technology Sydney, Australia & founder KES International.
Dr. A. Govardhan, Rector, JNTUH, Hyderabad, India.

Editorial Board

Dr. C. Kiran Mai, *VNR VJIET, Hyderabad, India.*
Dr. B. V. Kiranmayee, *VNR VJIET, Hyderabad, India.*
Dr. Margarita N. Favorskaya, *Siberian State Aerospace University, Russian federation.*
Dr. Suresh Chandra Satapathy, *KIIT University, Bhuvaneswar, India.*
Dr. K. Srujan Raju, *CMRTC, Hyderabad, India.*

Programme Committee

Dr. G. Ramesh Chandra, CSE, VNR VJIET.
Dr. S. Nagini, CSE, VNR VJIET.
Dr. P. V. Siva Kumar, CSE, VNRVJIET.
Mrs. V. Baby, CSE, VNR VJIET.
Dr. T. Sunil Kumar, CSE, VNR VJIET.
Dr. Y. Sagar, CSE, VNR VJIET.
Mr. M. Gangappa, CSE, VNR VJIET.
Dr. P. Subhash, CSE, VNR VJIET.

Advisory Committee

Dr. Aynur Unal, Director, Member of the Executive Team, UK.
Dr. Pawan Lingras, SAINT MARY'S University, Canada.
Dr. Raghava Rao Mukkamala, CBDA, CBS, Denmark.
Dr. Rajeev Srivastava IIT BHU, India.

Dr. C. Krishna Mohan, IIT, Hyderabad, India.
Dr. N. S. Choudary, IIT, Indore, India.
Dr. P. Premchand, Osmania University.
Dr. Rajkamal Formar-VC, DAVV, Indore, India.
Dr. R. B. V. Subramanyam, NITW, Warangal, India.
Dr. O. B. V Ramanaiah, JNTUH, Hyderabad, India.
Dr. Supreethi K. P., JNTUH, Hyderabad, India.
Dr. Sujatha Banothu, Osmania University, Hyderabad, India.
Dr. I. L. Narasimha Rao, Cyber Peace Foundation, New Delhi, India.
Dr. Y. Padmasai, VNR VJIET, Hyderabad, India.
Dr. G. Ramesh Chandra, VNR VJIET, Hyderabad, India.
Dr. G. Suresh Reddy, VNR VJIET, Hyderabad, India.
Dr. K. Anuradha, VNR VJIET, Hyderabad, India.
Dr. Poonam Upadhyay, VNR VJIET, Hyderabad, India.
Dr. R. Manjula Sri, VNR VJIET, Hyderabad, India.
Dr. A. Mallika, VNR VJIET, Hyderabad, India.
Dr. G. Srinivasa Gupta, VNR VJIET, Hyderabad, India.
Dr. Srinivasa Rao T., VNR VJIET, Hyderabad, India.
Dr. T. Jayashree, VNR VJIET, Hyderabad, India.
Dr. Sankar K. Pal, Indian Statistical Institute, Calcutta, India.
Dr. KrithiRamamritham, IIT Bombay, India.
Dr. Abdul Sattar, Griffith University, Australia.
Dr. N. Mangathayaru, VNRVJIET, Hyderabad, India.
Anil Sukheja, Scientist “E” ISRO Ahmadabad, India.

Technical Committee

Dr. Umesh Kumar Singh, Vikram University, Ujjain, India.
Dr. Padmaja Joshi, CDAC, India.
Dr. Pilli Emmanuel Subhakar, MNIT, Jaipur, India.
Dr. M. B. Srinivas, BITS Pilani, Hyderabad, India.
Dr. K. Seetha Ram Babu, Chairman CSI-Hyderabad Chapter, India.
Dr. A. V. Krishna Prasad, Secretary CSI-Hyderabad Chapter, India.
Dr. D. Vasumathi, JNTUH, Hyderabad, India.
Dr. A. P. Siva Kumar, JNTUCEA, Andhra Pradesh, India.
Dr. T. V. Rajanikanth, SNIST, Hyderabad, India.
Dr. Dolly Sharma, Shiv Nadar University, UP, India.
Dr. Virander Saraswat, Dr. B. R. Ambedkar University, Agra, India.
Dr. T. Kishore Kumar, NIT Warangal, India.
Dr. B. Vishnu Vardhan, JNTU Mantani, Peddapalli, India.
Dr. N. Sandhya, VNRVJIET, Hyderabad, India.
Dr. P. Neelakantan, VNRVJIET, Hyderabad, India.
Dr. K. Vijaya Kumar, CMRCET, Hyderabad, India.

Dr. P. Vijaya Pal Reddy, Matrusri Engineering college, Hyderabad, India.
Dr. M. Raja Sekar, VNRVJIET, Hyderabad, India.
Dr. G. L. Prajapat, IET, Devi Ahilya University, Indore.
Dr. Parag Parandkar Reva University, Bangalore, India.
Dr. S. Nagini, VNRVJIET, Hyderabad, India.
Ms. V. Baby, VNRVJIET, Hyderabad, India.
Dr. P. V. Siva Kumar, VNRVJIET, Hyderabad, India.
Dr. D. Raman, VCE, Shamshabad, Hyderabad, India.
Dr. T. Sunil Kumar, VNRVJIET, Hyderabad, India.
Dr. Y. Sagar, VNRVJIET, Hyderabad, India.
Dr. P. Subhash, VNRVJIET, Hyderabad, India.
Dr. K. Srinivas, VNRVJIET, Hyderabad, India.
Dr. A. Kousar Nikhath, VNRVJIET, Hyderabad, India.
Dr. Niranjan Shrivastava IMS DAVV, Indore, India.
Dr. Uttam Kumar Roy, Jadavpur University, Kolkata, India.
Dr. D. Srinivasa Rao, VNRVJIET, Hyderabad, India.
Dr. G. Madhu, VNRVJIET, Hyderabad, India.
Dr. K. Srinivas, VNRVJIET, Hyderabad, India.
Dr. Chalumuru Suresh, VNRVJIET, Hyderabad, India.
Dr. P. Kishore, VNRVJIET, Hyderabad, India.
Dr. S. Rajendra Prasad, VNRVJIET, Hyderabad, India.
Dr. C. Kiran, VNRVJIET, Hyderabad, India.
Dr. S. Pranavanand, VNRVJIET, Hyderabad, India.
Dr. D. Ravi Kumar, VNRVJIET, Hyderabad, India.
Ms. D. N. Vasundhara, VNRVJIET, Hyderabad, India.
Mr. G. S. Ramesh, VNRVJIET, Hyderabad, India.
Ms. Y. Bhanusree, VNRVJIET, Hyderabad, India.
Mr. G. Nagaraju, VNRVJIET, Hyderabad, India.
Ms. K. Jhansi Lakshmi Bai, VNRVJIET, Hyderabad, India.
Mr. R. Ranthi Kumar, VNRVJIET, Hyderabad, India.
Ms. L. Indira, VNRVJIET, Hyderabad, India.
Ms. Priyabhatnagar, VNRVJIET, Hyderabad, India.
Ms. Krithi Ohri, VNRVJIET, Hyderabad, India.

Organizing Committee

Dr. A. Brahmananda Reddy, CSE, VNR VJIET.
Dr. Deepak Sukheja, CSE, VNR VJIET.
Dr. P. V. Siva Kumar, CSE, VNR VJIET.
Dr. P. Subhash, CSE, VNR VJIET.
N. V. Sailaja, CSE, VNR VJIET.
D. N. Vasundara, CSE, VNR VJIET.
G. S. Ramesh, CSE, VNR VJIET.

P. Radhika, CSE, VNR VJIET.
G. Nagaraju, CSE, VNR VJIET.
R. Kranthi Kumar, CSE, VNR VJIET.
Priya Bhatnagar, CSE, VNR VJIET.
Kriti Ohri, CSE, VNR VJIET.
V. Hareesh, CSE, VNR VJIET.

Publication and Proceedings Committee

Dr. G. Ramesh Chandra, CSE, VNR VJIET.
Dr. P. Subhash, CSE, VNR VJIET.
Dr. K. Srinivas, CSE, VNR VJIET.
Chenna Basamma, CSE, VNR VJIET.

Review Committee

Dr. Mohammed Shafiul Alam Khan, Royal Holloway University of London.
Dr. Hikmat Ullah Khan, COMSATS Institute of Information Technology, Wah Campus, Islamabad.
Dr. Radhakrishnan Palanikumar, King Khalid University, Saudi Arabia.
Dr. KanwalinderjitGagneja, Florida Polytechnic University, US.
Dr. Ravi Vadapalli, HPCC, Texas Tech University, US.
Dr. Jagdish Shivhare, Distinguished Scientist & Member-IRBA, USA.
Dr. Junyuan Zeng, The University of Texas at Dallas, US.
Dr. Neeraj Mittal, The University of Texas at Dallas, US.
Dr. G. L. Prajapati, IET, Devi Ahilya University, Indore India.
Dr. Vivek Tiwari, IIIT Raipur, India.
Dr. Muralidhar Kulkarni, NIT-Suratkal, Mangalore, Karnataka, India.
Dr. Raghavan S., NIT-Tiruchirapalli, India.
Dr. T. Kishore Kumar, NIT Warangal, India.
Dr. Supreethi K. P., JNTUH, Hyderabad. India.
Dr. Sujatha Banothu, Osmania University, Hyderabad, India.
Dr. T. V. Rajini Kanth, Sreenidhi Institute of Science and Technology, Hyderabad, India.
Dr. AmjanShek, BVRIT, Hyderabad.
Dr. P. Vijyapal Reddy, Matrusri Engineering College, Hyderabad, India.
Dr. Aynur Unal, Palo Alto, CA, USA.
Dr. Manu Pratap Singh, DR. BHIMRAO AMBEDKAR UNIVERSITY, AGRA, India.
Dr. V. K. Saraswat, DR. BHIMRAO AMBEDKAR UNIVERSITY, AGRA.
Dr. Halavath Balaji, Sreenidhi Institute of Science & Technology, Hyderabad, India.

Dr. D. Stalin Alex, Guru Nanak Institute of Technology, Hyderabad, India.
Dr. Nidhi Arora, SAGE university, Indore, India.
Dr. N. Rajasekhar, GRIET, Hyderabad, India.
Dr. Sanjeev Tokeykar, IET, Devi Ahilya Vishwa Vidyalaya, Indore, India.
Dr. B. T. Krishna, UCEV, JNTUK, AP, India.
Dr. A. Narendra Babu, LBRCE, Mylavaram, AP, India.
Dr. S. P. Singh, Mahatma Gandhi Institute of Technology, Hyderabad, India.
Dr. Yatendra Nath Singh, IIT Kanpur, India.
Dr. Atul Negi, University of Hyderabad, India.
Dr. Ganapati Panda, IIT Bhubaneswar, India.
Dr. Kuldeep Kumar, Birla Institute of Technology and Science, Pilani, India.
Dr. N. Rama Subramanian, NIT Tiruchirapalli, India.
Dr. L. Anjaneyulu, NIT Warangal, India.
Dr. P. Sreehari Rao, NIT Warangal, India.
Dr. Saugata Sinha, VNIT-Nagpur, India.
Dr. Prabhat Kumar Sharma, VNIT Nagpur, India.
Dr. Dipankar Pal, BITS Pilani, K K Birla Goa Campus, India.
Dr. Ravindhar Reddy B., Annamacharya Institute of Technology and Sciences, India.
Dr. M. Chinna Rao, Srinivasa Institute of Engineering and Technology.
Mr. Sachin Chirgaiya, SVVV, Indore, India.
Dr. Jitendra Choudhary, Medicaps University, Indore, India.
Dr. Dharmendra Choukse, IPS Academy, Indore, India.
Dr. Delshi Howsalya, Bharat Institue of Engineering and Technology, Hyderabad.
Dr. Dhirendra Pandey, Babasaheb Bhimrao Ambedkar University, Lucknow, India.
Dr. Jayesh Gangrade, IPS Academy, Indore, India.
Dr. S. Govinda Rao, GRIET, Hyderabad, India.
Reetu Gupta, SDBCT, Indore, India.
Dr. Chanchala Joshi, Vikram University, Ujjain, India.
Dr. Purna Chand K., BVRIT Narsapur, Hyderabad, India.
Dr. Sreekala K., Mahatma Gandhi Institute Of Technology, Hyderabad, India.
Dr. Kamakshaiah, Geethanjali College of Engineering and Technology, Hyderabad.
Dr. Priyesh Kanungo, DAVV, Indore.
Dr. KousarNikhath, VNRVJIET, Hyderabad, India.
Dr. B. Krishna, Vaagdevi College of Engineering, Warangal.
Dr. Sheo Kumar, CMR Engineering College, Hyderabad.
Dr. Korra Lakshman, NIELIT, Aurangabad, India.
Dr. Srinivas Madana, Jyothishmathi Institute of Technological Sciences, Karimnagar, India.
Dr. M. Mahalakshmi, CMR College of Engineering and Technology, Hyderabad, India.
Dr. Bharti Malukani, prestige institute of management and research, Indore.
Dr. Maya Rathore, Christianeminent college, Indore.
Dr. Merugu Suresh, CMR College of Engineering and Technology, Hyderabad, India.
Dr. Vijay N., Malla Reddy College of Engineering & Technology, Hyderabad, India.
Dr. Sitanshu Sekhar Sahu, BITS-Mesra, Ranchi, India.

Dr. B. Anuradha, S V University College of Engineering, Tirupati, India.
Dr. M. Asha Rani, JNTUH, Hyderabad, India.
Dr. D. Sreenivasa Rao, JNTUH, Hyderabad, India.
Dr. G. Nagaraju, VNR VJIET, Hyderabad, India.
Dr. Nagaveni, Acharya Institute of Technology, Bangalore, India.
Dr. Y. Jeevan Nagendra Kumar, GRIET, Hyderabad, India.
Dr. A. Narayana Rao, NBKRIST, Nellore, India.
Dr. Neelima Vontela, Jyothishmathi Institute of Technology & Science, Karimnagar, India.
Dr. Bhupendra Pandya, Vikram University, Ujjain, India.
Dr. Parag Parandkar, REVA University, Bangalore, India.
Dr. Subhash Parimalla, VNR VJIET, Hyderabad, India.
Dr. Krishna Prasad Ponnekanti, SV College of Engineering, Tirupathi, India.
Dr. K. Prasanna Lakshmi, GRIET, Hyderabad, India.
Dr. Adiraju Prashanth Rao, Anurag University, Hyderabad, India.
Dr. V. Prashanthi, GRIET, Hyderabad, India.
Mohd. Qayyum, King Khalid University, Abha, Saudi Arabia.
Dr. K. Prabhakar Nayak, MIT, Manipal, India.
Dr. J. Ramesh, PSG College of Technology, Coimbatore, India.
Dr. K. V. Raghavendra, Mallareddy College of Engineering and Technology, Hyderabad, India.
Dr. Vijaya Latha, GRIET, Hyderabad, India.
Dr. K. Vijayalakshmi, Sreenidhi Institute of Science & Technology, Hyderabad, India.
Dr. Rohit Raja, SREYAS Institute of Engineering and Technology, Hyderabad.
Dr. E. Raju, Vaagdevi Engineering College, Warangal, India.
Dr. Rakesh Kumar Tripathi, BM College, Indore, India.
Dr. V. Raman, Vardhaman College of Engineering, Hyderabad, India.
Dr. V. Ramesh, Presidency University, Bangalore, India.
Dr. Manish Sahajwani, IPS Academy, Indore, India.
Dr. Seethramulu, ICFAI University, Hyderabad, India.
Dr. M. Shanmukhi, Mahatma Gandhi Institute of Technology ,Hyderabad.
Dr. M. Sharadha Varalakshmi, St. Peters Engineering College, Hyderabad, India.
Dr. Rashid Sheikh, Acropolis Institute of Technology and Research, Indore.
Dr. Niranjan Shrivastava, DAVV, Indore.
Dr. K. Srinivas, VNR VJIET, Hyderabad, India.
Dr. Y. Suresh, BallariInstitue of Technology and Management, Ballari, India.
Dr. M. Suresh Kumar, Vaagdevi College of Engineering, Warangal, India.
Dr. G. Suresh Reddy, VNRVJIET, Hyderabad, India.
Dr. R. S. Thakur, MNIT, Bhopal, India.
Dr. Ugrasen Suman, DAVV, Indore, India.
Dr. Umesh Kumar Singh, Vikram University, Ujjain, India.
Dr. Subhashini Valluru, MLRIT, Hyderabad, India.
Dr. V. Venkateshwarlu, Vaagdevi College of Engineering, Warangal, India.
Dr. M. Venugopala Chari, CBIT, Hyderabad, India.

Dr. V. Akhila, GRIET, Hyderabad, India.
Dr. A. Swarna Bai, Scientist E, RCI, DRDO, Hyderabad, India.
Dr. Parag Parandkar, REVA University, Bangalore, India.
Dr. Niranjan Shrivastava IMS DAVV, Indore, India.

Web and Publicity Committee

G. S. Ramesh, CSE, VNR VJIET.
P. Radhika,CSE, VNR VJIET.
R. Kranthi Kumar, CSE, VNR VJIET.
S. Kranthi Kumar, CSE, VNR VJIET.
N. Sandeep Chaitanya,CSE, VNR VJIET.

Preface

Computer Engineering and Communication Systems are entwined more now than any other time in the history. The interplay of Information and Communication Technologies, the rise of Internet of Things (IoT) applications, smart computing and the inroads that technology has taken into personal lives through wearable https://didacindia.com/works, and so on have significant roles to play in the marriage of computer engineering with communication systems. The International Conference on Advances in Computer Engineering and Communication Systems (ICACECS-2020) is themed around smart innovations, Industry 4.0 technologies, data analytics, networks and communication systems, thereby celebrating the emerging technology trends in Computer Engineering and Communication Systems. The Conference is organised as six parallel tracks, viz. Mezzanine Technologies; Big Data and Data Analytics; Cloud, IoT and Distributed Computing; Smart Systems; Network and Communication Systems; and Education Technology and Business Engineering.

The primary objective is to encourage National and International communication and collaboration, promote professional interaction and lifelong learning on emerging technologies. The conference consists of keynote lectures, tutorials, workshops and oral presentations on all aspects of advance computing and communications. The aim of this International Conference on Advances in Computer Engineering and Communication Systems (ICACECS-2020) is to present a unified platform for advanced and multidisciplinary research towards the design of smart computing, information and communication systems.

It is being organized specifically to help the academicians and professionals from IT and communication organizations/industries to derive benefits from the advances of next generation computer and communication technology. The conference provides a platform for the experts and researchers from all over the world to discuss the contemporary developments and novel concepts in the fields of advanced computing and communication technologies. This International Conference is being organised ONLINE by the Computer Science and Engineering Department, VNR VJIET, Bachupally, Nizampet, Hyderabad, from 13th–14th August 2020.

Quality articles are published in this volume after performing the plagiarism check and review process by three eminent experts in the respective domain. Eminent Academicians & Researchers are delivering Keynote addresses on contemporary thrust areas. The resource pool is drawn from IITs, NITs, IIITs, IDRBT and Universities along with Software companies like TCS, CAP Gemini, Microsoft, Cyber Jagrithi, etc.

A galaxy of 32 eminent personalities is chairing and acting as Conference Jury to review the presentations. The papers are classified into 7 tracks which will be delivered ONLINE in 2 days using MICROSOFT TEAMS. The Pre-conference Workshops on 'Machine Learning & Artificial Intelligence', 'Block Chain Technologies' and 'Cyber Security' will be organised through MICROSOFT TEAMS from 12th–14th August 2020.

We wish our Participants a Healthy and Fruitful Learning.

Hyderabad, India
July 2020

Dr. C. Kiran Mai
Dr. B. V. Kiranmayee
Suresh Chandra Satapathy
Margarita N. Favorskaya
K. Srujan Raju

Acknowledgements

We would like to acknowledge the support extended by AICTE, by partially funding the event. It furthered and accelerated our thoughts to practice.

We thank all the authors for their contributions and timely response. A special thanks to our reviewers who invested their valuable time and scrupulously evaluated the submissions for the best outcomes.

We express our profound gratitude for the inspiring and informative presentations of our Keynote speakers on the frontline technologies, creating the curiosity to explore more.

The co-operation extended by the Sessions Chair is immense, elevating the presentation skills amongst the participants.

We would like to thank our Chief Guest and Guest of honour for making it to the e-conference, boosting up our spirit to achieve more for providing technology solutions to the Society—Involve and Evolve.

Thanks to the founding members of Vignana Jyothi and their wisdom and social responsibility, we firmly believe that all technology must serve the urgent need of advancing the society, at the micro and/or macro levels.

Our sincere thanks are extended to all the Patrons, Chairpersons, members of the Editorial Board, eminent members of the Programme Committee and Advisory Committee for their guidance and support and to the enthusiastic people amongst the Technical Committee for their coordination and help in execution.

We would like to extend our appreciation for the amazing work done by our self-reliant and motivated team of VNR VJIET. The amazing dedication and effort of the team enabled us to reach our goal, in spite of the Pandemic.

A special acknowledgement to Ms. Sharmila Mary Panner Selvam, Project Coordinator, Books Production SPRINGER for the prompt communications and Support.

Vallurupalli Nageswara Rao Vignana Jyothi Institute of Engineering and Technology (VNRVJIET), Hyderabad, India

Vallurupalli Nageswara Rao Vignana Jyothi Institute of Engineering and Technology (VNRVJIET), an Autonomous Institute, approved by AICTE, affiliated to JNTUH is accredited by **NAAC with A++, a score of 3.73** and also 7 UG & 4PG programmes are accredited by NBA. The institute, from its inception, has charted distinct pathways to academic excellence. It is one of the most distinguished and premier institutions of higher education in the State of Telangana. Its complexity, diversity and comprehensiveness is a fountainhead of creativity and innovation.

VNR VJIET, a modern world-class academic institute, certified as **COLLEGE WITH POTENTIAL FOR EXCELLENCE** by **UGC**, has a strong inclination towards sustainable development through research and expansion of innovative technologies. In an era of global progression propelled by technology, research at academic institutes will foster economic growth and helps in attaining self-reliance in technology and innovation. The Institute is committed to fundamental long-term research and innovation in leading edge technologies, performs a diverse and expanded set of activities like,

- Producing high quality engineers with the required skills and knowledge at different levels (undergraduate, postgraduate).
- Exploring new horizons through fundamental research.
- Continuous improvement of knowledge repository and domain expertise.
- Encouraging new ideas and proposals through research awards and remuneration.
- Becoming a source for innovation, addressing societal needs and developing new products leading to revenue generation.

About the Department—Computer Science and Engineering

Computer Science and Engineering (CSE) department, was started in the year 1995, and is significant for its ability of research and development activities and excellence in academics. The department is NBA Accredited and is recognised as a Research

Center by JNTUH, the affiliating University. It offers two courses at the Undergraduate level, Computer Science & Engineering (CSE) with an intake of 240 and Computer Science and Business Systems (CSBS) with an annual intake of 60. It also offers two PG Courses in Software Engineering (SE) and Computer Science Engineering (CSE) with an intake of 18 each. The state-of-the-art infrastructure of the department is commendable.

The Department is dynamic and has several publications to its credit listed in SCOPUS, Web of Science, SCI indexed, etc. It is involved in pioneering research in Advanced Research areas like Image Processing, Data Mining, Networks, Artificial Intelligence, Databases and Wireless Area Network. Computer Science and Engineering Department executes several funded projects from reputed agencies like DST, DRDO, AICTE, UGC, etc. The student involvement in the research projects nurtures their involvement in handling societal problems, thus making them the technology solution providers. Involve and evolve with the changing technology keeps the fraternity updated. Committed faculty with strong aptitude to learn and engage in research enable to bridge the industry–academia gap leading the department to number one position.

Contents

About the Editors

Dr. C. Kiran Mai working as Professor in the Department of Computer Science& Engineering, VNR VJIET, has over 31 years of experience in the field of academic research and technological education. She has a Multidisciplinary approach due to varied roles taken up, including but not limited to teaching, research and administration. She was awarded as 'Best teacher in Computer Science' in the year 2010, by the professional body—InternationalSociety for Technology in Education (ISTE). She also worked at various administrative positions in the institute (Principal, Dean Academics, and Head of the Department) and has an extensive experience in internal administrative tasks and communication. While serving in the administrative positions as Vice-Principal and Principal, she was instrumental in designing the policies and strategies for the institute and also the administrative manuals. Dr. Kiran Mai, also administered the processes and could get five departments of the institute recognised as Research centers by the JNTUH, the affiliating university. As Dean Academics, she played a key role in curriculum revision, enhanced the Learning by doing component for Practical courses, introduced the concept of WIT & WIL (Why I am Teaching, What am I teaching and Why I am Learning, What am I learning). With her industrial experience, where she headed ISO 9002 division—document control, she could frame and document the processes and procedures in the institute with ease. Being a member of the Internal Quality Assurance Cell (IQAC) administers the quality procedures in the institute and performs a periodical audit of the academic and administrative processes. Under her leadership, the institute was recognised by UGC as College with Potential for Excellence, got the UG and PG courses re-accredited. She published 40 papers in various reputed National and International journals, Conducted Faculty development programs in the cutting edge technologies with the funding from AICTE. Actively participated in the research projects and guided nearly 75 UG projects and 20 PG projects. Currently, four research scholars are working under her guidance. She had co-chaired many international conferences. Her research paper on Data mining for deforestation using Polyanalyst, presented at the IEEE conference held at Seoul, South Korea, in 2005, was selected as the best paper. She was on the Editorial Board for two Korean journals. Her areas of interest are Network communications, Data engineering and Blockchain Technologies.

Dr. B. V. Kiranmayee is currently working as Associate Professor and Heading Department of Computer Science & Engineering at VNR VJIET. She obtained her Ph.D. from JNTU Hyderabad in the area of Data Mining, M.Tech in 2007, and B.Tech in 1995, in Computer Science and Engineering. She has a rich teaching and research experience of 22 years in VNR Vignana Jyothi Institute of Engineering and Technology, Hyderabad. She has more than 20 research papers published in various reputed National/International Conferences and Journals are listed in Scopus, IEEE, and Springer Proceedings, etc. She is a member of IEEE, and lifetime member of ISTE and CSI. Dr. Kiranmayee's research interests include Data Mining, Algorithms, Image processing, Machine Learning, Deep Learning and Blockchain technologies. She has rich experience in generating funds from various funding sources like UGC, Consultancies such as organizing Virtual Tour for Warangal Tourism, Developing Educational Videos for School Children, TCS Online examinations. She is the convener for conducting many seminars/workshops/FDPs for the benefit of students and faculty in and out by the eminent personalities from various reputed institutions and industries. She is the chairman of the Board of Studies for the department and also served as BOS member in various other engineering colleges. She has played a crucial role in establishing Center of Excellence in Data Science, Big Data, Virtual Reality and Augmented Reality. In her leadership, the department has undergone MoUs with various prestigious software industries and academic institutions for reducing the gap between Industry and Academia. She has contributed more towards in successions of getting NBA accreditation, NAAC A++, QS Diamond rating and NIRF ranking. She got Best Paper award for 'Eye state Detection and Analysis for Driver's Fatigue' in International Conference of Systemics, Cybernetics and Informatics. She has guided many UG and PG projects in various mezzanine technologies. Under her mentorship, the students have bagged various ISTE and CSI awards like: 'Best Student Innovator', 'Best CSE Student', 'Best Accredited Student Branch', 'Highest Committed Student Branch Activist'. Under her guidance and motivation, many students have excelled and won awards in various coding competitions like: Smart Indian Hackathon-2018 and 2019, Code for Good (JPMC), Swish Hackathon, ACM/ICPC, VesAithon, etc.

Dr. Margarita N. Favorskaya is a Professor and Head of Department of Informatics and Computer Techniques at Siberian State Aerospace University, Russian Federation. Professor Favorskaya is a member of KES organization since 2010, the IPC member and the Chair of invited sessions of international conferences. She serves as a reviewer in international journals (Neurocomputing, Knowledge Engineering and Soft Data Paradigms, Pattern Recognition Letters, Engineering Applications of Artificial Intelligence), an associate editor of Intelligent Decision Technologies Journal and Computer and Information Science Journal. She is the author or the co-author of 160 publications and 20 educational manuals in computer science. She co-edited three books for Springer recently. She supervised eight Ph.D. candidates and presently supervising five Ph.D. students. Her main research interests are digital image and videos processing, remote sensing, pattern recognition, fractal image processing, artificial intelligence and information technologies and the applications of artificial intelligence techniques in many disciplines including mechanical engineering, electrical engineering, healthcare and so on.

Suresh Chandra Satapathy is currently working as Professor, KIIT Deemed to be University, Odisha, India. He obtained his Ph.D. in Computer Science Engineering from JNTUH, Hyderabad and Master degree in Computer Science and Engineering from National Institute of Technology (NIT), Rourkela, Odisha. He has more than 27 years of teaching and research experience. His research interest includes machine learning, data mining, swarm intelligence studies and their applications to engineering. He has more than 98 publications to his credit in various reputed international journals and conference proceedings. He has edited many volumes from Springer AISC, LNEE, SIST and LNCS in past and he is also the editorial board member in few international journals. He is a senior member of IEEE and Life Member of Computer Society of India. Currently, he is the National Chairman of Division-V (Education and Research) of Computer Society of India.

Dr. K. Srujan Raju is currently working as Dean Student Welfare and Heading Department of Computer Science & Engineering at CMR Technical Campus. He obtained his Doctorate in Computer Science in the area of Network Security. He has more than 20 years of experience in academics and research. His research interest areas include Computer Networks, Information Security, Data Mining, Cognitive Radio Networks and Image Processing and other Programming Languages. Dr. Raju is presently working on 2 projects funded by Government of India under CSRI & NSTMIS, has also filed 7 patents and 1 copyright at Indian Patent Office, edited more than 14 books published by Springer Book Proceedings of AISC series, LAIS series and other which are indexed by Scopus also authored books in C Programming & Data Structure, Exploring to Internet, Hacking Secrets, contributed chapters in various books and published more than 30 papers in reputed peer-reviewed Journals and International Conferences. Dr. Raju was invited as Session Chair, Keynote Speaker, Technical Programme Committee member, Track Manager and reviewer for many National and International conferences also appointed as subject Expert by CEPTAM DRDO—Delhi & CDAC. He was undergone specific training conducted by Wipro Mission 10X & NITTTR-Chennai which helped his involvement with students is very conducive for solving their day to day problems. He has guided various student clubs for activities ranging from photography to Hackathon. He mentored more than 100 students for incubating cutting edge solutions. He has organized many conferences, FDPs, Workshops and Symposiums. He has established the Centre of Excellence in IoT, Data Analytics. Dr. Raju is a member of various professional bodies and received Significant Contributor Award and Active Young Member award from Computer Society of India also served as Management Committee Member, State Student Coordinator & Secretary of CSI-Hyderabad Chapter.

Smart Dustbin: A Reward Provider

Ankitraj Gaddam and A. Kousar Nikhath

Abstract Today, an average resident produces seven and a half pounds of garbage every day and that is buried down in landfills and litters lands costing a great amount of money and environmental damage. Nowadays, people face no more critical trouble than the need to save the weakening environment, mainly in urban areas, where solid wastes are uselessly dumped. Although an intrinsic part of our everyday routines, the dustbin's role as a mediator of changing waste practices has rarely been considered. The idea to exalt simple acts in civil behaviour may not set a good example, but is a start nevertheless towards fostering a culture where these actions are given the attention they deserve. This paper presents an IOT innovation in the form of a smart dustbin with real-time waste monitoring system that integrates multiple technologies, such as mobile technologies, sensors and wireless communication technologies. The main idea of the smart dustbin evolves from the traditional basic dustbins to have multiple smart features, like: to have two bins, one for the wet waste and another for the dry waste. As soon as the person throws the trash into the smart bin, the smart dustbin provides a reward as an incentive. Rewards include internet facilities, coupons for different payment driven apps, wallet money. Smart dustbin also provides the location of the bins, so as to notify the concerned department to come and collect the trash when the smart bin is full with the shortest path distance navigation with data analytics and also to provide the garbage collectors with rewards for successfully emptying the smart dustbin with the rewards for encouraging them. The aim of this paper is to provide an efficient and cost-effective waste collection management system, while encouraging the individuals to put the trash where it belongs, hence providing clean, healthy and green environment.

Keywords Real-time waste monitoring system · Rewards · Smart dustbin · Waste collection management system · A reward provider · Reward points

A. Gaddam (✉) · A. Kousar Nikhath
VNR Vignana Jyothi Institute of Engineering and Technology, Hyderabad, India
e-mail: ankitraj.g13@iiits.in

A. Kousar Nikhath
e-mail: kousarnikhath@gmail.com

C. Kiran Mai et al. (eds.), *Proceedings of International Conference on Advances in Computer Engineering and Communication Systems*, Learning and Analytics in Intelligent Systems 20,
https://doi.org/10.1007/978-981-15-9293-5_1

1 Introduction

Accumulation of garbage is so high that if left uncollected, it becomes a problem. If the garbage collector does not appear, a household will probably direct their maid-servant to pick up the trash bags, as it would be too much to carry the bags from inside the house. Perhaps the servant will dump the garbage at the end of the lane. Others would have followed suit after seeing that. The negligence of people throwing the garbage in their surroundings mainly due to the laziness and lack of enthusiasm to throw the garbage in the dustbin. The site will gradually turn into a garbage dump yard, transforming into a refuge for diseases of health. Inefficient waste collection systems lead to pollution of the environment, resulting in animal scavengers and rodents, insect breeding and a range of other diseases. If not obtained in time, the traditional method involves burning the waste. Waste burning causes large amounts of air pollution. The aerobic decomposition of burnt waste also adds unregulated release of methane to social health problems. Accumulation of waste is an important issue that needs to be cleverly addressed. Wisely, for ease of processing and recycling, we segregate the waste in our homes. We observed that trash vans/trucks arrive irregularly in homes that create household havoc. Because of this, many people in open spaces clear their overflowing dustbins. It, in effect, raises emissions from the atmosphere. Hence a robust waste management system with analytics is needed with an intent to make every individual a habit for proper cleanliness.

2 Related Work

The paper [1] proposes a system where if the bin has been filled up to the brim, then it will send an automated message to the system's server end. And the workers have to accept the notification sent to him/her and if he/she is not available then he/she will reject it then the work to collect the garbage is assigned to another worker.

The paper [2] suggests a design that consists of an integrated system with sensors and microcontroller to detect Bins Sending the information to the workstation, located at the municipal office "for discovering the shortest path."

The paper [3] suggests IR camera, microcontroller and Wi-Fi smart garbage management framework where dustbins are washed quickly as the amount of garbage reaches its limit. If the dustbin is not washed in a reasonable period, the report is submitted to the higher authority concerned. This program also helps track the false results and "may eliminate abuse in the management system as a whole."

In paper [4], authors introduced an internet of things (IOT) program that will deliver a warning to the waste department about the dustbins overload and toxicity rates with ultrasonic sensor and gas sensor for assessment, respectively. A database is established to monitor the dustbins relevant info. Message is sent to the cell phone using GSM module through Arduino microcontroller, and the dustbin status data is changed on the website. Citizens may also file reports about dustbin or waste control at this website.

In this paper [5], authors propose a two-axis robot that has IR, gas (to detect harmful gas within the holder) and rain sensor (to deter rainwater from flooding the bin) with microcontroller. IR sensor after reaching a threshold level, closes the bin's top door automatically and enters the environment it must open the bin's side door for a limited time to dispose of the waste in the container. After the bin gets emptied, it shuts the bin's side door immediately and returns to its position and reopens the top door again.

This paper [6] suggests a system that has ultrasonic and MQ4 sensors with microcontroller. The ultrasonic sensor checks the amount of garbage, MQ4 sensor test odour levels in the biodegradable dustbin. For non-biodegradable smart bin, if the bin is full, then the information is given to the waste department for disposal. For biodegradable container, if it is exceeded, then the resulting chamber is poured onto a cover slides and the wastes.

In paper [7], the authors suggested a smart Wi-Fi dustbin device where the microcontroller shows the random number as someone drops the trash or waste into the bin and is used to achieve free Wi-Fi connectivity.

In paper [8], the authors surveyed and proposed that if someone puts the garbage in the dustbin, they gets a temporary Wi-Fi code that allows the user to connect to free Wi-Fi for a short time span. A single person is permitted to access the Wi-Fi twice a day.

3 Existing System

- Few of the papers were just offered a system to detect the object and open the lid of the dustbin.
- While most of the papers offered a system to detect how much amount of the waste is filled inside a dustbin and if it is full then send the notification to the waste collection department.
- Finally, only some of the papers offered a system of giving the internet facilities through Wi-Fi (which is up to a certain distance around the bin) for throwing waste into dustbins, but those systems did not include waste management system in them.

4 Proposed System

All the proposed systems until now have a Wi-Fi router in them and a Wi-Fi router works only for a fixed place (not everywhere). What if we do not keep a Wi-Fi router at all for users? Instead, we give the trash thrown users, the rewards like the internet facilities or wallet money or coupons for various service-based apps like bookmyshow, swiggy, etc. Then the user can travel anywhere far from the bin and still can get the reward of getting the internet facilities for throwing the trash or wallet money or coupons. The garbage collectors can also get the rewards (as same rewards as given for the users for throwing the trash) if they successfully empty the smart dustbin.

4.1 Reward System

A new system is introduced to encourage the individuals to throw the trash into the smart dustbin. Whenever a user throws the trash into the smart dustbin, the bin automatically detects the trash and generates a unique code, that the user needs to type on his/her smartphone application, then the application gives out the random reward points, these reward points are like trophies the more you collect, the more exciting rewards you get in terms of either in coupons or wallet money or mobile data recharges and the same goes to the garbage collectors, as soon as they empty the smart dustbin a unique code is generated and that is then typed into the same field provided in the application for the users, with its uniqueness in the code the application will identify that the garbage collector is giving the code and accordingly the rewards are provided.

Suppose you get the reward points as 20 in one throw, then you need to throw the trash into the smart dustbin a few more times in order to get to the 100 reward points mark to collect the reward, say a wallet money of 50 rupees or mobile data recharges or coupons. A reward timeline is introduced with each milestone rewards in between.

4.2 Hardware Sensors

Hardware sensor components used in this working flow diagram connected to a microcontroller are with two sensors, one is IR sensor and another is an ultrasonic sensor, IR sensor will emit the infrared ray from an IR emitter LED, when an object is thrown into the smart dustbin the IR receiver LED will detect the object thrown and sends the sensor data to the microcontroller (Fig. 1).

Whereas another sensor is the ultrasonic sensor, it is used to sense the level of the trash in the smart dustbin. Here the ultrasonic sensor uses ultrasonic waves, these waves are emitted by the head of the sensor and collected back those same waves from the target (i.e. the present level of the trash present in the smart dustbin). The sensor has a transducer that is used for transmitting and capturing the ultrasonic waves. This ultrasonic sensor measures the distance to a target (top level of the trash present in the smart dustbin) by measuring time-lapses between transmitting and receiving the ultrasonic pulse. After calculating the distance, the sensor data of calculated distance is sent to the microcontroller (Fig. 2).

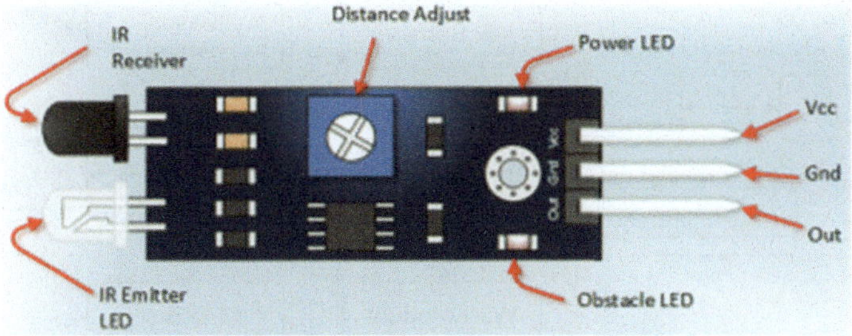

Fig. 1 IR sensor

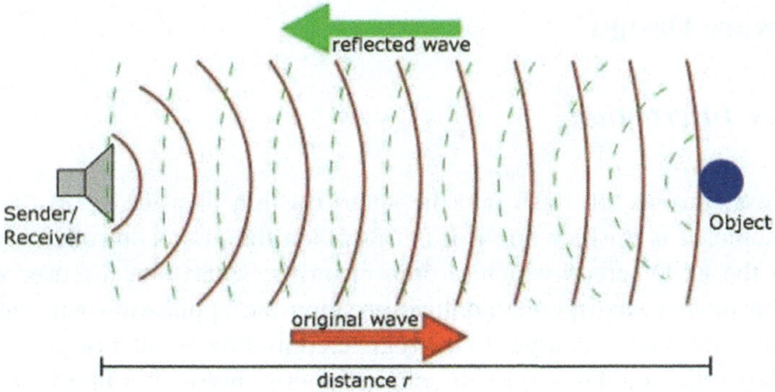

Fig. 2 Working of ultrasonic sensor

4.3 Working Flow Diagram

Block Diagram Starts with the mobile application with its database and that is connected to the microcontroller over the internet with the help of Wi-Fi module connected to the microcontroller. Power is supplied to the microcontroller through the solar panel, where power is stored into a battery or direct power supply. The two sensors named IR sensor and ultra-sensor are connected to the microcontroller, where IR sensor helps in detection of the trash object and ultra-sensor here will give the present level of the trash present in the dustbin. LED bulb is used to indicate if the bin is full or not. Finally, LCD display is used to display the code and necessary instructions (Fig. 3).

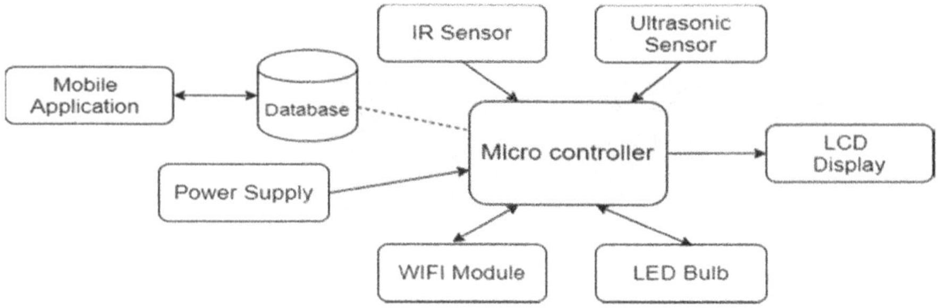

Fig. 3 Block diagram of smart dustbin: a reward provider

5 Software Design

5.1 User Interaction

First, the user throws the trash into the smart dustbin then the smart dustbin will checks whether it is trash or not, if it is trash then the microcontroller will display a code on the LCD screen which is present on the smart bin, the user will enter the code on his/her smartphone application, then the application will validate the code and then does the recharges to the registered mobile number or gives any other rewards, depending on the rewards timeline. Suppose he/she is a first time user then the WI-FI module helps the user to get the initial installation and registration to the mobile application. And if it is not trash that is thrown in the smart bin, then the smart dustbin will show the written message on the LCD display that "Incorrect trash, please throw some trash."

5.2 Process Execution

In process execution, first, the user throws the trash into the smart dustbin then the IR sensor will check whether it is a trash or not, if it is trash then the IR sensor sends the sensor data to the microcontroller then the microcontroller will send the random code from the database on to the LCD screen of the smart dustbin. Then that code is entered into the application and will be checked if it is the same code given to the microcontroller or not and if it is same code then the application will recharge the data balance to the registered mobile number or gives any other rewards, depending on the rewards timeline. And if it is not same, then the application will ask the user to enter properly. And if it is not the trash that is thrown into the smart dustbin, then the smart dustbin will show the written message on the LCD display that "incorrect trash, please throw some trash."

5.3 *Waste Management Execution*

In waste management system, the microcontroller with the help of ultrasonic sensor will check whether the bin is filled or not, if it is filled then the LED outside will be ON and a notification will be sent to the waste collection department then the department sends the truck to the notified location with the optimised path included in the application, the collector person after collecting the trash, will get the code which he/she enters into the application and gets rewards for emptying the dustbin and also updates the LED light signal, it's an indication that the bin is empty again and the process repeats.

6 Hardware Parts and Design

The hardware parts required for the complete connections in Fig. 4 are breadboard—half size, Arduino Uno, LCD Display screen 16 × 2 I2C, USB cable A to B, ESP8266-01—Wi-Fi Module, Logic Level Converter—Bi-Directional, IR Obstacle Avoidance Sensor, Ultrasonic Sensor—HC-SR04, Jumper Wires Pack—M/M, Jumper Wires

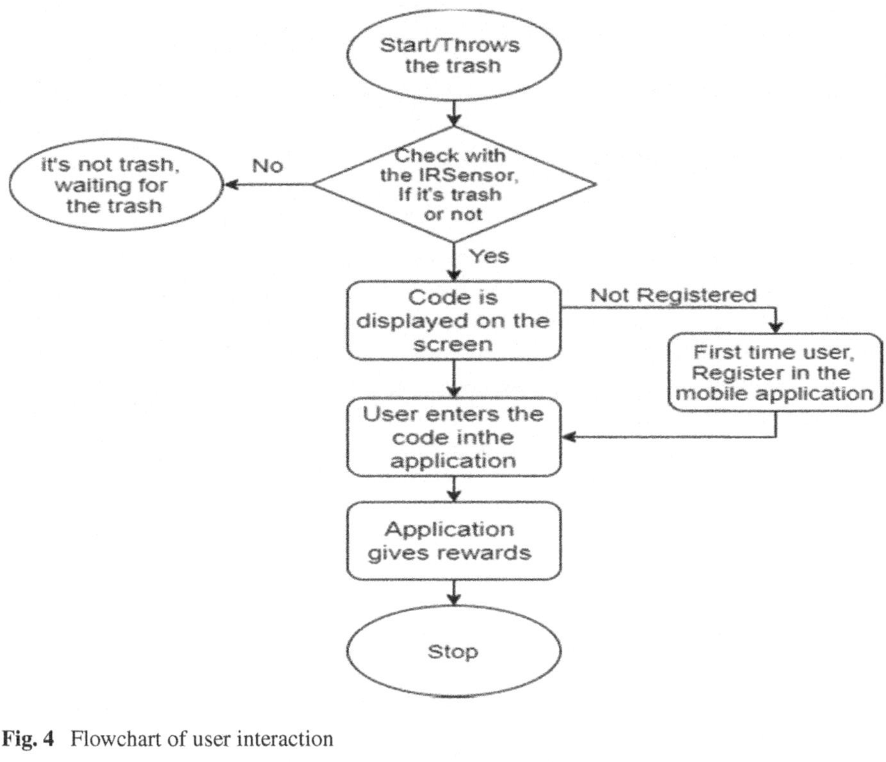

Fig. 4 Flowchart of user interaction

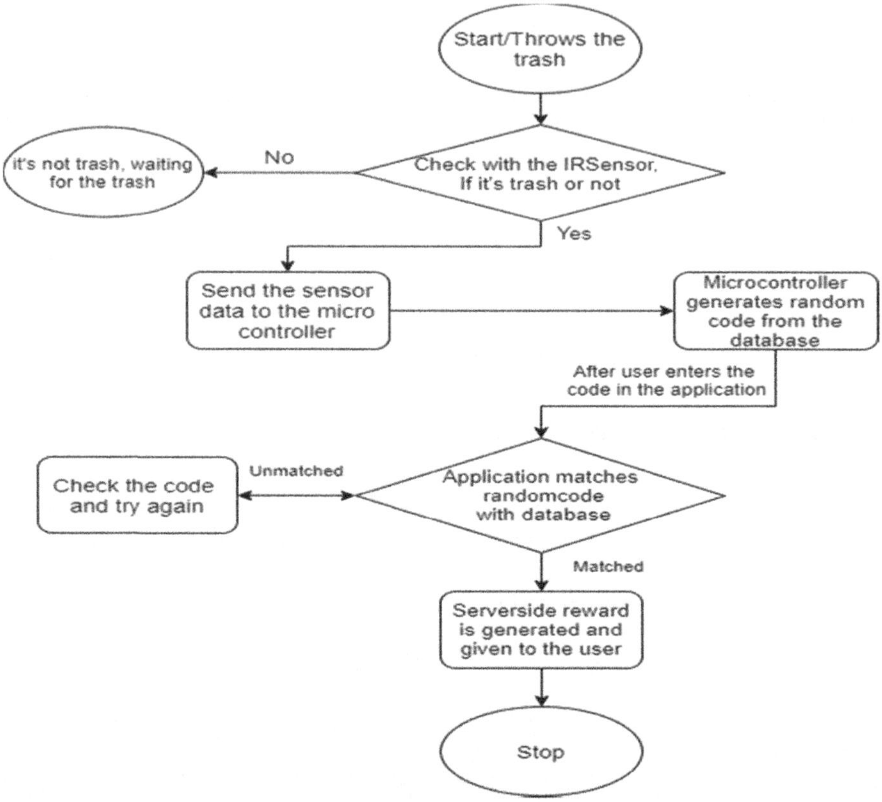

Fig. 5 Flowchart of process execution

Pack—M/F, Male Headers Pack-Break-Away. The cost for all the devices in Fig. 4 is approximately costs to 1500 rupees. In Fig. 4 with all the connections and required software code, the process is 99% success rate with the immediate response time. The energy taken for the complete process is very low as Arduino or any other microcontroller requires a very low power supply to function. Even if the many user joins or uses simultaneously, the server can handle large amounts of data and can simultaneously handle hits that is because the servers are maintained by the third-party software, such as amazon web services (Fig. 5).

7 Results

The final result would be in terms of a reward, which is given through the mobile application, the following picture shown in Fig. 5 has the reward points page of the mobile application, where the reward points are unlocked as the user throws into the

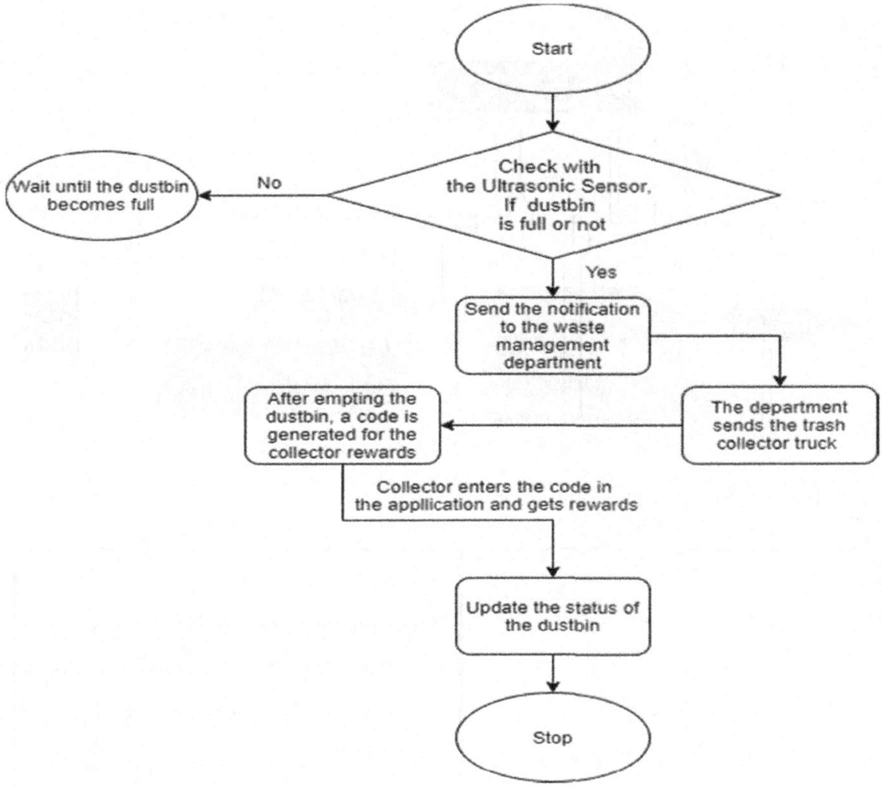

Fig. 6 Flowchart of waste management execution

smart dustbin and the another page is the rewards timeline page where the user can see what are all the rewards are present ahead, if they reach certain milestone reward points, say for 200 reward points then you get the wallet money of 50 rupees into your account (Figs. 6, 7, 8).

8 Conclusion

This proposed system has two main modules, the first module is for the customers and for the garbage collectors, the customers can get the rewards for their action of throwing the trash into the smart dustbin, whereas the garbage collectors can get the rewards if they successfully empty the smart dustbin. The second module is the robust waste management system, that is, for the waste collection department to come and collect once the smart dustbin is full, the bin triggers the notification to the department with its location and status information from which date the smart dustbin is full. The application takes care of the second module and by also providing

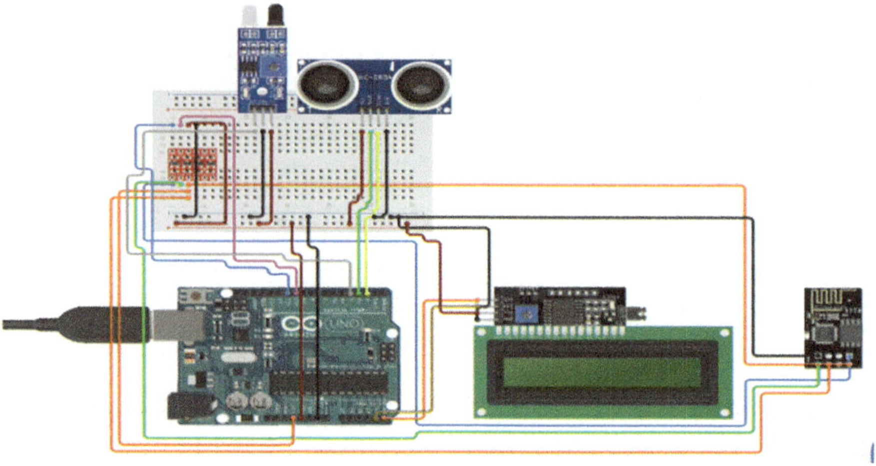

Fig. 7 Hardware design with complete connections

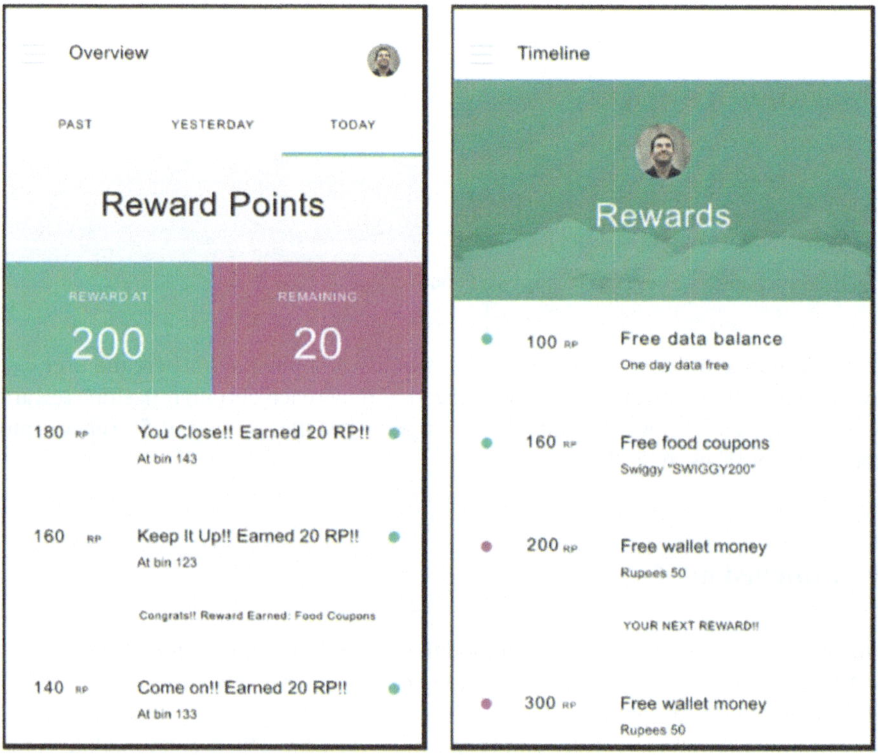

Fig. 8 Rewards points mobile application page and rewards timeline mobile application page

the schedules and also optimises the waste collection routes, which is based on historical data and real-time data collected regularly, by doing data analytics, It also offers predictive analytics to allow early decision making and advises on waste bin collections and allocations. The ability to take less time and get fewer trucks on the track, based on schedules and optimising dynamic routing with the shortest distance from the smart dustbin to its disposal, results in fewer emissions of greenhouse gases, less surface degradation and less noise pollution. This smart way of waste management solution significantly decreases waste disposal frequency production, which in heavily developed cities often contributes to overflowing garbage containers and unsightly intersections. The system lets waste disposal staff scan real-time fill-levels and provides excess warnings, enabling the waste department staff to be able to save on diesel, energy and cost of vehicle maintenance. This real-time analysis by continuously monitoring the amount of trash is filled into the bin with percentage values incorporated to each and every bin throughout the city results in a clean green environment for the safety of an individual and the society, which in turns saves the nature and the future generations.

References

1. M. Prasanth, P. Srinivasan, M. Pendyala, Automated waste clearance: street-wise cleanliness! ICHCI, Chennai **2013**, 1–5 (2013). https://doi.org/10.1109/ICHCI-IEEE.2013.6887798
2. S. Dugdhe, P. Shelar, S. Jire, A. Apte, Efficient waste collection system, in *2016 IOTA, Pune* (2016), pp. 143–147. https://doi.org/10.1109/iota.2016.7562711
3. S.S. Navghane1, M.S. Killedar, Dr. V.M. Rohokale, IoT based smart garbage and waste collection bin. IJARECE, **5**(5), 1576–1578 (2016)
4. R.K. Singhvi, R.L. Lohar, A. Kumar, R. Sharma, L.D. Sharma, R.K. Saraswat, IoT based smart waste management system: India prospective, in *2019 4th International Conference on IoT-SIU, Ghaziabad, India* (2019), pp. 1–6. https://doi.org/10.1109/iot-siu.2019.8777698
5. S. Sreejith, R. Ramya, R. Roja, A.S. Kumar, Smart bin for waste management system, in 2019 5th *ICACCS, Coimbatore, India* (2019), pp. 1079–1082. 10.1109/ICACSS.2019.8728531
6. P. Chowdhury, R. Sen, D. Ray, P. Roy, Sarkar S.: Garbage monitoring and disposal system for smart city using Iot, in 2018 ICGCIoT, Bangalore, India (2018), pp. 455–460. https://doi.org/10.1109/ICGCIoT.2018.8753060
7. Bandal, A. et al., Smart Wi-Fi dustbin system. Int. J. Adv. Res. Ideas Innov. Technol. **2**(5) (2016)
8. S. Bhuravane, M. Panindre, S. Patole, T. Pooja, Survey of Wi-Fi trash bin. IOSR J. Comput. Eng. (IOSR-JCE), 52–55 (2016)

Cyclic Exploration-Based Whale Optimization to Linear Discriminant Regression Classification for Face Recognition

T. Syed Akheel and V. Usha Shree

Abstract Generally, face recognition determines the judgment of whether or not a specific face is "known." Moreover, the face identification denotes the retrieval of data or information about the "owner" of face. Under this concept, a number of researches are there in progression. Still, the research works are not yet up to the mark as human brain recognition. The intelligence in face recognition should be enhanced with high accuracy rate, and this approach tends to present a novel face recognition pattern with a concept of feature extraction and classification. The features are extracted using Active Appearance Model (AAM). Then the classification is done via linear collaborative Discriminant regression classification (LCDRC) model proposed by Xiaochao Qu. In the LCDRC classifier, the most important evaluation is the projection matrix that might get multiplied to the features while classification. The projection matrix must be optimal, so that the recognition accuracy can be greatly attained. In order to select the optimal projection matrix, this paper presents a Cyclic Exploration-based Whale Optimization model (CEWO), which is the modified form of Whale Optimization Algorithm (WOA). The comparison of the proposed face recognition model is done with the performance across the additional conventional techniques with regard to measures such as Accuracy, Precision, False positive rate (FPR), False negative rate (FNR) of the proposed model is proven.

Keywords Face recognition · Active appearance model · Linear collaborative discriminant regression categorization · Whale optimization

---

T. Syed Akheel (✉)
Department of ECE, K.V. Subbareddy Institute of Technology, Kurnool, A.P, India
e-mail: tsaece@gmail.com

V. Usha Shree
Department of Electronics and Communication Engineering, Joginipally BR Engineering College, Hyderabad, T.S, India

C. Kiran Mai et al. (eds.), *Proceedings of International Conference on Advances in Computer Engineering and Communication Systems*, Learning and Analytics in Intelligent Systems 20,
https://doi.org/10.1007/978-981-15-9293-5_2

1 Introduction

In this modest network world, the need for information or data security is becoming a vital role, and it is progressively more tedious process. Recently, to eradicate the conventional exploitation of unsecured approach and to substantiate the right to use the image, the users can employ various solutions pooled with biometrics to access the data [1, 2]. Here, the security of those biometric data is measured as the most important trouble and that should be confined from a variety of attacks. A biometric authentication is thus described to recognize individuals depending upon the respective physical characteristics or behavioral qualities like fingerprint, face, voice, iris, etc., rather than other conventional approaches such as using passwords and PIN codes [3–5]. Face recognition has various beneficiaries over other biometrics like fingerprint, iris, voice, etc. The face can be considered as a complex configuration and should need high-quality evaluation approaches for recognition [6–8]. The face can be regarded as the major and primary focus of concentration in social life playing an imperative role in the uniqueness of individual. Several numbers of faces can be recognized throughout the lifespan, and it can be identified even after years. More variations are there and that occur in faces just because of causes like aging and distractions including beard, glasses, or hairstyle changes [9].

Generally, recognition of face is a fundamental aspect of biometric strategy. In biometrics, the matching takes place for the human's indispensable characteristic to the existing data, and the human is traced as per the results. The features of face are extracted and implemented via different approaches and that are proficient. Further modifications have been needed to improve the conventional recognition approaches [10–12]. Normally, face recognition model is subdivided into two groups on the basis of face illustration. This comprises of both appearance-based and feature-based approach. In the first category, it utilizes holistic features and after it is applied either on total face or some specific regions in face image. In the next category, it utilizes more facial features including mouth, eyes, eyebrows, cheeks and so on and along with the geometric relationship among the features [13, 14, 16]. Computerized approaches can be implemented to detect and recognize faces, which can be applied to some broad application ranges including identification of criminal case, security systems, verification (identity), etc. Moreover, the detection of face and recognition is deployed in various areas [15, 18]. Numerous approaches are there for recognizing and detecting the face. Feature extraction approach can be done in which the extracted features from the face can be processed and compared to several similarities processed faces available in the database [17, 19, 20].

Once the face is recognized, it is identified or the system may demonstrate the same face present in database. In the respective surveillance system, if an unidentified face appears for many times, then it is accumulated in database for future recognition [21, 22]. The steps are more efficient in criminal identification. Several methods can be introduced for the classification of data, and the main technique used is feature learning technique. In this technique, the training image data can be constructed by reducing the number of features. Thus the performance can be improved in the classification technique [23]. Hence, several active learning approaches have been proposed with the combination of information and representation about the training samples and thus better classification accuracy can be achieved [24].

This paper presents a novel face recognition model by usual phases like feature extraction as well as classification. Initially, the features remain obtained using AAM model. Then the classification is done via LCDRC model. In LCDRC, the projection matrix is multiplied with extracted features. Moreover, it is very important to identify the optimal projection matrix for better recognition results. Hence, this paper introduces a new optimization model, CEWO, which is the improved version of WOA. The proposed face recognition model matches the respective efficiency compared with the additional conventional approaches.

2 Literature Review

2.1 Related Works

In 2017, Lin and Chiu [1] have formulated an approach for face detection using contour-based binary descriptor. Face recognition can be considered as an accepted theme with respect to its security significance, vital surveillance, etc. The existing local approaches including the local binary pattern (LBP) can have better performance than the holistic approaches due to its stability on local variations including misalignment, occlusion, but their elevated computational difficulties, which leads to several limitations. While LBP can be regarded as an excellent feature approach, the Scale Invariant Feature Transform (SIFT) was extensively established by means of the effective characteristics for incarcerating edge information and shape data. Yet, SIFT-built models were susceptible to illuminating changes. Thus, an LBP edge-mapped describer with maximum levels of gradient magnitude items was proposed. From the simulation results, it is obvious that this technique attains an enhanced recognition rate even in the uncontrolled condition while comparing with the other approaches.

In 2017, Ghasemzadeh and Demirel [2] have implemented a face recognition technique based on wavelet transform. Facial hyperspectral image examination was considered as a trendy topic as it grants spectral information on topics. Thus, it has ability on enhancing the accuracy of face recognition. They have evaluated three new models on feature extraction, which was evaluated for classifying the image. The models have utilized three-dimensional discrete wavelet transform (3D-DWT) for extracting characteristics out of the facial hyperspectral images. All the horizontal, vertical, as well as spectral information was parallel processed. The vital 3D-DWT features were decomposing the hyperspectral images within a sequence of spatiospectral frequency subbands. Finally, compared with additional approaches, superior outcomes are produced by the computation results.

In 2017, Angadi and Kagawade [3] have developed an algorithm namely spectral domain feature extraction. The model was used for extracting the features, where the initial face part was extracted via Viola-Jones algorithm. Here in this model, the mined face part was converted within 120×120 pixels size gray scale image. In order to attain dominant magnitude spectral feature, they have formulated polar Fourier transform. As the very first process, 2D-DFT was carried out on the preprocessed image for generating DFT spectrum. The developed model was not only effective but also more accurate over conventional methods, which was on the basis of Cartesian Fourier Transform.

In 2016, Chakraborty et al. [4] have developed an approach for the detection and retrieval of face using local descriptors. Within it, a local gradient hexa pattern (LGHP) was further evaluated to identify the relations among reference pixel in addition to the respective adjoining pixels. Discriminative data are present within the local neighborhood and within dissimilar derivation directions. The developed descriptor efficiently converts those relations within binary micropatterns. Finally, the performance in terms of both recognition and retrieval has been compared over other approaches and found to be more accurate.

In 2017, Yang et al. [5] have formulated a technique on behalf of face recognition. The facial images obtained from the surveillance videos typically have an inadequate resolution. In recent times, multidimensional scaling (MDS) has been engaged to resolve these issues. Here, a discriminative MDS approach can be employed to discover a mapping matrix that shows the higher resolution images as well as lower resolution images to an ordinary subspace. This approach can be considered to be more discriminative depending on both interclass distances and intra-class distances. Out of the experimental results, it has been proven that approach attains enhanced performance when comparing by additional techniques.

In 2016, Wang et al. [6] have developed a new image processing approach namely Adaptive Singular Value Decomposition within a 2D discrete Fourier domain (ASVDF) toward recognizing face under different lighting. The approach could make clear color-face images, which was more natural and smoother. This could be achieved even the face image has undertaken lateral lighting. The model transfers the color-face image to the Fourier domain. The investigational outcomes have proven the betterments of developed model across other techniques with regard to real-time applications.

In 2017, García et al. [7] have introduced a face recognition approach. Here, face recognition system using capability of attaining higher recognition rates, although the face image produces essential occluding, based on a wireframe pattern in addition to a Face fit tool that employs a coordinated method using the fiducial points which differentiate parts of facial. Thus, the estimation of the occlusions is done with the selection of the items which correspond to four circumstances of incomplete blockings of the face like blockage within the area of eyes, blockage within the mouth region as well as blockage within left as well as right area of face independently. From the simulation outcomes, it is obvious that this approach attains better stability and recognition rate for the identification system than the other approaches.

In 2016, Xu et al. [8] have developed a sparse representation approach for face recognition using l2 regularization in which a noticeable performance can be attained in face recognition. The discriminative presentation could be attained by correlation reduction of presentations. In sparse representation approaches, the presentation correlation of the test sample generated from various classes could be minimized in terms of sparsity and collaboration. The corresponding idea could be implemented for enhancing the class's differences. However, the optimal value of parameters was not yet satisfied with the proposed model, which will be considered in future.

(a) **AAM-based Feature Extraction**

AAM [25] is a computer vision algorithm for structuring the facial features: shape and appearance. The extraction is done by automated locating landmark points, which directly determine the texture and shape of statistically modeled image objects. AAM is a broadly used facial expression analysis.

Shape Model: This is the form of geometric data and it is stable from corner to corner of image class. Scientifically, the shape that is determined by landmark points within dimensional space is denoted using vector. In 2D images, landmarks define vector as in Eq. (1).with statistical validity of T.

$$\bar{x} = \left(\bar{x}_{1,}\bar{x}_{1,}\bar{x}_{2,}\bar{x}_{2.} \ldots\ldots \bar{x}_{z}, \bar{x}_{z}\right)^{\mathrm{T}} \tag{1}$$

The representation of shape with respect to same referential shape is more vital for attaining statistical validity. The eradication of location impact, scale, and rotation impact is done, and Generalized Procrustes Analysis (GPA) is executed for placing the entire shapes within a mutual frame. This GPA comprises of consecutively alignment shape pairs that are by mean shape. The process continues until there have no changes in mean shape within the iterations up to N. Moreover, the recomputation of aligned shape is defined in Eq. (2).

$$\tilde{x}_w = \frac{1}{N}\sum_{i=1}^{N}\bar{x}_i \tag{2}$$

Then, Principal Components Analysis (PCA) is done for minimizing the sizes of data by data direction exploration with highest variance and data are projected onto the direction. Each point is evaluated as the sum of mean and Eq. (3) gives the orthogonal linear transformation. Here, $\tilde{x}$ indicates the mean shape vector, shape parameter is denoted by $\emptyset_i$ with its b_i coefficients.

$$\bar{x}_i = \tilde{x} + \sum_{i=1}^{t'} \emptyset_i b_i \tag{3}$$

Appearance Model: The construction of full model of face image needs both the texture and shape model. Thus, the construction of statistical texture model is the next step, where the texture sample alignment is needed for a reference surface frame and the appearance is composed of texture information. As the structuring of statistical appearance model is needed to warp the color channels, initially the controlling points are paralleled toward the mean dimension. The piece-wise affine warping (i.e., dividing the convex outlines of the mean dimension by triangle sets) is done for matching the texture. The appearance model is attained by concerning PCA for texture vectors as defined in Eq. (4), in which AP_O denotes mean appearance vector, δ_i indicates appearance parameter, as well as the synthesized appearance vector out of affine wrapping is indicated by $\boldsymbol{AP}_i(\bar{x})$.

$$AP(\bar{x}) = AP_0(\bar{x}) + \sum_{i=1}^{mm} \delta_i AP_i(\bar{x}) \tag{4}$$

3 LCDRC-Based Classification [26]

This paper uses LCDRC [26] classifier to recognize the face image from the training image. Let the training matrix is represented as $X = [X_{1\ldots}, X_i \ldots X_P] \in \Re^{pxqi}$ where p indicates the dimension of respective image of training face, q_i refers to counting training face image out of class. Consider that matrix of subspace projection that must be drawn is denoted a $U \in \Re^{p\times d}$ & d < p. Every x_{ij} can be mapped for learnt subspace using $g_{ij} = U^T \cdot x_{ij}$, where $1 \leq j \leq q_i$. The matrix of whole image of training face is then mapped by means of $G = U^T.X \in \Re^{d\times qi}$ and for each class $G_i = U^T.X_i \in \Re^{d\times qi}$. The collaborative between-class reconstruction error (CBCRE) and within-class reconstruction error (WCRE) are defined as in Eqs. (5) and (6).

$$CBCRE = \frac{1}{q}\sum_{i=1}^{c}\sum_{j=1}^{q_i} \|g_{ij} - \hat{g}_{ij}^{inter}\|_2^2 \tag{5}$$

$$WCRE = \frac{1}{q}\sum_{i=1}^{c}\sum_{j=1}^{q_i} \|g_{ij} - \hat{g}_{ij}^{intra}\|_2^2 \tag{6}$$

where $\hat{g}_{ij}^{inter} = G_{ij}^{inter}\alpha_{ij}^{inter}$ and $\hat{g}_{ij}^{intra} = G_{ij}^{intra}\alpha_{ij}^{intra}$, $G_{ij} is$ the G with G_i eliminated and G_{ij}^{intra} is the G_i with g_{ij} eliminated. α_{ij}^{inter} Andand α_{ij}^{intra} is are attained by Eq. (7).

$$\hat{\alpha}_i = \left(X_i^T X_i\right)^{-1} X_i^T g, i = 1, 2, \ldots c \tag{7}$$

Prior obtaining the U, the value of α in learned subspace is unknown for us. However, the $\hat{\alpha}$ is evaluated within primitive space as well as $\hat{\alpha}$ is used by means of α approximation. Out of Eqs. (5) and (6), variation among CBCRE and BCRE is seen. As per relationship among X as well as G, CBCRE and WCRE are rewritten by means of Eqs. (8) and (9). This is again rewritten as in Eqs. (10) and (11).

$$CBCRE = \sum_{i=1}^{c}\sum_{j=1}^{q_i} \| U^T x_{ij} - U^T X_{ij}^{inter} \alpha_{ij}^{inter} \|_2^2. \tag{8}$$

$$WCRE = \sum_{i=1}^{c}\sum_{j=1}^{q_i} \| U^T x_{ij} - U^T X_{ij}^{intra} \alpha_{ij}^{intra} \|_2^2 \tag{9}$$

$$CBCRE = \sum_{i=1}^{c}\sum_{j=1}^{q_i} \left(x_{ij} - X_{ij}^{inter} \alpha_{ij}^{inter}\right)^T UU^T \left(x_{ij} - X_{ij}^{inter} \alpha_{ij}^{inter}\right) \tag{10}$$

$$WCRE = \sum_{i=1}^{c}\sum_{j=1}^{q_i} \left(x_{ij} - X_{ij}^{inter} \alpha_{ij}^{intra}\right)^T UU^T \left(x_{ij} - X_{ij}^{intra} \alpha_{ij}^{intra}\right) \tag{11}$$

In this, CBCRE as well as WCRE possess $\frac{1}{P}$ as factor and hence, it is prudent for eradicating $\frac{1}{P}$ out of CBCRE and WCRE correspondingly excluding the effect of ratio value of CBCRE across WCRE. In this situation, the CBCRE and WCRE are expressed as in Eqs. (12) and (13) where $tr(\cdot)$ indicates the trace operator.

$$CBCRE = \sum_{i=1}^{c}\sum_{j=1}^{q_i} tr\left(U^T \left(x_{ij}\_X_{ij}^{inter} \alpha_{ij}^{inter}\right) \left(x_{ij} - X_{ij}^{inter} \alpha_{ij}^{inter}\right)^T U \right) \tag{12}$$

$$WCRE = \sum_{i=1}^{c}\sum_{j=1}^{q_i} tr\left(U^T \left(x_{ij}\_X_{ij}^{intra} \alpha_{ij}^{intra}\right) \left(x_{ij} - X_{ij}^{intra} \alpha_{ij}^{intra}\right)^T U \right) \tag{13}$$

The Eigen vectors EI_b and EI_w is are determined as in Eqs. (14) and (15).

$$EI_b = \frac{1}{p} \sum_{i=1}^{c}\sum_{j=1}^{qi} \left(x_{ij} - X_{ij}^{inter} \alpha_{ij}^{inter}\right) \left(x_{ij} - X_{ij}^{inter} \alpha_{ij}^{inter}\right)^T \tag{14}$$

$$EI_w = \frac{1}{p} \sum_{i=1}^{c}\sum_{j=1}^{q_i} \left(x_{ij} - X_{ij}^{intra} \alpha_{ij}^{intra}\right) \left(x_{ij} - X_{ij}^{intra} \alpha_{ij}^{intra}\right)^T \tag{15}$$

Finally, the CBCRE as well as WCRE are rewritten as in Eqs. (16) and (17)

$$CBCRE = tr\left(U^T EI_b U\right) \tag{16}$$

$$WCRE = tr\left(U^T EI_w U\right) \tag{17}$$

For maximizing the CBCRE as well as to reduce the WCRE instantaneously, the Maximum Margin Criterion (MMC) is assumed for maximizing a given standard in Eqs. (18) and (19)

$$\max_U J(U) = \max_U (CBCRE - WCRE) \quad (18)$$

$$\max_U J(U) = \max_U \big(tr\big(U^T (EI_b - EI_W)U\big)\big) \quad (19)$$

This can be solved with the identification of a largest d Eigen values in addition to the permitting Eigen values by means of the following Eq. (20).

$$(EI_b - EI_w)u_k = \lambda_k u_k, k = 1, 2 \ldots d \quad (20)$$

where $\lambda_1 \geq \ldots \geq \lambda_k \ldots\ldots \lambda_d$ and $U = [u_1, \ldots u_k, \ldots, u_d]$ MMC solve the small sample size problem (SSP), in which the face image dimensions are greater compared with the images of count of training face.

4 The Role of Proposed Face Recognition Model

As declared above, the major steps of LCDRC classification go out with the projection matrix U. The projection matrix U gets multiplied with the features extracted, which is the part of LCDRC classification. It is planned to find the optimal U, so that the recognition accuracy may increase. The optimal selection of U is achieved by the introduction of a new CEWO algorithm, which is explained in the upcoming section.

5 CEWO Algorithm for Projection Matrix Optimization Using Objective Function

The objective model of the proposed face recognition model is given in Eqs. (21), and (22). At first, the error, *error*, between actual value, *act* and predicted value, *pred,* is evaluated followed by fitness function. In Eq. (22), λ indicates the regularization and the minimization of total error along with λ is considered as the major intensive of this proposed work.

$$error = (act - pred) \quad (21)$$

$$FT = Min\left(Sum(error) + \lambda * \sum_{i=1}^{NU} (U)^2\right) \quad (22)$$

5.1 Solution Encoding

The solution given to the proposed CEWO is the projection matrix, U that is to be optimally chosen. Figure 2 illustrates the diagrammatic solution encoding. Here U_{NU} denotes the total number of projection matrix (Fig. 1).

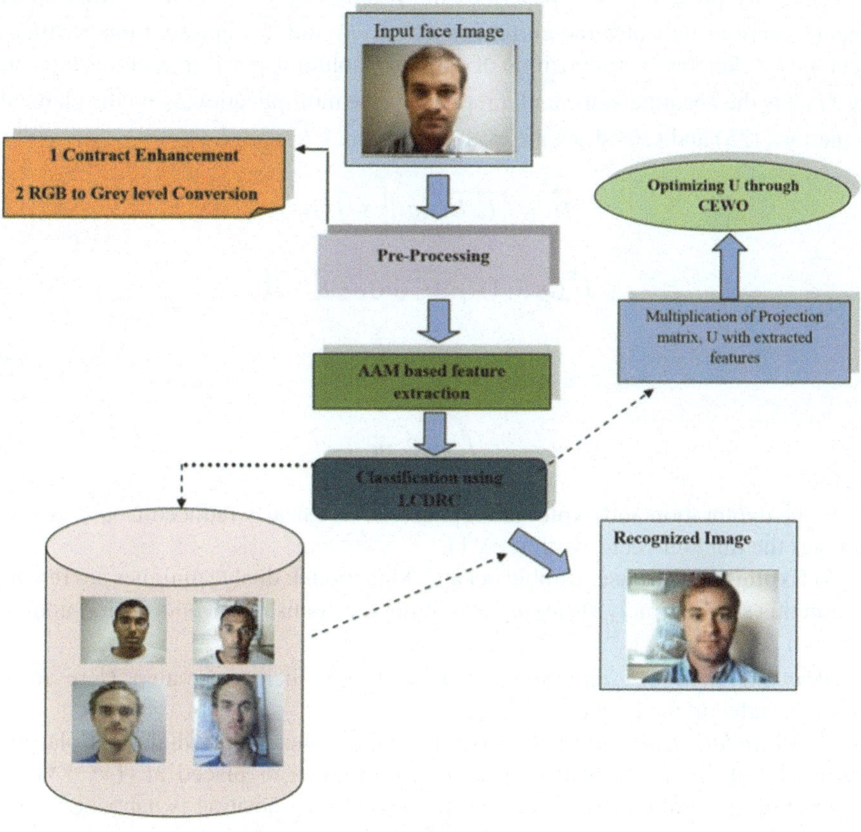

Fig. 1 Architecture of the proposed face recognition

Fig. 2 Solution Encoding

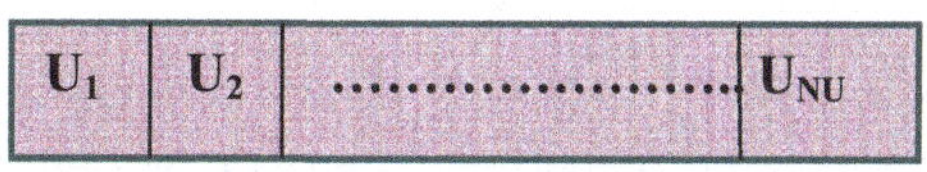

5.2 Conventional WOA

Different phases are there in WOA algorithm [28] developed by Seyedali Mirjalili and Andrew Lewis, in this, they are encircling prey, spiral bubble-net feeding maneuver, and search for prey. An improvement is made with the conventional WOA to get the optimal U, which is explained below.

Enclosing prey: Eq. (23) describes the position update for the optimal search agent, where tn indicates the current iteration, $\vec{K}$ and $\vec{L}$ represent the coefficient vectors, U^* denotes vector position of optimal resolution, position vector is indicated by $\vec{U}$, || is the absolute value and it represents the multiplication using the elements. Equations (25) and (26) show the evaluation of and $\vec{K}$ and $\vec{L}$.

$$\vec{H} = \left|\vec{L}.\vec{U}^*(tn)\right| - \vec{U}(tn) \tag{23}$$

$$\vec{U}(tn+1) = \vec{U^*}(tn) - \vec{K} \cdot \vec{H} \tag{24}$$

$$\vec{K} = 2.\vec{a}.ru - \vec{a} \tag{25}$$

$$\vec{L} = 2.\vec{ru} \tag{26}$$

In the exploration and exploitation phase, $\vec{a}$ is linearly reduced from 2 to 0, $\vec{ru}$ denotes the random vector within [0, 1].

(i) **Exploitation phase**: Bubble net attacking model: dual techniques are resolved within this stage, namely shrinking encircling mechanism as well as spiral updating position.

Shrinking encircling approach is achieved by reducing $\vec{a}$ value. Here $\vec{K}$ is a arbitrary value in the interval $-\vec{a}, \vec{a}$.

Spiral updating position first computes the distance amid whale that placed or positioned at (U, Y) as well as prey that positioned or placed at (U^*, Y^*). The generation of spiral equation between whale and prey position is happened, which gives the humpback whale's helix-shaped movement and it is defined in Eq. (27).

$$\vec{U}(tn+1) = \vec{H} \cdot e^{kn} \cdot \cos(2\pi n) + \vec{U^*}(tn) \tag{27}$$

where $\vec{H'} = \left|\vec{U^*}(tn) - \vec{U}(tn)\right|$, determines the distance of ith whale toward prey, k represents a constant value, n represents a random integer within [− 1, 1] and · represents the multiplication using the elements. Equation (28) gives a mathematical model of whale's position update, where $\tilde{m}$ denotes a random number in [0, 1].

$$\vec{U}(tn+1) = \begin{cases} \vec{U^*}(tn+1) - \vec{K} \cdot \vec{H} \; if\, \tilde{m} < 0.5 \\ \vec{H''} \cdot e^{kn} \cdot \cos(2\pi n) \; if\, \tilde{m} \geq 0.5 \end{cases} \tag{28}$$

(ii) **Exploration phase**: In this exploration phase, investigation agent position is upgraded according to an arbitrarily selected investigation agent. Equations (29) and (30) show its mathematical model, in which U^{ra} denotes a vector of random position. Algorithm 1 presents a pseudo code of whale optimization:

$$\vec{H} = \left| \vec{L} \cdot \overrightarrow{U^{ra}} - \vec{U} \right| \tag{29}$$

$$\vec{U}(tn + 1) = \overrightarrow{U^{ra}} - \vec{K} \cdot \vec{H} \tag{30}$$

Algorithm 1: WOA algorithm [28]

Fitness Calculation of every search agent

U^* is the best search agent

While $(tn < MAX^{tn})$

  For each search agent

  Update, $\vec{a}, \vec{K}, \vec{L}$, n and $\tilde{m}$

    If 1 $(\tilde{m} < 0.5)$

      If $2(|\vec{K}| < 1)$

        Position update as per Eq. (23)

      Elseif $2(|\vec{K}| \geq 1)$

        Select X^{ra}

        Position update by Eq. (30)

      End if 2

    Else if1 $\tilde{m} < 0.5$

        Position update of current search by Eq.(27)

    End if

  End for

Check for the search agent that goes beyond the search space

  Fitness evaluation of each search agent

  Update U^* if there is a better solution

  $tn=tn+1$

End while

Return U^*

6 Proposed CEWO Model

This research framework proposes a novel CEWO procedure for attaining an optimal projection matrix, U, which is the improvement of conventional WOA algorithm. The pseudo code of proposed CEWO is offered within Algorithm 2. The enhancement is made in case of no improvement in fitness evaluation. Further section in Algorithm 2 is executed as per the number of cycle N_{cycle}. This paper fixes $N_{cycle} = 3$, and finds the best solution as the optimal projection matrix U^*.

Algorithm 2: CEWO based optimal Projection matrix selection

Input: Projection matrix U

Output: Optimal Projection matrix U^*.

Fitness Calculation of each search agent

U^* is the best search agent

While($tn < MAX^{tn}$)

For each search agent

Update, $\vec{a}, \vec{K}, \vec{L}$, n and $\tilde{m}$

If 1($\tilde{m} < 0.5$)

If 2 ($|\vec{K}| < 1$)

Position update as per Eq. (23)

Else if2($|\vec{K}| \geq 1$)

Select X^{ra}

Position update by Eq. (30)

End if 2

Else if 1$\tilde{m} \geq 0.5$

Position update of current search by Eq.(27)

End if

End for

Check for the search agent that goes beyond the search space

Fitness evaluation of each search agent

Update U^* if there exist better solution

$tn = tn + 1$

End while

If (Fitness is not improved)

For $(cycle = 1, \ldots N_{cycle})$

$\tilde{m} < 0.5$

if $|\vec{K}| \geq 1$

Execute Eq. (30)

End

ReturnU^*

7 Results and Discussions

7.1 Simulation Setup

The presented face recognition model was implemented within MATLAB 2015a, i3, 2.10 GHz Processor. The image database was downloaded from URL: http://cswww.essex.ac.uk/mv/allfaces/index.html. A database includes both male and female images. There are many optimization techniques that provide the better solution form the random samples such as [27–30]. The proposed method is superior to conventional methods such as the fruit fly optimization algorithm [27], conventional Whale Optimization Algorithm (WOA) [28], Grey Wolf Optimization (GWO) [29], Firefly (FF) [30]. The performance of the proposed model was analyzed in terms of measures like Accuracy, Precision, FPR, FNR with the consideration of varying learning percentages to 40, 50, 60, 70, and 80%.

7.2 Performance Analysis of the Proposed Recognition Model

The efficiency of presented face recognition with regard to different learning percentages is illustrated in Fig. 3. Percentage is varying to 40%, 50%, 60%, 70%, and 80%, respectively. The study has proven the superiority of the presented technique with respect to positive as well as negative measurements. Figure 3a shows the resultant graph of accuracy measure, and through the graphical representation, the performance rate of the proposed method is clearly evident with high accuracy. It is also shown that for 40% learning, the proposed method enhanced compared with GWO as well as FF, correspondingly. The presented approach is efficient compared with

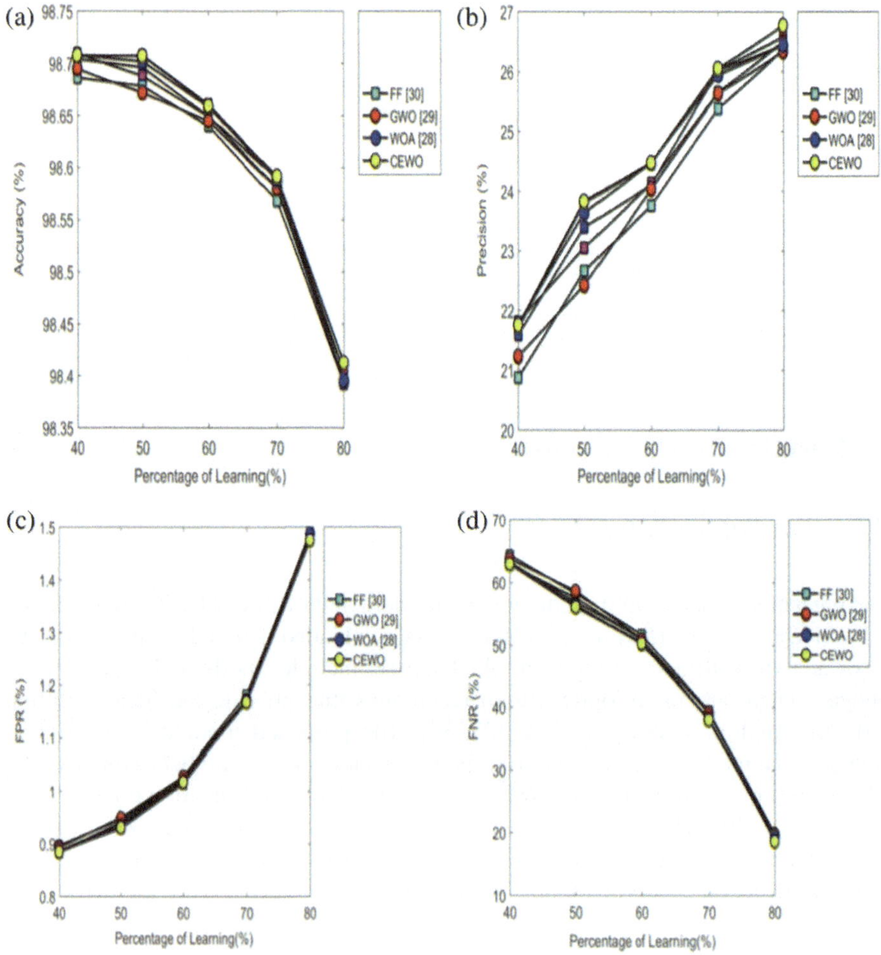

Fig. 3 Performance analysis of proposed and conventional methods. **a** Accuracy, **b** precision, **c** FPR, and **d** FNR

Table 1 Overall performance of the proposed and conventional methods

Measures	FF [30]	GWO [29]	WOA [28]	CEWO
Accuracy	0.984007	0.983877	0.983823	0.984169
Precision	0.265625	0.263666	0.261943	0.268544
FPR	0.014846	0.014966	0.014977	0.014715
FNR	0.189145	0.190789	0.197368	0.184211

other conventional methods like WOA, GWO, FF on behalf of 50% of the learning. Compared with GWO and FF, the proposed method is enhanced on behalf of 60% learning. Figure 3b gives the resultant graph of precision which is measure of quality illustrated in the graph, compared with the existing approaches, the precision rate is enhanced by this approach. Figure 3c gives the FPR which is false positive rate and is less compared with other methods. Similarly, Fig. 3d gives FNR status and presents the better performance of the proposed work. Hence for 40, 50, 60% learning attains less FPR values and it is better from WOA, GWO, and FF. Hence the enhancement of presented face recognition model is proven in terms of various learning proportions.

Table 1 shows the complete performance of presented face recognition model. From Table 1, it is clearly shown that the presented method is enhanced compared with conventional approaches with regard to all respective measures. Here, the accurateness of the presented approach is enhanced from WOA, and GWO, and superior to FF correspondingly. The accuracy of the presented approach is also better than WOA, GWO, FF, respectively. Similarly, the FPR, FNR of this approach are quite less than additional models like WOA, GWO, FF, therefore, the entire study has shown the capability of this work in recognizing face images.

8 Conclusion

This research framework has developed a novel face recognition model with a new contribution on feature extraction phase. The feature extraction was depending upon AAM as well as on behalf of categorization; this paper has used the LCDRC classifier. The major contribution was in LCDRC classifier, in which the projection matrix was multiplied with extracted features, the issue of choosing the optimal projection matrix was fulfilled by proposing a new CEWO. The efficiency of the presented method, the CEWO, and additional conventional methods are compared with regard to both positive as well as negative measurements, and from results, it is proved that the proposed method is superior to WOA, GWO, FF. Hence, the satisfactory results have shown the better performance of the proposed model with high accuracy.

References

1. J. Lin, C.T. Chiu, Low-complexity face recognition using contour-based binary descriptor. IET Image Process. **11**(12), 1179–1187 (2017).https://doi.org/10.1049/iet-ipr.2016.1074
2. A. Ghasemzadeh, H. Demirel, 3D discrete wavelet transform-based feature extraction for hyperspectral face recognition. IET Biometr. **7**(1), 49–55 (2018).https://doi.org/10.1049/iet-bmt.2017.0082
3. S.A. Angadi, V.C. Kagawade, A robust face recognition approach through symbolic modeling of Polar FFT features. Pattern Recogn. **71**, 235–248 (2017). https://doi.org/10.1016/j.patcog.2017.06.014
4. S. Chakraborty, S.K. Singh, P. Chakraborty, Local gradient hexa pattern: a descriptor for face recognition and retrieval. IEEE Trans. Circuits Syst. Video Technol. **28**(1), 171–180 (2018). https://doi.org/10.1007/s11042-019-07908-3
5. F. Yang, W. Yang, R. Gao, Q. Liao, Discriminative multidimensional scaling for low-resolution face recognition. IEEE Signal Process. Lett. **25**(3), 388–392 (2018)
6. J.-W. Wang, N.T. Le, J.-S. Lee, C.-C. Wang, Color face image enhancement using adaptive singular value decomposition in fourier domain for face recognition. Pattern Recogn. **57**, 31–49 (2016). https://doi.org/10.1016/j.patcog.2016.03.021
7. E. Garcia Rios, E. Escamilla Hernandez, M. Nakano Miyatake, H. Perez Meana, Face recognition with occlusion using a wireframe model and support vector machine. IEEE Latin Am. Trans. **15**(10), 1960–1966 (2017). https://doi.org/10.1109/tla.2017.8071241
8. Y. Xu, Z. Zhong, J. Yang, J. You, D. Zhang, A new discriminative sparse representation method for robust face recognition via l2 regularization. IEEE Trans. Neural Netw. Learn. Syst. **28**(10), 2233–2242 (2017). https://doi.org/10.1109/TNNLS.2016.2580572
9. Y. Peng, L. Spreeuwers, R. Veldhuis, Low-resolution face alignment and recognition using mixed-resolution classifiers. IET Biometr. **6**(6), 418–428 (2017). https://doi.org/10.1049/iet-bmt.2016.0026
10. Z. Xia, X. Peng, X. Feng, A. Hadid, Scarce face recognition via two-layer collaborative representation. IET Biometr. **7**(1), 56–62 (2018).https://doi.org/10.1049/iet-bmt.2017.0193
11. K. Grm, V. Štruc, A. Artiges, M. Caron, H.K. Ekenel, Strengths and weaknesses of deep learning models for face recognition against image degradations. IET Biometr. **7**(1), 81–89 (2018).https://doi.org/10.1049/iet-bmt.2017.0083
12. J. Liang, M. Wang, Z. Chai, W. Qin, Different lighting processing and feature extraction methods for efficient face recognition. IET Image Process. **8**(9), 528–538 (2014). https://doi.org/10.1049/iet-ipr.2013.0006
13. H. Boughrara, C.B. Amar, L. Chen, Face recognition based on perceived facial images and multilayer perceptron neural network using constructive training algorithm. IET Comput. Vis. **8**(6), 729–739 (2014). https://doi.org/10.1049/iet-cvi.2013.0294
14. W. Chen, Y. Gao, Face recognition using ensemble string matching. IEEE Trans. Image Process. **22**(12), 4798–4808 (2013). https://doi.org/10.1049/iet-cvi.2013.0294
15. Dagher, R. Nachar, Face recognition using IPCA-ICA algorithm. IEEE Trans. Pattern Anal. Mach. Intell. **28**(6), 996–1000 (2006). https://doi.org/10.1109/tpami.2006.118
16. J. Lu, K.N. Plataniotis, A.N. Venetsanopoulos, Face recognition using kernel direct discriminant analysis algorithms. IEEE Trans. Neural Netw. **14**(1), 117–126 (2003).https://doi.org/10.1109/tnn.2002.806629
17. J. Lu, K.N. Plataniotis, A.N. Venetsanopoulos, Face recognition using LDA-based algorithms. IEEE Trans. Neural Netw. **14**(1), 195–200 (2003).https://doi.org/10.1109/tnn.2002.806647
18. W. Deng, J. Hu, Z. Wu, G. Jun, From one to many: pose-aware metric learning for single-sample face recognition. Pattern Recogn. In press, corrected proof. Available online (2017). https://doi.org/10.1016/j.patcog.2017.10.020
19. T. Larrain, J.S. Bernhard, D. Mery, K.W. Bowyer, ace recognition using sparse fingerprint classification algorithm. IEEE Trans. Inf. Forensics Secur. **12**(7), 1646–1657 (2017). https://doi.org/10.1007/978-3-030-304652

20. N. Hao, H. Liao, Y. Qiu, J. Yang, Face super-resolution reconstruction and recognition using non-local similarity dictionary learning based algorithm. IEEE/CAA J. Autom. Sinica **3**(2), 213–224 (2016)
21. C. Galea, R.A. Farrugia, Forensic face photo-sketch recognition using a deep learning-based architecture. IEEE Signal Process. Lett. **24**(11), 1586–1590 (2017). https://doi.org/10.1109/LSP.2017.2749266
22. A.A.S. Syed, M. Bennamoun, F. Boussaid, Iterative deep learning for image set based face and object recognition. Neurocomputing **174**, 866–874 (2016). https://doi.org/10.1016/j.neucom.2015.10.004
23. X. Zhao, G. Evangelopoulos, D. Chu, S. Shah, I.A. Kakadiaris, Minimizing illumination differences for 3D to 2D face recognition using lighting maps. IEEE Trans. Cybern. **44**(5), 725–736 (2014). https://doi.org/10.1109/TCYB.2013.2291196
24. M. De Marsico, M. Nappi, D. Riccio, H. Wechsler, Robust face recognition for uncontrolled pose and illumination changes. IEEE Trans. Syst. Man Cybern. Syst. **43**(1), 149–163 (2013). https://doi.org/10.1109/TSMCA.2012.2192427
25. Y.H. Lee, W. Han, Y. Kim, B. Kim, Facial feature extraction using an active appearance model on the iPhone, in *2014 Eighth International Conference on Innovative Mobile and Internet Services in Ubiquitous Computing, Birmingham*, pp. 196–201 (2014).https://doi.org/10.1109/imis.2014.24
26. X. Qu, S. Kim, R. Cui, H.J. Kim, Linear collaborative discriminant regression classification for face recognition. Visual Commun. Image Retrieval **31**, 312–319 (2015). https://doi.org/10.1016/j.jvcir.2015.07.009
27. Y. Wang, Y. Bai, Y. Hao, Image restoration based on structure and fruit fly optimization algorithm, in *2016 7th IEEE International Conference on Software Engineering and Service Science (ICSESS), Beijing*, pp. 622–626 (2016).https://doi.org/10.1109/icsess.2016.7883146
28. S. Mirjalili, A. Lewis, The whale optimization algorithm. Adv. Eng. Softw. **95**, 51–67 (2016). https://doi.org/10.1016/j.advengsoft.2016.01.008
29. S. Mirjalili, S.M. Mirjalili, A. Lewis, Grey wolf optimizer. Adv. Eng. Softw. **69**, 46–61 (2014). https://doi.org/10.1016/j.advengsoft.2013.12007
30. A.H. Gandomi, X.-S. Yang, S. Talatahari, A.H. Alavi, Firefly algorithm with chaos. Commun. Nonlinear Sci. Numer. Simulat. **18**, 89–98 (2013). https://doi.org/10.1155/2013/125625

Auction System in Food Supply Chain Management Using Blockchain

A. N. Shwetha and C. P. Prabodh

Abstract Blockchain has become very popular these days because of its decentralized nature. The applications of blockchain are spread out in all domains from agriculture to medical field. Food supply chain can be made transparent using blockchain. The need for transparency in food supply chain is increasing day by day because nowadays people are more cautious about the food they buy and consume. So every individual should be able to get complete details about the food they eat from farming to consumer end. In this area, blockchain is playing a vital role. Along with giving complete information of food products to end user, it is also required for a farmer to get good prices for the food products they had grown. Currently, food products are sold in centralized auction system with the involvement of a third party. Because of the biased nature of third party, farmers are not able to get valid price for food items they have even though there is a demand in market. So, this paper is discussing an auction system for food products without a third party using blockchain technology, which avoids biasing.

Keywords Auction · Biasing · Blockchain · Supply chain management

1 Introduction

1.1 Blockchain

Blockchain is a distributed ledger that has its inherent properties like decentralization, transparency, security and trust. Blockchain became quite popular after its usage in cryptocurrency applications like Bitcoin. Blockchain allows peers to negotiate and

A. N. Shwetha · C. P. Prabodh (✉)
Department of Computer Science and Engineering, Siddaganga Institute of Technology, Tumakuru, India
e-mail: prabodh@sit.ac.in

A. N. Shwetha
e-mail: shwethaan@sit.ac.in

C. Kiran Mai et al. (eds.), *Proceedings of International Conference on Advances in Computer Engineering and Communication Systems*, Learning and Analytics in Intelligent Systems 20,
https://doi.org/10.1007/978-981-15-9293-5_3

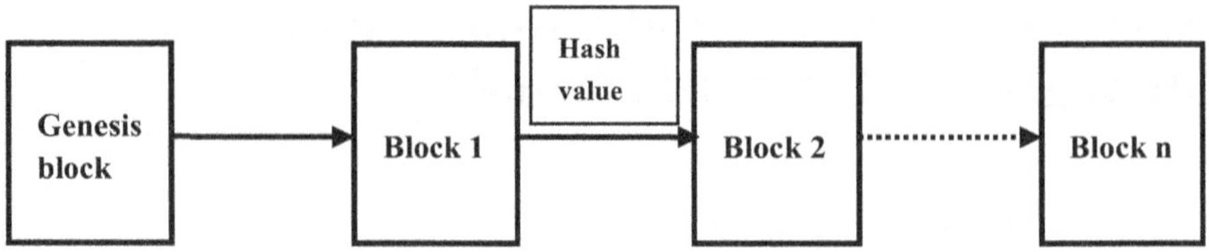

Fig. 1 Illustration of a blockchain

record transactions in a transparent ledger, which is publicly viewable and verifiable. Blockchain is basically a decentralized database where data will be stored in the form of blocks as shown in Fig. 1. In each block, a set of transactions will be stored. Each block is immutable, once the block is attached to a chain and stored in database, no one can alter the data present in a block. When a transaction is completed in a blockchain network, all nodes present in that network need to validate a transaction depending on the information of previous transactions. There are two available mechanisms to select the node for mining namely proof of work and proof of stake.

1.2 Auction System

The most common auction system that is in use currently is a public auction system where auction/bidding will be done face to face. The highest bidder will win the bidding and he will own the bidding products after payments. Here bidding information is not available to everyone because it is centralized and which can also be modified.

In the traditional market, crops are auctioned using the Ascending Auction where the auction of crop produce starts from an initial bid specified by the auctioneer. He then increases the bid amount in small increments, bidders can choose to continue or drop out as the price increases. Finally, when everyone except one bidder has dropped out, the left out bidder then purchase the crop at the final bid price. In the other end, some of the farmers do not know about the bidding value or demand of food products in the market. A third party or broker will go to a farm land and buy the food products for the rate of his benefit.

This problem can be solved by using blockchain in the auction system which will make the bidding transparent to everyone and farmers will get benefited. Because of usage of blockchain in bidding, the bidding price will be immutable and that information will be available to everyone. So here third party is not required, because it can be replaced by smart contract of blockchain that is trustworthy. The bidding happens online directly between a buyer and a seller which is profitable to both parties. Table 1 illustrates the comparison between the existing and proposed systems.

Table 1 Comparison of auctioning solutions in the existing and proposed systems

	Traditional centralized auction system	Blockchain-based decentralized auction system
Third party	Required	Not required
Security to data	Less	More
Availability of bidding data	Available to few people	Available to everyone in blockchain network
Profit to farmers	Less	More
Mode of auction	Offline	Online
Trust on system	Less	More
Transparency	Less	More

2 Literature Survey

Zheng et al. [1] have given the overview of blockchain technology, explained the architecture of blockchain, consensus algorithms used to mine a block, block structure and future challenges in blockchain. Beck et al. [2] have developed a proof of concept prototype to replace coffee shop payment system. They had also evaluated strengths and weaknesses of blockchain technology in replacing it with a trust-based centralized system. Apart from its many advantages, there are a few hindrances also like scalability, cost and volatility in currency value.

Sonnino et al. [3] proposed a sealed bid auction system using blockchain. Here workers will submit a sealed bid containing a minimum price for their work without disclosing a bid value to their peers. Then auction system will determine optimal price of services and assign suitable workers to clients. Xiong and Wang [4] proposed sealed bid auction protocol for consortium blockchain. Digital certificate-based identity mechanism is used to identify authorized participants. To prevent intruders and to encrypt the bids, public key encryption algorithm is used.

Baki [5] describes auctioning, traditional auction system, three categories of blockchain and also analyzed the benefits of using blockchain in auction mechanism. Caro et al. [6] present a decentralized, blockchain-based traceability system for agricultural food supply chain management. They have used various IoT sensors from farm to fork to collect data at each level and stored in blockchain. This traceability system is implemented using ethereum blockchain frameworks.

Motivated by the food scandals in Taiwan such as contaminated food and drainage oil, Lin et al. [7] have proposed a conceptual framework for food traceability system using ethereum blockchain framework. They had also analyzed the system by considering effectiveness of blockchain in food traceability system. Leng et al. [8] have proposed a public blockchain on double chain architecture for agriculture supply chain. Mainly discussed about structure of dual chain, procedure of creation, its data storage mode, resource availability on rent basis and consensus algorithms of blockchain used to mine a block. The double chain architecture of blockchain in agriculture food supply chain is able to provide transparency, security to transactions, privacy, resource allocation based on rent basis and credibility.

Kim et al. [9] have proposed a farm to fork network for food traceability using IoT devices. To track food products at every stage, all physical assets are represented digitally and a token is assigned to each asset. As the assets move from one place to other like processing to package, IoT devices attached to it will track the movement of assets and tracking information will be stored as transactions in blockchain. The stored transactions will be processed by smart contract. The end user can see details of each product starting from farming till purchase using blockchain network.

Lin et al. [10] have developed a prototype model for food safety traceability system based on blockchain, IoT devices and EPCIS. IoT devices are used to collect data of food products at each stage and then the collected data are categorized as on chain data and off chain data. On chain data will be stored in blockchain. Off chain data will be stored on cloud for further analysis.

Biswas et al. [11] have proposed a traceability system for wine supply chain using blockchain. Every transaction of wine is stored in block and is transparent to users. Hua et al. [12] have developed an agricultural provenance system based on blockchain, which helps to build a reliable community among different stakeholders. Hong et al. [13] have proposed agricultural products reliability system based on IoT and blockchain to provide reliability and trust. Mao et al. [14] provide credit evaluation system using blockchain to improve the management of food supply chain. The credit data are analyzed using deep learning network. Marin et al. [15, 16] have developed e-learning platform where users can post information about their methodology of farming, crops they grow, etc.

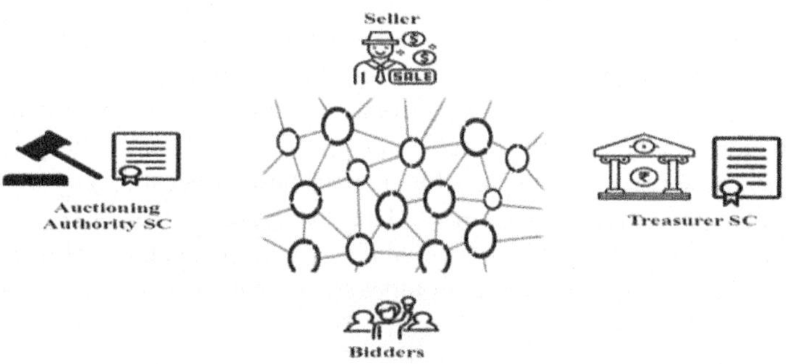

Fig. 2 System architecture

3 Design

The system architecture proposed in this paper is illustrated in Fig. 2. Generally, there are four types of auctions. They are

1. First price auction: Here all bidders submit their bid value in sealed envelope to auctioneer, who verifies bid values and announces the winner.
2. Second price auction: It is similar to previous one but the winner needs to pay the second-highest price.
3. Open ascending auction: Bidders will gradually increase bidding value and stop when they do not want to pay more than the current highest bid value.
4. Open descending auction: Auctioneer initially decides the highest bid value and gradually decreases until one of the bidders ready to pay the current bid value.

This paper is implemented open ascending auction to sell and buy food products online, which will help both buyers and sellers using blockchain. The smart contract of a blockchain is going to replace third party or auctioneer and makes the bidding system transparent and secure. Smart contract is a set of rules that perform various transactions involved in auction. The smart contract will decide the bidding winner without any bias. The auction system is implemented using Ethereum blockchain.

4 Implementation

The following roles exist in an Auction BlockChain Network.

Seller—anyone who wants to sell his produce

Bidder—anyone who wants to bid in order to purchase an auctioned item

An Auctioning Authority—which validates bids and chooses the winner

Treasurer—deals with collecting stake amount and settle monetary transaction between winning bidder and the seller.

Due to the very decentralized nature of the blockchain network, the last two roles, Auctioning Authority and the Treasurer, are coded as smart contracts that implement a business logic required by the Auction System.

For conducting an auction, the participant entities exchange the following message types to complete the auction.

- SaleMessage—<SellerId, ProductId, MinBidAmt, StTime, EndTime>
- BidMessage—<BidderId, ProductId, BidAmt, BidTime>
- WinNotice—<BidderId, ProductId, BidAmt>

First the Person, who intends to sell his produce, the Seller, creates SaleMessage with the following information minimum bid amount, bidding increment, Start time and End Time. This is then notified on the Blockchain network by the Auctioning Authority by recording this on the Blockchain as intent to sale transaction. For a Bidder to take part in the bidding process he has to stake a predefined amount of money in the form of ether with the treasurer. A possible choice for the stake value could be the same as minimum bid amount. The treasurer verifies the stake and identity of the Bidder and upon approval allows the Bidder to take part in the auction. The approved Bidders send BidMessages in response to SaleMessage notification.

The first such message is then appended to the blockchain. Later, whenever a new BidMessage arrives, the Auction authority checks if the bid amount is greater than the last bid amount recorded in the BlockChain, if yes then the BidMessage is accepted and appended to the blockchain. When the end time expires, the Auctioning Authority sends a WinNotice containing the details of the winning Bid. Any node on the network can verify the details by referring to the blockchain contents and contest this claim if it is a spurious claim within a ClaimGrievancePeriod. Then if this claim is validated, the auction is invalidated. Otherwise, the Auctioning Authority intimates the Treasurer about the successful bid. The treasurer notifies the winning bidder to transfer the winning bid amount to the Seller using Cryptocurrency.

After the Seller notifies the treasurer of successful transfer, the Treasurer then reimburses the staked money for the failed bidders and indicates the Auctioning Authority to mark the Auction as complete. In case, the bid winner backtracks to pay the highest bid price, his staked money is transferred to the seller and the next highest bidder is offered to pay his bid amount. On successful transfer, the agricultural produce is sold to him. To safeguard the interest of the Seller will not be allowed not more than three higher bids to back off, if not then all the bid deposits are pooled and paid to the Seller, which shall not exceed half the amount of the highest bid amount after which the auction sale is considered as failed.

5 Results

The results of our system are shown below. First, a seller registers his produce to be auctioned by performing a SaleMessage operation as shown in Fig. 3. Once this is notified on the blockchain network, the buyers can place their bid or back out by performing BidMessage operation as shown in Fig. 4.

After the bidding time has elapsed, the smart contract executes the WinNotice operation notifying the bidders in the blockchain network as to who has won the bid. For transparency, bid information of other bidders along with the winner is displayed as illustrated in Fig. 5. Finally, several successful auction details that have been recorded and displayed in the blockchain network are depicted in Fig. 6.

6 Conclusion

The usage of blockchain in food supply chain auction will provide transparency, security to data, increased profit to both buyers and sellers and avoids third party in auction system. The information about bidders, bidding and winner details will be

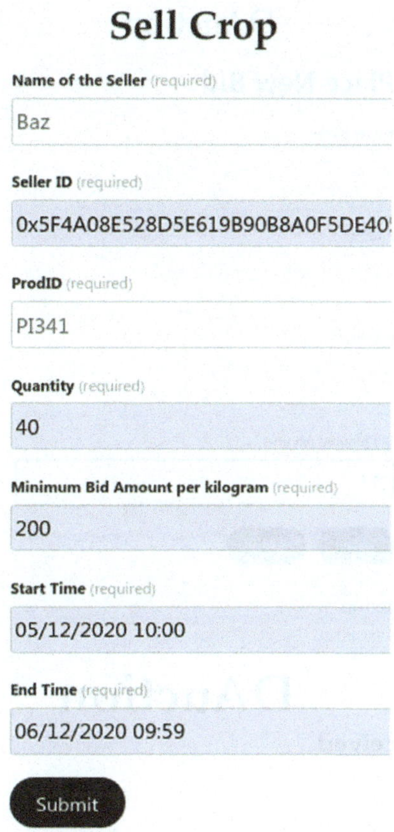

Fig. 3 Seller placing produce for bidding

stored in blockchain as transactions. So anyone can verify the transactions stored in blockchain, through which it provides transparency.

Because the transactions stored in blockchain are immutable, the data stored in blockchain are secure. If a third party is involved in auction system, some percentage of transaction fees needs to be paid to a third party. Here third party is replaced by smart contract. By usage of blockchain, smart contract is responsible to execute auction. So extra amount needs to be paid for third party is avoided, which will make both buyers and sellers profitable and they can do auction directly online without the involvement of third party.

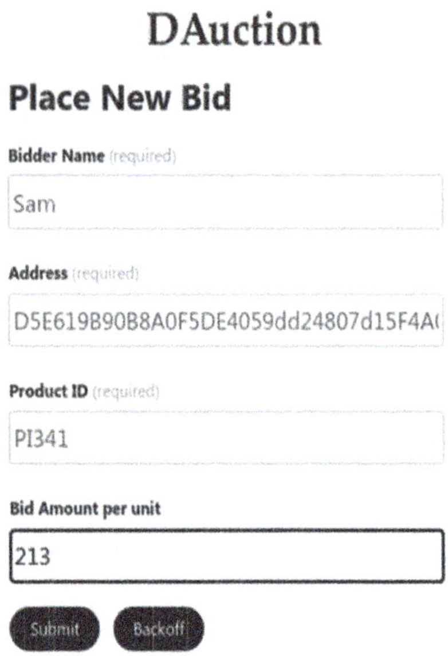

Fig. 4 Bidder placing his auction

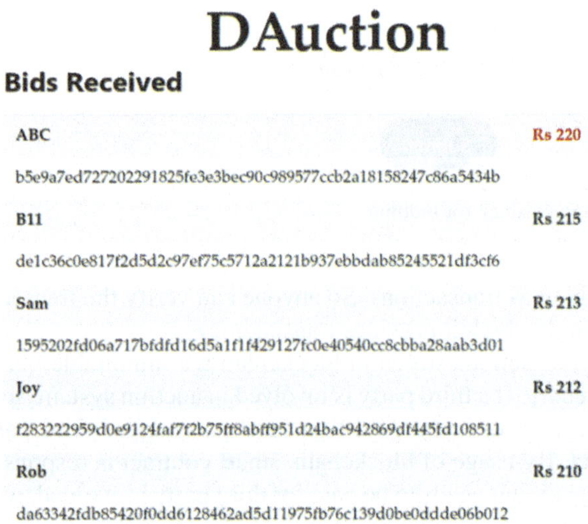

Fig. 5 Winner notification

Fig. 6 List of successful auctions

References

1. Z. Zheng, S. Xie, H. Dai, X. Chen, H. Wang (2017) An overview of blockchain technology: architecture, consensus, and future trends, in *IEEE 6th International Congress on Big Data; Honolulu*. https://doi.org/10.1109/bigdatacongress.2017.85, pp. 557–564
2. R. Beck, J. StenumCzepluch, N. Lollike, S. Malone, Blockchain—the gateway to trust-free cryptographic transactions, *Twenty-Fourth European Conference on Information Systems (ECIS)* (Turkey. Springer Publishing Company, İstanbul, 2016), pp. 1–14
3. A. Sonnino, M. Król, A.G. Tasiopoulos, I. Psaras, ASTERISK: auction-based shared economy resolution system for blockchain, in *Conference on Cryptography and Security* (2019)
4. J. Xiong, Q. Wang, Anonymous auction protocol based on time-released encryption atop consortium blockchain. Int. J. Adv. Inf. Technol. **9**(1) (2019). https://doi.org/10.5121/ijait.2019.9101
5. M.N. Baki, Auctioning using blockchain advantage analysis. Int. J. New Technol. Res. **5**(4), 109–113 (2019). ISSN: 2454-4116
6. M.P. Caro, M.S. Ali, M. Vecchio, R. Giaffreda, Blockchain based traceability in agri-food supply chain management: a practical implementation, in *IoT Vertical and Topical Summit on Agriculture, Tuscany* (2018). https://doi.org/10.1109/iot-tuscany.2018.8373021
7. I.C. Lin, H. Shih, J.C. Liu, Y.X. Jie, Food tracebility system using blockchain, in *Proceedings of 79th IASTEM International Conference, Tokyo* (2017)
8. K. Leng, Y. Bi, L. Jing, H.-C. Fu, I. Van Nieuwenhuyse, Research on agricultural supply chain system with double chain architecture based on blockchain technology. Future Gener. Comput. Syst. **86**, 641–649 (2018). ISSN 0167-739X. https://doi.org/10.1016/j.future.2018.04.061
9. M. Kim, B. Hilton, Z. Burks, J. Reyes, Integrating blockchain, smart contract-tokens, and IOT to design a food traceability solution, in *IEEE 9th Annual Information Technology, Electronics and Mobile Communication Conference (IEMCON), Vancouver, BC*, pp. 335–340 (2018). https://doi.org/10.1109/iemcon.2018.8615007
10. Q. Lin, H. Wang, X. Pei, J. Wang, Food safety traceability system based on blockchain and EPCIS, in Smart *Caching, Communications, Computing and Cyber-security for Information-Centric Internet of Things, IEEE Access* (2019). https://doi.org/10.1109/access.2019.2897792
11. K. Biswas, V. Muthukkumarasamy, W.L. Tan, Blockchain based wine supply chain traceability system, in *Future Technologies Conference (FTC), United Kingdom*, pp. 56–62 (2017)

12. J. Hua, X. Wang, M. Kang, H. Wang, F. Wang, Blockchain based provenance for agricultural products: A distributed platform with duplicated and shared bookkeeping, in *IEEE Intelligent Vehicles Symposium (IV), Changshu*, pp. 97–101 (2018). https://doi.org/10.1109/ivs.2018.8500647
13. W. Hong, Y. Cai, Z. Yu, X. Yu, An agri-product traceability system based on iot and blockchain technology, in *1st IEEE International Conference on Hot Information-Centric Networking (HotICN)*, pp. 254–255 (2018)
14. D. Mao, F. Wang, Z. Hao, H. Li, Credit evaluation system based on blockchain for multiple stakeholders in the food supply chain. Int. J. Environ. Res. Pub. Health **15**(8), 1627 (2018). https://doi.org/10.3390/ijerph15081627
15. M. Marin, I. Marin, Iuliana, L. Vidu, Learning about the reduction of food waste using blockchain technology, in *13th annual International Technology, Education and Development Conference* (2019). https://doi.org/10.21125/inted.2019.0856
16. M. Satish, P. Srinvasa Rao, M. Ramakrishna Murty et al. Identification of natural disaster affected area using twitter, in *International Conference and Publish the Proceedings in AISC Springer ICETC-2019 at Osmania University, Hyderabad*, pp. 782–801, July 2019

The Mechanism of Generating the Automated Java Unit Test Cases by Achieving Maximum Code Coverage

Bindu Avadhani, Saumya Ranjan Giri, and Venkateswara Rao Pulipati

Abstract The code coverage, being the factor of focus on unit testing that tells us which part of the source code is tested or untested. If the code consists of abstract classes or interfaces, they are also to be tested to ensure the maximum code coverage as generating the high coverage test suites play an important role in realistic systems like airbag systems, autopilot systems, etc. Thus, the correctness in the functionality is properly tested. Mocking is the process of constituting an imitation of the object. It is the technique of unit testing where the virtual objects are created to mock the behavior of the actual object, which helps us to bypass faults to some other class or system like database operation, REST API calls, etc. Hence, this proposed work has been designed to generate highly covered test cases in the aspects of Java. This work helps us to produce the test cases along with the skeleton, which forms a syntactically correct JUnit. It helps to test the subclasses implementing the Abstract class's methods and classes implementing interfaces, to generate the test cases for no access modifier methods. The generated test cases support mocking as well. Mockito framework is being used to do so.

Keywords Code coverage · Mocking · JUnit · Mockito · Abstract class · Interfaces

---

B. Avadhani (✉) · V. R. Pulipati
VNR Vignana Jyothi Institute of Engineering and Technology, Hyderabad, India
e-mail: binduavadhani302@gmail.com

V. R. Pulipati
e-mail: pvenkat2004@gmail.com

S. R. Giri
ITC Infotech Ltd., Bengaluru, India
e-mail: saumya.giri07@gmail.com

C. Kiran Mai et al. (eds.), *Proceedings of International Conference on Advances in Computer Engineering and Communication Systems*, Learning and Analytics in Intelligent Systems 20,
https://doi.org/10.1007/978-981-15-9293-5_4

1 Introduction

Testing, the key to the process of development of a product lets us know the efficiency of the software where the functionality is being tested thoroughly to avoid the bugs' existence in the code. In complex systems, testing manually requires longer time. Hence, automation is the process that makes the process of testing easy in lower time to identify the existence of the bugs, errors, exceptions without any human intervention. Unit testing is one of the types of testing targets, the unit to be tested for the correctness of the functionality.

In the object-oriented programming concepts, abstraction and inheritance are the one among the concepts play an important role. In java, both Abstract class and interfaces help us to implement the abstraction in the code. Abstraction in Java shows the functionality instead of exhibiting how it is being implemented.

1.1 Abstract Class

Abstract class contains both abstract and non-abstract methods where the implementation of the interfaces is provided. An abstract class is extended by a concrete class in which the abstract methods are defined in the concrete class. To test the functionality of the abstract classes, subclasses extending the abstract class are tested to check the correctness of the program. The example of abstract class and its concrete class extending the abstract class are shown in Figs. 1 and 2, respectively.

1.2 Interfaces

Interfaces have only method declarations, which are the abstract methods. A concrete class is defined, which implements the interface, i.e., interfaces are implemented but not extended. Testing of interfaces is possible by testing the concrete classes implementing that interface. The example of interface and its implementing class are shown in Figs. 3 and 4.

1.3 Mocking

The efficient way of testing includes a substantial challenge of handling the dependencies of the software. It is not always possible to test the software effectively with all its dependencies. It has the following three reasons. One is the invocation of databases, REST API's that reduces the speed of the process and that is not encouraged in unit testing process. Second, the existence of the bugs in the dependencies

Fig. 1 Example of abstract class

```
package com.vnr.bindu.abcls;

public abstract class Person {

    private String name;
    private int age;

    public abstract String getSocialStatus();

    public boolean isSeniorCitizen() {
        return age > 60;
    }


    public String getName() {
        return name;
    }
    public void setName(String name) {
        this.name = name;
    }
    public int getAge() {
        return age;
    }
    public void setAge(int age) {
        this.age = age;
    }

}
```

resulting the current unit under test resulting in bugs and fixing all of them together is complex, which takes longer amount of time. Generally, the development of a software is done module wise and integration is done in the latter part. If the current unit of test is to be tested with the other that has not been developed is the third challenge to be handled. For the effective testing of the software, to handle these challenges, the concept of mocking helps us to substitute the dependencies with the unreal(mock) objects to perform the process of testing, i.e., the dummy values are inserted at the places of dependencies and the unit is tested. There are certain mocking frameworks for Java-like Easy Mock, Mockito, etc.

1.4 Code Coverage

Code coverage, the systematic process of determining which part of code is exercised by the test cases, which part of the code is to be exercised by generating the additional test cases. It helps us to know the wellness of the test cases and part of the code

```
package com.vnr.bindu.abcls;
public class Bindu extends Person { //concrete class
    private int salary;
    private String company;

    @Override
    public int hashCode() {
        final int prime = 31;
        int result = 1;
        result = prime * result + ((company == null) ? 0 : company.hashCode());
        result = prime * result + salary;
        return result;
    }
    @Override
    public boolean equals(Object obj) {
        if (this == obj)
            return true;
        if (obj == null)
            return false;
        if (getClass() != obj.getClass())
            return false;
        Bindu other = (Bindu) obj;
        if (company == null) {
            if (other.company != null)
                return false;
        } else if (!company.equals(other.company))
            return false;
        if (salary != other.salary)
            return false;
        return true;
    }
    public int getSalary() {
        return salary;
    }

    public void setSalary(int salary) {
        this.salary = salary;
    }

    public String getCompany() {
        return company;
    }

    @Override
    public String getSocialStatus() {
        return "Celebrity";
    }

}
```

Fig. 2 Example of a concrete class extending the abstract class in Fig. 1

Fig. 3 Example of Interface

```
package com.vnr.bindu.in;

public interface Animal {
    public int getYearOfBirth();
    public int getNumberOfLegs();
    public boolean canSwim();
}
```

```java
package com.vnr.bindu.in;

public class Dog implements Animal {
    private String kindOfSound;

    public String getKindOfSound() {
        return kindOfSound;
    }

    public void setKindOfSound(String kindOfSound) {
        this.kindOfSound = kindOfSound;
    }

    @Override
    public int hashCode() {
        final int prime = 31;
        int result = 1;
        result = prime * result + ((kindOfSound == null) ? 0 : kindOfSound.hashCode());
        return result;
    }

    @Override
    public boolean equals(Object obj) {
        if (this == obj)
            return true;
        if (obj == null)
            return false;
        if (getClass() != obj.getClass())
            return false;
        Dog other = (Dog) obj;
        if (kindOfSound == null) {
            if (other.kindOfSound != null)
                return false;
        } else if (!kindOfSound.equals(other.kindOfSound))
            return false;
        return true;
    }
    @Override
    public int getYearOfBirth() {
         return 1999;
    }

    @Override
    public int getNumberOfLegs() {
         return 4;
    }

    @Override
    public boolean canSwim() {
         return true;
    }

}
```

Fig. 4 Example of implementing class of interface example in Fig. 3

tested. By the determination of code coverage circuitously quantifies the quality of the software.

Though there are many testing types, Ref. [1] tells us about the concolic testing which is a combination of both symbolic and concrete execution. In this concolic testing, the formation of an execution tree takes part which helps us to generate the test cases for every possible path in tree covering all the lines, branches of the code in order to achieve maximum code coverage.

Example of a method in a class Concolic Demo as shown in Fig. 5 has its code coverage as shown in Fig. 6, by one test case and the complete code coverage is shown in Fig. 7.

Fig. 5 Example of a method

```
int test (int x)
{
  if(x > 10)
      return 1;
  else
      return -1;
}
```

Fig. 6 Part of the code covered by one Test case

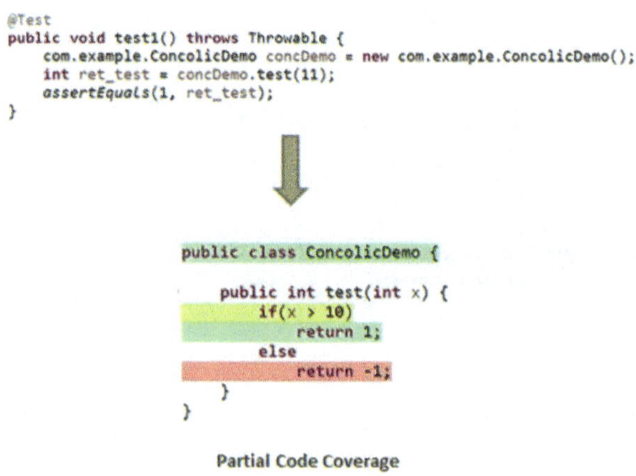

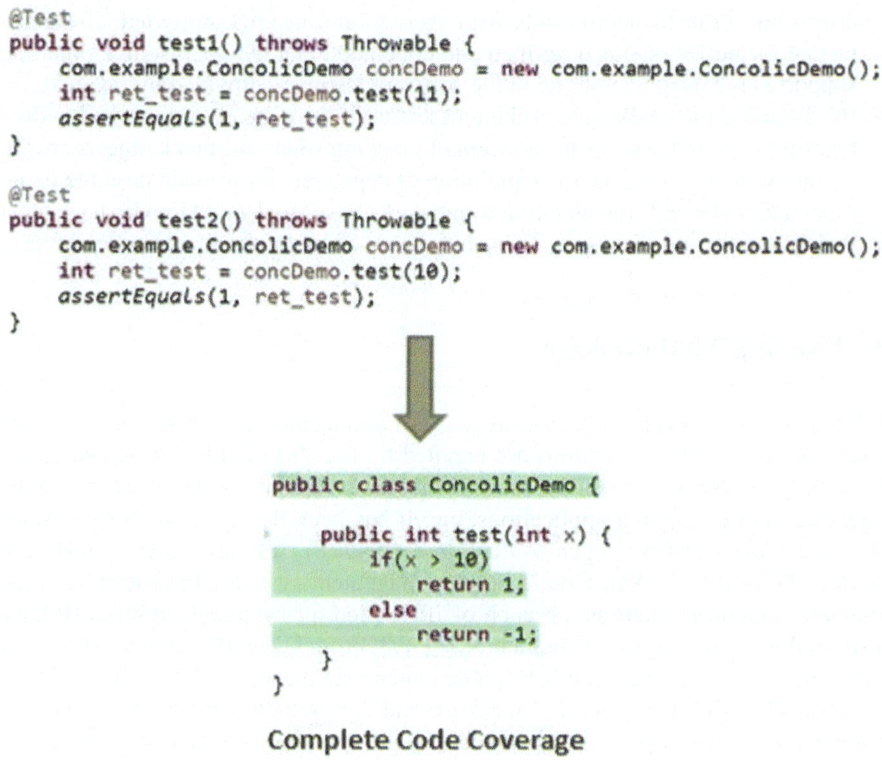

Fig. 7 Complete code covered by the remaining optimum number of test cases

2 Background of the Work

- Li et al. [2] summarizes that any of the tools studied during the work are not implementing all the branching structures available in Java, branch coverage is improperly implemented, byte code instrumentation is an invalid procedure to implement branch coverage. The statement coverage, tool that produces qualitative test cases are the two to be focused for future enhancements.
- From the comparison tables in [3] tells us that the JaCoCo is the code coverage tool that implements the byte code instrumentation, measures the statement, decision and method coverages. Evosuite, JaCoCo (Java Code Coverage), DCC (Dynamic Code Coverage) are the tools that support Java, implement byte code implementation and can be used for automation.
- Dhareula and Ganpati [4] states that the test case effectiveness can be observed by the code covered in the System Under Test. This paper tells us that the EclEmma is the most effective tool than CodeCover and implements byte code instrumentation.
- In [5], the comparison table of the Coverage Tool tells us that Java Code Coverage (JaCoCo), Emma, are the open-source tools with byte instrumentation which

covers line, branch, method and class coverage, and are GUI supported, The architecture of the JaCoCo is described and concluded that the open-source database support is the most useful one in the software testing is provided by JaCoCo.
- [6] discusses the challenges of the unit testing if the dependencies to be handled appropriately. It discusses the concept of mocking where the mock objects are the mimic of the real objects, the simulation of dependencies is made possible using mocking. It also tells the mocking frameworks EasyMock and Mockito have most widely used unique features. These help in mocking on the source code classes.

3 Existing Methodology

JPF (Java Path Finder) is the custom java virtual machine that works on Java byte code. The byte code instructions are handled by the VM of JPF. JPF is known for its usage in test-case generation using symbolic execution, model checking of the interfaces and distributed applications, etc. It has been developed at NASA Ames Research Center and was open sourced in 2005 [7, 8]. JPF has a plugin to eclipse and can be run using Ant, Gradle build management systems. In order to run the symbolic execution, there is a branch of JPF called jpf-symbc. To run the JPF, we also need jpf-core-master, the branch of JPF [7]. From [1], with examples in Figs. 4 and 5, it is already proved that JPF is able to generate the test cases for the primitive data types but with the examples in Figs. 6 and 7, it is shown that the test cases are not generated using JPF. It is also stated that JPF can generate the test cases for the Strings.

Bindu et al. [1] Upon experimentation, it is seen that PIT is the fast mutation testing tool that is able to catch the logical errors within the code by performing the mutation testing. There is also a plugin available for eclipse. [1] also tells us that mutation testing is the process of introducing the faults into the code with the help of some operations like arithmetic, logical, etc.

3.1 EclEmma

EclEmma, the Java code coverage tool plugin for eclipse, which gets the code coverage analysis into the workbench of eclipse. The JaCoCo library is used as the base for EclEmma, but EclEmma was originally made as plugin using EMMA, the one developed by Vlad Roubstov [9].

3.2 *Mockito*

The method mock () is used to create the mock objects, when () and given () methods definition within the code tells how the mock has to act. spy() is the method used for stubbing though the real methods invocation is done. @InjectMocks is an annotation used to mock the fields with annotations as @Spy or @Mock automatically [10].

4 Proposed Methodology

The proposed work uses jpf-core, jpf-symbc along with the constraint solver Z3 to solve the equations within the code. As discussed in the proposed work of [1], the prerequisites of the work are config.xml file with its contents including classpath, sourcepath, JUnit sourcepath (where the outputs i.e., generated JUnits are stored). The inputs to the system comprise of config.xml file, System Under Test's source path and class path. The Intermediate Java file is been created, if there exist the complexed types of the data as the arguments such as objects, lists, set, etc. The Intermediate file is shown in Figs. 8, 9 of [1]. The breakage of the complexed data structures is also shown in Fig. 10 of [1]. The generated outputs are stored into the given JUnit source folder. The process of verification of the test suites is done using the method called mutation testing [1] which is able to find the logical errors within the code. The open source software PIT [1] is used here for the quality determination of the test cases. Cyclomatic complexity is the software metric and quantitative measure

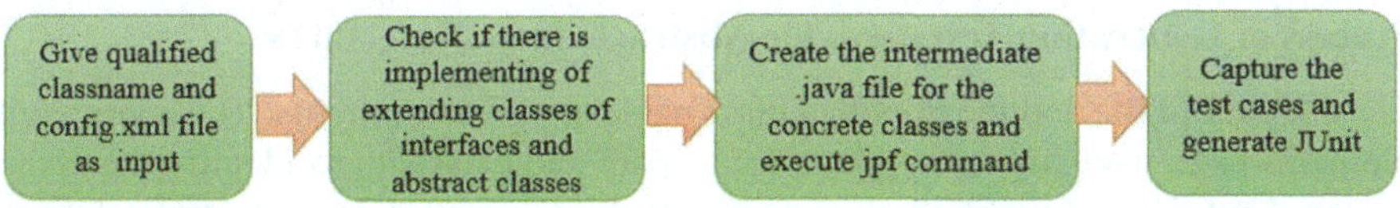

Fig. 8 The process flow

```java
public class TestAbClass {

    //example #79
    public static Person abstractTest(Person p) {
        Person p1 = new Bindu();
        if(((Bindu)p).getCompany().equals("HP") && ((Bindu)p).getSalary() > 30000){
            p1.setAge(25);
            p1.setName("Henry");
        }else if(((Bindu)p).getCompany().equals("Accenture") && ((Bindu)p).getSalary() < 90000){
            p1.setAge(30);
            p1.setName("Patrick");
        }
        return p1;
    }
}
```

Fig. 9 Example SUT using abstract class (shown in Fig. 1) as an argument

```java
public class TestInterface {

    public static Animal testInterface(Animal a){
        Animal a1 = new Dog();
        Dog d = new Dog();
        Cow c = new Cow();
        a1.getNumberOfLegs();
        if(((Dog)a).getKindOfSound().equals("Bark")){
            d.setKindOfSound("Bark");
            d.getYearOfBirth();
            d.getNumberOfLegs();
            return d;
        }
        return a1;
    }
}
```

Fig. 10 Example SUT using interface (shown in Fig. 3) as an argument

that helps us to calculate the number of linearly independent paths where the control flow graph (execution tree in concolic approach [1]) of the program is developed. The optimum number of test cases generated by the system is greater than or equal to the cyclomatic complexity.

Process of Generating Test cases for Abstract classes/Interfaces:

The config.xml file and the qualified class name are given as inputs. The config.xml file helps us to navigate to the given sourcepath and classpath to identify the given System Under Test (qualified class name). If the given SUT is an abstract class or interface, the system tries to fetch the subclasses that are extending the given abstract class or the class implementing the given interface. The intermediate Java file is generated. .jpf file is created and the jpf command is executed. The jpf generated test cases are captured and a complete JUnit with generated test cases is given as the output. The process of verification of the test cases is same as discussed in Fig. 11 of [1]. If there are any additional test cases generated, they are appended to the existing JUnit upon user requisition.

The proposed methodology in the above example tries to fetch the details of the argument. As the argument in the above example is an abstract class shown in Fig. 3, the system tries to fetch the details of the concrete classes extending the abstract class Person. As there exists a concrete class extending Person, it is picked up.

Similarly, the below example in Fig. 10 is the SUT (System Under Test) has an interface as an argument, in turn the proposed system tries to fetch the details of the implementing classes of the specified interface Animal as shown in Fig. 4. If

```java
@Test
public void testAbstractTest_1() throws Throwable {
    com.vnr.bindu.abcls.Person ret_abstractTest = com.vnr.bindu.examples.TestAbClass.abstractTest((com.vnr.bindu.abcls.Person)null);
    //TODO assert here
    assertNotNull(ret_abstractTest);
}

@Test
public void testAbstractTest_2() throws Throwable {
    com.vnr.bindu.abcls.Bindu bindu_ConcCls_0 = new com.vnr.bindu.abcls.Bindu();
    bindu_ConcCls_0.setCompany(" ");
    com.vnr.bindu.abcls.Person person_AbstractCls_0 = bindu_ConcCls_0;
    com.vnr.bindu.abcls.Person ret_abstractTest = com.vnr.bindu.examples.TestAbClass.abstractTest(person_AbstractCls_0);
    //TODO assert here
    assertNotNull(ret_abstractTest);
}


@Test
public void testAbstractTest_3() throws Throwable {
    com.vnr.bindu.abcls.Bindu bindu_ConcCls_0 = new com.vnr.bindu.abcls.Bindu();
    bindu_ConcCls_0.setSalary(0);
    com.vnr.bindu.abcls.Person person_AbstractCls_0 = bindu_ConcCls_0;
    bindu_ConcCls_0.setCompany("Accenture");
    com.vnr.bindu.abcls.Person ret_abstractTest = com.vnr.bindu.examples.TestAbClass.abstractTest(person_AbstractCls_0);
    //TODO assert here
    assertNotNull(ret_abstractTest);
}


@Test
public void testAbstractTest_4() throws Throwable {
    com.vnr.bindu.abcls.Bindu bindu_ConcCls_0 = new com.vnr.bindu.abcls.Bindu();
    bindu_ConcCls_0.setSalary(90000);
    com.vnr.bindu.abcls.Person person_AbstractCls_0 = bindu_ConcCls_0;
    bindu_ConcCls_0.setCompany("Accenture");
    com.vnr.bindu.abcls.Person ret_abstractTest = com.vnr.bindu.examples.TestAbClass.abstractTest(person_AbstractCls_0);
    //TODO assert here
    assertNotNull(ret_abstractTest);
}
```

Fig. 11 Test cases for the example SUT shown in Fig. 9

the system is not able to find the details of the implementing classes or extending classes of the interface and abstract class, respectively, it gives us the warning that no concrete classes are found.

5 Outcomes of the Work

5.1 *For Abstract Classes*

For the SUT (System Under Test) shown in Fig. 9, the proposed system generates the JUnit for testing the abstract class and for the SUT, seven test cases are generated but a portion of JUnit can be seen in Fig. 11.

The test cases here generated by the system are the optimum number of test cases, which helps us to cover almost all the paths of the code, which gives the maximum code coverage and the JUnit generated is syntactically and semantically correct JUnit. Even if the arguments are collections (List for example) in Java, arrays, other complexed data structures, the system tries to evaluate the condition available and generates the optimum number of test cases. If there are no conditions in the code, it generates the random test cases. The output of the SUT, the example method in Fig. 12, having one of the complexed structures as an argument is shown in Fig. 13.

```java
    public Set<Person[]> testSetOfArrayOfAbCls(Set<Person[]> personSet){
        Set<Person[]> set = new HashSet<>();
        Person[] person = new Person[2];
        Person b = new Bindu();
        for(Person[] arr: personSet){
            if(((Bindu)arr[1]).getSalary() >= 30000 && ((Bindu)arr[1]).getCompany().equals("Capital First")){
                b.setName("Harish");
                person[0] = b;
                set.add(person);
            }else if(((Bindu)arr[2]).getSalary() < 60000 && ((Bindu)arr[2]).getCompany().equals("Google")){
                b.setName("Srinivas");
                person[1] = b;
                set.add(person);
            }
        }
        return set;
    }

}
```

Fig. 12 Example SUT with complexed structure of abstract class as an argument

```java
public class TestSetOfArrayOfAbClsTest {

    // Launch the test
    public static void main(String[] args) {
        new org.junit.runner.JUnitCore().run(TestSetOfArrayOfAbClsTest.class);
    }
    @Test
    public void testTestSetOfArrayOfAbCls_1() throws Throwable {
        java.util.Set ret_testSetOfArrayOfAbCls = new com.vnr.bindu.examples.TestSetOfArrayOfAbCls().testSetOfArrayOfAbCls((java.util.Set<com.vnr.bindu.abcls.Person[]>)null);
        //TODO assert here
        assertNotNull(ret_testSetOfArrayOfAbCls);
    }

    @Test
    public void testTestSetOfArrayOfAbCls_2() throws Throwable {
        com.vnr.bindu.examples.TestSetOfArrayOfAbCls testSetOfArrayOfAbCls_obj = new com.vnr.bindu.examples.TestSetOfArrayOfAbCls();
        com.vnr.bindu.abcls.Bindu bindu_ConcCls_0 = new com.vnr.bindu.abcls.Bindu();
        bindu_ConcCls_0.setSalary(920);
        com.vnr.bindu.abcls.Person person_AbstractCls_0 = bindu_ConcCls_0;
        com.vnr.bindu.abcls.Person[] person_Arr_0 = new com.vnr.bindu.abcls.Person[2];
        person_Arr_0[0] = person_AbstractCls_0;
        java.util.Set<com.vnr.bindu.abcls.Person[]> person_Set_0 = new java.util.HashSet<>();
        person_Set_0.add(person_Arr_0);
        bindu_ConcCls_0.setCompany("9V3GT9CNLF");
        com.vnr.bindu.abcls.Bindu bindu_ConcCls_1 = new com.vnr.bindu.abcls.Bindu();
        bindu_ConcCls_1.setSalary(564);
        com.vnr.bindu.abcls.Person person_AbstractCls_1 = bindu_ConcCls_1;
        person_Arr_0[1] = person_AbstractCls_1;
        bindu_ConcCls_1.setCompany("K0MSABJ38E");
        com.vnr.bindu.abcls.Bindu bindu_ConcCls_2 = new com.vnr.bindu.abcls.Bindu();
        bindu_ConcCls_2.setSalary(20);
        com.vnr.bindu.abcls.Person person_AbstractCls_2 = bindu_ConcCls_2;
        com.vnr.bindu.abcls.Person[] person_Arr_1 = new com.vnr.bindu.abcls.Person[2];
        person_Arr_1[0] = person_AbstractCls_2;
        person_Set_0.add(person_Arr_1);
        bindu_ConcCls_2.setCompany("5X7aNFB8LX");
        com.vnr.bindu.abcls.Bindu bindu_ConcCls_3 = new com.vnr.bindu.abcls.Bindu();
        bindu_ConcCls_3.setSalary(819);
        com.vnr.bindu.abcls.Person person_AbstractCls_3 = bindu_ConcCls_3;
        person_Arr_1[1] = person_AbstractCls_3;
        bindu_ConcCls_3.setCompany("T7W92aPN92");
        java.util.Set ret_testSetOfArrayOfAbCls = testSetOfArrayOfAbCls_obj.testSetOfArrayOfAbCls(person_Set_0);
        //TODO assert here
        assertNotNull(ret_testSetOfArrayOfAbCls);
    }

}
```

Fig. 13 JUnit for the example SUT shown in Fig. 12

5.2 For Interfaces

As shown for the abstract class for the complexed data structures, the output for example SUT for the complexed structure of the interface as an argument shown in Fig. 15 is shown in Fig. 16.

The total number of test cases generated is four but here only a portion of JUnit, i.e., three test cases are shown for the good visuality (Fig. 14).

```java
public class TestInterfaceTest {


    // Launch the test
    public static void main(String[] args) {
        new org.junit.runner.JUnitCore().run(TestInterfaceTest.class);
    }


    @Test
    public void testTestInterface_1() throws Throwable {
        com.vnr.bindu.in.Animal ret_testInterface = com.vnr.bindu.examples.TestInterface.testInterface((com.vnr.bindu.in.Animal)null);
        //TODO assert here
        assertNotNull(ret_testInterface);
    }

    @Test
    public void testTestInterface_2() throws Throwable {
        com.vnr.bindu.in.Dog dog_ConcCls_0 = new com.vnr.bindu.in.Dog();
        dog_ConcCls_0.setKindOfSound(" ");
        com.vnr.bindu.in.Animal animal_Interface_0 = dog_ConcCls_0;
        com.vnr.bindu.in.Animal ret_testInterface = com.vnr.bindu.examples.TestInterface.testInterface(animal_Interface_0);
        //TODO assert here
        assertNotNull(ret_testInterface);
    }


    @Test
    public void testTestInterface_3() throws Throwable {
        com.vnr.bindu.in.Dog dog_ConcCls_0 = new com.vnr.bindu.in.Dog();
        dog_ConcCls_0.setKindOfSound("Bark");
        com.vnr.bindu.in.Animal animal_Interface_0 = dog_ConcCls_0;
        com.vnr.bindu.in.Animal ret_testInterface = com.vnr.bindu.examples.TestInterface.testInterface(animal_Interface_0);
        //TODO assert here
        assertNotNull(ret_testInterface);
    }

}
```

Fig. 14 JUnit for the example SUT shown in Fig. 10

```java
    public List<Animal[]> testListOfArrayOfInterface(List<Animal[]> list){
        List<Animal[]> animalList = new ArrayList<>();
        Animal[] animalArr = new Animal[2];
        Dog d = new Dog();
        Cow c = new Cow();
        for(Animal[] arr: list){
            if(((Dog)arr[0]).getKindOfSound().equals("Bark")){
                d.setKindOfSound("Bark");
                d.getNumberOfLegs();
                d.getYearOfBirth();
                animalArr[0] = d;
                animalList.add(animalArr);
            }else if(((Cow)arr[1]).getKindOfSound().equals("Moo")){
                c.setKindOfSound("Moo");
                c.getNumberOfLegs();
                c.getYearOfBirth();
                animalArr[1] = c;
                animalList.add(animalArr);
            }
        }
        return animalList;
    }
}
```

Fig. 15 Example SUT of complexed structure of interface as an argument

```
@Test
public void testTestListOfArrayOfInterface_1() throws Throwable {
    java.util.List ret_testListOfArrayOfInterface = new com.vnr.bindu.examples.TestListOfArrayOfInterface().testListOfArrayOfInterface((java.util.List<com.vnr.bindu.in.Animal[]>)null);
    //TODO assert here
    assertNotNull(ret_testListOfArrayOfInterface);
}
@Test//(expected = java.lang.ClassCastException.class)
public void testTestListOfArrayOfInterface_2() throws Throwable {
    com.vnr.bindu.examples.TestListOfArrayOfInterface testListOfArrayOfInterface_obj = new com.vnr.bindu.examples.TestListOfArrayOfInterface();
    com.vnr.bindu.in.Dog dog_ConcCls_0 = new com.vnr.bindu.in.Dog();
    dog_ConcCls_0.setKindOfSound(" ");
    com.vnr.bindu.in.Animal animal_Interface_0 = dog_ConcCls_0;
    com.vnr.bindu.in.Animal[] animal_Arr_0 = new com.vnr.bindu.in.Animal[2];
    animal_Arr_0[0] = animal_Interface_0;
    java.util.List<com.vnr.bindu.in.Animal[]> animal_List_0 = new java.util.ArrayList<>();
    animal_List_0.add(animal_Arr_0);
    java.util.List ret_testListOfArrayOfInterface = testListOfArrayOfInterface_obj.testListOfArrayOfInterface(animal_List_0);
    //TODO assert here
    assertNotNull(ret_testListOfArrayOfInterface);
}
@Test//(expected = java.lang.ClassCastException.class)
public void testTestListOfArrayOfInterface_3() throws Throwable {
    com.vnr.bindu.examples.TestListOfArrayOfInterface testListOfArrayOfInterface_obj = new com.vnr.bindu.examples.TestListOfArrayOfInterface();
    com.vnr.bindu.in.Dog dog_ConcCls_0 = new com.vnr.bindu.in.Dog();
    dog_ConcCls_0.setKindOfSound("Bark");
    com.vnr.bindu.in.Animal animal_Interface_0 = dog_ConcCls_0;
    com.vnr.bindu.in.Animal[] animal_Arr_0 = new com.vnr.bindu.in.Animal[2];
    animal_Arr_0[0] = animal_Interface_0;
    java.util.List<com.vnr.bindu.in.Animal[]> animal_List_0 = new java.util.ArrayList<>();
    animal_List_0.add(animal_Arr_0);
    com.vnr.bindu.in.Dog dog_ConcCls_1 = new com.vnr.bindu.in.Dog();
    dog_ConcCls_1.setKindOfSound(" ");
    com.vnr.bindu.in.Animal animal_Interface_1 = dog_ConcCls_1;
    com.vnr.bindu.in.Animal[] animal_Arr_1 = new com.vnr.bindu.in.Animal[2];
    animal_Arr_1[0] = animal_Interface_1;
    animal_List_0.add(animal_Arr_1);
    java.util.List ret_testListOfArrayOfInterface = testListOfArrayOfInterface_obj.testListOfArrayOfInterface(animal_List_0);
    //TODO assert here
    assertNotNull(ret_testListOfArrayOfInterface);
}
```

Fig. 16 Test cases for the example SUT in Fig. 15

5.3 *For Database Calls and Private Constructors in the Code*

As discussed, if there is a database call or REST API calls in the code, it is most essential to avoid the time consumption and testing the code functionality appropriately. Here, the proposed method can also implement the mocking concept using Mockito where the when() and then() methods are to written by the user to guide what action is to be performed. The example method is shown in Fig. 17 and its respective output in Fig. 18.

If the constructor is made private, the system shows us the warning that the method is inaccessible as shown in Figs. 17 and 18.

6 Conclusion and Future Enhancements

The system is able to generate the test cases for the extended classes of the respective abstract classes and implementing classes of the respective interfaces. It also generates the proper comments in some cases when code generation is impossible such

```java
public class Demo {
    private Demo(){
    }
    private Demo(int i){

    }

    public int testMe(Employee e){
        if(e.getAge()> 20 && e.getName().equals("abcd"))
            return 1;
        else
            return 0;
    }

    //Fetch data from a data base, employee table, having salary more than the argument
    public List<Employee> getData(int salary) {
        String query = "select * from EMPLOYEE where SALARY > "+salary;
        List<Employee> result = new ArrayList<>();
        // database call with the above query and result is stored in 'result'
        return result;
    }

}
```

Fig. 17 Example SUT having database call in the method

```java
public class DemoTest {

    // Launch the test
    public static void main(String[] args) {
        new org.junit.runner.JUnitCore().run(DemoTest.class);
    }
    @BeforeClass
    public static void setUp() throws Exception {
        // write set up code here
    }
    @Test
    public void testTestMe_1() throws Throwable {
        com.vnr.constr.bindu.Demo demo_ConcCls_0 = mock(com.vnr.constr.bindu.Demo.class);
        int ret_testMe = demo_ConcCls_0.testMe((com.vnr.constr.bindu.Employee)null);
        //TODO assert here
        //int expectedOutput = 0;
        //assertEquals(expectedOutput, ret_testMe);
    }
    @Test
    public void testTestMe_2() throws Throwable {
        com.vnr.constr.bindu.Demo demo_obj = mock(com.vnr.constr.bindu.Demo.class);
        com.vnr.constr.bindu.Employee employee_ConcCls_0 = mock(com.vnr.constr.bindu.Employee.class);
        int ret_testMe = demo_obj.testMe(employee_ConcCls_0);
        //TODO assert here
        //int expectedOutput = 0;
        //assertEquals(expectedOutput, ret_testMe);
    }
    @Test
    public void testTestMe_3() throws Throwable {
        /* !! Warning !!  Cannot call the method as it is not accessible. */
    }
    @Test
    public void testTestMe_4() throws Throwable {
        com.vnr.constr.bindu.Employee employee_ConcCls_0 = null; //TODO default constructor is not found or accessible
        employee_ConcCls_0.setEmpNo(44);
        employee_ConcCls_0.setName("ZY7ZPEITGQ");
        employee_ConcCls_0.setDept("60RJUsHSc8");
        employee_ConcCls_0.setAge(484);
        /* !! Warning !!  Cannot call the method as it is not accessible. */
    }
}
```

Fig. 18 JUnit for the example SUT in Fig. 17

as inaccessibility of default constructor and method, etc. Keeping the performance in view, the system is robust in test-case generation for the derived data types by generating the random test cases if the system fails to generate the optimum number of test cases. Though the arguments in the method are of complexed structures, the system generates the optimum number of test cases and tries to give the maximum code coverage, which can be seen using EclEmma (the plugin for Eclipse), which internally implements the JaCoCo libraries. The optimum number of test cases can be greater than or equal to the cyclomatic complexity of the code. Mockito, the framework for mocking in Java is used to mock whenever the user asks the system to mock by enabling the option. The system is loosely coupled such that any open source generating the test cases can be integrated as per the requirement or the unavailability of the existing sources.

JPF, the open source software has a flexibility of developing the own listeners but this procedure would become a dependency of the JPF.

Acknowledgements We would like to express our gratitude to **Dr. Corina S. Pasareanu**, Researcher, NASA Ames Research Center for the utmost support provided during the implementation of the system.

References

1. A. Bindu et al., Automatic JUnit generation and quality assessment using concolic and mutation testing. Int. J. Innov. Technol. Explor. Eng. (IJITEE) **8**(10) (2019). ISSN: 2278-3075
2. N. Li, X. Meng, J. Offutt, L. Deng, Is bytecode instrumentation as good as source code instrumentation: An empirical study with industrial tools (Experience Report), pp. 380–389 (2013). https://doi.org/10.1109/issre.2013.6698891. Author, F., Author, S., Author, T.: Book title. 2nd edn. Publisher, Location (1999)
3. S. Shelke, S. Nagpure, The Study of various code coverage tools. Int. J. Comput. Trends Technol. (IJCTT) **V13**(1):46–49 (2014). ISSN:2231-2803, www.ijcttjournal.org. Published by Seventh Sense Research Group. LNCS
4. P. Dhareula, A. Ganpati, Open source code coverage tools for Java: a comparative analysis. Indian J. Sci. Technol **9**(32) (2016)
5. R. Lingampally, A. Gupta, P. Jalote, A multipurpose code coverage tool for Java, in *2007 40th Annual Hawaii International Conference on System Sciences (HICSS'07), Waikoloa, HI, 2007, pp. 261b–261b*
6. S. Mostafa, X. Wang, An empirical study on the usage of mocking frameworks in software testing, in *Proceedings of the 2014 14th International conference on quality software (QSIC'14)* (IEEE Computer Society, USA, 2014), pp. 127–132. https://doi.org/10.1109/QSIC.2014.19
7. S. Anand, C.S. Pasareanu, W. Visser, JPF-SE: a symbolic execution extension to Java PathFinder, in *TACAS'07* (2007)
8. Java Path Finder. https://github.com/javapathfinder/jpf-core/wiki
9. EclEmma. https://www.eclemma.org/
10. Mockito framework. https://site.mockito.org/

Line Stability Index-Based Voltage Stability Assessment Placing Series Compensation of TCSC

S. Ravikumar, Poonam Upadhyay, and B. Anjan

Abstract Nowadays, power system security is the main concern for the continuous transfer of power supply, which has to be maintained under steady-state condition when comes to the line outages and the transformer outages. So, severe stress on the lines causes line outages, which involves voltage instability leads to the collapse of voltage at the entire system. In this paper, IEEE-5 and 6 bus system is considered for the performance of voltage stability condition whereby giving the contingency ranking of the severe line using Active Power performance index/Real power performance index only for 5 bus for which the series compensation is placed as the optimal location providing reactive power based on Line voltage stability factor to reduce the vulnerability of the bus and the optimal size is determined based on the maximum compensation provided from the TCSC.

Keywords APPI · LVSF · Optimal location · Reactive power compensation

1 Introduction

Security for the power system should be maintained for the continuous supply of power to the longer distances. Checking the security on the power system makes proper planning for the operation. Dylicco gave five working states of the power system [1] as shown in Fig. 1. Generally, longer the distance of power supply means, the proper security of the power system is needed for the continuous power supply. So the proper planning is needed for maintaining security of the power system.

S. Ravikumar (✉) · P. Upadhyay · B. Anjan
Department of EEE, VNR Vignana Jyothi Institute of Engineering and Technology, Hyderabad, India
e-mail: suramravikumar94@gmail.com

P. Upadhyay
e-mail: upadhyay_p@vnrvjiet.in

B. Anjan
e-mail: anjan_b@vnrvjiet.in

C. Kiran Mai et al. (eds.), *Proceedings of International Conference on Advances in Computer Engineering and Communication Systems*, Learning and Analytics in Intelligent Systems 20,
https://doi.org/10.1007/978-981-15-9293-5_5

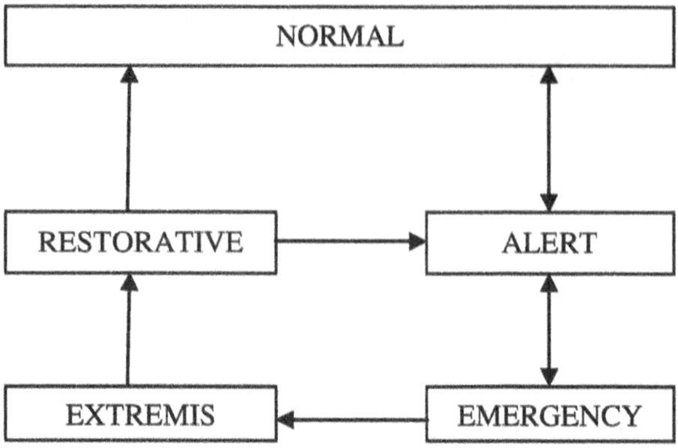

Fig. 1 Operating states of the power system

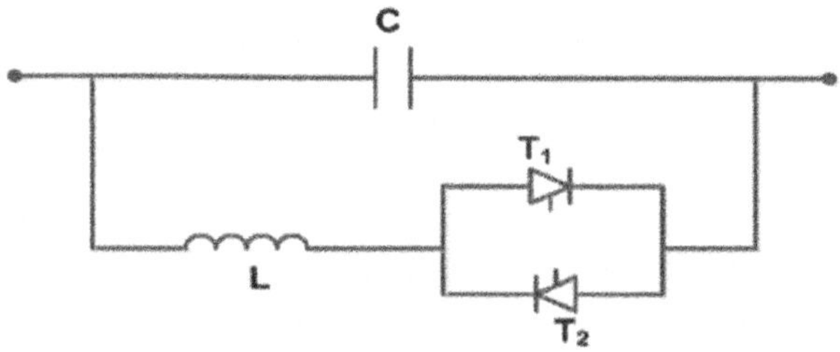

Fig. 2 Schematic diagram of TCSC

From Fig. 1, we can observe that power system is operated under normal state but it goes to the respective states when security constraints are violated. Power system security is classified into the steady state, transient and dynamic stability security where line outages come under steady-state stability security. Voltage instability arises due to the violation of the security constraints like bus voltages and magnitudes [2]. In this paper, voltage instability problem can be solved by placing series compensation as TCSC (Thyristor controlled series capacitor) at a suitable location supplying reactive power of definite size [3]. This paper uses APPI for ranking the critical line of having the highest value and LVSF for the placement of TCSC at the critical line corresponding to the collapsing bus having value equal or nearer to the one [4]. The entire analysis was carried out in the MATLAB program using Newton–Raphson method for the load flow of IEEE Test bus system [5].

2 Thyristor Controlled Series Capacitor

It is one of the series facts, device [1] is used for the reactive power compensation. It has a series controlled capacitor that is shunted by the Thyristor controlled reactor. It will be run under capacitive and inductive mode of operations.

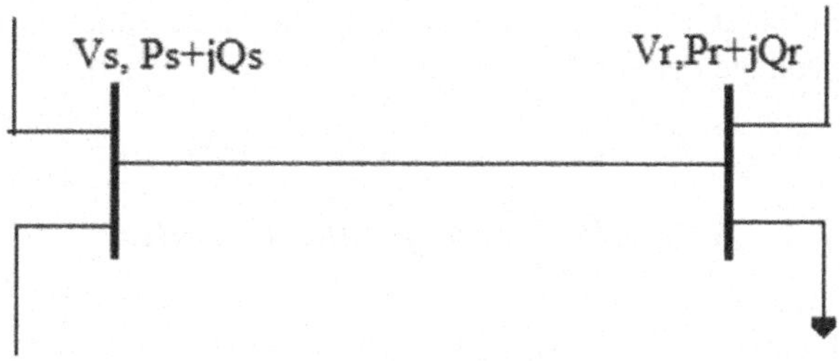

Fig. 3 Single-line diagram of the transmission network

Figure 2 depicts the schematic diagram of TCSC operates at the capacitive mode for the injection of the reactive power at the vulnerable bus associated with the critical line. It operates at different firing angles like at delay angle (α), the reactance is 0.055, at α = 30, the reactance is 0.1090, at α = 45, the reactance is 0.1857, at α = 60, the reactance is 0.6228, at α = 65, it will be resonance condition, at α = 75, the reactance is −0.4605, at α = 80, the reactance is −0.2915, at α = 900 the reactance is −0.1687 to be operated for the required injection of the reactive power within the limits of the reactance for the line. Figure 3 represents the single line representation of the transmission network for to solve the line voltage stability factor [6].

3 Predicament Detection

Under steady-state, power system works under secure state by considering all security constraints in the limits. Whenever the limits violated due to the severe overloads, which is the condition that arises caused by the uneven seasonal or day-to-day increase of the loads makes the voltage instability at the particular bus. The bus having voltage instability leads to the voltage collapse that makes the line connected to the bus goes to the critical state of disconnection which in turn stops the delivery of the power to the customer domestic applications. Because of overstress on the lines by the increases loads causes the reduction of reactive power will be injected at the location where LVSF value shows nearer to one of the particular bus compensating equipments like ULTC also varied because of voltage instability. Contingency ranking is given by the real power index/active power index due to the congestion created after the load flow of the IEEE Test bus system where LVSF provides the optimal placement of the series compensation [6].

4 Method Implemented

Here contingency ranking is given by active power performance index and the voltage instability problem can be identified by using line voltage stability factor where vulnerable bus can be known by the value which is varied from 0 (No load) to 1 (voltage collapse).

4.1 Active Power Performance Index

Contingency ranking of the lines was depicted by the APPI. The most severe line of having highest value will be ranked accordingly. The explanation of the active power performance index is shown at [7].

$$\mathrm{APPI} = \sum_{i=1}^{L} (Pi/Pi\mathrm{max})^{2n}$$

where L = number of lines in system, n = specific exponent, W_i = weighting factor, P_i = real power in line i, Pimax = max active power in line i.

4.2 Line Voltage Stability Factor

Power system is a big network with different devices and equipments were connected to it. So for the simplification purpose entire system is taken as single line representation of having single load. The derived method is shown [5].

$$L_{QP} = 4\left(\frac{X}{V_i^2}\right)\left(\left(\frac{X}{V_i^2}\right) + P_i^2 + Q_j\right)$$

where x-line reactance, P_i = real power at receiving end, Q_j = reactive power at end, V_i = voltage at the front end. The line voltage stability factor in the system at which the line with greatest stability factor index value ranges zero (no load) to one (voltage stability) shows the proximity to voltage collapse. The line stability factor can be depicted as the static indicator for the collapsing of voltage condition [8].

Reactive power supplied using normal conventional method.

From Fig. 4, we can explain the process for the injection of the reactive power with the series compensation TCSC device. Load flow analysis has done using Newton–Raphson method for IEEE 5 and IEEE 6 bus system to know the power flows in the buses then create a congestion on the load by maintaining within the limits of the real power to which active power performance index calculates the highest amount power passing through by the line indicates the critical line there by creating a single line outage as a N-1 contingency. Solve the load flow again after creating the contingency and the LVSF should be calculated for the remaining lines. If the LVSF is within the limits then we can allow the system to work, if not then consider it is the vulnerable line associated with load bus for which the injection of the reactive power from

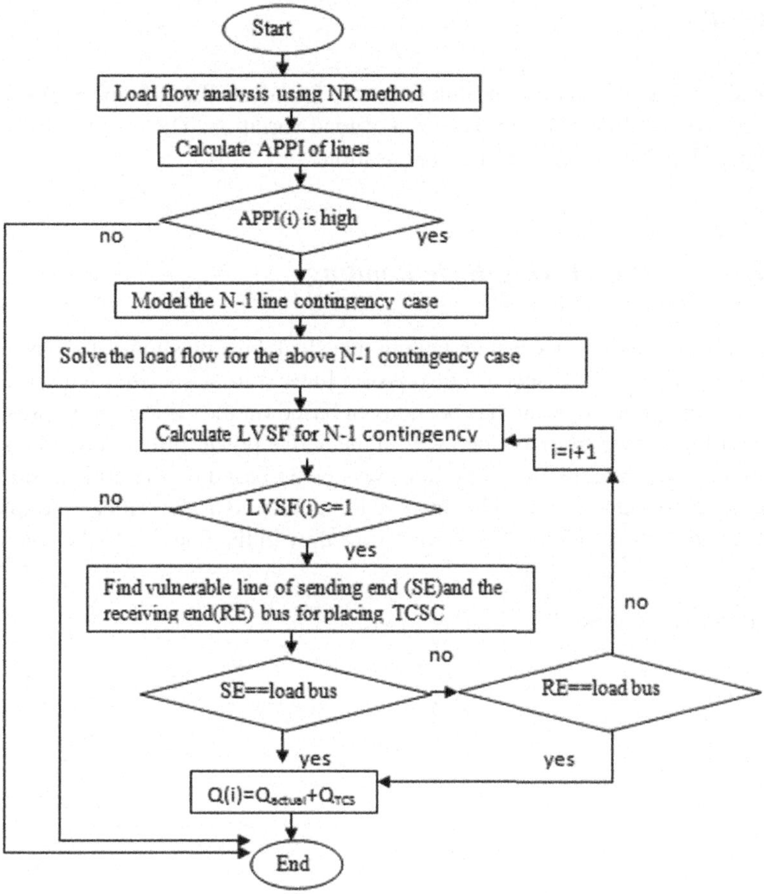

Fig. 4 Reactive power injection using TCSC

the compensating device is to be done to attain the line to its normal condition. Repeat the same procedure for the next lines creating different contingencies. Here the reactive power is injected by considering the reactance limits of the particular line at the vulnerable bus. MATLAB code is developed in such a way that injection will be done in view of not violating the limits of the line considering the maximum compensation value of the TCSC done by the firing angles at capacitor and inductor values [8], which considers as the optimal size of the TCSC.

5 Results

Finding the optimal location and the size was found on IEEE 5 and 6 bus system with base case loading having no criteria of load variation. The results were shown before and after the placement of compensation.

5.1 IEEE 5 Bus Having Base Loading

From Fig. 5, the IEEE5 bus system having one slack bus, one generator bus and the three load buses to which seven transmission lines were associated with it. For all the lines, contingency ranking has been given based on the critical stress present on the line with the congestion created within the active power limits of the lines.

From Table 1, the ranking of the lines was given based on the critical line using real power performance index. The line 1–2 is considered as the critical line since the rank associated is one based on the real power flow in the line due to the congestion

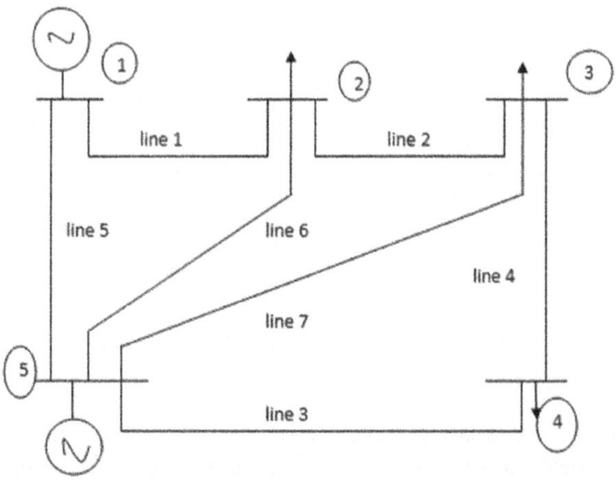

Fig. 5 IEEE 5 bus system

Table 1 Line contingency ranking using APPI

Lines	APPI value	Rank
1–2	0.8233	1
2–3	0.7134	2
4–5	0.6642	3
3–4	0.3246	4
3–5	0.3133	5
1–5	0.2234	6
2–5	0.2112	7

created and the remaining lines were ranked accordingly. The LVSF is calculated for all the lines after creating contingency analysis based on APPI. Here base loading was considered for the base case and the contingency case.

From the TCSC, 90% of compensation is chosen considering the limits of the line reactance for particular critical line which is stated as the optimal size of the device. Reactive power needed to be injected for the base case is of 31.98 MVAR and for the contingency 1–2 is of 35.02 MVAR.

From Table 2, the reactive power compensation at load buses of base case loading considering line contingency is depicted. At base case, bus 5 is the vulnerable bus in which the reactive power to be injected is 31.98. Bus 3 is the vulnerable bus at line 1–2 which the reactive power to be injected is 35.02.

From Fig. 6, voltage performance of different five buses is depicted for base case condition. At every bus, the voltage has been enhanced after placing the series compensation of TCSC based on the LVSF index within the limits not exceeding ±0.05 above 1.

From Fig. 7, The LVSF index value has been decreased for the base case at all the lines which are achieved by placing the series compensation of TCSC indicating the normal lines to be operated in the system.

Line Contingency 1–2 case

Table 2 Reactive power compensation at critical bus of 5 bus system

Loading	Contingency	Compensation	Vulnerable bus	Compensation (MVAR)
Base case loading	Base case	Before	BUS 5	31.98
		After		
	Line 1–2	Before	BUS 3	35.02
		After		

Base case

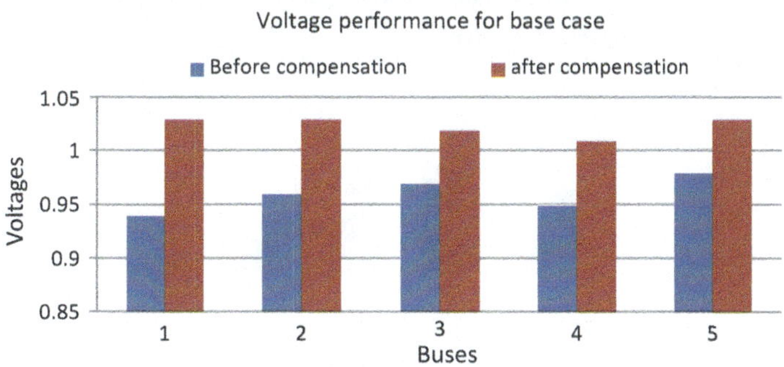

Fig. 6 Voltage performance bar chart for base case

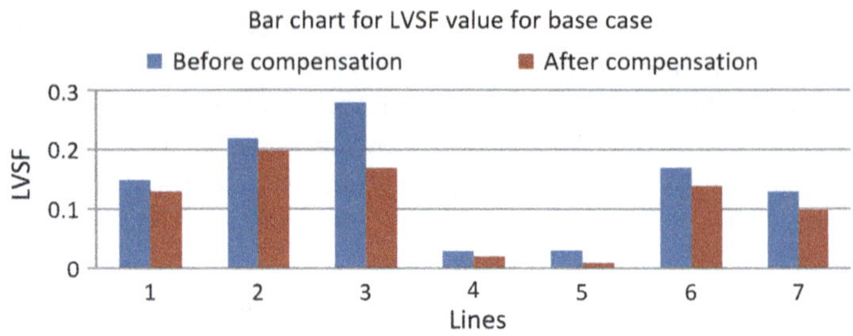

Fig. 7 Bar chart of LVSF values

From Fig. 8, voltage performance of different five buses is depicted for line contingency 1–2 case condition. At every bus, the voltage has been enhanced after placing the series compensation of TCSC based on the LVSF index within the limits not exceeding ±0.05 above 1.

From Fig. 9, the LVSF index value has been decreased for the contingency 1–2 case at all the lines except for the first line since, it is the critical line required for the

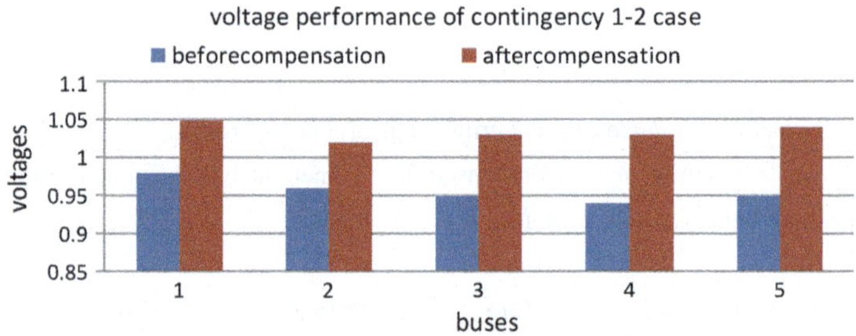

Fig. 8 Voltage performance bar chart for line contingency 1–2 case

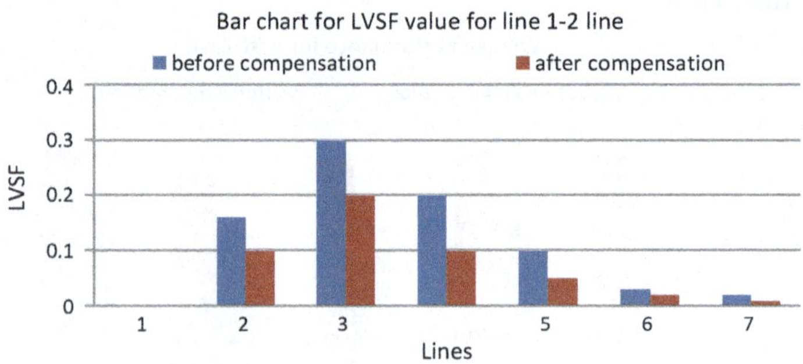

Fig. 9 Bar chart values of LVSF values

contingency condition by placing the series compensation of TCSC indicating the normal lines to be operated in the system.

5.2 *IEEE 6 Bus Having Base Loading*

Figure 10 is the IEEE6 bus system having one slack bus, three generator buses and the three load buses to which 11 transmission lines were associated with it. Here the load flow analysis was done for base case and then contingency analysis was done by selecting random line for which the LVSF value is known having the value nearer or equal to the one of the highest index is best for the placement of compensation device with 90% available within the reactance of the line limits.

From Table 3, the reactive power compensation at load buses of base case loading is considered with different contingency of lines 1–4 are depicted. At base case, bus 4 is the vulnerable bus in which the reactive power to be injected is 70.3. Bus 5 is the vulnerable bus at line 1–4 in which the reactive power to be injected is 90.73.

Line Contingency 1–4 case

From Fig. 11, voltage performance of different six buses is depicted for contingency 1–4 case condition of IEEE 6 bus system. At every bus, the voltage has been enhanced after placing the series compensation of TCSC based on the LVSF index within the limits not exceeding ±0.05 above 1.

Table 3 Injected reactive power at critical bus of 6 bus system

Contingency	Compensation	Critical bus	Reactive power (MVAR)
Base case	Before	BUS4	70.3
	After		
Line 1–4	Before	BUS 5	90.73
	After		

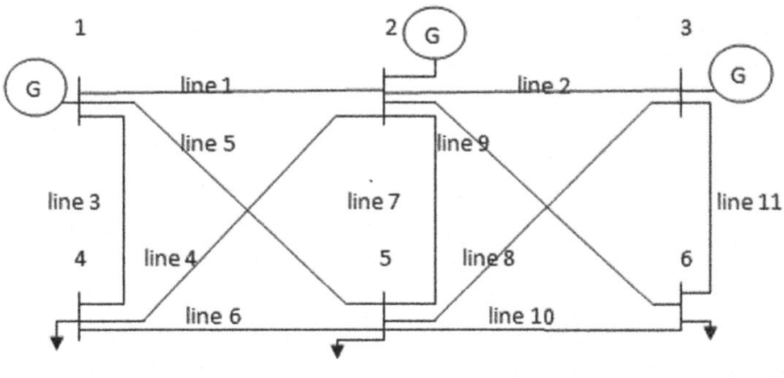

Fig. 10 IEEE 6 bus system

Line Contingency 1-4 case

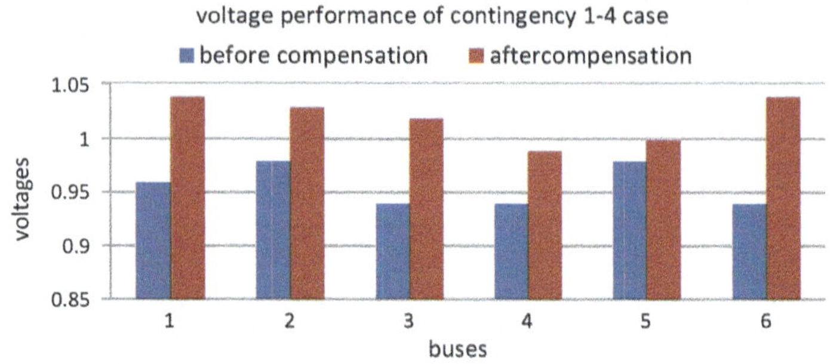

Fig. 11 Voltage performance bar chart for line contingency 1–4 case

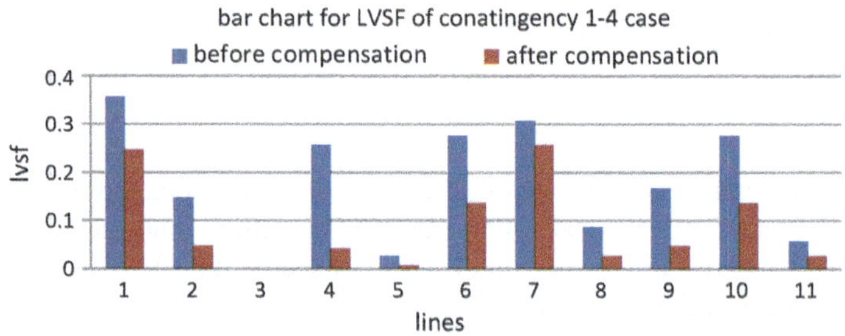

Fig. 12 Bar chart for LVSF values

From Fig. 12, the LVSF index value has been decreased for the contingency 1–4 case at all the lines except for the third line since, it is the critical line required for the contingency condition by placing the series compensation of TCSC indicating the normal lines to be operated in the system.

6 Conclusion

Load flow analysis was done by using Newton–Raphson method, which is considered as the base case values and the ranking is given by APPI for 5 bus based on the highest value which is used for the contingency analysis for each line then using LVSF index compensating device is placed as the optimal location to reduce the vulnerability of the bus by increasing the voltage whereas for the 6 bus system random line contingency was selected with no loadings based on LVSF for the placement of the series compensation. Optimal size is determined from the reactance of the line that depicts the maximum compensation to be provided, which reduces the size that operated in extreme condition.

References

1. N.G. Hingorani, L. Gyugi, *Understanding Facts: Concepts and Technology of Flexible AC Transmission Systems* (IEEE Press, New York, 2000)
2. C.W. Taylor, *Power System Voltage Stability* (McGraw-Hill, New york, 1994), p. 273
3. C.-F. Yang, G.G. Lai, C.-H. Lee, C.-T. Su, G.W. Chang, optimal setting of reactive compensation devices with an improved voltage stability index for voltage stability enhancement. Int. J. Electr. Power Energy Syst. **37**, 50–57 (2012)
4. A.A. Abou El-Ela, A.M. Kinawy, T. Mouwafi, R.A. El Sehiemy, Efficient performance indices for voltage detection, in *IEEE International Energy Conference and Exhibition (EnergyCon) 2010*, Manama 18–22, pp. 178–182 (2010)
5. P. Naik, Power system contingency ranking using Newton Raphson load flow method and its prediction using soft computing techniques. Thesis of Master of Technology in Industrial Electronics, National Institute of Technology, Rourkela, India (2014)
6. S.D. Venkataramana, TCSC incorporated voltage stability assessment under contingency condition. Int. J. Grid Distrib. Comput. **10**(7), 27–40 (2017)
7. S. Burada, D. Joshi, K.D. Mistry, Contingency analysis of power system by using voltage and active power performance index, in: *1st IEEE International Conference on Power Electronics, Intelligent Control and Energy Systems (ICPEICES-2016)*, pp. 1–5 (2016)
8. P. Rajyalakshmi, A comparision of transmission line voltage stability indices, in: *2nd IEEE Conference Electrical Electronics, Information, Communication and Bioinformatics*, pp. 44–48 (2016)

Preventing Crime Using Advanced Artificial Intelligence Techniques

Saikiran Gogineni, Anjusha Pimpalshende, Poongodai Arumugham, and Porika Dhanrajnath

Abstract Crime is increasing significantly from day to day. Every crime is associated with a certain pattern. The modern AI techniques can be implemented in crime scenarios to obtain valuable insights. In contrast to the existing works on crime analysis, we used the latest and much more efficient algorithms to handle crime. Our research work is segregated into three independent modules—prevention, detection and investigation. Prevention phase uses AI strategies to predict regions with high probability of crime occurrence in specified time and preventing crime from happening. Detection phase uses Faster RCNN to detect the kind of crime happening in an area through a security camera and alerting the police for instant response. If crime happens, AI Investigation can predict criminals through clues left in the crime spot. Our work also classifies police complaints. Criminal complaints registered by victims are automatically classified as an appropriate type such as robbery or murder. Our work achieved significant performance compared with earlier works. For prevention stage, SAN FRANCISCO crime dataset is considered. For detection stage, crime images dataset is constructed and annotated by extracting frames from UCF crime dataset (1900 videos, 128 hours of videos). Text complaints are manually constructed.

Keywords Crime prevention · Crime detection · Faster RCNN · NAÏVE BAYES · Random forest · Recurrent Neural Network (RNN) · Long–short-term memory

S. Gogineni (✉) · A. Pimpalshende · P. Arumugham · P. Dhanrajnath
CSE, CMR College of Engineering & Technology, Hyderabad, India
e-mail: goginenisaikiran31677@gmail.com

A. Pimpalshende
e-mail: anjushap@cmrcet.org

P. Arumugham
e-mail: apoongodai@cmrcet.org

P. Dhanrajnath
e-mail: porikadhanrajnath@gmail.com

C. Kiran Mai et al. (eds.), *Proceedings of International Conference on Advances in Computer Engineering and Communication Systems*, Learning and Analytics in Intelligent Systems 20,
https://doi.org/10.1007/978-981-15-9293-5_6

1 Introduction

Crime rate is increasing significantly from day to day. The two characteristics of crime are—it is not systematic and not random in nature as it follows a pattern. In this digital-driven world, criminals are using modern technologies to do misdeeds. We cannot come up with a strategy that prevents crime 100% from happening but we need to mitigate the crime ratio as far as possible. Artificial intelligence can be employed in crime investigation scenarios to find insights. As per statistics, every 22 out of 100 citizens of India are taking part in crime through some means. Crime can take several forms such as Robbery, Kidnapping, Burglary, Assault, Murder, fight and rapes. Detecting them accurately and finding proper insights from these crime scenarios is quite an important task as it provides a mechanism to handle them in the nearby future. Federal bureau of investigation department has classified top crimes majorly into four categories that are listed below:

1. Robbery 2. Sexual abuse 3. Murder 4. Assault.

2 Related Work

Bogahawatte and Adikari [1] proposed a strategy named (ICSIS) intelligent crime investigation system. It used clustering, data mining and classification techniques efficiently to find the criminal as per the clues collected from crime location. In this approach, crime patterns are identified using clustering techniques as each and every crime is associated with a certain unique pattern. They implemented NAÏVE BAYES algorithm for achieving the classification. Agarwal et al. [2] made use of a rapid miner tool for visualization and analysis of increasing crime rates. They implemented k means algorithms for achieving the task. Mahdavi et al. [3] used genetic algorithm approach and optimization of outlier detection to analyze crime rates. This work also included clustering and classification techniques based on available culprit data. Sathyadevan et al. [4] used different strategies to detect frequent crime locations. They used NAÏVE BAYES, support vector machine and logistic regression classifiers for crime pattern classification. Crime spot is detected with support of the decision tree. But their performance is quite less. Bruin et al. [5] proposed a quite interesting strategy. It is designed to achieve the clustering of lawbreakers based on readily available lawbreaker profiles. The lawbreakers profile per crime per year is obtained from the stored database and a profile distance matrix is calculated. Let the distance matrix be M. Immediately, matrix M in profile per year is calculated. Matrix M including the frequency value is used in creating clusters through standard NAÏVE BAYES clustering approach. It also created a lawbreakers profile. It is created in a particular manner—representing the offense profile of a lawbreaker for 1 (single) year. Using these established data, the huge group or cluster of offenders is easily analyzed and visualized. They also tried to predict the upcoming behavior and trends of individual suspects. It is widely useful in drawing the clear image of different existing kinds of offender careers. They efficiently tested the tool on Dutch National Criminal Record Database for obtaining the key factors for recognizing the offender careers of a person. [6, 7] worked on detecting the hotspots of a particular crime. Taha and Yoo [8] predicted the most influential members of crime in an organization. NavyaSri et al. [9] focused on future crimes likely to happen in a given city. As per [10], the following are the main four factors that we need to take into consideration:

- Huge volume of crime data and relevant records has to be stored for better analysis purposes.
- Accuracy of the algorithm developed will be proportional to the accurate training dataset
- Analyzing the data will become much more difficult and useless if data are inconsistent and incomplete
- Difficulty in gathering data. As we are dealing with crime scenes, the law enforcement and federal bureau investigation department may not readily provide data.

3 Proposed Work

We have segregated our work into three separate modules. The first module deals with the prevention of crime and the second module deals with detection of crime using modern AI. Third module is the investigation phase. Prevention tasks involve taking actions in order to control crime before it happens. This can be done by collecting previous year's crime records, analyzing them and making predictions through machine learning algorithms. They can predict the location that is more prone to crime as per earlier statistics and can allocate more police in that particular area in particular duration to arrest crimes. They can also analyze a particular criminal behavior and predict if he is willing to commit the crime in near future. If yes, police can take him into consideration and prevent him from doing any crime. So, the prevention phase will not let the crime happen. The detection module is quite opposite to this. The detection phase involves taking actions in order to control crime while it is happening. The trained AI algorithms might detect offense in an area through a security camera fitted. They can detect the type of crime happening in the video and send messages to nearby police for instant response. The crime might start but police can interfere and stop it before executing completely. This is also an attempt to reduce the crime ratio. The above discussed two modules are an attempt to arrest crime and restore peace. But all crimes cannot be interrupted from executing. Many of the crimes will be successful leaving a challenge for the police to find the criminal. Here, our third module crime investigation will come into the picture. Well-trained AI algorithms can predict criminals through clues left in the crime spot. This is quite a complex phase and out of scope of our proposed work. Our work mainly focuses on the prevention and detection of crime through advanced AI algorithms leaving the investigation part. Our work also implements advanced text classification techniques in the crime domain. So, police complaints or criminal complaints registered by victims are automatically classified as an appropriate type such as robbery or murder.

3.1 Prevention

Let us consider the SAN FRANCISCO crime dataset containing crime information from the year 2003–2015. This is quite a large dataset with 878049 records and with one output variable. Attributes of the dataset are: time stamp, category of crime, description about crime, day in which crime happened, police district responsible, resolution, crime spot, latitude and longitude. All are independent variables and the category of crime is a dependent variable that we need to predict accurately. Data preprocessing is a crucial step before analyzing data or making predictions on data. The following are important factors that we need to deal with before analyzing phase:

1. missing values 2. categorical features 3. numerical features 4. feature selection 5. feature scaling 6. Normalization.

The dataset considered is quite clean as there are no considerable missing values but only one missing value in y attribute. We need to fill the missing value with appropriate value with one of the given approaches—we opted to leave record:

1. If attribute is categorical, substitute with mode
2. If attribute is numerical, substitute with mean or median of attribute through some statistic tests
3. If missing ratio is less than 20%, leave records containing missing values.

Categorical features have to be encoded as numeric values. Suppose if attribute sex has two values male and female then male = 0 and female = 1. Our crime category attribute has 39 unique values. As we are interested in top crime categories, this count is reduced to 12 taking top 11 and renaming others as "other offenses."

In numerical features, discretization and binarization are two standard techniques. If the attribute is continuous in nature, then it can be divided into bins for better processing. Example, an attribute age can be divided into four bins: children (0–9 years), young (10–18 years), adults (19–40 years) and (41–80 years). Binarization is a way of transforming an attribute into a Boolean attribute with a certain threshold. For our crime dataset, discretization is applied on latitude and longitude variables as they are continuous in nature. The dataset may contain large values which will make them harder to get involved in computation. In such cases, the scaling or normalization can be applied to the attribute without any loss of data.

$$\text{Scaling}: x1\_new = (x1 - min(x))/(max(x) - min(x)) \quad (1)$$

$$\text{Normalization}: x1\_new = (x1 - u)/s \quad (2)$$

where u is the mean of x and s is the standard deviation of x.

Feature selection is crucial for a better output. All attributes of the dataset may not help in determining target variable. Sometimes, though, it helps in determining target variable it might be noisy and result in drop in accuracy if it is included in the dataset. So, we need to eliminate all such types of variables and be choosy in attribute selection. Feature transformation is a way of creating new attributes so that meaning and completeness are obtained for the dataset (Figs. 1 and 2).

Only in 10000 cases, criminals have been arrested and cited. In nearly 20000 cases, criminals have been arrested and booked. But no action has been taken in nearly 55000 cases which might erase fear in criminals. This helps police to rethink in regard to making laws too strict and to cultivate fear in criminals (Figs. 3 and 4).

The crimes such as assault, noncriminal and drug were minimum in the year 2005 but have increased dramatically with time. October month has been opted by all criminals irrespective of year. So, police have to consider these insights and plan accordingly to prevent crime in future. A large number of crimes are happening in the southern police district compared with Richmond police district. Mission police district is next to southern in entertaining crimes. This suggests to use more officials in southern districts and to make people more cautious living in these areas.

After analysis and proper data preprocessing, the dataset is aptly suitable for machine learning algorithms. Mostly, machine learning algorithms are either predictive or generative in nature. Still one attribute named as "crime description" in our

```
[47] x['Category'].value_counts()
     a=['LARCENY/THEFT','OTHER OFFENSES','NON-CRIMINAL','ASSAULT','DRUG/NARCOTIC'.
     def func(s):
       if s in a:
         return s
       else:
         return "OTHER OFFENSES"
     x['Category']=x['Category'].apply(func)
     x.groupby('Category').size().sort_values(ascending=False)[:12].plot('barh')

     /usr/local/lib/python3.6/dist-packages/ipykernel_launcher.py:9: FutureWarning
       if __name__ == '__main__':
     <matplotlib.axes._subplots.AxesSubplot at 0x7f6f7c915518>
```

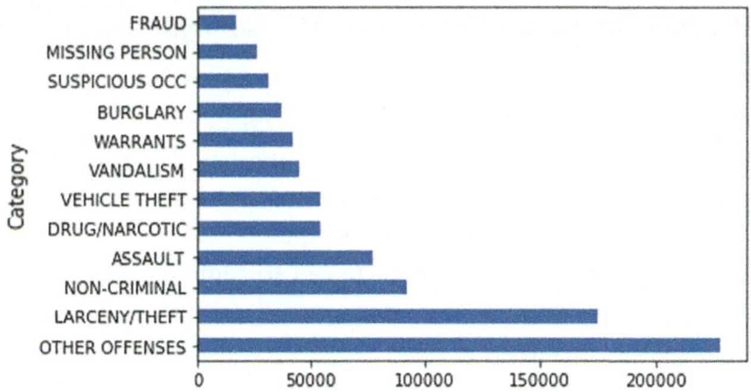

Fig. 1 Finding out the most frequent crimes through visualization

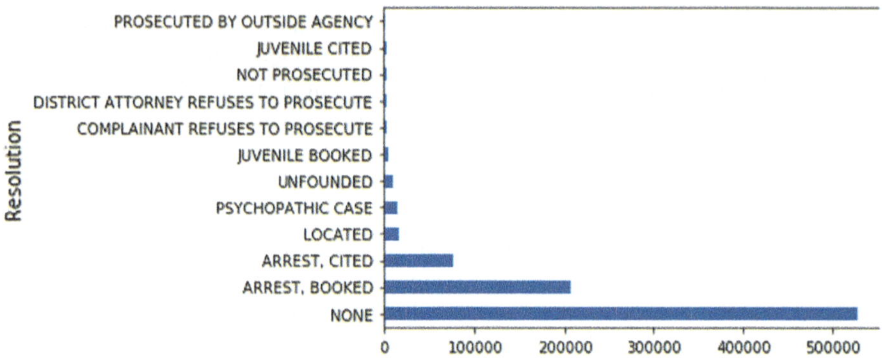

Fig. 2 Describing the count of different actions taken by police per crime

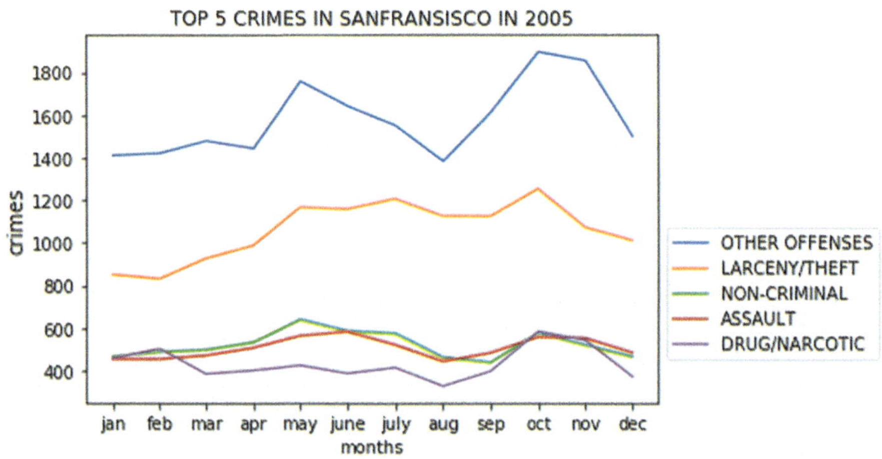

Fig. 3 Crime ratio in each month in the year 2005

dataset is not processed. The attribute has string values describing the crime of particular record. For understanding, let us take two records as a sample and process them (Table 1).

1. Conversion—all sentences are converted into lower case.
2. Lemmatization and stemming—the words in sentences are replaced with a common synonym. Words are transformed into root word by removing suffixes and prefixes as (locked and arrested in the above case)
3. Remove stop words—to remove noise we can opt to remove either stop words (such as [the, from]) or more frequent words in all sentences as they do not contribute to a particular sentence.

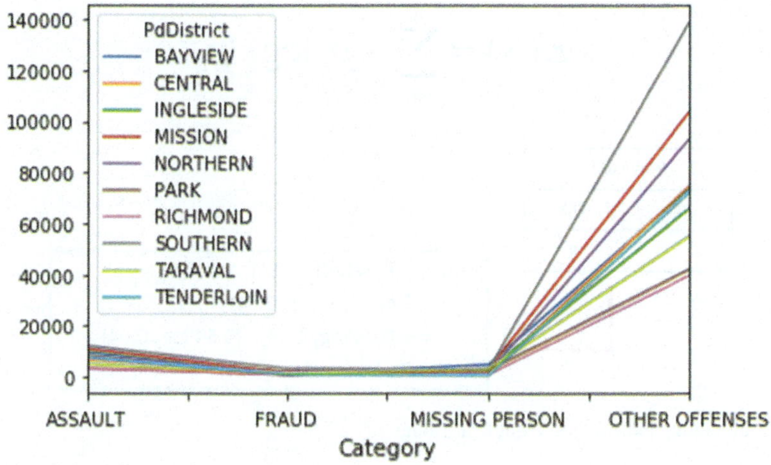

Fig. 4 Crime rate in each of the police district

Table 1 Sample crime description attribute before and after preprocessing

Crime description	Traffic	Violation	Arrest	Grand	Theft	Lock	Auto
Traffic violation arrest	1	1	1	0	0	0	0
Grand theft from locked auto	0	0	0	1	1	1	1
Arrested	0	0	1	0	0	0	0

Decision Tree

The algorithm is quite similar to a tree. The entire training data are fed as an input to the root of the tree. Here as per the rules and conditions mentioned in the node, an instance of training data chooses its branch and travels till the terminal node. A training instance passes through several intermediate nodes to reach a terminal node. All the terminal nodes will lead to one of the classes of the target variable. Here two terms entropy and GINI INDEX are quite important to understand. Lower entropy, it is easier to extract much information from the data. GINI INDEX is similar to a cost metric, which is adapted in evaluating the split of training sets. An attribute with low GINI INDEX is apt for splitting and calculated as:

$$G = 1 - \Sigma pi2 \tag{3}$$

If count of yes = 10 and no = 0 then entropy (play golf) = entropy (10, 0) = − (10/10 $\log_2 10$) − (0/10 $\log_2 0$) = 0. Such a variable can be a leaf node or terminating node. Else, the variable needs to split further (Figs. 5 and 6).

$$Informationgain\,(C, N) = entropy\,(c) - entropy\,(C, N) \tag{4}$$

$$E(S) = \sum_{i=1}^{c} - p_i \log_2 p_i$$

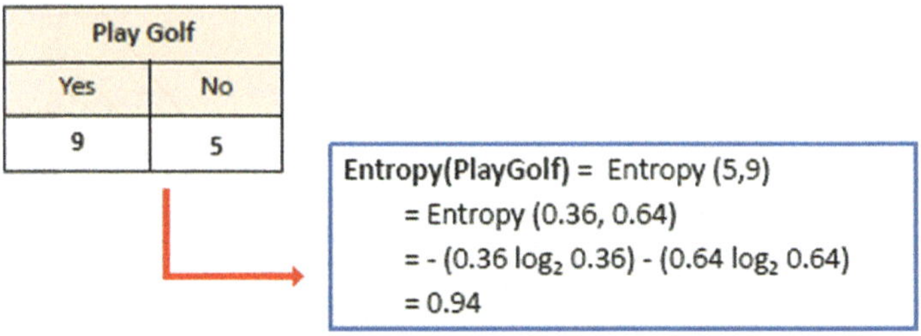

Fig. 5 Calculating entropy for a single attribute

$$E(T, X) = \sum_{c \in X} P(c)E(c)$$

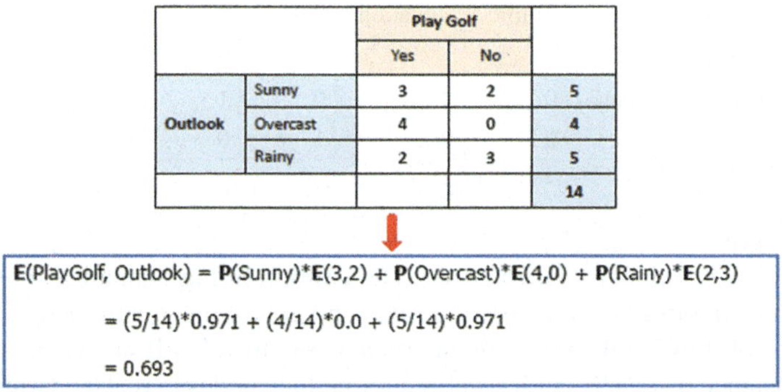

Fig. 6 Calculating entropy for multiple variables

C = current attribute N = next selected attribute.

For the next split, an unselected attribute with higher information gain is chosen. Decision trees are easily prone to overfitting. The two solutions to handle this are described below.

Pruning

Actually, the process of splitting keeps on happening till a terminal node with the least entropy or zero entropy is obtained. This can perform well on training data but result in poor accuracy on test data. So, further splitting of nodes is stopped after the tree reaches a certain depth.

Table 2 Accuracy of different classifiers in San Francisco crime dataset

Algorithm	Reported accuracy (%)	Precision
Decision tree	81	0.76
Random forest	84	0.81
ADABOOST	90	0.85
SVM	89	0.82
KNN	75	0.72
XGBOOST	85	0.81

Random Forest

Random forest algorithm falls under the bagging technique. An ensemble of trees is constructed through various training datasets obtained by sampling and replacement from original dataset. Generally, it is a collection of decision trees together. Each tree will make a prediction. The class of the test instance is decided by voting in case of classification task and output is decided by average in regression task. Each tree in the forest will be given equal importance in determining final class. This can reduce overfitting and can increase the accuracy on test data (Table 2).

ADABOOST

ADABOOST is based on decision stump and weak learner concept. In contrast to decision trees, ADABOOST constructs a group of trees, each of depth 2 usually known as decision stump. A single stump may not provide good accuracy as they do classification based on a single attribute. Hence, these are also known as weak learners. In contrast to random forest, in groups of stumps made by ADABOOST, some stumps have more say in predicting the final class of an instance. A stump with least error will get higher say in predicting the final class. Order of stumps generated is quite important. The errors made by the first stump impact the training set of the second stump. Misclassified training instances in the first stump are included in the training set of the second stump and so on.

At first, all the training instances in a dataset of size N are allocated with equal sample weight 1/N.

$$w^i = \frac{1}{N} fori = 1\, to\, N \tag{5}$$

A random training set for the first stump is generated. We can use earlier decision tree concepts in determining the root attribute of stump. Before the second iteration, the sample weights of training instances get updated.

Step 1: calculate error of Mth stump. M (x) is stump output

$$E(M) = \frac{\sum W(i) where\, class(i) = M(X)}{\sum \mathrm{W\ (i)}} \tag{6}$$

Step 2: calculate amount of say of Mth stump

$$A(M) = amountofsay(Mthclassifier) = \frac{1}{2}log\frac{1 - \text{E (M)}}{\text{E (M)}} \quad (7)$$

Step 3: update weights of training instances after Mth stump.
If ith training instance is correctly classified:

$$w\_new^i = w^i * e^{A(M)} \quad (8)$$

If ith training instance is misclassified:

$$w\_new^i = w^i * e^{-A(M)} \quad (9)$$

Step 4: normalize weights of all instances.

Step 4 will decrease sample weights of instances, which are correctly classified and increase weights of misclassified instances. Misclassified instances with larger sample weights are more likely to be included in the training set of (M + 1) stump.

Step 5: predicting the final class of the test instance after all stumps are made.

If classes are [i, j]. Total there are M stumps and A (M) is say of stump final class of test instance x = j if

$$\left(\sum_{m=1}^{M} A(m)whereM(x) = j\right) > \left(\sum_{m=1}^{M} A(m)whereM(x) = i\right) \quad (10)$$

3.2 Detection

Ren et al. [11] proposed Faster RCNN Object detection which has increasing applications in real life. Faster RCNN is an advancement of fast RCNN. Selective search technique in fast RCNN is successfully replaced by region proposal network in faster RCNN. Region proposal network (RPN) aims at proposing regions. A standard input image is passed through blocks of convolutional layers like in vgg16 to extract the final feature map. Every point in this feature map of size 37 * 50 * 256 (assumed as output feature map for explanation) represents an anchor. With respect to the size of objects to be detected, the anchor ratios [1:1, 1:2, 2:1] and anchor sizes [128 * 128, 256 * 256, 512 * 512] are chosen (Fig. 7).

Region proposal network is further connected to a convolutional layer of 3 * 3 kernel size, padding = 1 and output features = 512. The output of this layer is passed to 1*1 fully convolutional layer to perform regression (to obtain 4 * n values-denoting object coordinates where n is no of anchors) and classification tasks (2 * n values—0 or 1 denoting the presence of object). Every point in 37 * 50 * 256 with given anchor ratios and sizes creates 3 * 3 = 9 boxes in the input image. So, there are totally 37 * 50 * 256 * 9 = 16650 proposed regions by RPN in input image (Figs. 8 and 9).

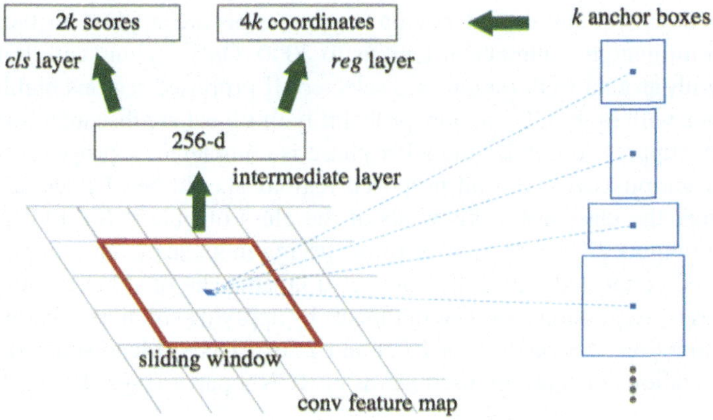

Fig. 7 Each cell of this feature map will have k anchor boxes of different shapes

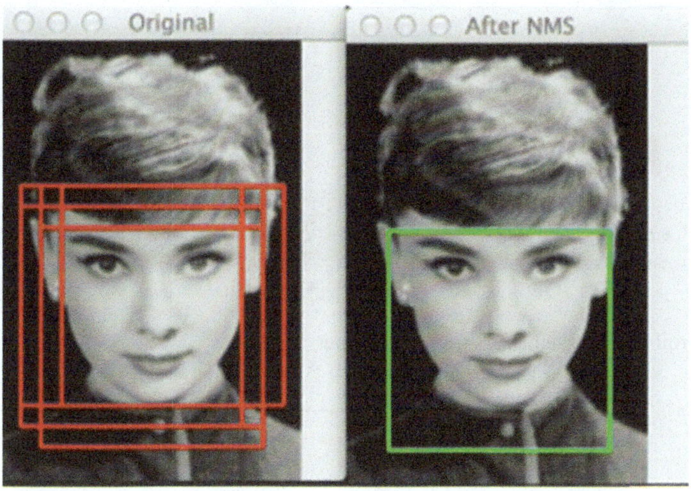

Fig. 8 Before and after applying nonmax suppression

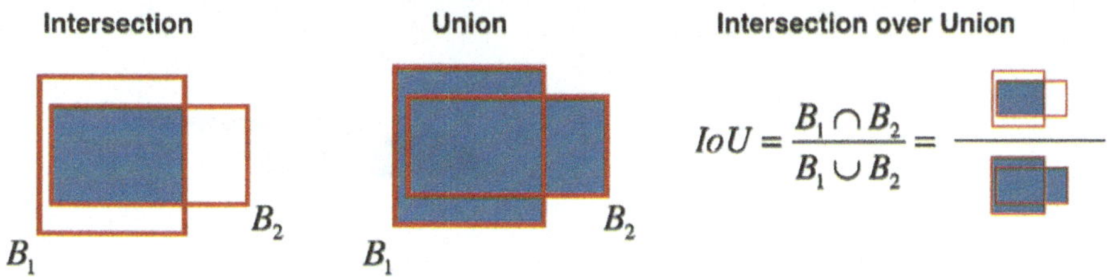

Fig. 9 Calculating intersection over union

Intersection over union and nonmax suppression are applied to these 16650 regions to mitigate the number of regions to 2000. Only regions with IOU greater than 0.5 with ground truth regions are selected. If proposed regions highly overlap then regions with high IOU and low probability of object are discarded in the name of nonmax suppression. After the RPN phase is completed, it proposes regions of interest of various sizes. But all proposed regions should be of fixed size as they pass through the classifier to know about the class of object. So, ROI pooling is applied. It outputs 7 * 7 * 512 tensor from varying input sizes. This output tensor is flattened and connected with fully connected layers. The final layer outputs object coordinates by regression and class of objects by applying soft max. Finally, we have four loss functions defined—2 for RPN and 2 for RCNN. Mean squared error and L2 regularization concepts are used in Faster RCNN paper (Figs. 10 and 11).

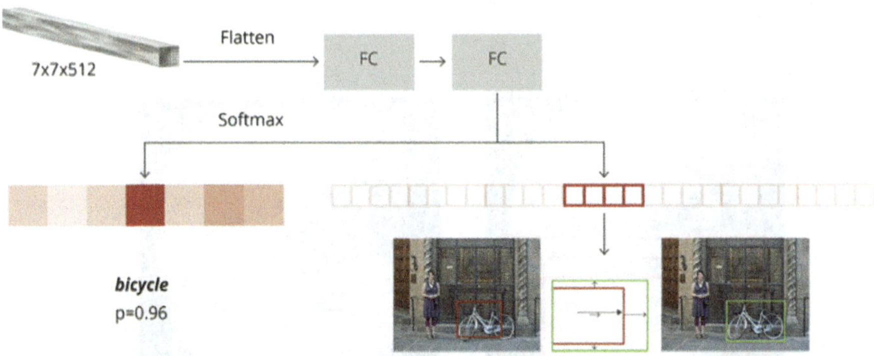

Fig. 10 Forming connected layers from ROI output

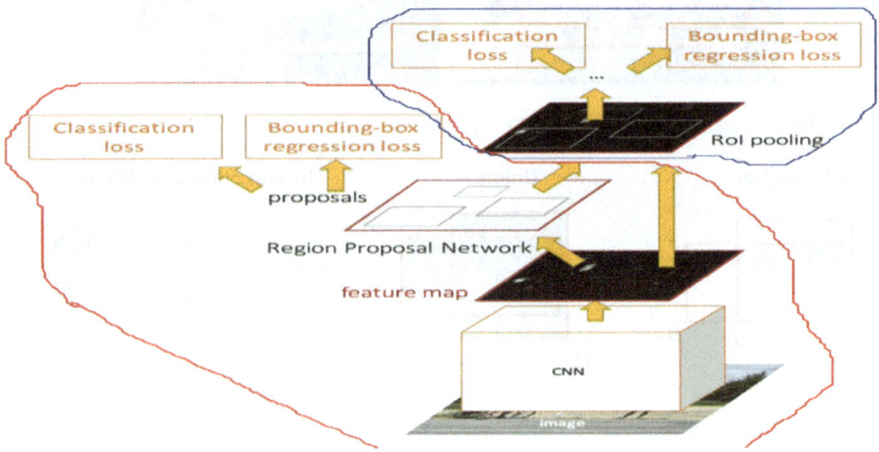

Fig. 11 Full faster RCNN architecture

Faster RCNN Detecting Crime

We trained Faster RCNN on crime images dataset covering top crimes such as Assault, burglary, Sex abuse, Murder, etc., crimes such as fraud and forgery are not taken into consideration as they are quite complex to detect and differentiate. PYTORCH and TENSORFLOW frameworks are used. Faster RCNN performance on test images is reported here. We have also experimented with other standard object detection models such as YOLO, SSD whose performance is compared below (Figs. 12, 13, 14, 15 and 16, Table 3).

Crime Complaints: Text Classification

Crime complaints in the form of text given either in police stations or through online application by victims are increasing day by day. Manpower is needed in reading all these text complaints and categorizing them so that they are passed to respective departments for investigation. Through machine learning algorithms, without human intervention, the complaint documents can be categorized by decreasing the time needed and speeding up investigation. Let us compare machine learning algorithms and advanced deep learning techniques in this domain. The algorithm working on sample small dataset is described below (Table 4):

NAÏVE BAYES

It is a popular supervised machine learning classifier constructed on Bayes formula.

$$P(category/complaint) = \frac{P(compliant/category) * P(category)}{p(compliant)} \tag{11}$$

$$P(A/B) = \frac{P(B/A) * P(A)}{p(B)} = P(B/A) * P(A) \tag{12}$$

Fig. 12 Faster RCNN detecting sex abuse

Fig. 13 Faster RCNN detecting burglary

Fig. 14 Faster RCNN detecting murder

Fig. 15 Faster RCNN detecting robbery

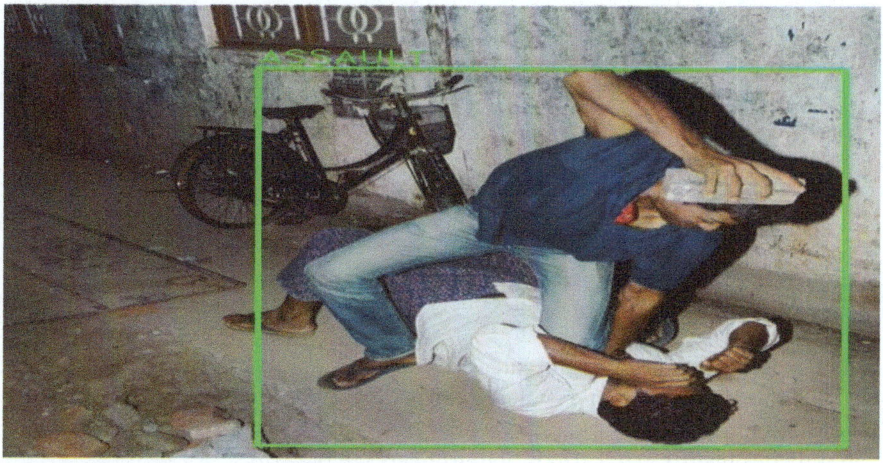

Fig. 16 Faster RCNN detecting assault

Table 3 Comparing different object detection algorithms performance in detecting crime

Algorithm	MAP	FPS
FASTER RCNN	75.6	12
YOLO 448 * 448	71.3	45
SSD 300	69.5	46

Table 4 A sample crime classification dataset

Category	Complaint	Document
Murder-0	My brother was killed with knife	D1
Robbery-1	Brother's wallet—stolen	D2
Robbery-1	Someone stolen—television	D3
???	He killed wife	D4

P (B) is neglected as it is evidence and is common for all the classes in computation.

$$\text{Category} = \text{argmax}\ (\text{P}\ (\text{murder}/\text{d4}), \text{p}(\text{robbery}/\text{d4})) \tag{13}$$

$$\text{P}\ (\text{murder}/\text{d4}) = \text{p}\ (\text{d4}/\text{murder}) \times \text{p}(\text{murder}) \tag{14}$$

$$p\left(\frac{murder}{d4}\right) = p\left(\frac{hekilledherwife}{murder}\right) * p(murder) \tag{15}$$

$$\text{P}\left(\frac{\text{murder}}{\text{d4}}\right) = \text{p}\left(\frac{\text{he}}{\text{murder}}\right) \times \text{p}\left(\frac{\text{killed}}{\text{murder}}\right) \times \text{p}\left(\frac{\text{wife}}{\text{murder}}\right) \times \text{p}(\text{murder}) \tag{16}$$

$$P\left(\frac{robbery}{d4}\right) = p\left(\frac{he}{robbery}\right) \times p\left(\frac{killed}{robbery}\right) \times p\left(\frac{wife}{robbery}\right) \times p(robbery) \quad (17)$$

Step 1: calculate prior probabilities

$$\text{P (murder)} = 1/3 \quad (18)$$

$$\text{P (robbery)} = 2/3 \quad (19)$$

Step 2: remove stop words and calculate conditional probabilities for test sentence:
P (he/murder) = 0 + 1/6 p (killed/murder) = 1 + 1/6
p (wife/murder) = 0 + 1/6 p (he/robbery) = 0 + 1/6
p (killed/robbery) = 0 + 1/6 p (wife/robbery) = 0 + 1/6
P (murder/d4) = 0.16 * 0.33 * 0.16 * 0.33 = 0.002787
P (robbery/d4) = (0.16)4 * 0.66 = 0.00270
Category = murder

We have created crime complaints dataset through available datasets, internet and social media. We have applied NAVIE BAYES, RNN, GRU, LSTM techniques and compared their performance on the dataset with crimes—murder, assault, fraud, sex abuse and robbery (Table 5).

RNN (Recurrent Neural Network) RNN is quite an effective model in which timestamps of input are considered.

For the text—"let kill her" timestamp values are [let = t_0, kill = t_1, her = t_3]. All the unique words in the crime complaints corpus are expressed in one hot encoding format. For example, let = [1, 0, 0] kill = [0, 0, 1] her = [0, 1, 0]. The size of the vector is equal to the number of unique words in the corpus. If there are N words in the complaint text corpus. Then x[t] will be [1 * N] vector. If we choose k hidden neurons in hidden layer, then h[t] will be [1 * k] vector. U, w and v are weights that are randomly initialized at beginning but gradually updated through backpropagation (Fig. 17).

$$\text{Shape of vector } U = [N * K] \quad (20)$$

Table 5 Comparing the performance of algorithms on crime complaint dataset

Algorithm	Classification accuracy (%)
NAIVE BAYES	81
RNN	84
GRU	86
LSTM	87

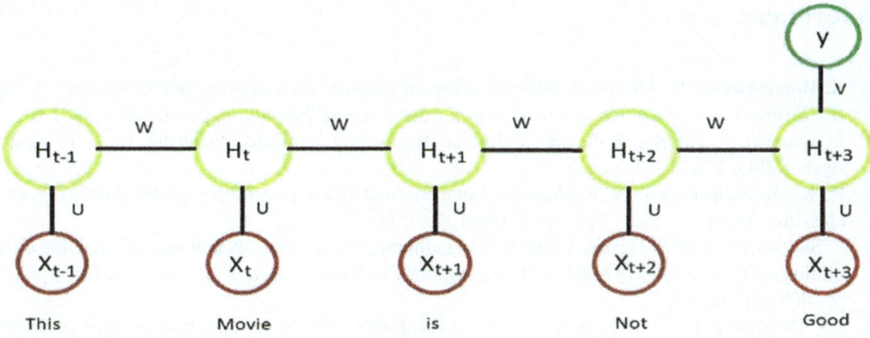

Fig. 17 Complete structure of many to one RNN

$$W = [K * K] \tag{21}$$

$$V = [k * c] \tag{22}$$

V is a vector where c is no of final classes.

$$x[t-1] = [1, 0, 0]x[t] = [0, 0, 1]x[t+1] = [0, 1, 0] \tag{23}$$

$$h[t] = \tan h(x[t] * u + w * h[t-1]) \tag{24}$$

$$y = soft - max(H[t] * v) \tag{25}$$

Cross entropy loss function is used while updating weights through backpropagation.

4 Conclusion

The proposed work is a significant contribution to crime prevention and detection using advanced artificial intelligence methods. The analysis and visualization provided fruitful insights regarding crime patterns. Faster RCNN performed better compared with YOLO in terms of accuracy in the object detection process. Long–short-term memory has achieved good accuracy compared with other models in crime text classification. The techniques used can be easily deployed on the cloud for maximum usage. The proposed crime detection can be done in the real world through video surveillance and processing. We would also like to carry on further research in crime investigation through AI.

References

1. K. Bogahawatte, S. Adikari, Intelligent criminal identification system, in *Proceedings of 8th IEEE International Conference on Computer Science and Education*, pp. 633–638 (2013)
2. J. Agarwal, R. Nagpal, R. Sehgal, Crime analysis using K-means clustering. Int. J. Comput. Appl. **83**(4), 1–4 (2013)
3. R. Kiani, S. Mahdavi, A. Keshavarzi, Analysis and prediction of crimes by clustering and classification. Int. J. Adv. Res. Artif. Intell. **4**(8), 11–17 (2015)
4. S. Sathyadevan, M.S. Devan, S. Surya Gangadharan, Crime analysis and prediction using data mining, in *Proceedings of IEEE 1st International Conference on Networks and Soft Computing*, pp. 406–412 (2014)
5. J.S. De Bruin, T.K. Cocx, W.A. Kosters, J.F.J. Laros, J.N. Kok, Data mining approaches to criminal career analysis, in *Proceedings of 6th IEEE International Conference on Data Mining*, pp. 1–7 (2006)
6. D. Wang, W. Ding, H. Lo, T. Stepinski, J. Salazar, M. Morabito, Crime hotspot mapping using the crime related factors-a spatial data mining approach. Appl. Intell. **39**(4), 772–781 (2013)
7. E. Hamdy, A. Adl, A. Ella Hassanien, O. Hegazy, T.-H. Kim, Criminal act detection and identification model, in *Proceedings of 7th International Conference on Advanced Communication and Networking*, pp. 79–83 (2015)
8. K. Taha, P.D. Yoo, SIIMCO: a forensic investigation tool for identifying the influential members of a criminal organization. IEEE Trans. Inf. Forensics Secur. **11**(4), 811822 (2016)
9. M. Navya Sri, M. Ramakrishna Murty et al., Robust features for emotion recognition from speech by using Gaussian mixture model classification, in *International Conference and Published Proceeding in SIST Series*. Springer, vol. 2, pp. 437–444, August-2017
10. S. Sathyadevan, S. Gangadharan, Crime Analysis and Prediction Using Data Mining
11. S. Ren, K. He, R. Girshick, J. Sun, Faster R-CNN: towards real-time object detection with region proposal networks

Issue Resolution Process: Salesforce

Ch. Rishika and J. Sirisha Devi

Abstract The world we live in today is fully synchronized with the data being generated every second using cloud. One such company that delivers cloud services is "Salesforce." Salesforce is a cloud-based software organization providing Customer-Relationship Management (CRM) services. This paper proposes the case escalation concept and how it can solve the issue mentioned by the customers in the specified time. Case Escalation, also known as Issue Resolution Process, is done when an issue reported by the customer is left unresolved. This mechanism will be crystal clear when a sample of data is taken. The data project the issues arriving at a pediatric hospital with different departments. An interesting feature of Salesforce will take a big hand in this application to resolve a problem regarding the purchase of licenses for each and every manager in each department, on a whole containing more than 50 departments. This feature helps to provide access of the application to the managers, executives, directors, president, vice president, to solve the case without having a licensed account on Salesforce. The cases will be solved depending on its priority whether high, moderate or low. This paper also defines the outline of stimulating traits for further improvements.

Keywords Salesforce · Apex · Case escalation · Customer-Relationship management · Unauthenticated sites · Priority of a case

---

Ch. Rishika (✉) · J. Sirisha Devi
Department of Computer Science and Engineering, Institute of Aeronautical Engineering, JNTUH, Hyderabad, India
e-mail: chigurupatirishika@gmail.com

J. Sirisha Devi
e-mail: siri.cse21@gmail.com

C. Kiran Mai et al. (eds.), *Proceedings of International Conference on Advances in Computer Engineering and Communication Systems*, Learning and Analytics in Intelligent Systems 20,
https://doi.org/10.1007/978-981-15-9293-5_7

1 Introduction

Case is a problem or a conflict that is not solved even after a given time. This time is treated as the deadline for the case to be resolved. Generally, the deadline will be specified depending upon the type of issue and its priority. The priority of the case can be categorized into three, namely:

1. High priority
2. Moderate priority
3. Low priority.

Salesforce is the platform that provides an automatic routing for important events, which are nothing but the issues. It allows customers to track their own issues, report them, and get it solved. It also promises the automatic escalation of the issue, when delayed.

The aim of this paper is to bring up one of the key elements of Salesforce called "Unauthorized sites" where a link is generated by the Salesforce team to the department concerned with the issue, which enables them to fix the issue. A single license is way better than acquiring hundreds of them.

In Sect. 2, we will discuss the previous research done by various authors regarding the domain and the research topic. Section 3 defines the problem statement, workflow, and its solution. Section 4 depicts the results of the fully developed system. The major details are summarized in Sect. 5.

2 Literature Survey

The below survey done by a few researchers mainly focuses on the domain used in the paper that is "Salesforce." It also indicates the programming language and the key feature of Salesforce used to expand the functionality. The following subsections describe each of them using the references.

2.1 Salesforce: An Emerging Platform

Cloud, today is on demand and considered to be most desired from two companies whether it might be small or large scale. Services provided by it are being used daily in one or the other form like Gmail, Google Docs and Dropbox. One such organization that contributes cloud as a service is Salesforce. It is mostly used for the business background. Major social networking sites like Twitter, Facebook, and so on use cloud to store personal and business data. Author Jelena Jeknic, Bojan Kraut in their study stated that all the data saved are accessible for several purposes [1]. But they also mentioned that users need to give up their privacy. Since the data are made public, anyone can use this personal information for their own needs and profits. The marketers are the most benefited with all the fresh and updated information about consumers, which can be used wisely to develop products for customer satisfaction.

Author Matt D'Angelo in his writings praised Salesforce to be the renowned and best software for Customer-Relationship Management Services [2]. It came into light 20 years ago to take over sales, marketing, and customer service. Salesforce brings hope, power, and functionality to small businesses to scale up and upgrade over time. Salesforce has the ability to get a startup to its enterprise level. Today, Salesforce is recognized as one of the top-five fastest-growing enterprise software companies in the world. One far-reaching feature of it is that it offers clients a broad marketplace for easy scalability of their business.

Salesforce takes the upper hand in CRM market and becomes a risk to other companies in the CRM field because it is powerful, scalable, and customizable. It is easy to use and very less training is required for nonadmin users. The page is separated into tabs or fields such as Home, Contacts, Accounts, Leads, Campaigns, Opportunities, Forecasts, and Files. These categories make it easy to access any task when needed. Also, permissions are separated for individual users and user groups, which are under admin control. It is a product (Saas) that can be accessed anywhere and anytime, even on a mobile.

2.2 Case Escalation and Its Definition

Authors Wil M.P. Van der Aalst, Michael Rosemann, Marlon Dumas, in their writings, proposed that decision-making in information systems includes build-time and run-time [3]. He defines both of them and how important they are. Process models always assume a time for all the activities planned to be accomplished in a certain period. When these activities are not fulfilled in that period, users must take decisions to reach the goal. Such decisions are called escalations. Their paper proposed a framework for escalations and also escalation mechanisms. It identifies and classifies these mechanisms such as routing the task, distributing the task, and so on.

2.3 Key Feature in Salesforce

Salesforce has a special feature called "unauthenticated access" to a user. The user may not be having a licensed account but he can still access the data shared using these pages [4]. A guest user profile is created to manage the access of data, content, and objects for public without authentication. This profile acts as a community where public can get access to its data. This feature is available in all five editions. After creation of community, Salesforce creates a profile, user record, sharing mechanisms only for the guest users.

Community -> Guest User Profile -> Community Site Guest User.

Guest user has access to certain pages in the community as long as the community is active in your organization.

2.4 Apex as a Programming Language

Apex Code is a strongly typed programming language that executes on the Salesforce platform-Force.com [5–7]. Apex's functionality is to add business logic to applications, writing database triggers, and also program controllers in the user interface layer.

Jon Mountjoy, in his article, initially, introduced the Apex language [8]. After delivering the overview of the syntax, he focused on the database support, testing capability, web services provided by it, and finally the dynamic Apex.

3 Problem Definition

3.1 Process Flow

The below Fig. 1 depicts the workflow of the escalation of a case. It starts from the customer who opens a case to the department and then further escalated when left unsolved to the higher officials of the same department.

3.2 Problem Statement

The client wanted a system to improve the time and efficiency in issue resolution process for the cases logged by users. They wanted a system that allows automated escalation of cases to upper management if it is not resolved in the specified time. Upper management should be able to access the case details and give resolution steps and next steps to the users from outside without the need to login to Salesforce system.

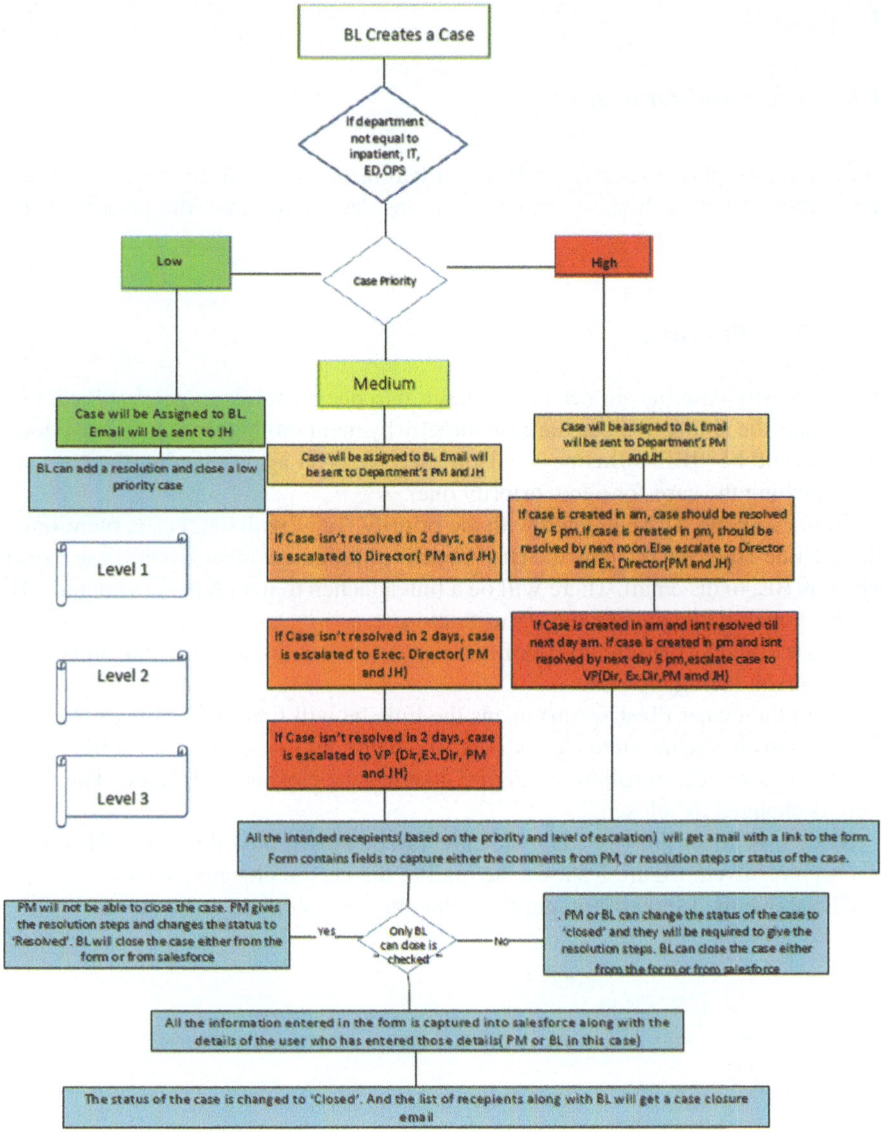

Fig. 1 Process flow of the system

4 Results

4.1 Input and Output

The issue resolution process can be described in one go with the case assignment flow depicted below. It shows three variations depending upon the priority of the case.

4.1.1 Low Priority

A low priority case has no need to undergo into deeper levels to get resolved. One person has the ability to close the case directly by mentioning the steps to solve them and it would be "BL"—Business liaison user. He is an agent opening the case and also finishing the case for a low priority one.

Figure 2 shows a case created with the priority "low" with the reason mentioning about clinic-staff Svc issue by the BL. An acknowledgment is sent to the case owner, which is BL, to his email. There will be a link attached to it with the case details. He needs to open it and change the status to close it.

Figure 3 Shows the screenshot of the case owner's email with the link and details of the person facing the issue.

When the owner (Test varun) opens the link, he will find the details of the case. He needs to change the status to "closed" and mentions few steps to solve the case in the text area beside "resolution." Figure 4 shows the closing of the case. The status there is changed to "closed."

Since it is a low priority case, the case can be closed directly by the BL once it has been resolved. Figure 5 shows the mail of the person closing the case.

The above mail screenshot mentions that the case is closed at L0 level.

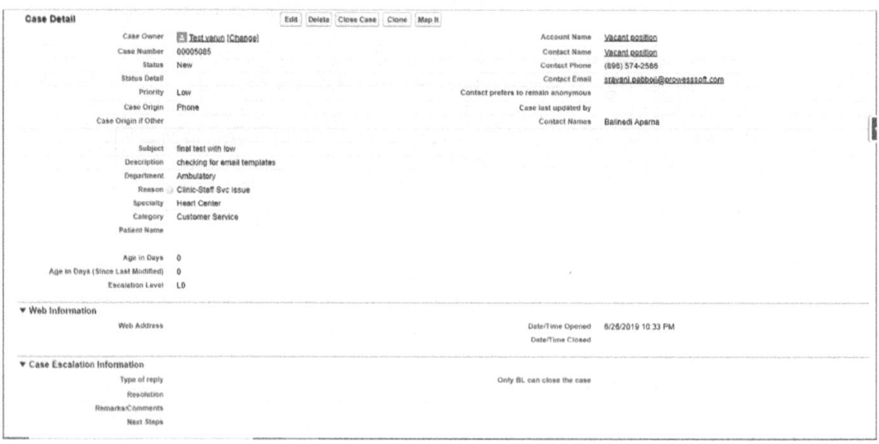

Fig. 2 Creating case with low priority

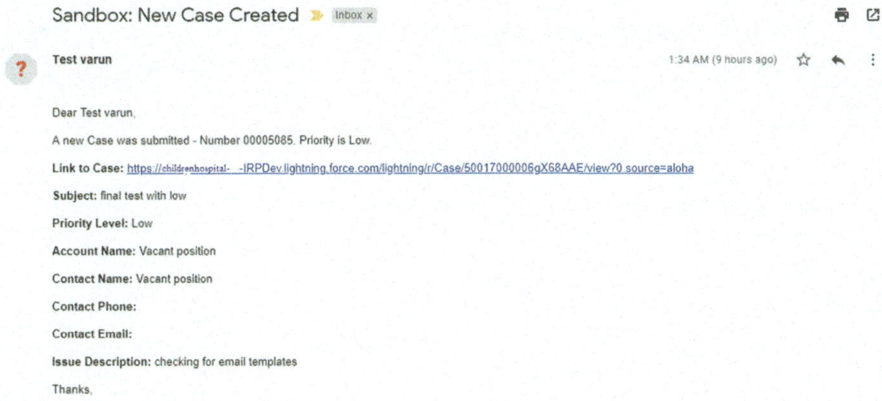

Fig. 3 Case owner mail

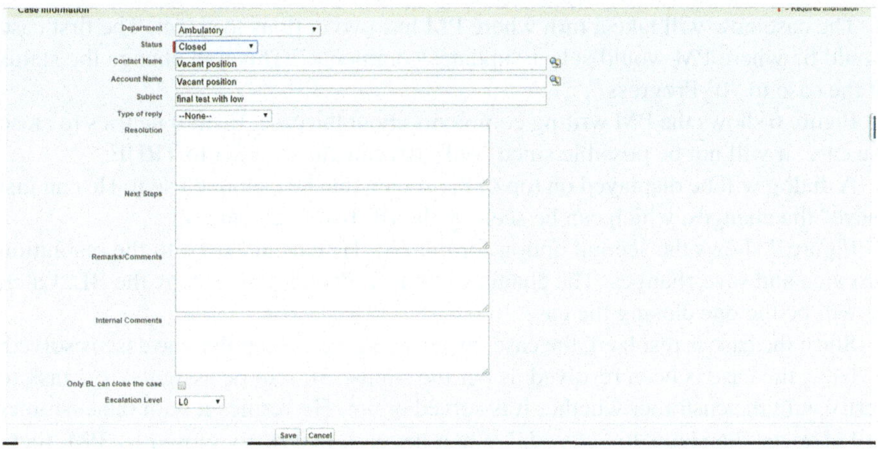

Fig. 4 Case closure

4.2 Medium Priority

A medium priority case will be created but here the case details will also be sent to the PM and JH. PM indicates “Practice manager” and JH is an “Executive assistant.” These two persons will be able to handle the case, open it, make changes, and add resolutions. Closing the case will be done by the BL or the PM.

While creating the case, if the box mentioning “only BL can close” is checked, then PM and JH cannot close the case what so ever. But if the box is unchecked, either PM or JH can close the case after mentioning the steps for resolution.

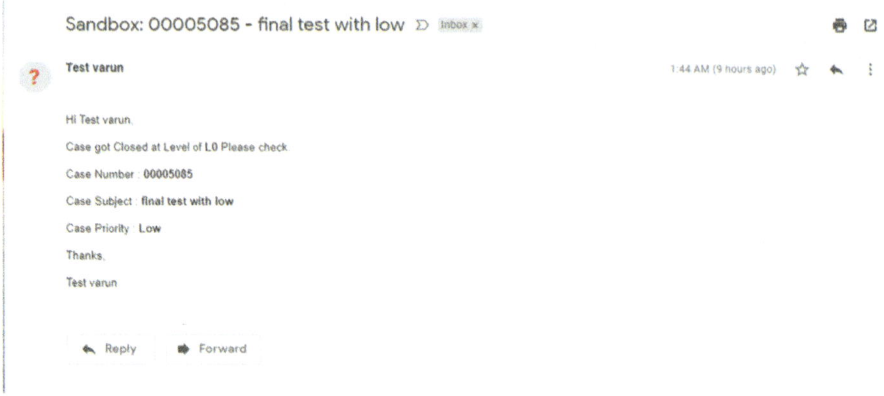

Fig. 5 Case closure mail

The case now will take a turn where PM has two options to select. The first case would be where PM would select "updates/comments." This will change the status of the case to "In-Progress."

Figure 6 shows the PM writing comments about the case. Even if he tries to close the case, it will not be possible since "only BL can close" is set to TRUE.

A dialog will be displayed on top of the screen that he cannot close it. He can just "save" the changes, which can be seen by the BL before closing it.

Figure 7 shows the second option for the PM. He can add steps to the resolution text area and save changes. The changes will be reflected and seen by the BL. Later, he will be the one closing the case.

Since the case is resolved, the case owner will get an email that the case is solved.

Now, the case is been resolved as per the status. BL will be assigned to a task to verify with the customer whether it is solved or not. He verifies it with the customer and changes the status to "closed." When the case is directly closed by PM, there will be no closure mail sent neither any task will be created.

Medium and high priority cases are much similar but the only difference seen in high priority is that the case will be escalated to the next higher official.

Fig. 6 Status as in-progress

Fig. 7 Status as resolved

4.3 *Escalation Level*

What if medium priority case is left unsolved for a longer time? It will be escalated to next-level executives who will receive the case details via mail with a link. Those executives could be president, vice president, director, etc.

5 Conclusion

The past few years were very challenging to the industries relating to all sectors for bringing up something innovative into action. The on-going innovations are making Salesforce as their backbone since it provides the best cloud services. Not only cloud, it also introduces marketing world to those who are new to it and makes it easier to them. This paper involves such an astonishing platform to solve issues by the customers in a very soothing way. Force.com platform's main mark is case routing and escalation. The issues are routed automatically to their respective departments and solved by its managers. If they are not done at specified time, it can be escalated to higher officials. This paper mainly introduces the masterpiece of Salesforce and how it solved the needs of the client. A major challenge is taken up and solved which resulted in a better system in hospitals. The finished product introduced benefits in terms of cost-efficiency. Many more enhancements can be made into healthcare using Salesforce in the upcoming years.

References

1. J. Jeknic, B. Kraut, Cloud services and marketing's, in: *38th International Convention and Communication Technology*, Electronics and Microelectronics, May 2015
2. M. D'Angelo, Contributing Writer, Salesforce.com, June 2019
3. W.M.P. Van der Aalst, M. Rosemann, M. Dumas, Decision Support Systems **43**(2) (2007)
4. Trailblazer Community (Give secure access to unauthenticated users with guest user profile), Salesforce.com.
5. Developer Force provides access to free developer edition accounts. It also provides links to documentation, forums, and more, "Trailblazer community", June 2012
6. The Apex Language Reference provides a comprehensive introduction to the Apex language, "Trailblazer community ", June 2012
7. An Introduction to the Force.com Database provides a much more comprehensive introduction to the Force.com Database, "Trailblazer community", June 2012
8. J. Mount, Community manager and editor-in-chief. In: An Introduction to force.com Apex code", Salesforce.com, June 2012

Automated Member Enrollment: Health Insurance Agency

A. Sai Bhavana and P. L. Srinivasa Murthy

Abstract Popular arrangements emerge with anticipated cutting edge innovations during this cloud world. Salesforce advancement, which is an exceptional float these days, has been extraordinarily helpful to small, medium, and huge estimated associations. Being an unmistakable industry, medicinal services like a few unique enterprises must advance to fulfill the requests of the computerized age in every area, especially health care coverage, patient data. To accomplish this assignment, health insurance has been inviting Salesforce for its assistance. This paper examines a plan created utilizing Salesforce to ensure uncomplicated techniques inside the medical coverage segment. Starting with the point of the errand, the paper diagrams the strategies, highlights, and edges of the anticipated framework. To acknowledge less manual work, effective procedure management and to allow direct assistance to every person, a programmed part enlistment plan is exhibited in detail. The paper incorporates the structure plan and a top-level perspective on some eye-catching alternatives for additional upgrades.

Keywords Salesforce in health care · Challenges to insurance agencies · Automated enrollment · Individual assistance · Stripe payment gateway · Framework plan

1 Introduction

There are enormous changes astir inside the medicinal services industry. Mechanical developments are changing how care is analyzed, treated, and oversaw. These progressions are inciting the medical coverage framework for new improvements.

A. Sai Bhavana (✉) · P. L. Srinivasa Murthy
Department of CSE, IARE, Hyderabad, Telangana, India
e-mail: akkineni.saibhavana@gmail.com

P. L. Srinivasa Murthy
e-mail: pl.srinivasamurthy@iare.ac.in

C. Kiran Mai et al. (eds.), *Proceedings of International Conference on Advances in Computer Engineering and Communication Systems*, Learning and Analytics in Intelligent Systems 20,
https://doi.org/10.1007/978-981-15-9293-5_8

The crucial problem depicted when clients are surveyed is the exhausting methodology to get took a crack at the strategies or to turn into a part in a predefined protection organization. Medical coverage is dauntingly troublesome, and numerous customers who have not recently had inclusion need assistance to hack their methods through the wilderness of protection details and making sense of which plan is best for them. As uninsured ordinarily needs assistance through the strategy, the new results make it direct for each person. This paper expects to build up a structure that empowers computerized enlistment, plans, payments, and security for data in medical coverage. We expect to abstain from burning through the time required to comprehend the strategies and to make availability clear.

1.1 Existing System

Insurance providers comprehend that numerous countries face a wide cluster of the present moment. Clients likewise will in general overlook that while getting protection may include tedious procedures, for example, enrollment or document obtaining, the result merits their difficulty. Build up an open digital stage that takes the client's very own conditions, gives simple access to a consultant, maker or mentor who gives custom-fitted direction and noteworthy answers for different monetary well-being concerns have been developed.

1.2 Limitations

To state that the insurance industry has intentionally held off digitization is not totally valid. Truth be told, the industry has been experiencing its own computerized change for as far back as 5 years. There are two significant ways insurance providers are trying to beat obstructions. One is to put resources into select abilities and the other is to take an advanced first outlook. These kinds of enhancements in the insurance process have been developing for many years. In any case, a framework giving the answer for numerous difficulties did not advance.

1.3 Software Requirements

Salesforce.com user license can get to all the custom applications in the association. We require a developer license in order to design the proposed system with the corresponding features.

Apex is a specific, object-situated programming language that permits engineers to execute stream and exchange control proclamations on Salesforce servers.

HelloSign is the easiest e-Signature answer for Salesforce is HelloSign, which helps in faster contract signing and is increasingly beneficial.

Stripe payment gateway comes with a smart payment page permitting in-person payments.

1.4 Hardware Requirements

Electronic device. Any gadget with Windows, iOS, or Android operating systems can be utilized. The system functions admirably on each operating system with no particular necessities.

In Sect. 2, we talk about past research on this point. Issue definition and framework plan advancement are talked about in Sect. 3. In Sect. 4, the outcomes following the framework formed are investigated in detail. The outcomes got are talked about in Sect. 5.

2 Literature Review

Medical coverage is as of now ascending as a device to deal with the financial wants of people to chase well-being administrations. Dr Ramaiah Itumalla, G. V. R. K. Acharyulu, and Dr. L. Kalyan Viswanath Reddy directed an examination, that unequivocal that the first imperative issue in medical coverage is that the low degree of customer awareness [1]. This examination reasons that there is a need to make familiarity with items, plans, and advantages. The fundamental dispersion of medical coverage has been finished by the business power, deals specialist, TPAs. What drawback will raise here? This is replied by Binny and Dr. Meenu Gupta in their examination paper regarding "Medical coverage in India—Opportunities and Challenges." [2]. Additionally, the absence of straightforwardness is the greatest test for protection companies [3]. The level of India's national spending plan assigned to well-being, the area stays one of the most minimal on the planets, and medicinal service consumptions are to a great extent out-of-pocket (OOP) [4].

The key explanation is referenced by Tanja Ahlina, Mark Nichter, and Gopukrishnan Pillai, in their article covering the subject of what do we think about medical coverage and ethnography. There is a developing consciousness of the way that evil well-being propagates monetary conditions. Maximillian Kolbe Domapielle, in his paper, which provides information on "Medical coverage and access to social insurance benefits in developing nations," contends that, regardless of the model of health care coverage being upheld these components of access oversee poor people and the most unfortunate family choices concerning taking on a medical coverage plot and using human services [5, 6]. India has constrained involvement with medical coverage. Dileep Mavalankar and Ramesh Bhat discussed a challenge with regard to the healthcare insurance industry in their project report "Health Insurance in India Opportunities, Challenges and Concerns." [7].

Salesforce, the overall head in CRM, today conveyed developments to Salesforce Health Cloud that improve human services through associated, shrewd patient engagement [8]. This Cloud brings the world's #1 CRM to social insurance, empowering suppliers like Piedmont Healthcare and 180 Health Partners to customize persistent communications and improve results. In a present circumstance, IT businesses are developing with the help of suitable utilization of accessible assets. Arockia Panimalar.S, Priyadharshan.R, Mithun Kumar.R, Visweshwaran.G in their paper, "Salesforce.com—A Cloud Provider," attempted to focus on basic and regular features of salesforce.com [9].

In the course of recent years, the methodology of building a "sufficient" installment stage without any preparation will no longer suffice [10]. Jordan Jewell and Matthew Marden found that stripe addresses all the difficulties with an API-based installment stage, applications, and administrations that permit clients to streamline the sending and support of their installments framework. Most applications on the Salesforce AppExchange have overseen bundles and the ISVs that create them can use all the devices. Ryan Huff's article talks through the utilization and usage of the SDK [11].

3 Problem Definition

3.1 Problem Statement

The client needed a system that enables automated enrollment of members through agents and sends policy documents to the customers. The system should also take care of getting signatures from the customers, giving them online access to their policies and making payments.

3.2 Process Flow

The flow diagram in Fig. 1 speaks to the main considerations associated with the framework.

Figure 2 is a flow diagram depicting the procedure stream with the customer necessities.

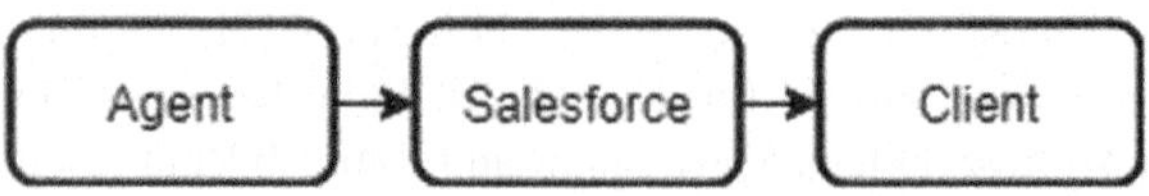

Fig. 1 Focal details in the proposed system

Fig. 2 Procedure stream

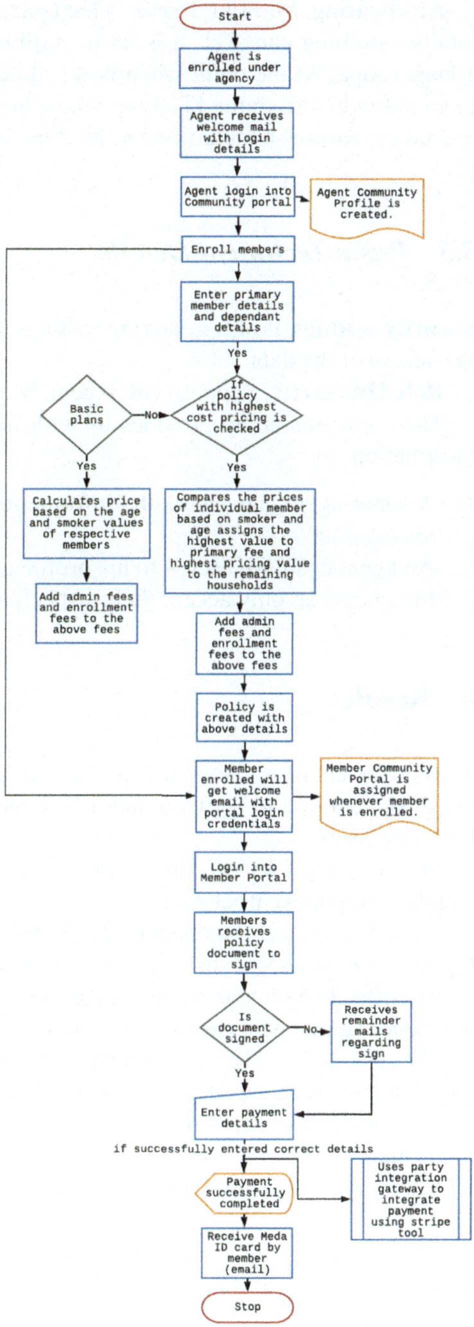

Eradicating Manual Error. Mechanization will never expel human mistake totally—nothing can. Yet, it is an incredible method to guarantee consistency for a huge scope. At the point when issues do emerge, an alert is consequently hailed to an individual from the IT group who can rapidly recognize the underlying driver and take measures to determine it. Nothing is missed and the fix is a lot snappier.

3.3 *Basic Technical Details*

Security settings. Profiles, Sharing Settings, and Role Hierarchy are used to maintain the access of the data.

Role Hierarchy. Super-agent, Agent, Member.

The above hierarchy provides us with information about who can access what information.

1. A super-agent can access the information of the agent working under him and his members.
2. An agent can have access to his profile and the member's profiles he enrolled.
3. Members can only access their respective profiles.

4 Results

The enrollment form shown in Fig. 3 is built using Lightning Component, we display the fields required to be filled and action like plan price calculation is also done in lightning itself.

The next step after enrolling is picking an appropriate plan. Figure 4 shows plans and their respective prices.

Immediately, the customer receives the policy document through mail to sign. Figure 5 is the message displayed asking to sign the document once received to mail.

After the document is signed, the page is redirected to the payment gateway with the choice of payments as in Figs. 6 and 7.

The customer receives a "payment successful" message once the payment is done, enrolling him in the agency as shown in Fig. 8.

Once the process is done, customer automatically receives the policy document to his mail as shown in Fig. 9.

If the payment is done, the customer is issued with a MedaID as in Fig. 10.

5 Conclusion

Most likely medical coverage is one of the developing fragments of the protection business. This paper set out to exhibit a framework, which is equipped for explaining different difficulties and gives an improved system to the clients and offices. We

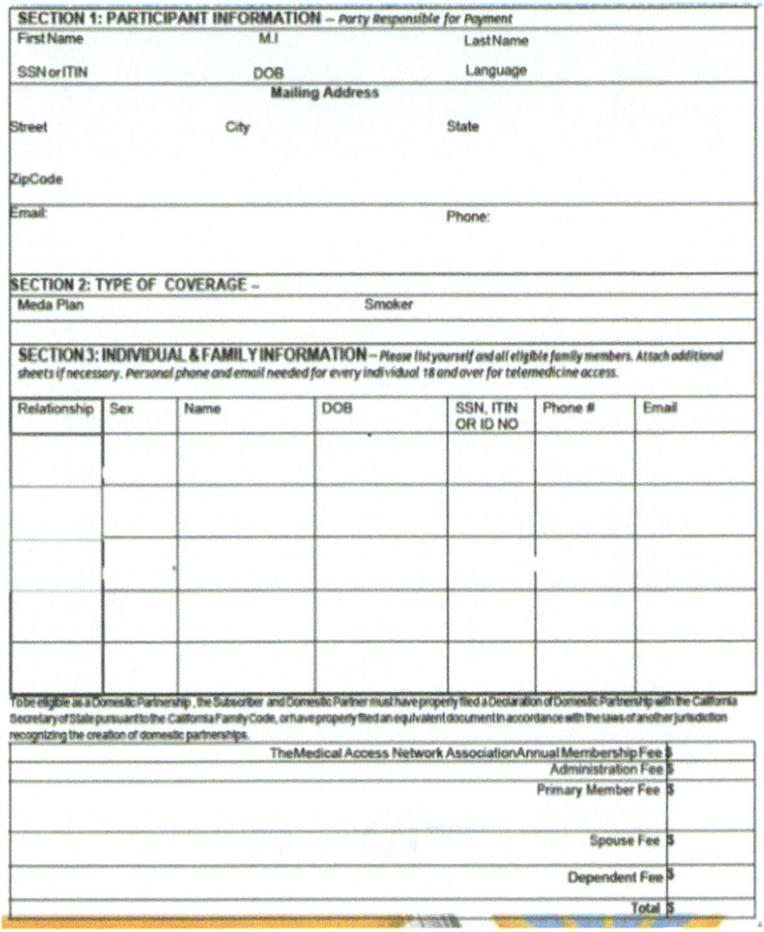

SECTION 1: PARTICIPANT INFORMATION – *Party Responsible for Payment*

First Name | M.I | Last Name

SSN or ITIN | DOB | Language

Mailing Address

Street | City | State

ZipCode

Email: | Phone:

SECTION 2: TYPE OF COVERAGE –

Meda Plan | Smoker

SECTION 3: INDIVIDUAL & FAMILY INFORMATION – *Please list yourself and all eligible family members. Attach additional sheets if necessary. Personal phone and email needed for every individual 18 and over for telemedicine access.*

Relationship	Sex	Name	DOB	SSN, ITIN OR ID NO	Phone #	Email

To be eligible as a Domestic Partnership, the Subscriber and Domestic Partner must have properly filed a Declaration of Domestic Partnership with the California Secretary of State pursuant to the California Family Code, or have properly filed an equivalent document in accordance with the laws of another jurisdiction recognizing the creation of domestic partnerships.

The Medical Access Network Association Annual Membership Fee	$
Administration Fee	$
Primary Member Fee	$
Spouse Fee	$
Dependent Fee	$
Total	$

Fig. 3 Enrollment form

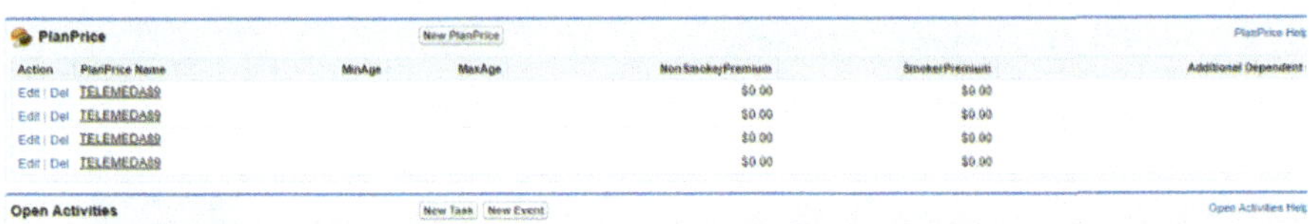

Fig. 4 Plan pricing

have actualized a procedure for operator enlistment and part payments. Clients have the adaptability to pick various plans via mechanizing the system, other than the procedure getting simpler, control and permeability are improved alongside consistency and security. A superior item is conveyed as far as highlights, advantages, and

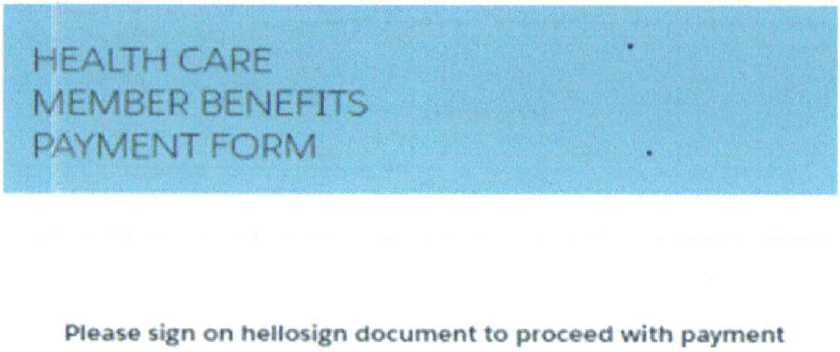

Fig. 5 Mail requesting signature

Fig. 6 Payment page

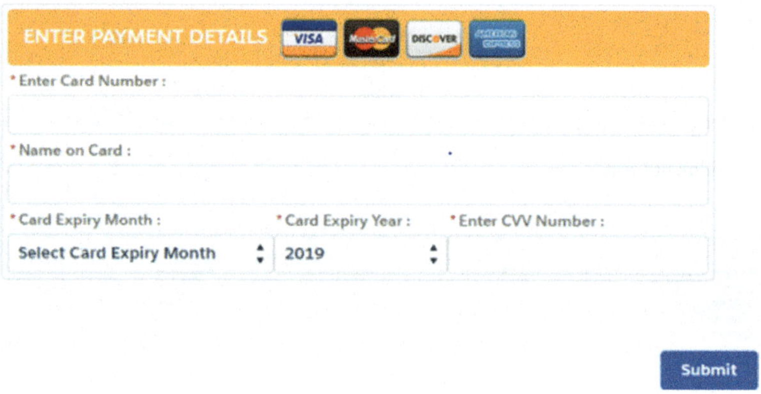

Fig. 7 Payment details

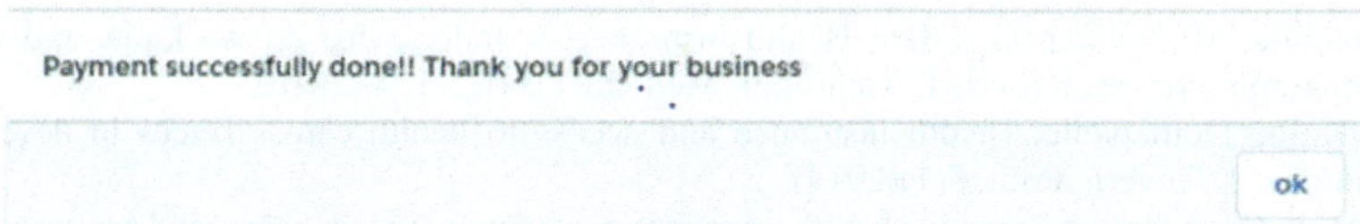

Fig. 8 Payment successful page

Fig. 9 Policy document in PDF format

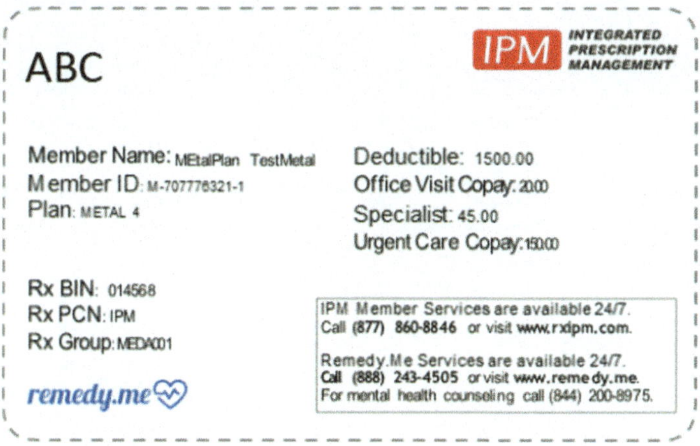

Fig. 10 Unique Identification Card

choices to the clients. A few territories for additional improvements are stretching out this framework to other protection segments alongside medical coverage.

References

1. R. Itumalla, G.V.R.K. Acharyulu, L. Kalyan Viswanath Reddy, Health insurance in India: issues and challenges. Int. J. Current Res **8**(02), 26815–26817 (2016)
2. Binny, M. Gupta, Health insurance in India – Opportunities and challenges. Int. J. Lat. Technol. Eng., Manag. & Appl. Sci. (IJLTEMAS) **VI**(VIII) (2017)
3. A. Aggarwal, N. Kapoor, G. Anchal (2013) Health insurance: innovation and challenges ahead. Glob. J. Manag. Bus. Stud. **3**(5), 475–780

4. T. Ahlina, M. Nichter, G. Pillai, Health insurance in India: what do we know and why is ethnographic research needed. Anthropol. Med. **23**(1), 102–124 (2016)
5. M. Kolbe Domapielle, Health insurance and access to health care services in developing countries. J. Govern. Polit. **5**(1)(2014)
6. C. Atim, Contribution of mutual health organizations to financing, delivery, and access to health care. Synthesis of Research in Nine West and Central African Countries. PHR Report N 18., July 1998
7. D. Mavalankar, R. Bhat, Health insurance in India opportunities, challenges and concerns. Report, November 2000
8. Salesforce.com, Experiences to Improve Outcomes, Available from:ti, Feb 11, 2019, 08:00 ET
9. S. Arockia Panimalar, R Priyadharshan, R. Mithun Kumar, G. Visweshwaran, Salesforce.com – a cloud provider. Int. Res. J. Eng. Technol. (IRJET) **04**(09) (2017)
10. J. Jewell, M. Marden, The business value of the stripe payments platform. https://stripe.com/files/payments/IDC_Business_Value_of_Stripe_Platform_Full%20Study.pdf (2018)
11. R. Huff, Simplify recurring billing with the stripe SDK for salesforce", https://developer.salesforce.com/blogs/developer-relations/2014/04/recurring-billing-with-the-stripe-sdk-for-salesforce.html. Accessed 25 April 2014

Comprehensive Analysis of State-of-the-Art Techniques for VQA

Sachin Chirgaiya and Anand Rajavat

Abstract The visual question answering is an upcoming field in the domain of computer vision and NLP, i.e., natural language processing community. In this domain, a person will be giving any image and can answer certain questions using common sense about the given image, this combines various visual and linguistic processing using the understanding of the user by analyzing the image. This discussion concerned the various datasets that are used in these studies and using the various question and the answer format and which increases the robustness and efficiency of different machine learning models. Also, here, various methods are discussed, which are classified using the machine learning mechanism taking the help of popular textual and visual modalities. The latest convolutional and neural networks of recurring nature are used to map several questions and images in the feature space, which is very common in deep learning. Also, numerous question answers are examined from the diverse depth and the project of visual modalities to understand the relevance of the annotations of structured images and photographs using a visual question answering method in this study.

Keywords Deep learning · Artificial intelligence · VQA · Models

1 Introduction

Sentiment Visual Question Answering (VQA) aims to answer about the questions relevant to a given image [1–3]. VQA is an emerging field of computer vision and multimedia as in past decade, a lot of researches have witnessed [4–11] related to VQA. With the evolution of deep learning in the area of AI, VQA systems now can be very useful in many real-world applications. For example, providing assistance to

S. Chirgaiya (✉) · A. Rajavat
SVVV, Indore, India
e-mail: sachin.chirgaiya@gmail.com

A. Rajavat
e-mail: anandrajavat@yahoo.co.in

C. Kiran Mai et al. (eds.), *Proceedings of International Conference on Advances in Computer Engineering and Communication Systems*, Learning and Analytics in Intelligent Systems 20,
https://doi.org/10.1007/978-981-15-9293-5_9

Fig. 1 What woman is doing?

visually impaired people by capturing the current scene as an image and answering the natural questions about it in real time.

Given an image I and a question Q, state-of-the-art approaches [1, 12–14] first retrieve image feature embedding from I by using pretrained model. Next, language feature embedding is extracted from Q. Now the joint embedding of image and language is learned through the multimodal fusion. At last, the joint embedding is inputted to a sequential network in order to form the possible answers for Q. However, using only image and language features does not yield accurate results. Therefore, some VQA techniques [12–17] introduced semantic information like entity objects, object color, etc.

Combination of the objects and the relation in between them can provide substantial semantic capacities, for example, given the question "what woman is doing?" and the image shown in Fig. 1, following are possible semantic information in the form < *subject, relation, object* >. *(1)* < *women, riding on, cycle* >, *(2)* < *woman, on, road* >,and < *person, in, red-suit* >.

The most difficult task in VQA systems lies in bridging up the semantic gap between image and language. The researches that deal with semantic gap challenge can be classified into two categories. Researches lie in the first category extract high-level semantic information [12–14], for example, entities, and entity features or even information from Knowledge bases [13] such as YAGO [18] and DBpedia [19]. Researches belong to the second category exploit visual attention [16, 20–22] to select important regions in the image by using salient visual features. However, the use of semantic knowledge for VQA task remains limited in two aspects. On one side, entities and features are used as high-level semantic information which only encompasses limited knowledge information. Whereas on the other side, off-the-shelf computer vision techniques are used to extract the concepts in other tasks and therefore, the candidate concepts may not be relevant to the VQA task. Hence the knowledge present in the visual form in the image is not fully exploited in the existing researches. This study attempts to efficiently use the available semantic information for the VQA task.

2 Related Works

In the computer vision field, researchers have made extensive efforts to construct the Visual Question Answering systems.

Xu et al. [23] proposed a novel architecture for VQA based on exploring question-guided spatial attention. They evaluate their model on two DAQUAR and VQA and achieved the promising results. Zhou et al. [24] suggested the bag-of-words baseline model for VQA. In this approach, they get the word features from question and image features from CNN to predict the answer after the evaluation from VQA dataset. Fukui et al. [25] proposed the Multimodal Compact Bilinear Pooling (MCB) model. They combine image and text to represent the results for visual question answering, their architecture with attention and multiple MCBs gives significant improvements on two VQA datasets COCO and Visual7W compare to state-of-the-art with effective accuracy. Ilievski et al. [26] proposed a novel Focused Dynamic Attention (FDA) model for VQA. This model uses the LSTM unit to provide global feature and questions to VQA. Agrawal et al. [27] suggested a new area of research that is Visual Question Answering (VQA). They worked on the real-world scene images dataset with questions and answers. They give a large comparison of the state-of-the-art approaches for VQA with human performance. Wu et al. [12] proposed a model for VQA which is the combination of an image internal content with information extracted from a general knowledge base. They find the answers up to the mark with respect to the image contents from the standard dataset. Hu et al. [28] presented a Compositional Modular Networks (CMNs) for the VQA. It is a novel architecture that takes the language analysis and images from the multiple referential expression datasets for the evaluation purpose and presents the outperformed state-of-the-art approach. Kim et al. [29] proposed the Multimodal Residual Networks (MRN) for the VQA. They use the back-propagation algorithm over the Open-Ended and Multiple-Choice tasks Visual QA dataset to achieve the state-of-the-art results. Yang et al. [20] proposed a Multiple-Layer Stacked Attention Networks (MLSAN) for VQA. The experiments conducted on four VQA data sets to demonstrate the performance of the SANs effectively. Noh et al. [30] proposed a recurrent deep neural network-based algorithm for VQA, where every module of the method in the network corresponds to a complete answering unit with an attention mechanism by itself. Younes et al. [31] introduced a multimodal tensor-based Tucker (MUTAN) for fusion-based VQA. With MUTAN, they control the complexity of the merging scheme text and image concepts while keeping nice interpretable fusion relations. Teney et al. [32] proposed Graph-Structured Representations (GSR) for VQA to build graphs from the scene objects and over the question words. They use a deep neural network for the structure representations of the graph-based VQA. Lu et al. [33] proposed the Hierarchical Question-Image co-attention model for VQA. The model presents the reasons about image and question attention jointly and for this, they use CNN and test it with COCO-QA dataset. Goyal et al. [34] worked for Visual Question Answering (VQA) and make vision (the V in VQA) is a very important part. However, they manage the VQA dataset for their work by considering the balanced images with balanced questions associated with the images. Agrawal et al. [2] worked with the modified dataset a new compositional split of the standard VQA v1.0 dataset. They called this dataset as a Compositional VQA (C-VQA). They had tested it with the existing VQA methods and get the promising results related to the original data set. Teney et al. [35] give the deep survey of the VQA methods from the state-of-the-art model for VQA from 2017 VQA challenge. Hu et al. [36] proposed the End-to-End Module Networks (N2NMNs) for VQA. This model is equipped with a learning network parameter to learn and generate network structures. They use CLVER dataset for their experiment purpose and achieved the results with nearly 50% of an error reduction rate over from the existing one. Wang et al. [3] suggested FVQA (Fact-based VQA). FVQA uses deeper reasoning to contain questions with external knowledge to answer in the dataset to support it with the promising results in the state-of-the-art strategies of VQA. Jiang et al. [37] suggested the updated bottom-up top-down (up-down) model with learning rate schedule and fine-tuning image feature addition with data augmentation for the VQA v2.0. Liu et al. [4] proposed the inverse problem of VQA (iVQA). They use iVQA for the generation of a question that corresponds to a given image and answer pair related to the image. They also suggested how iVQA work effectively in a reinforcement learning framework with VQA model. Li et al. [5] proposed a new model called VQA-E (VQA with Explanation). Their model is tested over the VQA 2 dataset for the generation of an explanation about the predicted answer. Singh et al. [6] proposed a model worked with 300 GPU hours of extensively with hyper-parameter. Their model was evaluated effectively with a score of 64.78% accuracy for the VQA. Lu et al. [7] suggested a novel framework to learn visual relation facts for VQA by multi-step semantic attention network (MSAN). They refined the VQA dataset with Relation-VQA (R-VQA) dataset and tested it on MSAN with promising results. Huang et al. [8] proposed a compressive dataset for VQA to solvable in the compressed domain. They suggested the series of deep-network architectures for the testing and training purpose with the compressive data to get the increased accuracy in VQA. Anderson et al. [38] proposed a joint bottom-up (based on Faster R-CNN) and top-down attention approach. They allow considering the salient objects and other salient image regions for VQA. The proposed mechanism of bottom-up for feature vector and the top-down mechanism determine (feature weightings). Their results are achieved from the MSCOCO test server establish good score 117.9, 36.9 and 21.5, tested on CIDEr/BLEU-4/SPICE, respectively. Yu et al. [39] proposed a combined MFB and multimodal factorized high-order (MFH) approaches to achieve better performance on the VQA. Using the Kullback–Leibler divergence (KLD) as the loss function with complex correlations between multimodal features they achieved better results. Lao et al. [9] proposed a model called Cross-modal Multi-step Fusion (CMF). They generate multistep features to achieve multiple interactions for two modalities. Yang et al. [10] proposed CAQT Model for the mechanism of coattention combine self-attention-based question-guided visual attention from text and image. They performed the experiments on VQA v1.0 and v2.0 datasets to show outperform state-of-the-art approach for CAQT model for VQA.

Table 1 shows the comparative study of several methods, datasets and performance issue of the previously used approaches for VQA.

Table 1 Comparative analysis of state-of-the-art techniques for Visual Question Answering

S. No	Authors	Method	Percentage of accuracy	Year	Dataset
1	Xu et al. [23]	SMem-VQA	58.2	2015	COCO-(VQA) real images 1.0 open-ended
2	Zhou et al. [24]	iBOWIMG baseline	55.9	2015	COCO-(VQA) real images 1.0 open-ended
		iBOWIMG baseline	62		COCO-(VQA) real images 1.0 multiple choice
3	Fukui et al. [25]	MCB 7 att	66.5	2016	COCO-(VQA) real images 1.0 open ended
		MCB 7 att	70.1		COCO-(VQA) real images 1.0 multiple choice
		MCB + Att	62.2		Visual7W
4	Ilievski et al. [26]	FDA	59.5	2016	COCO-(VQA) real images 1.0 open ended
		FDA	64.2		COCO-(VQA) real images 1.0 multiple choice
5	Agrawal et al. [27]	LSTM Q + I	58.2	2016	COCO-(VQA) real images 1.0 open ended
		LSTM Q + I	63.1		COCO-(VQA) real images 1.0 multiple choice
		HDU-USYD-UNCC	68.2		COCO-(VQA) real images 2.0 open ended
		DLAIT	68.1		COCO-(VQA) real images 2.0 open ended
		Dualnet ensemble	71.2		COCO-(VQA) abstract 1.0 multiple choice
		LSTM + global features	69.2		COCO-(VQA) abstract 1.0 multiple choice
		LSTM blind	61.4		COCO-(VQA) abstract 1.0 multiple choice

(continued)

Table 1 (continued)

S. No	Authors	Method	Percentage of accuracy	Year	Dataset
		Dualnet ensemble	69.7		COCO-(VQA) abstract images 1.0 open ended
		LSTM + global features	65		COCO-(VQA) abstract images 1.0 open ended
		LSTM blind	57.2		COCO-(VQA) abstract images 1.0 open ended
6	Wu et al. [12]	CNN-RNN	59.5	2016	COCO-(VQA) real images 1.0 open ended
7	Hu et al. [28]	CMN	72.5	2016	Visual7W
		CMN	44.2		Visual Genome (subjects)
		CMN	28.5		Visual Genome (pairs)
8	Kim et al. [29]	MRN + global features	61.8	2016	COCO-(VQA) real images 1.0 open-ended
		MRN	66.3		COCO-(VQA) real images 1.0 multiple choice
9	Yang et al. [20]	SAN	58.9	2016	COCO-(VQA) real images 1.0

3 Current Challenges and Research Opportunities

Visual Question Answering is distinct from other traditional computer vision tasks such as object detection in an image; classify objects present in image, etc. As these traditional tasks only extract the object present in the image, however, these tasks do not provide answers to the natural question related to the image content. The current state-of-the-art VQA systems are developed on the assumptions that the question asked will always be relevant to the image and therefore this system confidently provides incorrect answers to the irrelevant questions. However, before answering a question, VQA systems should first check whether the asked is relevant to the image or not. To the best of our knowledge, there is much more work need to do. Also, current VQA techniques are largely focused on the image processing tasks, however, less effort is made on the understanding the language semantics of the question.

The existing researches in VQA not much focus on the available semantic information present in the image. Also, Lu et al. [7] suggest that existing available datasets for VQA tasks are too much sparse and biased. These opportunities motivate us to pursue our research on the VQA task. There are some of the research gaps in this area, as:

1. The domain-specific datasets are required for effective VQA.
2. Salient object detection from the image is still a challenging work for VQA.
3. The semantic relationship is missing in image objects due to that relevant answers are uncertain in nature.
4. Joint learning of visual question answering (VQA) and question relevance detection (QRD) is required to get efficient results.
5. In the present dataset, questions and answers are biased in nature due to the efficiency of the present VQA methods and models are imperfect.
6. Present VQA tasks answer confidently to the questions which are irrelevant to the given image.
7. More precise and accurate results are an important requirement for VQA.

4 Conclusion

This paper provides a sketch of the state-of-the-art on visual question answering. Some important and popular approaches have been taken into consideration for making the questions and their related images into a feature species using a common and comprehensive representation. Various advanced concepts also are taken into consideration like the memory augmented architecture, the modular pigmentations, and the attenuation mechanism. A large number of data sets have been taken into consideration, which highlights the difficulty and complexity of the research work in the field of VQA. A suggestion has been also made for including the structural knowledge and external knowledge bases for the exploration of a continuous dataset in natural language processing. The recent trend justifies about the implementation of Deep learning methods in visual and utilization cues for textual and structural modalities to sample the dataset, which ultimately used errors of textual mining applied at different algorithms. The current trend also shows that the best among several machine learning models does not achieve much significant accuracy level in the field of VQA. The current study displays that different strategies like deep learning models, segmentation analysis, and sentiment analysis show much more potential for implementing the next-generation VQA model.

References

1. D. Teney, Wu. Qi, A. van den Hengel, Visual Question Answering: A Tutorial. IEEE Signal Process. Mag. **34**(6), 63–75 (2017)
2. A. Agrawal, A. Kembhavi, D. Batra, D. Parikh, C-VQA: A compositional split of the visual question answering (VQA) v1.0 Dataset. arXiv:1704.08243 (2017)

3. P. Wang, Q. Wu, C. Shen, A. Dick, A. van den Hengel, FVQA: fact-based visual question answering. In: IEEE Transactions on Pattern Analysis and Machine Intelligence, vol. 40, no. 10, pp. 2413–2427 (2018)
4. F. Liu, T. Xiang, T.M. Hospedales, W. Yang, C. Sun, Inverse visual question answering: a new benchmark and VQA diagnosis tool. arXiv:1803.06936 (2018)
5. Q. Li, Q. Tao, S. Joty, J. Cai, J. Luo, VQA-E: explaining, elaborating and enhancing your answers for visual questions. arXiv:1803.07464 (2018)
6. J. Singh, V. Ying, A. Nutkiewicz, Attention on attention: architectures for visual question answering (VQA) arXiv:1803.07724, 9. (2018)
7. P. Lu, L. Ji, W. Zhang, N. Duan, M. Zhou, J. Wang, R-VQA: learning visual relation facts with semantic attention for visual question answering. arXiv:1805.09701 (2018)
8. L-C. Huang, K. Kulkarni, A. Jha, S. Lohit, S. Jayasuriya, P. Turaga, CS-VQA: visual question answering with compressively sensed images. arXiv:1806.03379 (2018)
9. Yu. Zhou, Yu. Jun, C. Xiang, J. Fan, D. Tao, B. Bilinear, Generalized Multimodal Factorized High-Order Pooling for Visual Question Answering. IEEE Trans. Neural Netw. Learn. Syst. **29**(12), 5947–5959 (2018)
10. M. Lao, Y. Guo, H. Wang, X. Zhang, Cross-modal multistep fusion network with co-attention for visual question answering. In: IEEE Access, vol. 6, pp. 31516–31524, (2018)
11. C. Yang, M. Jiang, B. Jiang, W. Zhou, K. Li, Co-attention network with question type for visual question answering. IEEE Access **7**, 40771–40781 (2019)
12. Q. Wu, C. Shen, A. van den Hengel, P. Wang, A. Dick, Image captioning and visual question answering based on attributes and external knowledge. arXiv:1603.02814 (2016)
13. FloydHub Blog. Asking questions to images with deep learning: a visual-question-answering tutorial. [online] Available at. https://blog.floydhub.com/asking-questions-to-images-with-deep-learning. Accessed 20 Jan 2020
14. Q. Wu, P. Wang, C. Shen, A. Dick, A. van den Hengel, Ask me anything: Free-form visual question answering based on knowledge from external sources. In: IEEE Conference on Computer Vision and Pattern Recognition, 4622–4630 (2016)
15. G. Li, H. Su, W. Zhu, Incorporating external knowledge to answer open-domain visual questions with dynamic memory networks arXiv preprint arXiv:1712.00733 (2017)
16. H. Xu, K. Saenko, Ask, attend and answer: Exploring question guided spatial attention for visual question answering. In: European Conference on Computer Vision, 451–466 (2016)
17. Z. Yang, X. He, J. Gao, L. Deng, A. Smola, Stackedattention networks for image question answering. In: IEEE Conference on Computer Vision and Pattern Recognition, 21–29 (2016)
18. F.M. Suchanek, G. Kasneci, G Weikum, Yago: a core of semantic knowledge in Proceedings of the 16th international conference on World Wide Web ACM, 697–706 (2007)
19. V. Annapurna, M. Ramakrishna Murty et al, Comparative analysis of frequent pattern mining for large data using FP-tree and CP-tree methods. International Conference FICTA-17 at KIIT University, Bhubaneswar, Springer , AISC series, 2017
20. Z. Yang, X. He, J. Gao, L. Deng, A. Smola, Stacked attention networks for image question answering. arXiv:1511.02274 (2016)
21. S. Auer, C. Bizer, G. Kobilarov, J. Lehmann, R. Cyganiak, Z. Ives, Dbpedia: A nucleus for a web of open data in the semantic web. Springer, 722–735 (2007)
22. J-H. Kim, K. Woon On, W. Lim, J. Kim, J-W Ha, B-T. Zhang, Hadamard product for low-rank bilinear pooling. In: International Conference on Learning Representations (2017)
23. H. Xu, K. Saenko, A. Attend, Answer, Exploring question-guided spatial attention for visual question answering. arXiv:1511.05234 (2015)
24. B. Zhou, Y. Tian, S. Sukhbaatar, A. Szlam, R. Fergus, Simple baseline for visual question answering. arXiv:1512.02167 (2015)
25. A. Fukui, D. Huk Park, D. Yang, A. Rohrbach, T. Darrell, M. Rohrbach, Multimodal compact bilinear pooling for visual question answering and visual grounding. arXiv:1606.01847 (2016)
26. I. Ilievski, S. Yan, J. Feng, A Focused dynamic attention model for visual question answering. arXiv:1604.01485 (2016)

27. A. Agrawal, J. Lu, S. Antol, M. Mitchell, C. Lawrence Zitnick, D. Batra, D. Parikh, VQA: visual question answering. arXiv:1505.00468 (2016)
28. R. Hu, M. Rohrbach, J. Andreas, T. Darrell, K. Saenko, Modeling relationships in referential expressions with compositional modular networks. arXiv:1611.09978 (2016)
29. J-H. Kim, S-W. Lee, D-H. Kwak, M-O. Heo, J. Kim, J-W. Ha, B.-T. Zhang, Multimodal residual learning for visual QA. arXiv:1606.01455 (2016)
30. H. Noh, B. Han, Training recurrent answering units with joint loss minimization for VQA. arXiv:1606.03647 (2016)
31. D. Teney, L. Liu, A. van den Hengel, Graph-structured representations for visual question answering. arXiv:1609.05600 (2017)
32. H. Ben-Younes, R. Cadene, M. Cord, N. Thome, MUTAN: multimodal tucker fusion for visual question answering. arXiv:1705.06676 (2017)
33. J. Lu, J. Yang, D. Batra, D. Parikh, Hierarchical question-image co-attention for visual question answering. arXiv:1606.00061 (2017)
34. Y. Goyal, T. Khot, D. Summers-Stay, D. Batra, D. Parikh, Making the V in VQA matter: elevating the role of image understanding in visual question answering. arXiv:1612.00837 (2017)
35. Damien Teney, Peter Anderson, Xiaodong He, Anton van den Hengel, Tips and Tricks for Visual Question Answering: Learnings from the 2017 Challenge arXiv:1708.02711
36. R. Hu, J. Andreas, M. Rohrbach, T. Darrell, K. Saenko, Learning to reason: end-to-end module networks for visual question answering. arXiv:1704.05526 (2017)
37. Y. Jiang, V. Natarajan, X. Chen, M. Rohrbach, D. Batra, D. Parikh, Pythia v0.1: the winning entry to the VQA challenge 2018. arXiv:1807.09956 (2018)
38. P. Anderson, X. He, C. Buehler, D. Teney, M. Johnson, S. Gould, L. Zhang, Bottom-up and top-down attention for image captioning and visual question answering. arXiv:1707.07998 (2018)
39. N. Ruwa, Q. Mao, L. Wang, M. Dong, Affective visual question answering network, conference on multimedia information processing and retrieval (MIPR), Miami, FL, IEEE, pp. 170–173.(2018)

Study of State of Arts Methods for Event Extraction

Nikhil Chaturvedi and Jigyasu Dubey

Abstract The amount of online published news articles increases day by day so readers need to find news articles related to his or her interest is very difficult. Lots of important events happed in every day but they are reported in different manner in different online published news articles. In the event extraction technique, detection of main event from the news articles is the most important task including some other tasks like identifying the argument, trigger and named entities. Event extraction and categorization have a wide range of applications in various domains. This article discusses the various states of arts and current methods used in the area of event extraction from text as well as news articles. We also discuss the comparative study about the various methods used in event extraction field. After that, we will discuss the current challenges and research opportunities in the methods of event extraction and categorization.

Keywords Event extraction · Event arguments and triggers · Event corpus · Event categorization

1 Introduction

In new era of online published news articles, hundreds of articles published every day by the popular news agencies like CNN, BBC, etc. The amount of online published news articles increases day by day, so the reader needs to find news articles related to specific event type of his or her interest. In general, reader to find news articles of his or her interest, a reader filter out desired into related to event of interest from headlines and teaser by scanning various sources of news articles about the same news information he already knows about it. Therefore, people always want

N. Chaturvedi (✉) · J. Dubey
SVVV, Indore, India
e-mail: nikhilcse2015@gmail.com

J. Dubey
e-mail: jigyasudube@yahoo.co.in

C. Kiran Mai et al. (eds.), *Proceedings of International Conference on Advances in Computer Engineering and Communication Systems*, Learning and Analytics in Intelligent Systems 20,
https://doi.org/10.1007/978-981-15-9293-5_10

to extract comprehend a huge amount of information in very less time. A reader not willing to read through a collection of news articles needs a representation of the news in a compact from describing the event type briefly.

The amount of unstructured data enormously increases every day, so does require a process it automatically extracts different types of knowledge from data [1, 2]. A continuously occurred problem in processing these large amount of data in intelligent manner because most of the data are unstructured form, e.g., data written in human-understandable language. This information in the form of online broadcast news, blogs to comments and social communication data [3–5]. Thus, it may also consist of various types of unrelated data or noisy discussions. Event extraction is the task to identifying the key terms related to the specific events from unstructured data.

Event defined as a specific occurrence involving participants. An event is something that explained as a change of state, so in various fields is classified in terms of events.

Formally, the event extraction is to automatically identify events in collection of text and to find detailed information about them is known as event extraction. Event extraction from texts could be beneficial to various domains. In the online published news articles application, the ability to identifying events may enhance the performance of specific news recommendation. Since the online published news articles may be selected more efficiently based on the specific extracted events and user preferences [6–8]. Event extraction and two categorization of online published news articles have many applications such as government policymakers uses the information related to significant international incidents tracking to make better decisions regarding policies. In cyber security domain, ongoing cyber-attack incidents help to monitoring. Another most valuable application of event extraction in the area of stock market extremely sensitive to the breaking news for deciding the stock trading planning such as time, price and volume.

Events are extracted from text at sentence level, some past studies are shown that the extraction of events using relations, entity and document level [9]. Event extraction is crucial and quite challenging task, as the argument detection and classification heavily depend on it [10–12]. In the news articles it can be challenging to identify the same event that have different form, either present in single or multiword units.

Event extraction comes from the unstructured data, where a large number of nonevent examples are present, in the other, so making event and argument identification are more challenging tasks.

Reference [13] For such essential characteristics of online published news articles, document required more semantic information processing, so that requires building a larger scale news semantic knowledge base [14, 15]. After that such knowledge base apply, the event extraction and event categorization are more challenging tasks.

2 Related Works

In the field of event extraction and analysis, many researchers have made extensive efforts to develop several methodologies of event extraction from news articles. A number of efforts have been undertaken to address the analysis of narrative work in online news articles.

Chen et al. [1] proposed a dynamic multipooling conventional neural network model (DMCNN). This uses dynamic multipooling according to event trigger and argument to receive more information. Michal Tillo et al. [2] proposed a web toolkit for event extraction in which automatically evaluated the suggestions and exploration process after that shows how user related to one to another relation.

Ralph Grishman et al. [16] described a two-stage training algorithm for neural networks that effectively transfers knowledge from the other event types to the target type. They also investigated the effectiveness of the method compared with other state of art method. This method for extracting of knowledge may be more efficient use of real-time dataset for description.

Yagung et al. [17] presented a timeline that shows significant events about query word based on our corpus, which can inform user's important things they are interested in. In the procedure to cluster sentences by their dates, they considered the value of no accurate date sentences. Because it is now one date vector for one date, they can only extract the most important thing in each date vector. In reality, it is possible that a number of things with same type happened on same date.

Shulin Liu et al. [18] present the extraction of argument information explicitly for event detection use of supervised mechanisms. In this model, researchers use systematically approach for the investigation of proposed model under the guidance of various attention strategies.

Somayeh Keshavarz et al. [19] proposed a novel probabilistic inference framework for complex video event recognition using supervised action concepts. In this method author attempt to model for the conditional relationships between complex events and exhaustive set of intermediate concepts by constraining dependencies to pairwise joint distributions while avoiding the need of manually re encode new graph structures as the number of concepts increases.

Mahdi Namazifar [20] suggests a Named Entity Recognition (NER) method aims at locating and classifying named entities in text. Proposed NER method applies to find Named entity from Tweets. Researcher suggests that the NER is also used for the online news articles text.

Gills Jacoab et al. [21] presented a dataset and classification experiments for company-specific economic event detection in English news articles. There is still plenty of room for improvement: more annotated data and augmentative resources are needed to further offset ambiguous event expressions.

Felix Hamborg et al. [6–8] proposed open source system that retrieves answers in the questions form of what, when, who, where, why and how to describe online published news articles information in main events. In these, researcher identified only the information related to main event, they do not categorize all types of main events.

Quan Yang et al. [22] presented to obtain event profiles event extraction automatic framework using unsupervised learning method. Their proposed framework based on event profiling. For implementation of work, lots of challenges occur such as find out event types in given event initially unknown for users, lake of event types schemas in previously defined in model. So that event profiling to their event types is still demanding in the field of information extraction from text.

Haibo Ding et al. [3–5] presented a method based on word and contexts for Human Needs categorization. They develop a method for assign words to related concept in between of categorization process. Researchers are creating supervised model for labeling the words into their sematic concepts for human need information categorization. They use 10 types of semantic concepts in their work.

J. Walkar et al. [23] presented a new model that is based on GRU that combines the temporal structure information and syntactic information in that mechanism. The result shows that it is competitive with other state of art methods like neural network architectures using empirical evaluations under the split of ACE dataset. They suggest that event can be categorized using their types.

Shuo Yang et al. [10] proposed an efficient collapsed Gibbs sampling approach to infer the truths of news and the users' credibility without any labeled data. In this model, researcher shows through the experiment results on two different data sets outperforms the other compared methods.

Rui Wang et al. [11] researchers proposed a new approach on the basis of adversarial training for the extraction of the structured presentation of events from the online text. In this proposed model, researcher compared the model with the different state of arts method and showed that AEM improved the extraction performance especially for long text. In future work, we will explore incorporating external knowledge (e.g., word relatedness contained in word embedding) into the learning framework for event extraction.

Momna Nassem et al. [12] proposed approach uses two level of clustering. The first level of clustering identifies major events among diverse social media text, and the second level of clustering method subevents identifies to a given super event by using semantic relationship and other temporal information. After that, they compare their results in terms of accuracy compared with other state of arts methods.

A comparative study of state of arts method for event extraction is presented in Table 1.

3 Current Challenges and Research Opportunities

Event extraction and categorization from online news articles are distinct from other traditional event extraction tasks such as trigger and argument detection in text, classify the trigger and argument present in text, etc., however, these tasks do not provide categorization of event type from the events. The current state-of-arts methods [5,

Table 1 Comparative study of state of arts method for event extraction

S. No	Paper Title	Author	Year	Method used	Findings
1	Event extraction via dynamic multi polling's conventional neural network	Yubo Chen et al. [1]	2015	Conventional neural network	They did not focus on trigger and argument classification
2	A two stage approach for extending event detection in new types via neural network	Thien Huret et al. [16]	2016	Neural network	In this article, there is no method for relation extraction and categorization
3	An event timeline extraction method based on news corpus	Yaguang Wu et al. [17]	2017	Design algorithm for event data using date vector	In this work, they focus only timeline of work on date if same type of event happens on same date also failed
4	Exploiting argument information to improve event detection via supervised mechanism	Shulin hiu et al. [18]	2017	Improve basic neural network model using supervised attention of argument	The remarkable improvement in argument extraction but insignificant to event detection
5	Named entity sequence classification	Mahdi Namazifar [20]	2017	NESC use NER as well as neural network	This method applicable only on short text like tweets
6	Economic event detection in company specific news text	Gills Jacobs et al. [21]	2018	RNN-LSTM and linear kernel SVM	They focus only company specific news means only one domain use for analysis
7	Sentiment content analysis and knowledge extraction from news articles	Mohammad kamel et al. [24]	2018	NLP Techniques	It is very time-consuming process
8	NLP for information extraction	Sonit singh et al. [25]	2018	Various improve methods for information extraction	Various challenges in NLP like OpenIE, Business analytics and web harvestings

(continued)

Table 1 (continued)

S. No	Paper Title	Author	Year	Method used	Findings
9	Extraction of main event descriptor from news analysis by answering journalistic five W and one H questions	Felix Homberg et al. [6]	2018	Three phase analysis pipeline pre-processor, phrase extraction and candidate scoring	They do not give the appropriate result in respect to precision metrics
10	Giveme5W: Main event retrival from news analysis by extraction of the five 5 journalist W questions	Felix Homberg et al. [7]	2018	Use open source information extraction system so improve precision	They suggest that use coreference resolution, semantic measure and combine scoring
11	Automated identification of media bias in news articles	Felix Homberg et al. [8]	2018	NLP process it uses frame, content and meta-analysis	They use manual content analysis for media bias so we can use automated content analysis
12	Open schema event profiling for massive news corpus	Quen Yuan et al. [22]	2018	Non parametric Baysian Model	Need to improve efficiency with large data sets
13	Why is an affective? Classifying affective event based on human needs	Haibo Ding et al. [3]	2018	Rule based system using LIWC and Supervised classifies	They suggest that in future improve performance of categorizing affective event
14	Human need categorization of affective events using labelled and unlabelled data	Haibo Ding et al. [4]	2018	For labelled data use trained event classifier and unlabelled data use self training and co training	In these methods, categorization performance is still room for improvement
15	Event detection with neural network: A rigorous empirical evaluation	J Walkar et al. [23]	2019	Neural network model including trigger identification	Evaluation based on single random initialization and single training set then did not get significant value

(continued)

Table 1 (continued)

S. No	Paper Title	Author	Year	Method used	Findings
16	Improving human needs categorization of events with semantic classification	Haibo Ding et al. [5]	2019	Supervised model that learn word and semantic concepts using word embedding	These models did not work on large data set
17	Classification and event identification using word embedding	Anais Oilagnier et al. [26]	2019	Word embedding techniques with linear classification	Method often sensitive to the domain of application
18	Open event extraction from online text using a generative adversarial network	Rui Wang et al. [11]	2019	AEM(Adversarial event model)	They require more time for training respect to other model
19	Big data and quality data for fake news and misinformation detection	Maite Taboada et al. [27]	2019	Labelled data set for various classification method NLP & machine learning approach	They did not use any data set for experiment
20	Discovering events from social media for emergency planning	Momna Nasim et al. [12]	2019	Hierarchical clustering machine learning techniques	Missing data qualify issue regarding fake tweets and social news
21	Survey on events extraction technology in information extraction research area	Liying Zang et al. [28]	2019	Compare rule based, statistical machine learning and deep learning	Lake of large scale mature corpus for experiments
22	Whose story is if anyway? Automatic extraction of accounts from news articles	Hao Zhang et al. [29]	2019	Machine learning based model on morden NLP process	These method missing semantic knowledge of domain and narrative study

10–12, 30–32] are developed on the assumption all words in a document are generated from a single event besides long texts such news articles often describe multiple events, which clearly violate this assumption. However, before extracting the events, first check the given events are relevant to the argument and trigger or not and after that categorize these events to specific event type.

Above related work shows the study different methods and data set issue of the previously used method for event extraction and categorization. There are some research gaps in this area, as:

- They assume that all words in a document are generated from a single event However, long texts such news articles often describe multiple events, which clearly violate this assumption.
- The domain-specific dataset is required for effective event extraction.
- Salient trigger and argument detection form article is still a challenging task.
- The semantic relationship is missing in trigger and argument due to that relevant event type is uncertain in nature.
- The present event extraction and categorization task are biased in nature due to the categorization of event some specific type only for analysis of special purpose only so the models are imperfect to all users.
- More precise event extraction and categorization of event type are required.

4 Conclusion

This article provides the sketch of the state of arts method on event extraction from text. Some important and popular methods have been undertaken into consideration for making event extraction using a common and compressive representation. Various advanced concepts also consider like the argument identification, trigger identification, named entity reorganization mechanism in event extraction method. In the field of event extraction, most difficult task when taken the large data sets for consideration. In the current methods, till develop, most of them taken as all words in a document from a single event but in long data sets text such as news articles have described multiple events then violate this assumption. However before extracting the events, first check the given events are relevant to the argument and trigger or not and after that categorize these events to specific event type. In the current state of arts, methods through machine learning and NLP do not achieve a much significant level of efficiency and accuracy in terms of event extraction from the online news articles. The current study displays that different strategies like computation neural network, machine learning, NLP show much more potential for implementing the event extraction for online news articles.

References

1. Y. Chen, L. Xu, K. liu, D. Zeng, J. Zhao, Event Extraction via dynamic multi-pollingconvolutional neural networks. In ACL-IJCNLP2, pp. 167–176 (2015)
2. T. Michael, A. Akbik, A web toolkit for exploratory relation Extraction. In ACL- IJCNLB (2015)
3. H. Ding, T. Jiang, E. Riloff, Why is an even affective? Classifying affective events based on human needs. In: The AAAI-18 Workshop on Affective Content Analysis (2018)

4. H. Ding, E. Riloff, Human needs categorization of Affective events using labeled and unlabelled data. In: Proceedings of the 2018 Conference of the North American Chapter of the Association for Computational Linguistics: Human Language Technologies (2018)
5. D. Haibo, R. Ellen, F. Zhe, Improving human needs categorization of events with semantic classification. Proceeding of Eighth Joint Conference on Lexical and Computation Semantics by ACL (2019)
6. F. Hamborg, C. Breitinger, M. Schboutz, B. Gipps, Extraction of main event descriptor from news articles by answering the journalist five W and one H questions. FCDL'18 June 3–7 , Fort Worth, TX, USA (2018)
7. F. Hamborg, C. Breitinger, M. Schboutz, B. Gipps, Giveme5W: main event retrieval from news articles by answering the journalist five W. I Conference in Sheffield, UK, (2018)
8. F. Hamborg, B. Gipps, Automated Identification of media bias in news articles: an interdisciplinary literature review. International Journal on Digital Library, Springer 16 November 2018
9. P. Sarma, Y. Liang, W. Sethares, Domain adapted word embedding's for improved sentiment classification. In: Proceedings of the ACL Workshop on Deep Learning Approaches for Low-Resource NLP (2018)
10. S. Yang, K. Shu, S Wang, R. Gu, H. Liu, Unsupervised fake news detection on social media: A generative approach. Proceedings of 33rd AAAI Conferences on Artificial Intelligence (2019)
11. R. Wang, D. Zhou, Y. He, Open event extraction from online text using a generative advisable network. Proceedings of 33rd AAAI Conferences on Artificial Intelligence (2019)
12. Momna Nassem, Basit Shafiq Soon Chon, Shafay Shamail. "Discovering Events from Social Media for Emergency Planning." Association for Computing Machinery, Dubai, UAE, June 18, 2019.
13. Navya Yarrabelly and Kamalakar Karlapalem. "Extracting Predictive Statements with their Scope from News Articles." Proceedings of 12th International AAAI Conference on Web and Social Media (2018).
14. B. Tarakeswara Rao, M. Ramakrishna Murty et al., A comparative study on effective approaches for unsupervised statistical machine translation. International conference and published the proceedings in AISC Springer conference, Vol. 1076, pp. 895–905, April 2020
15. F.M. Suchanek, G. Kasneci, G. Weikum, Yago: a core of semantic knowledge. In: Proceedings of the 16th international conference on World Wide Web ACM, pp. 697–706 (2007)
16. T. Huu Nguyen, L. Fu, K. Cho, R. Grishma, A two-stage approach for extending event detection to new types via neural network. In WRepL4NLP, pp. 158–165 (2016)
17. Y. Wu, H. Sun, C. Yan, An event timeline extraction method based on news corpus. International Conference on Big Data Analytics (2017)
18. Liu, S. et al., Exploiting argument information to improve event detection via supervised attention mechanisms. Proceedings of the 55th Annual Meeting of the Association for Computational Linguistics 1(2017)
19. K. Somayeh, S. Imran, A. George, Exploiting probabilistic relationships between action concepts for complex event classification. IEEE Transaction and Information Theory (2017)
20. M. Namazifar, Named entity sequence classification. arXiv preprint arXiv:1712 (2017)
21. G. Jacobs, E. Lefever, Hoste, Economic Event detection in company specific news text. In the proceedings of first workshop on Economic and NLP, Melbourne Australia ACL pp. 1–10 (2018).
22. Q. Yang, X. Ren Wenqi He, C. Zhang, H. Ji et al., Open schema event profiling for massive news corpus." CIKM'18, Torina, Italy, October 22–26 2018
23. J. Walker Orr, P. Tadepalli, X. Fern, Event detection with neural network: a rigorous empirical evaluation. arXiv preprint arXiv:1808 (2018)
24. M. Kamel, N. Keyvani, Sentiment content analysis and knowledge extraction from news articles. Association of computer linguistics conference, pp. 13–18 (2018)
25. S. Singh, Natural language processing for information extraction. Association of Computer Linguistics Conference, pp. 13–18 (2018)

26. A. Ollagnier, H. Williams, Classification and Event Identification Using word Embedding. Trans. Associat. Comput. Linguist. **7**, 91–122 (2019)
27. F. Torabi, M. Toboada, Big data and quality data for fake news and misinformation detection. Big Data and Society, pp. 1–4, January–June 2019
28. Sören Auer, Christian Bizer, Georgi Kobilarov, Jens Lehmann, Richard Cyganiak, and Zachar Ives, "Dbpedia: A nucleus for a web of open data" in the semantic web. Springer, pp. 722–735, (2007)
29. L. Zhan, X. Jiang, *Survey on Events Extraction Technology in Information Extraction Research Area 3rd Informatation TechnologyInformatation Technology* (Networking, Electronic and Automation Control Conference on IEEE, 2019)
30. P.K. Choubey, K. Raju, R. Huang, Identifying the most dominant event in a news article by mining event coreference relations. Proceedings of the 2018 Conference of the North American Chapter Of the Association For Computational Linguistics: Human Language Technologies, vol. 2, pp. 340e345. Short Papers (2018)
31. Y. Belinkov, J. Glass, Analysis methods in natural language processing: A Survey. Transaction of the Association for Computational Linguistics, 2019, vol. 7, pp. 49–72 (2019)
32. H. Zhang, F. Boons, R. Batista, Whose story is if anyway? Automatic extraction of accounts from news articles. Inf. Process. Manage. **56**, 1837–1848 (2019)

A Survey on Identification of Illegal Wildlife Trade

Sravani Nalluri, S. Jeevan Rishi Kumar, Manik Soni, Soheb Moin, and K. Nikhil

Abstract From the last two decades, there was a huge climb in the web where there are various changes in the criminal activities, through which various opportunities have risen and one of them is wildlife trafficking. Wildlife trafficking is increasing rapidly across the world and becoming a threat to various survival of species as well as to the security of the whole world. The researchers have examined this illegal wildlife trade using various theoretical frameworks, but the attention is not drawn over the ways of trade that is being happened on the web for a very long time. Thus, this paper gives a brief review of the present literature to check these gaps in the web as well as recommends empirical research in the difficulty of animal trafficking in the future. This paper also highlights the online wildlife trafficking, which is happening to date and the methods that were used to identify this illicit trade.

Keywords Wildlife trafficking · Illicit trade · Animal trafficking · Online trade framework

S. Nalluri (✉) · S. J. R. Kumar · M. Soni · S. Moin · K. Nikhil
Department of Computer Science, VNR VignanaJyothi Institute of Engineering and Technology, Hyderabad, India
e-mail: sravani22me@gmail.com

S. J. R. Kumar
e-mail: jeevanrishikumar@hotmail.com

M. Soni
e-mail: maniksoni653@gmail.com

S. Moin
e-mail: moinsoheb02@gmail.com

K. Nikhil
e-mail: kamarthapuneekill142@gmail.com

C. Kiran Mai et al. (eds.), *Proceedings of International Conference on Advances in Computer Engineering and Communication Systems*, Learning and Analytics in Intelligent Systems 20,
https://doi.org/10.1007/978-981-15-9293-5_11

1 Introduction

The extinction of species has an increased rate, which is 100 times above normal. Where 1/4th of biodiversity is currently endangered including species of mammals and birds, which are about 25 and 14%. Some scientists believe that the world is moving toward extinction and others believe it is really happening. Wildlife trafficking is one of the activities of extinction, namely agriculture activities, pollution, and global climate change due to which various animals are becoming extinct [1].

According to "The Convention on International trade of endangered species of wild Plants and Animals (CITES)" is that it addresses the vital agreement with the wildlife trafficking [2] and also divides the species that are being threatened based on the protection they have and also provided a framework that needs to be implemented by each party at the national level. Moreover, wildlife trafficking remains prominent and it is considered among all the illegal activities that are profitable behind drugs, human trafficking, counterfeiting, and so on. Researchers have examined repeatedly the illegal wildlife trade online through various frameworks related to criminology and more importantly using a harm-based approach. Since the last two decades, there was a huge climb in the web where there are various changes in the criminal activities, through which various opportunities have risen and one of them is wildlife trafficking. Regarding this issue, there needs to be an understanding of how wildlife is being trafficked over the internet [3] and insight needs to be developed so that it is not neglected by criminologists.

Wildlife trafficking is the least considered illegal activity over the internet, but it generates huge profits that are estimated from 5 billion dollars to 20 billion dollars, which includes an illegal fish market, the estimations for annual profits range between 15 billion dollars and 43.5 billion dollars [4]. These estimates are less likely considering that this kind of illegal trade is not seen by respective authorities, In the illegal economy generation, wildlife trafficking is the second largest across the world compared with drug trafficking, which stands first.

2 Wildlife Trafficking on Web

2.1 The Clear Web

Traffickers of wildlife enter into the web using legitimate platforms like advertisements, auction sites like eBay and forums (e.g., [5, 6]). The information related to wildlife trafficking is being sent to the web via advertisements mostly over websites and these ads are present only till a particular time, It is difficult to trace the ad back to the source. So using this method effectively, traffickers were able to target buyers on the web. The web has become a hub for illegal wildlife trade. IFAW (International Fund for Animal Welfare) has given a brief establishment of the Wildlife market on the web in 2005. It also reported that approximately 9000 ads were released based on animals as well as their derived products in a week [7].

There are tons of ads related to ivory because it is the only product that is taken up by the researchers to research [5]. However, animals that are living are also being advertised globally, for example, tigers and gorillas were kept on sale for 9000 dollars and 70,000 dollars, respectively. Based on the latest report from IFAW, it was found that within 6 weeks 33,006 ads are posted, which gives an estimate of 11 million dollars [7].

2.2 The Dark Web

Dark Web is an anonymous network that is invisible to the public internet [8]. It uses Onion routing or tunnel routing to overlay networks. TOR is one among them. It was originally developed to secure communication for the U.S Naval base, to safeguard as well as hide traffic by passing it through encrypted relays, which consist of multiple layers. Dark Web uses a P2P network and is accessed by the TOR browser. It is used for providing privacy and anonymity. So most cybercriminals use the TOR browser to hide their identity. According to Tor Project's Website, "Tor is free software and an open network that helps you defend against traffic analysis, a sort of network surveillance that threatens personal freedom and privacy, confidential business activities and relationships, and state security." [9].

It is difficult to analyze and investigate the crimes in Dark Web and there is a shortage of analytical tools and transparency because of the encryption techniques as well as anonymity of users. These encryptions are used by the vendors as well as customers. Technicians and law enforcement must often believe old-fashioned investigative techniques that are known to the general public. These traditional methods involve taking data from the Dark web and trying to correlate it with the instant usernames to identify their location on WWW [8].

3 Online Wildlife Trade

3.1 Online Trade

Since online transactions are faster as well as easier, they became much more "safe" to offenders. There is an underground black market website that is developed for cybercriminals to sell wildlife products on websites like eBay. Table 1 gives an insight into the illegal trafficking happening in recent years based on advertisements on the internet.

Table 1 The advertisement of wildlife on the internet [10]

	US	UK	China	France
Website tracked	28	22	5	11
Numbers of ads	5028	551	544	380
Ads on international shipping	260	66	28	69
Percentage of total ads tracked	70%	7.2%	7.8%	8.8%
Value of final sales recorded	3781	5087	12.58	223.91
Number of ads found on eBay	3610	217	17	10
Ads that were exotic birds	1025	217	17	10
Ads of electric products	3921	285	376	325

3.2 Secret Wildlife Species Trade

From January 2015 to May 2016, 52 species of prohibited wild animals were sold within the public sector. Ivory and its posts were almost one-third because of the ads that were totaled to 3047 in the report. Ivory is being smuggled from Africa and poaching of elephants is a bit high in Central Africa especially in Congo. Other states like Tanzania, Mozambique, Zimbabwe, Zambia and Kenya are also affected. "Serous Reserve" in Tanzania has a report of 50 elephants being killed on a monthly basis. Total number of species with their ads is shown in Fig. 1 [10].

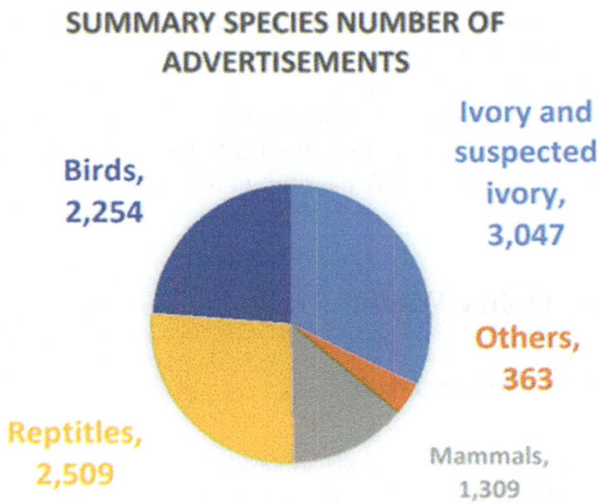

Fig. 1 Summary species number of advertisements [10]

4 Methods Used for Identifying Illegal Wildlife Trade Across the World

4.1 Deep Learning to Identify Illegal Trade on Social Media

Data collected from social media via APIs are getting used increasingly in conservation (e.g., [11]), but the restriction of automated classification is seen because it can filter the information which is not relevant to the wildlife trade and render the content with the help of cost-efficient classification techniques.

ML along with its subfields like deep learning, speech processing, and neural networks are mostly used to identify visual, verbal, and audiovisual content concerning illegal Wildlife trade [12] (Fig. 2b). To find the abstract representations of the information training of neural networks with a massive labeled set of data that contains multiple layers of neurons [13, 14].

Based on the information and knowledge about the products related to biodiversity, IUCN (International Union for Conservation of Nature) made a "Red List," so that all the products related to wildlife trade of various species that are being threatened by the traffickers are on the list [15].

With the help of analysis techniques and available information on the trade, it is easy to identify the buyers, sellers, and countries which allow this kind of trade. Using sentiment analysis, one can identify and categorize opinions mentioned in social media [16].

So a framework was designed by researchers in Fig. 2, [12] where they have trained a neural network to find the code words related to wildlife trade on Twitter by taking images of Rhino species, which contains 19 different languages that were used. Based on this method, they removed 94% of pictures in tweets. Using another application that is different from social media [6] developed an automatic system to find illegal ivory items that were sold on eBay.

4.2 Illegal Wildlife Trade e-Commerce Monitoring (China)

The total number of advertisements from 2014 October to 2016 December has a steep fall. There are around 1500 advertisements that have shown up in the first 14 months. From 2015 to 2016 December, there were 1000 advertisements per month. It is because of the new websites that are evolving from period to period and more key terms are to be used for searching [17], the statistics of monthly advertisements are mentioned in Fig. 3.

Since 2012, monthly advertisements related to wildlife products were calculated by TRAFFIC. The ivory products have the best share of advertisements about 63.2% and 18.1% of rhino products. These products have the highest demand in the Dark markets as well as wildlife trade, which accounts for 80% of all monthly ads. The remaining 20% were split between Hornbills, Tiger bones, Pangolin scales, hawksbill

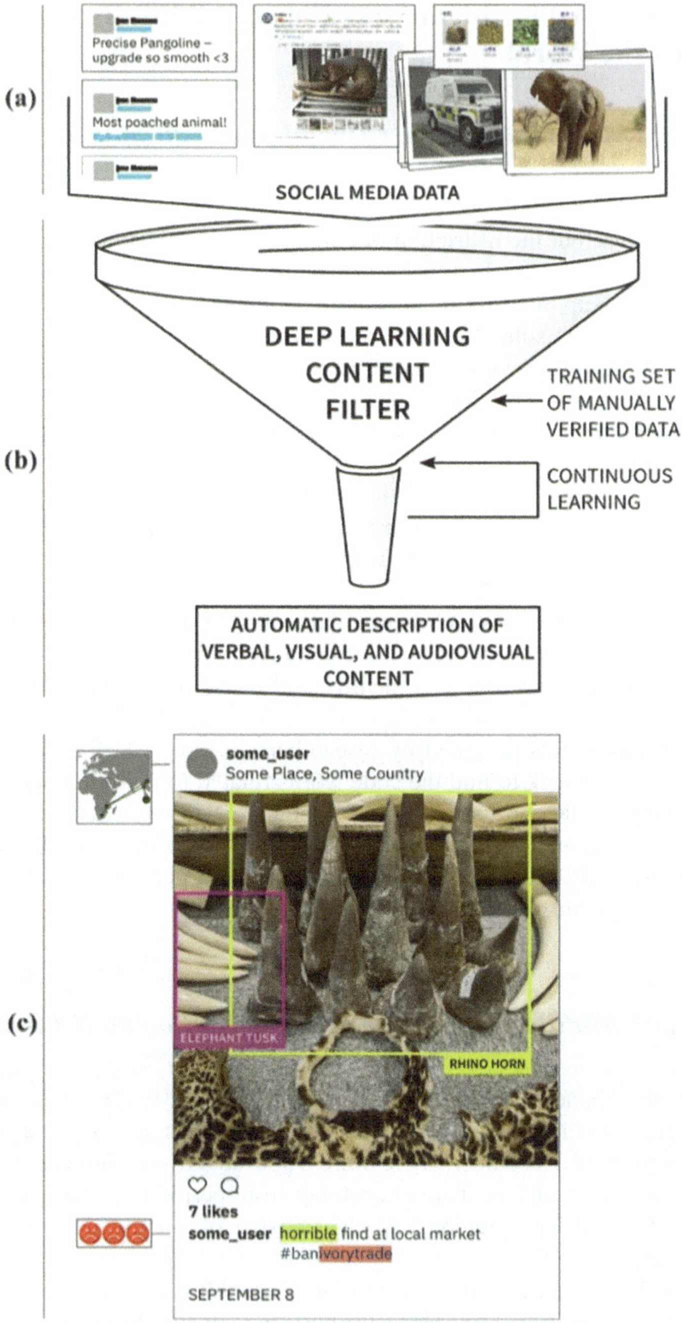

Fig. 2 Framework to **a** Mine **b** Filter and **c** identify relevant data on the illegal wildlife trade from social media platform with ML, photo in (**c**) is [16] taken from Twitter

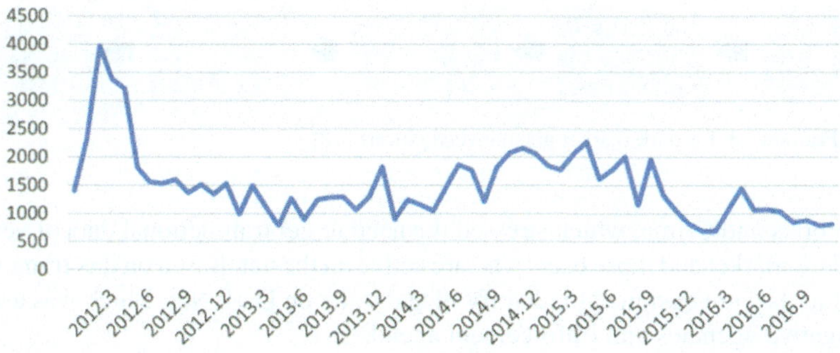

Fig. 3 Number of monthly advertisements (2012–2016) [17]

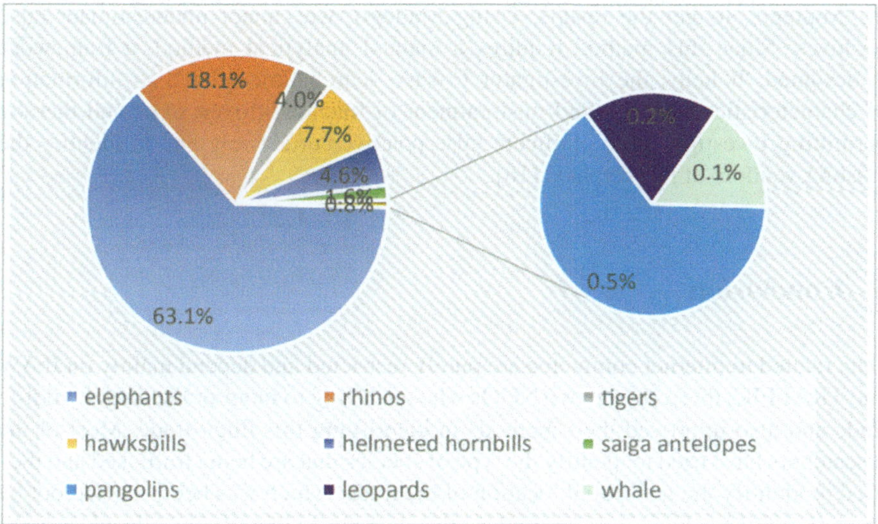

Fig. 4 Wildlife advertisements monthly [17]

shells, leopard bones, and whale products, respectively. TRAFFIC has analyzed ads related to rhino horns and ivory to understand the traits and features of internet-based wildlife cybercrime based on volume trade within the market [17] (Fig. 4.)

4.3 Effective Method for Investigating Dark Web Illegal Trade

To identify the sellers of the marketplace as well as their customers, it is somewhat difficult. So an analytical framework was developed (Fig. 5). This framework helped in identifying dark marketplaces effectively. It was first highlighted by the UN Office

Fig. 5 Framework for dark market place investigations [18]

of Narcotics and Crime, which stressed the need to get transactional data of vendors in the dark market and those users who are active on the net. Based on this framework, several tools were designed to identify illegal trade on Dark Net, which was used by Investigative agencies and Enforcement agencies.

The proposed framework consists of a Web crawler that was used on either Clear Net or Dark Net to find the HTML tags that exist on the website using a crawler script. Personal specifiable data are required manually and should be sent to "Maltego" to get the results. Using Maltego, we cannot underestimate false positives. Since this method requires a manual analytical method, a framework is designed. It helps the investigator to save search time related to information of characteristic relevance and mistreatment of machine-driven steps and thereby permitting investigators to eliminate false positives effectively and to identify the existing vendors of marketplace [18].

5 Conclusion

Data related to internet commerce are purely restricted and depend mainly on IFAW and TRAFFIC, these are the two NGOs who are trying to monitor the illegal wildlife trade and also improved their methods in identifying this illicit trade. Most of the researchers have tried to identify the type of species that are being trafficked and they tried to identify the geospatial location of the trade, which was being carried out by the dark market sellers. It is required for the agencies to identify this illegal trade for the conservation of wildlife.

So, one needs to take this as a responsibility to particularly use new methods and frameworks to identify this Dark Trade. Enforcements should work with website agencies and collect all the data related to the trade and report it to the higher authorities. Finally, online illegal wildlife trafficking is a crucial and important issue that needs to be considered and there are various challenges associated with it. Moreover, the current laws do not provide any protection to the wildlife species. New methods and frameworks are to be developed by researchers and enforcement agencies to stop this illegal trade.

References

1. S.L. Maxwell, R.A. Fuller, T.M. Brooks, J.E.M. Watson, Biodiversity: The ravage of guns, nets and bulldozers. Nature **536**(6515), 143–145 (2016). https://doi.org/10.1038/536143a
2. G. Ceballos, P.R. Ehrlich, A.D. Barnosky, A. García, R.M. Pringle, T.M. Palmer, Accelerated modern human-induced species losses: Entering the sixth mass extinction. Sci. Adv. **1**(5), 1–5 (2015)
3. R. Sollund, Wildlife trafficking in a globalized world: an example of motivations and modus operandi from a norwegian case study. In F.M. Angelici (ed), *Problematic Wildlife: A crossdisciplinary Approach* (Springer International Publishing, Switzerland, 2016)
4. E. Demeau, M. Vargas, K. Jeffrey, Wildlife trafficking on the internet: a virtual market similar to drug trafficking? Criminal. Mag. **61**(2), 101–112 (2019)
5. X. Yu, W. Jai, Moving targets: tracking online sales of illegal wildlife products in China (2015). Available at: https://static1.1.sqspcdn.com/static/f/157301/26245505/1432122394320/Chinamonitoring-report.pdf?token=Q8AuzEmc9lBrlrqSNkYIn6E%2FseI%3D
6. J. Hernandez-Castro, D. Roberts, Automatic detection of potentially illegal online sales of elephant ivory via data mining. Peer J. Comput. Sci. **0**(0), 1–11 (2015)
7. IFAW, Wanted Dead or Alive, Exposing Online Wildlife Trade (2014). Available at https://www.ifaw.org/sites/default/files/IFAW-Wanted-Dead-or-Alive-Exposing-Online-WildlifeTrade-2014.pdf
8. Article: A Framework for More Effective Dark Web Marketplace Investigations Darren R. Hayes 1, 2 ID , Francesco Cappa 3,* ID and James Cardon 1
9. Tor Project. Available online: https://www.torproject.org/
10. Survey: Wildlife Trade and Related Criminal Activities Over the Internet
11. E. Di Minin, H. Tenkanen, T. Toivonen, Prospects and challenges for social media data in conservation science. Front. Environ. Sci. **3**, 63 (2015)
12. E. Di Minin, C. Fink, H. Tenkanen, T. Hiippala, Machine learning for tracking illegal wildlife trade on social media. Nat. Ecol. Evol. **2**, 406–407 (2018)
13. A. Krizhevsky, I. Sutskever, G.E. Hinton, ImageNet classification with deep convolutional neural networks. Adv. Neural. Inf. Process. Syst. **1**, 1097–1105 (2012)
14. H. Liao, E. McDermott, A. Senior, Large scale deep neural network acoustic modeling with semi-supervised training data for YouTube video transcription. 2013 IEEE Workshop on Automatic Speech Recognition and Understanding, ASRU 2013—Proceedings, 368–373 (2013)
15. IUCN (International Union for Conservation of Nature), *The IUCN Red List of Threatened Species* (IUCN, Gland, Switzerland, 2016)
16. A framework for investigating illegal wildlife trade on social media with machine learning (2018) Enrico Di Minin, 1,2,3∗ Christoph Fink, 1,2 Tuomo Hiippala, 1,2,4 and Henrikki Tenkanen 1,2
17. WILDLIFE CYBERCRIME IN CHINA E-commerce and social media monitoring in 2016, Yu Xiao, Jing Guan and Ling XuMAY (TRAFFIC)
18. A Framework for More Effective Dark Web Marketplace Investigations 2018. https://www.mdpi.com/2078-2489/9/8/186

An Approach for Optimizing Algorithms Which Find k1-Most Demanding Products

Sireesha Chittepu, P. Yeshwanth Reddy, and M. Nikhil Reddy

Abstract The primary goal of this paper is on optimizing the process of finding k1-most demanding products using traditional algorithms and clustering. We have used four clustering algorithms over two traditional algorithms to optimize the time for finding k1-most demanding products. A synthetic data set was used for demonstration and the results are plotted in a graph for comparison. This approach can be used in various fields for optimization purposes.

Keywords Data mining · Decision support · Clustering set of rules · Performance comparison · Self organizing maps

1 Introduction

As the number of businesses is increasing and the competition between them for gaining more customers is growing [1]. Selecting the products that perform the best in the market is very crucial for their development. Also, the capability of a business to identify each of its clients, focusing on their purchases and customizing the advertisements will attract new customers [1]. So, expecting the performance of the brand new product (nominee products NP) in the market of a number of the clients (C) based on the demand of current products (EP) that are present in the market is very important. We have the Satisfaction (like/dislike) information of men or women consumers over the existing and nominee merchandise inside the market. This information may be proven in a tabular form called "Satisfaction bit

S. Chittepu (✉)
Vasavi College Of Engineering, Hyderabad, India
e-mail: sireesha@staff.vce.ac.in

P. Y. Reddy
NIIT University, Jaipur, India
e-mail: yashwanth.peddamallu@st.niituniversity.in

M. N. Reddy
VNR VJIET, Hyderabad, India
e-mail: maligi1998@gmail.com

C. Kiran Mai et al. (eds.), *Proceedings of International Conference on Advances in Computer Engineering and Communication Systems*, Learning and Analytics in Intelligent Systems 20,
https://doi.org/10.1007/978-981-15-9293-5_12

strings" (SBS) (Table 1) which may be built the usage of BMI index data. The table shows the consumer–product relationship. The ep1–ep3 is the present products (EP), np1–np9 is the nominee products (NP), and c1–c8 are the customers (C). Customer satisfaction about the product is represented with value as 1 and dissatisfaction with the value as 0, respectively. For example, the sequence 1, 1, 1, 1, 0, 1, 0, 0 in table [2] first row indicates that the customers c1, c2, c3, c4, c6 are satisfied through the ep2 and customers c5, c7 and c8 are not. The chance that consumer c1 will choose nominee product np7 is 1/7 (from column 1, Table 1). It is thought that clients pick any product with the same chance so as c1 likes seven products inside the set NP ⋃ EP the possibility of purchasing is 1/7 for every case. The predicted demand for np5 is found by including the possibilities for each customer ci ∈ C deciding on np5 as follows: (1/9 + 0/7 + 0/6 + 0/6 + 0/8 + 1/6 + 1/6 + 1/8) = 0.569444.

The company is expanding the business with a brand-new product (NP). Now to take diverse managerial selections, the goal is to perceive k1-most annoying merchandise (kNP) from the set EP ⋃ NP. For example, if k = 3, the expected quantity of total clients for np2, np3, and np7 are 1.0218, 1.1468, and 0.9802, respectively. The anticipated range of general customers for kNP = np2, np3, np7 is (1.0218 + 1.1468 + 0.9802 =) 3.1488. The cost 3.1488 is the best price among all of the combinations such as three (k1) applicants merchandise. The k1-most (right here, k1 is ready to a few) annoying merchandise are np2, np3, np7. The primary objective of this paper is to optimize the method of identifying the k1-maximum annoying products that combined with a few clustering sets of rules. First, we follow the clustering algorithm to organization comparable products right into a cluster and apply conventional algorithms on a particular cluster where the k1-most nominee products might also reside in. To evaluate the overall performance of our clustering-based k1-most demanding products [3, 4], we ran the conventional algorithms on the complete

Table 1 The SBS of the existing and nominee products

EP ⋃ NP	c1	c2	c3	c4	c5	c6	c7	c8
ep1	1	0	1	1	1	0	1	1
ep2	1	1	1	1	0	1	0	0
ep3	0	1	1	1	1	0	1	1
np1	0	0	0	0	1	1	0	0
np2	1	1	1	1	1	1	1	0
np3	1	1	1	1	1	1	1	1
np4	1	1	0	0	0	0	0	1
np5	1	0	0	0	0	1	1	1
np6	1	1	1	1	1	0	0	1
np7	1	1	0	1	1	1	1	1
np8	1	0	0	0	0	0	0	1
np9	0	0	0	0	1	0	0	0

dataset also. The resulting k1-merchandise selected in both the procedure has been the same.

We used synthetic dataset (car dataset) from UCI machine mastering repository (https://archive.Ics.Uci.Edu/ml/index.Php) [2].

This approach performs good on any type of datasets. In this paper, Sect. 2 is about the related work, Sect. 3 describes the proposed approach and Sect. 4 presents the results of the experiment.

2 Related Work

First, let us talk about the easy K-most demanding merchandise problem which is the primary consideration of our approach. Assume a fixed of clients C = {c1, c2, …, cnc}, wherein nc($\geq$1) is the number of customers for our company. Our company distributes some unique standard types of products. EP = {ep(1), ep(2), …, ep(nep)} where nep is the number of existing products in the market, which can be already popular in most of the customers C. The company is endeavor to release new products (nominee merchandise) NP = {np(1), np(2), …, np(nnp)} where nnp is the total number of nominee products which we have. Each product pi, pi $\in$ NP $\cup$ EP, consists of d range of functions that can be used to describe the products. Now, assuming that each client ci $\in$ C will without a doubt buy at least one of our products (from NP or EP or both) as per necessities. Now our goal is to pick out k1, 1 $\leq$ k1 $\leq$ nnp, products from NP, such that the anticipated range of the entire client for the k1, 1 $\leq$ k1 $\leq$ nnp, merchandise is maximized. Let kNP = knp1, knp2, …, knpk1, denote a set of k1 number of products, which can be selected from NP. Furthermore, N(EP, ci) and N(kNP, ci) represent the entire wide variety of products in EP and kNP which are liked by the customers ci $\in$ C, respectively.

To find out the products in set kNP from the set NP, traditional algorithms like Single Product Based Greedy Algorithm(SPG), Incremental Based Greedy Algorithm(IG), Upper Bound Pruning Algorithm(UBP) and Apriori algorithm(APR) [5] can be used. All these algorithms are explained very well in the reference papers [2, 6].

3 Proposed Optimized Clustering Approach

As we know that the traditional KMDP algorithms do not perform well on large datasets, we will try to increase the scalability of these existing algorithms proposed by Lin et al. [7] in the reference paper. Before making use of SPG, IG, APR and UBP algorithms, we use one of the clustering algorithms to preprocess the enter data. In this paper, we have done using four clustering algorithms K-Means [6], Mean Shift, Affinity Propagation, Agglomerative Hierarchical clustering. After applying, we select a particular cluster on which we apply the traditional algorithms assuming it contains the set kNP. To make sure the cluster which we selected contains the kNP set, we first apply the traditional algorithms over 10% sampled data and check which cluster contains the top products selected in this process that makes sure the cluster selected contains the kNP set.

The K-Means [6], Mean Shift, Affinity propagation and Agglomerative Hierarchical clustering algorithms are the clustering algorithms that are shown in this paper. The number of clusters that divides the data must be provided to K-means and Agglomerative Hierarchical clustering algorithms. The other two algorithms, Means shift and Affinity Propagation can determine this range on their own. So the evaluation is likewise made as a result. Our proposed model will work as follows.

4 Performance Evaluation

For evaluation of our proposed approach, a synthetic dataset (car data set) from UCI machine learning repository (https://archive.ics.uci.edu/ml/) has been used. The number of data points is 1728 and (we have used) 6 dimensions, respectively. To construct the satisfaction bit string table (SBS), we have randomly chosen if a customer would buy a product or not by setting the probability of the buying depending on the quality of the product. Also, the first 30% data from the SBS table are considered as EP and the rest of the data as NP. We implemented all the four clustering algorithms on the dataset. Later chosen the best algorithm based on the performance and SPG, IG is the traditional algorithms working on top of them. For the real data set, we considered.

K = 5 and clusters = 7. For all the cases, the k1 value was 5. About 10 trials were taken and the best result (lowest time) has been recorded.

5 Results and Discussion

The general execution of our proposed strategy particular conventional calculations has been plotted in the charts underneath. The overall performance (in phrases of time) of SPG and IG algorithms before clustering as compared with SPG and IG after clustering the dataset has a drastic difference (Table 2). The accuracy rates of the results in both cases are 100% (maximum of the times). But the AR for IG is not 100%. From the plot (the usage of synthetic datasets) (Fig. 1a, b), it is far visible that the times for Agglomerative Hierarchical clustering algorithm is the best (Table 3).

It is also visible that our approach scale nicely in terms of size (nnp) of the dataset. The approach solves the problem of large size, but large dimension datasets can still take a lot of time during clustering. To get around this problem, we can use Self Organizing Maps to scale down the dimensions and apply clustering over them. This approach of using SOM [9] can only be done on large datasets as competitive

Table 2 Experiments on data sets

Clustering algorithm	SPG	IG
None	4699	33,956
K-Means	339	2663
Agglomerative Hierarchical Clustering	444	3332
Affinity Propagation	328	1584
Mean Shift	741	5492

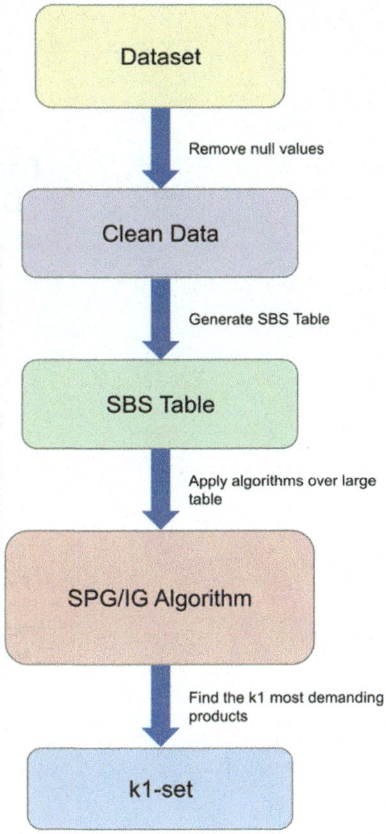

Fig. 1 Conventional approach

Table 3 Clustering time

Number of products in database	K-Means	Agglomerative	Affinity	Mean shift
1730	392	108	10,630	15,271
1000	384	107	9827	15,134
500	380	107	9849	14,366

learning can fit better when there is a large data size. We have tried applying SOM [Self Organizing Maps] over our dataset, but the accuracy was only 35% (Figs. 2, 3).

Fig. 2 Novel approach

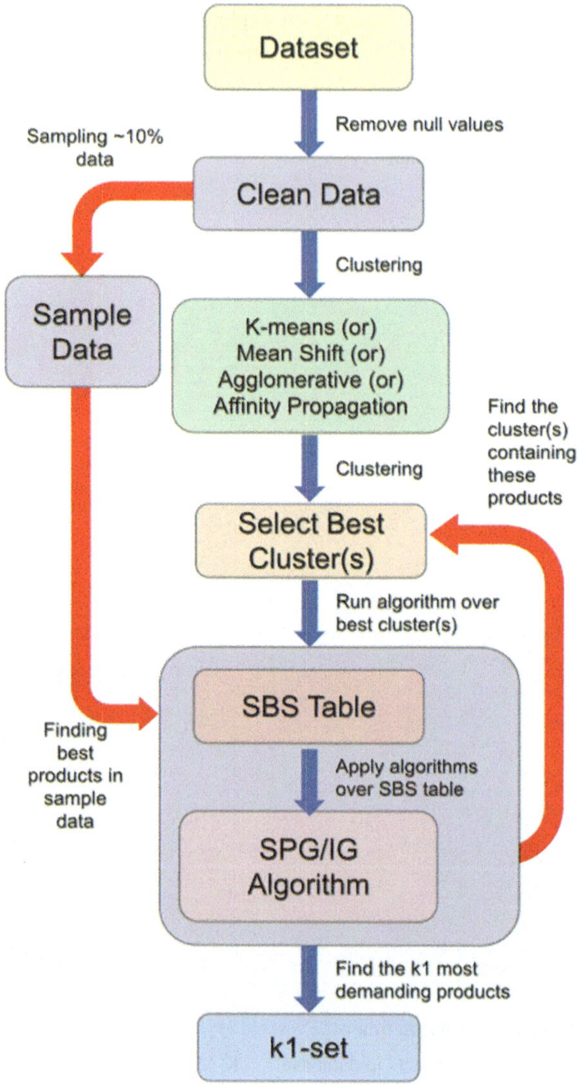

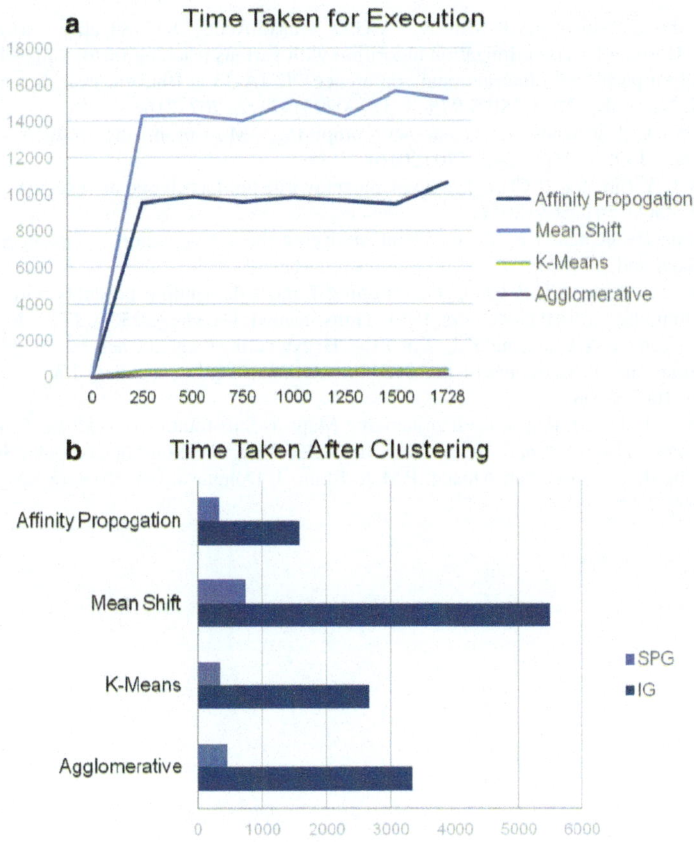

Fig. 3 **a** X—Number of products, Y—Running time of clustering. **b** X—Run time of SPG and IG on a group of products, Y—Various products grouped by clustering

6 Conclusion

The prevailing K-MDP algorithm to find the k1-most demanding merchandise has been proven [2]. The efficiency of the algorithm has been improved with the clustering algorithms, which have been shown. The experiments performed in contrast with current algorithms prove the effectiveness of our approach.

References

1. N.G. Mankiw, *Principle of Economics*, 5th edn. (South Western College Publication, New York, 2008)
2. R. Kumar, P. Bishnu, Identification of k-most promising features to set blue ocean strategy in decision making. Data Science and Engineering. 1–18. 10.1007/s41019-019-00106-z

3. M. RamakrishnaMurty, J.V.R Murthy, P.V.G.D. Prasad Reddy, S.C. Sapathy, Performance of teaching learning based optimization algorithm with various teaching factor values for solving optimization problems", International conference FICTA-13 at Bhuveneswar, Springer AISC, Vol. 247, November 2013, ISBN 978–3–319–02931–3, pp. 207–216.
4. M.S. Islam, C. Liu, Know your customer: computing k-Most promising products for targeted marketing. VLDB J. **25**(4), 545–570 (2016)
5. J.L. Koh, C.Y. Lin, A.L.P. Chen, Finding k most favorite products based on reverse top-t queries. VLDB J. **23**(4), 541–564 (2014)
6. PanelChing-HsueChengYou-ShyangChenClassifying the segmentation of customer value via RFM model and RS theory
7. C.Y. Lin, J.L. Koh, A.L.P. Chen, Determining k-most demanding products with maximum expected number of total customers. IEEE Trans. Knowl. Data Eng. **25**(8), 1732–1747 (2013)
8. W. Wu, F. Yang, C.Y. Chan, and K.L. Tan, FINCH: evaluating reverse k-nearest-neighbor queries on location data. Proceedings of the 34th International Conference Very Large Data Bases, pp. 1056–1067, 2008
9. F. Bação, V. Lobo, M. Painho, Self-organizing Maps as Substitutes for K-Means Clustering. In (eds) *Computational Science – ICCS 2005. ICCS 2005*. Lecture Notes in Computer Science, ed. by V.S. Sunderam, G.D. van Albada, P.M.A. Sloot, J. Dongarra, vol. 3516. (Springer, Berlin, Heidelberg 2005)

Homomorphic Analysis of Privacy-Preserving Psychological Medical Records

V. Kakulapati and S. Sai Teja

Abstract The mental health data and diagnosis notes are available online and accessible to all users, such as investigators, statisticians, data analysts, enduring, and psychiatric consultants. Conversely, electronic health records' accessibility, which contains hugely insightful patient data, threatens mentally diseased patients' privacy and secrecy. The objective of this work is to inspect privacy concerns about the psychological feelings of patients. Many investigators developed privacy-preserving data analysis methods. In the proposed approach, drawbacks of the accessible privacy-preserving techniques and analysis of the privacy-preserving data method allow investigators to provide mental illness patients' privacy permit the right of entry to psychological health records. In this work, we attain maximum privacy, preserving subjective data by applying the homomorphic encryption algorithm and our approach's performance analysis with other existing techniques. The implementation results demonstrated a substantial influence on numerous applications.

Keywords Privacy · Mental illness · Patients · Psychiatrists

1 Introduction

The improper leak of patient personal health information causes predicts employability or insurability; psychological conditions are frequently depending upon an added load of discrimination. The main features or characteristics of medical records are the type that can be useful for the record-based investigation. Healthcare providers combined medical data from heterogeneous resources and release the data provided to third parties, including researchers; the investigator must take care of the growth

V. Kakulapati (✉) · S. Sai Teja
Sreenidhi Institute of Science and Technology, Yamnampet, Ghatkesar, Hyderabad 501301, Telangana, India
e-mail: vldms@yahoo.com

S. Sai Teja
e-mail: mrsaiteja2@gmail.com

C. Kiran Mai et al. (eds.), *Proceedings of International Conference on Advances in Computer Engineering and Communication Systems*, Learning and Analytics in Intelligent Systems 20,
https://doi.org/10.1007/978-981-15-9293-5_13

rate of accessibility of patient health data, and there is a possibility of intimidates enduring privacy. The major challenge is that providing privacy-preserving patient sensitive information protecting when clinical trials utilize data. The existing privacy-preserving techniques are based on psychological medical records, such as De-Identification techniques. In our proposed method, we use alternative technology to facilitate investigators to secure and preserving patient privacy when they utilize the medical data in the form of encryption.

In today's world, the data generated have more value, and if the data are misused or placed in the wrong hands or any malpractice is performed, it could affect the personal life of an individual. There is a necessity to take certain protections and provide privacy to the mental health records. Many individuals are not willing to share their data and afraid of privacy concerns like breaches from trusted third parties. So, to counter these problems, we propose the novel approach of using privacy algorithms to provide privacy of mental health records and encrypt the files for unauthorized users.

Two-level privacy applies, such as when the System received the data, which is known as the input, and pattern illustration is known as output. The system input data obtained by utilizing different techniques such as k-anonymity [1], l-diversity [2], and t-closeness [3], or by altering the data already specified as an input using data perturbation techniques [4], whereas output techniques usually selected by mining algorithms due to patterns are different from different methods and handled all these properly [5].

Our system is efficient to secure and provide privacy to the mental health records where there is no need to follow the traditional methods and depend on third parties to ensure that data are safe. Also, our system encrypts the data to avoid the usage of unauthorized use or any hackers.

2 Related Work

Here is the list or a group of works that have been mentioned in the field of privacy preserving. The significance of mining techniques for preserving the privacy of data [6] was investigated and deliberated. The use of specific techniques revealed their capacity to avoid the prejudicial utilization of mining strategies. A few methods proposed that any defamed group must not emphasize on the generalization of data than everybody. PPRL (Privacy-Preserving Record Linkage) [7] consented the database connections to organizations by privacy preserving, and this technique-based classification was developed to analyze in 15 aspects. The method [8] described numerous existing methods implemented in mining for the privacy preserving be contingent on sharing of data, deformation, several mining algorithms, and imperatives hiding. Concerning the sharing of data, at present, few algorithms are utilized for privacy-preserving mining on consolidated and shared data. The requirement of include or developing the procedure based on homomorphic encryption [9] beside the accessible perception of electronic packet method in achieving shared mining methods though maintaining the sensitive data undamaged between the data sharing parties.

The present techniques of privacy [10, 11] evaluated for cloud services exhibit the outcomes based on proceeding with cryptographic factors, and it shows the general admission, the uplink facility, and the preservation of discretion of shared data. Fuzzy mapping techniques [12] are comparing with the perspective of privacy-preserving features and the potential sustainability of the same correlation with remaining features.

The evaluation focused on:

1. The definition of fuzzy function modification
2. Several functional assessments of fussy information send to a solitary value.
3. Different similarity measures utilized for the comparison between the novel data and recorded data.
4. The mapping persuades the obtained relationship.
5. Develop P3HR framework, which is utilizing smart IC cards with robust authentication mechanisms. These cards hold patient privacy data in confidentiality, and the accumulated information is prepared anonymously with the intention that unknown persons cannot correlate a record.

3 The Proposed Framework for Privacy

After the classification of data using Image segmentation, then K and L anomaly algorithm is applied to make data more secure. In the K and L anomaly algorithm, we take the data and apply constraints on it to make data more secure. The system framework provides us how data are modeled and give a brief description of the entire architecture (Fig. 1).

Based on the classification of image segmentation where we implement the K and L anomaly technique to provide privacy to the mental health records, and also, we apply homomorphic encryption technique to provide encryption to the data.

3.1 K-Anonymity

The K-anonymized dataset is for any tuple having specific traits in the dataset, at least k-1 further records that coordinate those qualities, which may accomplish by utilizing concealment and speculation.

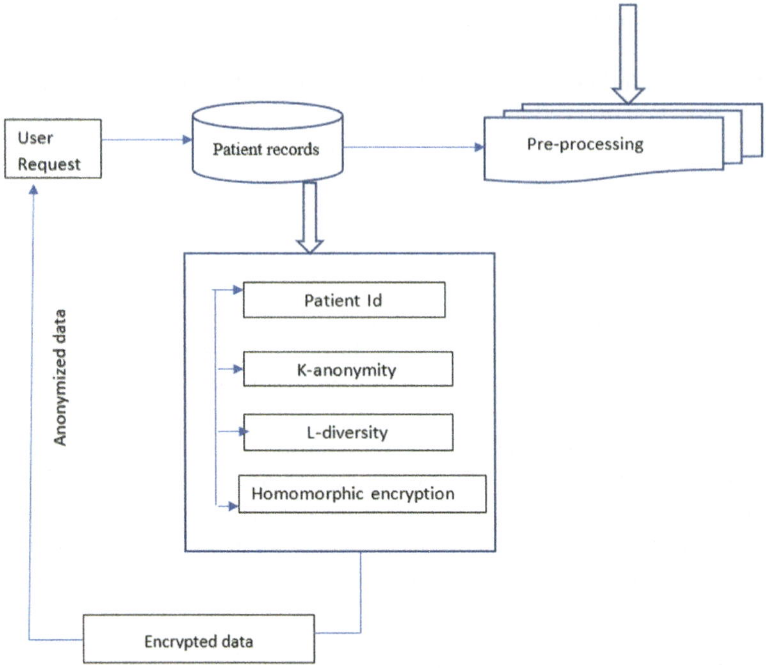

Fig. 1 Framework of privacy-preserving techniques

3.2 L-Diversity

This method avoids the consistency and contextual attacks of the k-anonymity technique. A correspondence class takes L-diversity once there are at least "L well signified" values for the delicate attribute.

3.3 Homomorphic Encryption

The estimation of massive volumes of data is an extremely modern, dissimilar, algorithm of privacy for explicit estimation typically cannot address the issues of massive data. Hence, the privacy calculation must be a function of a comprehensive scheme to ensure patients. Homomorphic encryption is an appropriate decision for massive data. For attaining privacy preserving in the context of massive data is difficult to assurance the communication security, and encryption methodology satisfies the demand. Therefore, the method of preservation of privacy based on data encryption frequently usage for sharing applications, for example, sharing data, conveyed security queries, geometric processing, and computing in scientific. In the shared data environment, the particular application typically relies upon the storage of data mode and site believability and its behavior.

For example, it is very complex to determine the steps using the special symbols of the algorithm; here is the algorithm explained with a small example of the given data encryption and decryption.

Following steps are the basic flow for the required security privacy systems.

- Initially, a dataset is taken from patient records preprocessed by removing unnecessary fields, remaining duplicate values, and no values.
- This dataset primarily encoded with relevant numerical integers for the field values of records to secure it more.
- These records are now ready for making secure systems with encrypted fields according to the system specification.
- Now we calculate K-anamolity and L-anamolity separately.
- K-anonymity is calculated for the dataset with some local suppression features to compress the given fields.
- The data fields now compressed with a particular compression factor.
- Now we calculate L-diversity, which is an extension to the above anonymity; this is a more generalized and suppressed method where the more fields are more specialized and made sensitive to the anonymous data, which makes it more secure.
- Apart from this now to fully encrypt the data, we are calculating using an encryption algorithm (Homomorphic encryption)
- Here we use this encryption technique, which entirely encrypts the data with special symbols.
- After encryption, entire data with unique symbols are not readable, which, if we need to read, we need to decode the data using the particular key, which is pre-defined.
- Here the encrypted fields are decrypted, and the required patient with keyed can see their data.
- The proposed method consists of three steps, one being k-anonymity, diversity, and homomorphic encryption (Fig. 2).

3.4 Suppression in K-anonymity

The above algorithm explains how homomorphic encryption works; homomorphic encryption is a type of encryption where a cipher used to encrypt the text. AES is the encryption that used here; AES is considered one of the most secure and hardest encryptions to crack (Figs. 3, 4 and Table 1).

128 bits of data are taken (16 bytes) and then converted to hex code using the above mapping table, and then a 128-bit key is used to encrypt the plain text.

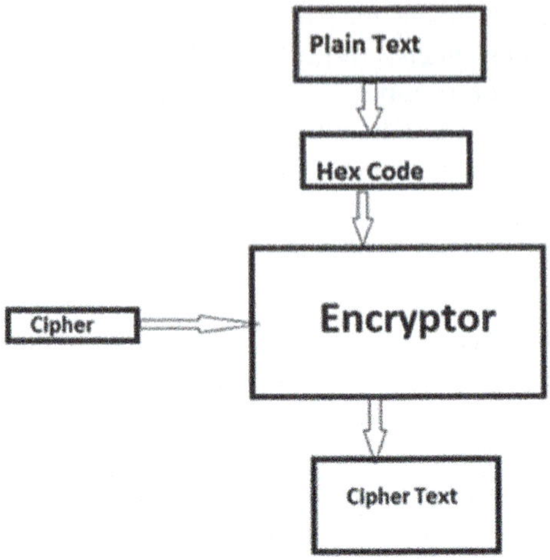

Fig. 2 Conversion of text into cipher text

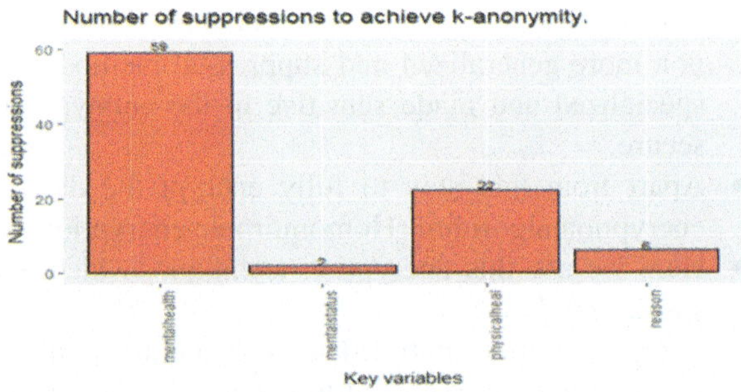

Fig. 3 Number of suppressions of K-anonymity

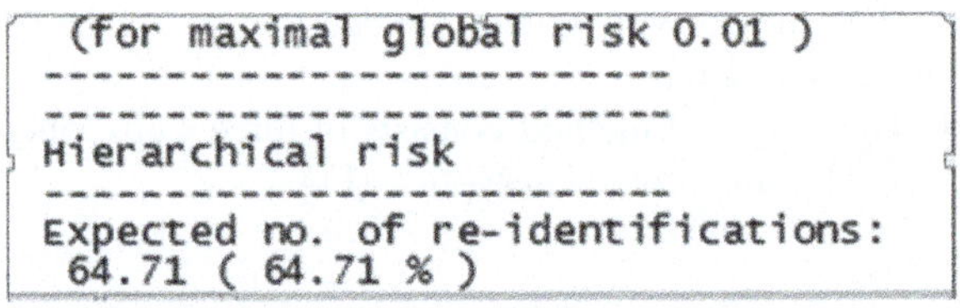

Fig. 4 Risk factor of data

The flow of encryption, the number 29, which is age, does not undergo homomorphic encryption as it is not considered as "sensitive data," instead, it undergoes k-anonymity (Figs. 5, 6 and Table 2).

Time complexity:

Time taken for encrypt: 21.715.

Time taken for decrypt: 16.457.

Table 1 The values of psychological data

ASCII		ASCII		ASCII	
Dec	Hex	Dec	Hex	Dec	Hex
32	20	48	30	64	40
33	21	49	31	65	41
34	22	50	32	66	42
35	23	51	33	67	43
36	24	52	34	68	44
37	25	53	35	69	45
38	26	54	36	70	46
39	27	55	37	71	47
40	28	56	38	72	48
41	29	57	39	73	49
42	2A	58	3A	74	4A
43	2B	59	3B	75	4B
44	2C	60	3C	76	4C
45	2D	61	3D	77	4D
46	2E	62	3E	78	4E
47	2F	63	3F	79	4F

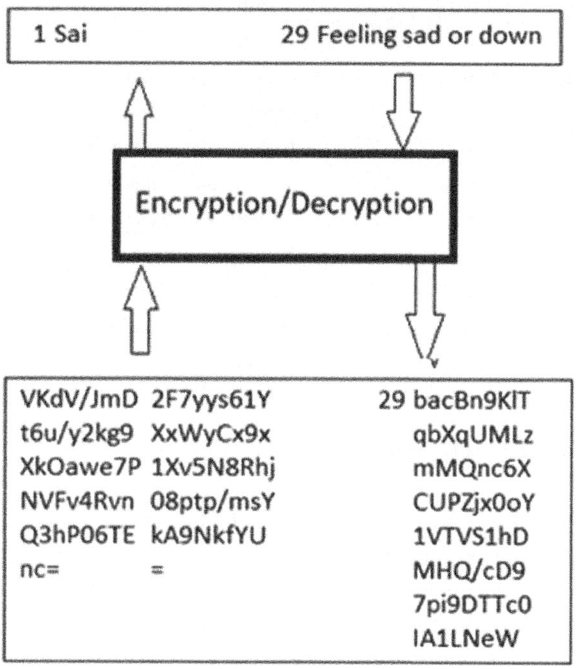

Fig. 5 The records after the encryption algorithm

Fig. 6 Psychological records after applying homomorphism

	ID	Name	Age	Disorder	salt
0	VKdV/JmD	2F7yys61Y	29	bacBn9KIT	d93d43d1dd157e0b6cc2a64116bc9a10
1	v1D2EY8g	SuL6YGbU	21	ZoE7RAK7	9c36612a523e2dd12364658c0b389d57
2	XV+6jidgu	NM//PE4I	28	0RwvNy1V	be73a83922bbebeb254dfb1f050b084a
3	nIBX2SoW	5QAYOdyY	23	psULYGOr	0163eb090dbcb8f64b1adfee3736dbb4
4	1+gKmRoN	ikTMZTJ43	43	jh9+xQOIK	3d88c1dcd7946b3854919a9e91bc76dd
5	FZBQdpmF	DNkVMMI	21	XAO93HvP	8ac1efd6c22b08381bff602b32a585c3
6	duvd0WL3	pUzdSLafC	19	XAFI03JhT	56b56520a1f922fcbb0daec0ac4c83f7
7	TxQnVm+	vPsMH0T6	22	1ZJOHOM	20060d9d58e076903a50ddabbec1bc57
8	WBlebmW	U/5RpTzY	23	NJ1DvRQq	2e80178e1ff2823d8ec5100b812875a7
9	oC7SdQXC	iSzanWeal	26	noS1v5mZ	f2d1f5465d4cc6cca44058fe6f931da9
10	PTZ3nZsDC	1NMFAEp	29	nuCMYxJC	93df2eb948f57690331bddd9a9365bea
11	MKUOUSV	bcvjJ4kz/3	31	HCQN4va	b0698857bba3f8632bf7d929d077a92d
12	nGxhTtnEF	NGC1kxDC	22	6I9hJI9KIJ	21a30829a2c993253e525c2918df7bdc
13	Ea/pDJW8	qu8tv15Ac	34	ibZISyDN0	859b369358ed80c7ba65e4d75707303e
14	tHsaP+4IU	88CdQ8yJ	21	OvS9FJSiq	22c0e2a42a3348023b5131cd9ba5eba3
15	uCyrSnHIE	XTjwN5ah	22	N2nCjx/XJ	f2370f9f1cbaf5dab87e653f940bbd33

Table 2 The records before homomorphic encryption algorithm

ID	Name	Age	Disorder
1	Sai	29	Feeling sad or down
2	Kashi	21	Confused thinking or reduced ability to concentrate
3	Vishwa	28	Excessive fears or worries, or extreme feelings of guilt
4	Ravi	23	Extreme mood changes of highs and lows
5	Teja	43	Withdrawal from friends and activities
6	Riteesh	21	Significant tiredness, low energy or problem sleeping
7	Reddy	19	Detachment from reality(delusions), paranoia or hallucinations
8	Sujan	22	Clinical depression
9	Shiva	23	Obsessive compulsion disorder
10	Srikar	26	Inability to cope with daily problems or stress
11	Eshwar	29	Trouble understanding and relating to situations and to people
12	Rohit	31	Problems with alcohol or drug use
13	Risheek	22	Social anxiety
14	Aahlad	34	Obsessive compulsion disorder
15	Charan	21	Hostility
16	Tejas	22	Suicidal thoughts

4 Conclusion

In this, we conclude that when the practical, systematic approach is built and appropriately organized, there is panic for theft and unauthorized use of data. So, the victims who feel uncomfortable sharing their data with the public feel free and grow their confidence. Also, we propose the Advanced Encryption Standard, which involves an analysis of data and provides high encryption standards.

5 Future Scope

In the future, in the context of medical perception, machine learning methods are addressing privacy issues, specifically psychological health. Develop a machine learning module that classifies psychological data into diverse protection levels. Then applying the churn methods to predict better accuracy of psychological data and provide better recommendation methods of diagnosis to reduce the psychological patient's stress and improve their behavioral therapy.

References

1. L. Sweeney, Achieving *k*-Anonymity privacy protection using generalization and suppression. Int. J. Uncertainty, Fuzziness Knowl.-Based Syst. **10**(5), 571–588 (2002)
2. H. Tian et al., Extending l-diversity for better data anonymization. In: *6th International Conference on Information Technology: New Generations, ITNG 2009, IEEE Xplore*, pp. 461–466, (2009)
3. N. Li et al., *t*-Closeness: privacy beyond *k*-anonymity and *l*-diversity. In: *23rd International Conference on Data Engineering, IEEE Xplore*, pp. 106–115 (2007)
4. Liu et al., The applicability of the perturbation based privacy preserving data mining for real-world data. Data. Knowl. Eng. **65**(1), 5–21 (2008)
5. Aggarwal et al., A general survey of privacy preserving data mining models and algorithms. In: *Privacy-Preserving Data Mining*, ed. By C.C. Aggarwal, P.S. Yu. (Kluwer Academic Publishers, Boston , 2008), pp. 11–52
6. S. Matwin, Privacy-preserving data mining techniques: survey and challenges. In: *Discrimination and Privacy in the Information Society. Studies in Applied Philosophy, Epistemology and Rational Ethics*, ed. by B. Custers, T. Calders, B. Schermer, T. Zarsky, Vol. 3 (Springer, Berlin, Heidelberg, 2013)
7. Vatsalan et al., A taxonomy of privacy-preserving record linkage techniques, Systems, Volume, pp. 946–969 September 2013
8. Qi, Zong, An overview of privacy preserving data mining. Procedia Environ. Sci. **12**(B), 1341–1347 (2012)
9. Raju et al., Privacy maintenance collaborative data mining. A Practical Approach' January 2009. https://doi.org/10.1109/ICETET.184
10. Malina et al., *Performance Evaluation of Primitives for Privacy-Enhancing Cryptography on Current Smart-Cards and Smart-Phones* (Springer, Berlin, Heidelberg, 2013) pp. 17–33
11. Sachan et al., A comprehensive review on privacy preserving data mining. Spring. Plus **4**, 694(1). https://doi.org/10.1186/s40064-015-1481-x, November 2015
12. R. Mukkamala et al., Fuzzy-based methods for privacy-preserving data mining. *2011 Eighth International Conference on Information Technology: New Generations,* Las Vegas, NV, pp. 348–353 (2011)

Prediction of Rice Plant Diseases Based on Soil and Weather Conditions

Akula Maneesha, Chalumuru Suresh, and B. V. Kiranmayee

Abstract India is an agriculture-dependent country as most of its inhabitants pivot on the cultivation of crops for their livelihood. The growth of crop mainly relies upon climatic conditions, water quality and water content in the soil. In recent years, the crop yield has been decreased due to the occurrence of diseases, variations in the climatic conditions and changes in the soil properties. The main challenge is to reduce the loss of crop due to diseases and increase the productivity of the crop. For this, it is very important to predict the disease in advance so that the necessary precautions will be taken. In order to achieve this, a system is proposed to detect the rice plant diseases based on Internet of Things and machine learning techniques where IOT is used to collect the sensor data and prediction of disease is done by Naïve Bayes machine learning algorithm. Climate and water data is obtained with the help of sensors and is sent to the ThingSpeak cloud and this data is extracted and utilized for predicting the plant disease and the accuracy obtained is nearly 92%.

Keywords Agriculture · Plant diseases prediction · Internet of things · Naïve bayes algorithm · ThingSpeak cloud

A. Maneesha (✉) · C. Suresh · B. V. Kiranmayee
Department of CSE, VNRVJIET, Hyderabad, India
e-mail: maneesha.akula39@gmail.com

C. Suresh
e-mail: suresh_ch@vnrvjiet.in

B. V. Kiranmayee
e-mail: kiranmayee_bv@vnrvjiet.in

C. Kiran Mai et al. (eds.), *Proceedings of International Conference on Advances in Computer Engineering and Communication Systems*, Learning and Analytics in Intelligent Systems 20,
https://doi.org/10.1007/978-981-15-9293-5_14

1 Introduction

Agriculture is one of the oldest and important activities of human being. Agriculture plays a significant part in the financial resources of India and it is the backbone of Indian economy. India is one of the top 3 countries in the production of various agricultural products like paddy, wheat, pulses, groundnut, natural products, vegetables, sugarcane, tea, jute, cotton, tobacco leaves and other products [1]. Paddy holds the topmost position in comparison to the other crops grown in India, as rice is the staple food and essential food of protein for most of the people in India. But the production of rice has been decreased as the plants get easily affected by the diseases due to changes in the climatic conditions. Climate variability plays a vital role in deciding the aggregate product yield as the variations in climatic changes can influence the plant growth at various stages. A small change in the temperature or humidity permits the growth of bacteria or virus or fungi that eventually promotes the occurrence of diseases. Diseases not only affect the quality of rice grains but also cause a decrease in the production. The loss of production of crop affects the financial status of the farmer as well as the country. Every year about 37% of the total crop is lost because of diseases. Thus, monitoring and accurate prediction of the occurrence of diseases are important. Early prediction of the disease is helpful to prevent the major damage to the crop [2].

Even though the technology was improved in the agriculture sector, it is still a major issue to predict the plant disease and reduce the yield loss. Hence, it has become a predominant issue to predict the rice plant disease based on the climatic conditions. In the previous years, the disease was predicted by the farmer at the starting stage manually based on his practical knowledge. Farmers used to monitor the plants at regular intervals of time and according to that, they used to spray the required quantity of fertiliser or pesticide [3]. But now-a-days, due to much variations in the climatic conditions, it is not feasible for the farmer to predict the crop disease and this may lead to loss of crop and application of wrong fertiliser may affect the properties of the soil. In recent years, the agriculture sector has a huge upgradation in order to survive the changing conditions of the Indian economy [4]. Many methods have come into existence for the prediction of crop disease based on various factors. The proposed system is developed using IOT and machine learning algorithms to predict the rice plant disease and helps to improve the crop yield.

At present, IOT has been used widely in improving the agriculture sector in various ways [5]. It plays a significant part not only in improving the agriculture sector, but also in the other sectors. In this system, we are using IOT as the main source to obtain the soil, water and weather data. For this, pH sensor, soil moisture sensor and DHT 11 sensor with Raspberry pi integration are used to get the required data. Machine learning technology uses various algorithms to analyse the data and to estimate the output values depending on the trained data. Various algorithms were used to anticipate the crop yield, diseases, to monitor the crop. In this, Naïve Bayes algorithm is used for the prediction of plant disease.

2 Literature Review

Agriculture is the main component of every country which provides food for the survival of the people. From the past few years, there has been a lot of development in the agricultural sector in order to improve the quality and quantity of the crop [6]. As the crop depends majorly on weather conditions, research was carried on forecasting the weather data. Various sensors were used to monitor the temperature, humidity, altitude, pressure, light intensity and level of rainwater in the field and the user can access the data easily [7]. Besides the weather conditions, a crop yield also depends on the type of soil it is cultivated. A system was developed that uses a machine learning algorithm on the soil and weather data to predict the crop. In this, the soil and weather past data are collected from the Indian Meteorological Department (IMD) and Multiple Linear Regression algorithm is applied to predict the best suitable crop. Through this, the farmer can get the maximum yield [8].

Even though the suitable crop is grown, plants might be attacked with diseases. In some cases, the farmers may not be aware of the measures that need to be taken and the pesticides that are ought to be used to prevent or reduce the plant diseases. In order to help such people, a system was developed that not only suggests the suitable crop, but also gives efficient information about the crop and fertilisers that should be used to get the maximum yield [9]. A complete smart field method was implemented to monitor the field condition from time to time. In this approach, various sensors were placed in the field to get information about water, soil and weather. This was developed in two modules where the first module provides complete information, advices and market prices about the crops. The second module allows the farmer to set reminders and plans about the crop and also monitor the condition of the crop at various stages [10].

Besides machine learning and IOT, some researchers have used Image Processing concepts to detect and classify the plant diseases. A method was developed to detect and classify the plant disease by considering leaf pictures as the primary input. In this, first the pictures of leaf are captured and Image Processing methods are applied on the images to extract the required features to detect the disease. After extracting the features, a recogniser is utilised to coordinate the diseased leaf with the images in the database. In this, SVM classifier with Back Propagation Neural Network is applied to detect and classify the leaf disease [11]. Sometimes, it is difficult to select the proper machine learning algorithm to predict the disease. A survey of all the algorithms with their advantages and issues along with a new approach was provided. A method is implemented that uses Fuzzy Conclusion Tree by optimizing the feature withdrawal using PSO-SVM algorithm [12]. Sometimes, there occurs a situation where it becomes difficult to find the fault area on a defect leaf. To overcome this, a method is implemented which uses K-means algorithm to find the ratio of the fault and normal portion of that leaf in a simple and effective way. Once the ratio is found, the disease of that particular portion is detected using image processing and Artificial Intelligence methods [13].

Table 1 List of papers with the algorithm used and accuracy results

Authors	Algorithm used	Disease	Accuracy (%)
[15]	NN	Rice blast, Sheath blight,	85
[16]	SVM classifier	Not specified	88
[17]	SVM	Rice blast	82
[18]	K-means clustering	Not specified	79
[19]	C4.5 and C5.0	Not specified	98

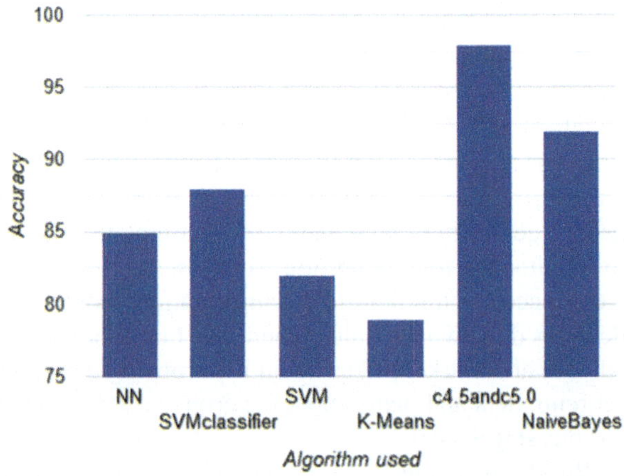

Fig. 1 Comparison of accuracy rate of output (in percentage) of existing algorithms with the proposed algorithm

In recent times, Convolutional Neural Network (CNN) has accomplished prominent results in the field of image classification. A method was developed in such a way that it can identify 13 distinct types of plant diseases out of healthy leaves. Caffe, a deep learning framework was used to train the dataset and recognise the disease [14]. Table 1 shows the rate of accuracy using existing algorithms and Fig. 1 represents the comparison of accuracy rates obtained using present algorithms and proposed one in graphical format.

3 Proposed System

The proposed work focuses on predicting the rice plant diseases based on pH, soil moisture, humidity and temperature values. An integrated system is developed that uses sensors with Raspberry Pi integration and machine learning algorithms. The system is implemented in two phases wherein the first phase, sensors are used to collect the data of water, soil and weather and this collected data is stored in 'ThingSpeak' cloud. In the second phase, the data are extracted from the cloud and machine learning algorithm is applied to this data to predict the plant disease.

The initial step is concerned with the collection of data. In this phase, the concept of IOT is used. Sensors such as pH sensor, DHT 11 sensor and soil moisture sensors are used to get the water, weather and soil data, respectively. The pH sensor is utilised to know the alkaline nature of water based on pH value. DHT 11 sensor is used to get the atmosphere's humidity and temperature values and the content of water in the soil is measured by a soil moisture sensor. These sensors are integrated with Raspberry pi-3 and it acts as a microcontroller. Here, eight GPIO pins are used for connecting the sensors.

All the data collected from the sensors are stored in 'ThingSpeak cloud', as shown in Fig. 2. ThingSpeak is an IOT analytics platform service that allows to aggregate, visualise and analyse live data streams in the cloud [20]. ThingSpeak provides instant visualisations of information posted by the devices to ThingSpeak [16].

ThingSpeak enables the sensors to send the data directly to the cloud where the user can continuously visualise and analyse the sensor data [21]. The user can observe the continuous information from sensors in the form of graphs as shown in Fig. 3. With the help of a unique API key, the user can analyse and download this data from any place and at any time. The data can be stored either in public or private channels. Here, public channel is used to store the humidity, temperature and pH and

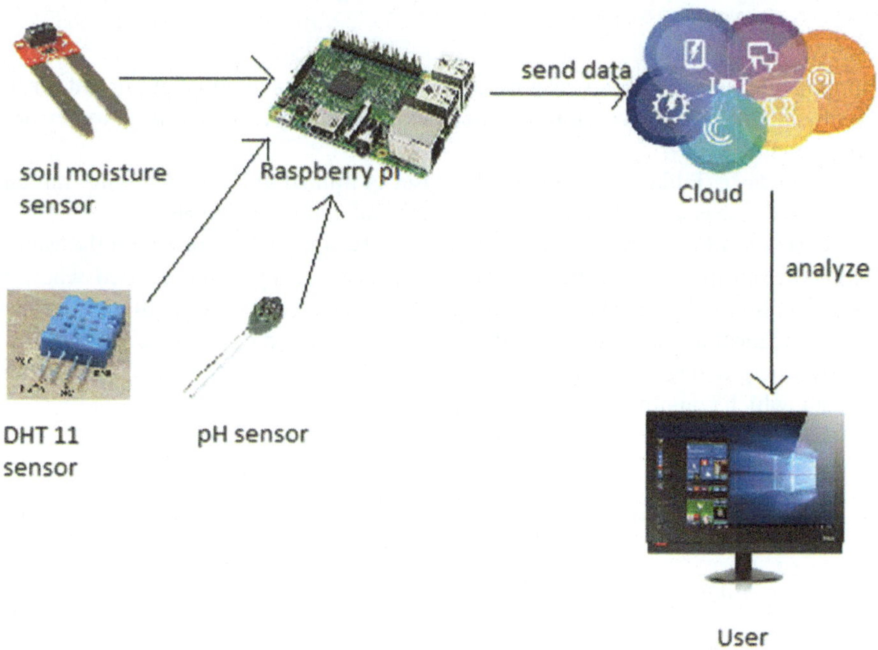

Fig. 2 Integration of sensors with Raspberry pi and sending the data to the cloud

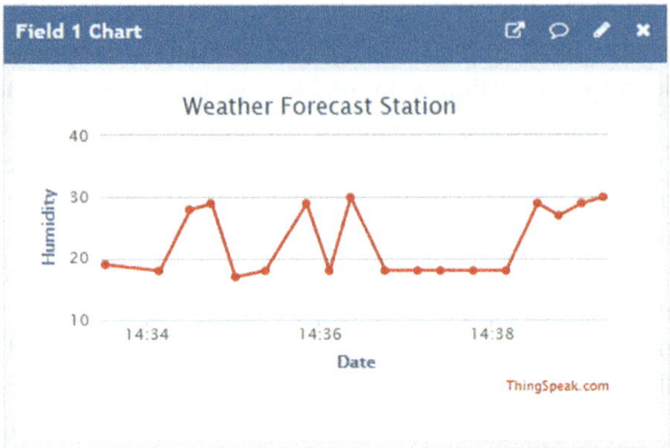

Fig. 3 Graphical representation of sensor values in ThingSpeak cloud

Table 2 Sample data of the dataset used

pH	Temperature	Humidity	Moisture	Disease
6	25	70	950	Sheath Blight
7	35	90	1035	Brown Spot
8	15	80	820	Rice Blast
5	21	85	1285	Bakanae
6	29	106	1020	Stack Burn
5	32	76	740	Rice Tungro

soil moisture data. It is also easy to retrieve the data or share the sensor data from this cloud. Once the data is stored in the cloud, the data is extracted from to perform further analysis [22].

As shown in Table 2, the obtained dataset is trained and based on the training dataset; the diseases are predicted for the new soil and weather values.

As shown in Fig. 4, the second step is to predict the disease based on the values obtained from the sensors. For this, machine learning algorithm is used where it helps to build models based on the sample data. In the field of agriculture, machine learning algorithms are used to predict the diseases based on the data that helps to improve the productivity of the yield. In this, Naïve Bayes algorithm is used to train the data, which contains sensor values, disease names.

Step 1: Loading the training and testing files.

$$\text{FileReader reader} = \text{new FileReader(trainfile.getPath());} \tag{1}$$

$$\text{FileReader reader} = \text{new FileReader(testfile.getPath());} \tag{2}$$

Step 2: Training is done using buildClassifier() method.

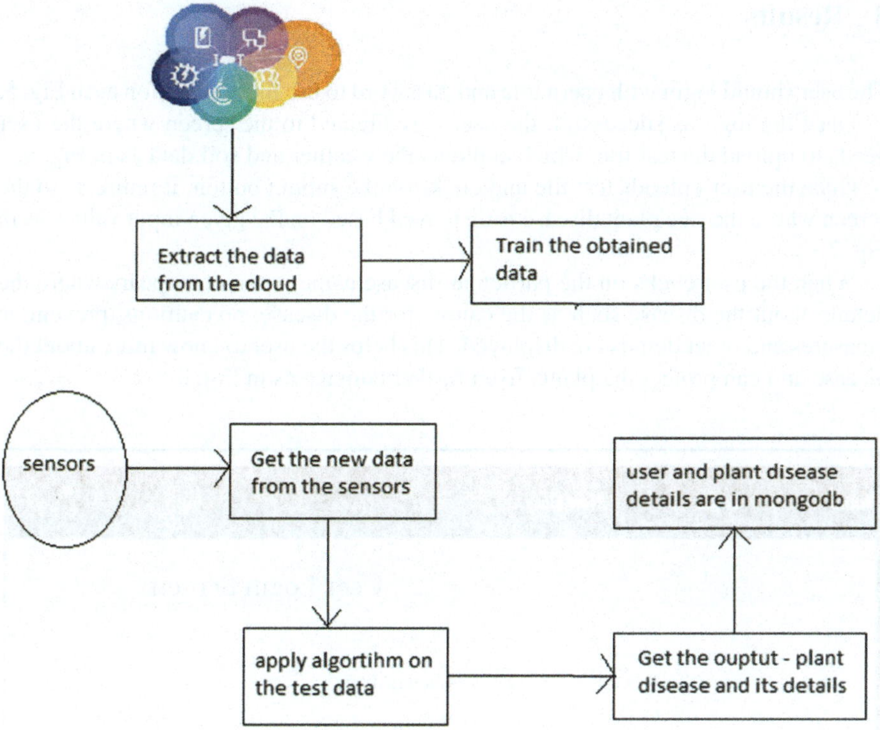

Fig. 4 Extracting the sensor data from the cloud and applying Naïve Bayes algorithm on the data to predict the disease

$$\text{nb.buildClassifier(train);} \tag{3}$$

Step 3: Evaluating the generated for test dataset.

$$\text{Eval.evaluateModel(nb, test);} \tag{4}$$

Step 4: Classifying the given test instance and displaying the output.

$$\text{double index} = \text{nb.classifyInstance(test.instance(i));} \tag{5}$$

$$\text{system.out.print(nb.toSummaryString());} \tag{6}$$

It gives the rice plant disease and its details as the final result. In this, 5 attributes and 400 instances are taken into consideration for training.

4 Results

The user should login with username and password to use the application as in Fig. 5.

Once the login is successful, the user is redirected to the screen where the user needs to upload the test file, which contains the weather and soil data as in Fig. 6.

Once the user uploads test file and clicks on the submit button, it redirects to the screen where the rice plant disease is displayed based on the given input values as in Fig. 7.

When the user clicks on the particular disease name, a screen appears where the details about the disease such as the causes for the disease, precautions, preventive measures and other details are displayed. This helps the user to know more about the disease and can protect the plants from further damage as in Fig. 8.

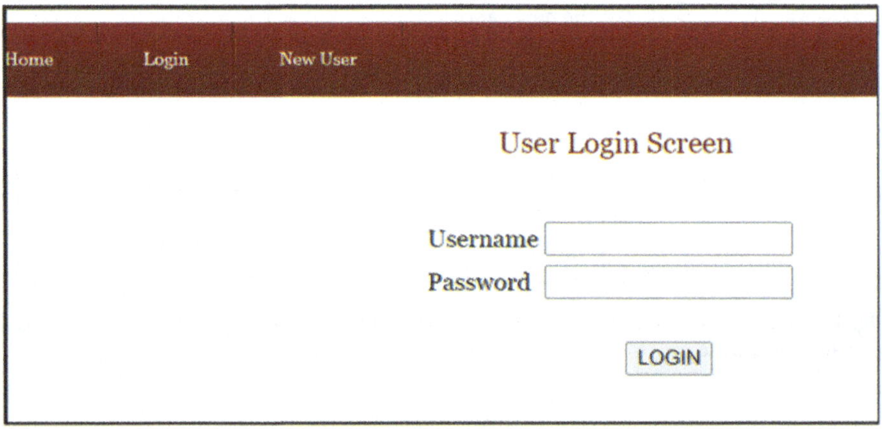

Fig. 5 Login for the user to use the application

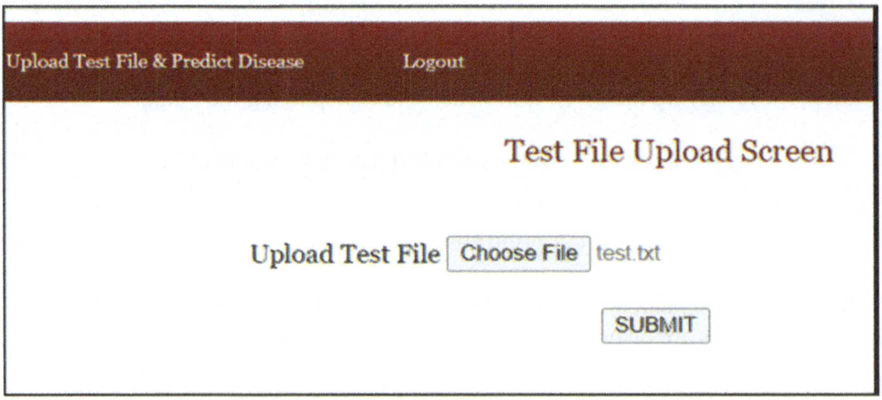

Fig. 6 Uploading test file that contains data of weather and soil

Test File Record No	Predicted Class
1	Crown Sheath Rot
2	Leaf smut
3	Grassy stunt
4	Crown Sheath Rot
5	Grassy stunt
6	Leaf smut
7	Crown Sheath Rot

Fig. 7 Prediction of disease based on given input values

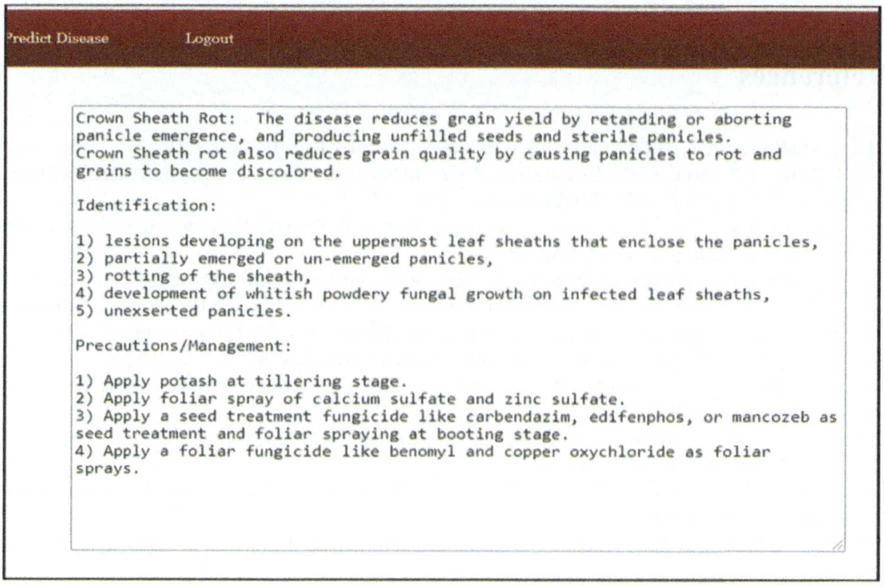

Fig. 8 Details about the disease

5 Conclusion

The proposed method uses Naïve Bayes algorithm to anticipate the plant disease based on sensor values and the accuracy of detecting the disease correctly is 92%. Predicting the disease based on temperature, humidity, pH and soil moisture values is the principal objective of the proposed approach. This method consists of two phases wherein first phase, Internet of Things technology is used to get the values from the sensors and in the following phase, machine learning algorithm is applied to predict the disease in rice plants. So by these stages, this proposed work provides the automatic system to predict the paddy diseases and overcome the loss of total harvest due to diseases and also gives safety measures that are ought to be taken to control the diseases. This helps to reduce the loss of crop due to diseases and increase the productivity of the crop.

6 Future Scope

The Proposed System is used to predict the diseases of the paddy crop and only few parameters (humidity, temperature, pH and moisture) are considered. In further, this work can be extended for other crops with more number of sensors so that more accurate results can be obtained and it can help many other farmers.

References

1. L. Madhusudhan, Agriculture role on Indian agriculture. Bus. Econ. J. (2015)
2. Y. Kim, J-H. Roh, Early forecasting of rice blast disease us-ing long short-term memory recurrent neural networks. MDPI (2017)
3. K.S. Archanal, A. Sahayadhas. Automatic rice leaf disease segmentation using image processing techniques. Int. J. Eng. Technol. (2018)
4. A. Badage, Crop disease detection using machine learning: Indian agriculture. IRJET (2018)
5. N. Vinayak, Malavade, P.K. Akulwar, Role of IoT in agriculture. IOSR Journal of Computer Engineering, National Conference on "Changing Technology and Rural Development (2016)
6. M.Y. Jyothi, Role of technology in advancement of Indian agriculture. Academia
7. S. D. Shewale, S. N. Gaikwad, An IoT based real-time weather monitoring system using raspberry Pi. IJAREEIE (2017)
8. D.S. Zingade, O. Buchade, N. Mehta, S. Ghodekar, C. Mehta, Crop prediction system using machine learning. IJAERD (2017)
9. R. Ghadge, J. Kulkarni, P. More, S. Nene, R.L. Priya, Prediction of crop yield using machine learning. IRJET (2018)
10. K.A. Mohanraj, J. Nare, Field monitoring and automation using IOT in agriculture domain. Elsevier (2016)
11. V. Ramya, M. Anthuvan Lydia, Leaf disease detection and classification using neural networks. Int. J. Adv. Res. Comput. Commun. Eng. (2016)
12. S. Chouhan, D. Singh, A. Singh, A survey and Analysis of various agricultural crops classification techniques. Int. J. Comput. Appl. (2016)
13. S. Sarkar, S. Biswas, A. Tapadar, P. Saha, AI based fault detection on leaf and disease prediction using K-means clustering. Int. Res. Eng. Technol (2018)
14. S. Sladojevic, M. Arsenovic, A. Anderla, D. Culibrk, D. Stefanovic1, Deep neural networks based recognition of plant diseases by leaf image classification. Research Article, Computational and Neuroscience (2016)
15. G. Anthonys, N. Wickramarachchi, An image recognition sys-tem for crop disease identification of paddy fields in Sri Lanka, International Conference on Industrial and Information Systems, IEEE (2009)
16. J.P. Shah, H.B. Prajapati, V.K. Dabhi, A survey on detection and classification of rice plant diseases. Conference: 2016 IEEE International Conference on Current Trends in Advanced Computing (2016)

17. N.N. Kurniawati, S. Abdullah, Texture analysis for diagnosing paddy disease. International Conference on Electrical Engineering and Informatics, IEEE (2009)
18. K. Sekhara, Trends in area, production and productivity of paddy crop: an overview. Int. J. Human. Soc. Sci. Inve. (2019)
19. R. Revathy, R. Lawrance, Comparative analysis of C4.5 and C5.0 algorithms on crop pest data. Int. J. Innov. Res. Comput. Commun. Eng. (2017)
20. https://thingspeak.com/pages/learn_more
21. K. Narsimha Reddy, B. Polaiah, N. Madhu, A literature survey: plant leaf diseases detection using image processing techniques. IOSR J. Electron. Commun. Eng. (2017)
22. B. Ramya, T. Tamilarasi, J. Tharani, E.A. Mary Anita, A survey on smart agriculture using internet of things. Int. J. Eng. Res. Technol. Special Issue (2018)

An Improved Web Information System Through Identification of Faculty Browsing Patterns in Affiliated Engineering Institutions

M. Swathi, K. C. Sreedhar, Meeravali Shaik, and V. Prabhakar

Abstract Now a days, the teaching methodology in engineering stream is extensively making use of internet resources. The concepts being taught in engineering institutions are very well presented in online resources, such as Wikipedia, Tutorialspoint etc. The faculty working in engineering institutions frequently visits the pages in the previously mentioned websites. As the syllabus the faculty has to teach is predetermined by the university to which they had been affiliated to. As a result, the pages those faculty browses are ordered by the topics in syllabus which is streamlined in a semester through the timetable of a particular year and branch of engineering. So we can know in advance the pages that can be browsed for all subjects, all branches, and all years of engineering courses. As a result, the web server efficiency can be improved by prefetching these pages into random access memory. We have implemented this idea with various well known page replacement algorithms like FIFO, LRU, LFU, OPT and analyzed the performance with proposed idea and without proposed idea in all four algorithms. Results suggest that there is a significant improvement in hit ratio and thus in access latency.

Keywords Web cache · Page replacement · Web prefetching · Faculty · Browsing patterns

1 Introduction

Web caching is a prominent technique for effective performance of web-based system by making different web objects available in the cache. The web caching mechanisms are enforced at three levels. They are client level, proxy level and the original server level. Proxy servers play an important role in reducing the response time and thus

M. Swathi
Department of CSE, VMTW, Hyderabad, India

K. C. Sreedhar (✉) · M. Shaik · V. Prabhakar
Department of CSE, Sreenidhi Institute of Science and Technology, Hyderabad, India
e-mail: simplykakarla@gmail.com

C. Kiran Mai et al. (eds.), *Proceedings of International Conference on Advances in Computer Engineering and Communication Systems*, Learning and Analytics in Intelligent Systems 20,
https://doi.org/10.1007/978-981-15-9293-5_15

bandwidth of network gets saved [2, 3]. The caching of data at client side has limited utility whereas at server we have greater utility. On the other hand, web prefetching extends the benefit of web caching by predicting the pages and prefetching them into cache and thus improving hit ratio [1]. The role of web cache replacement algorithms is to improve hit ratio. But unfortunately, hit ratio is not improved even with efficient web caching algorithms. Even though with a cache of infinite size, hit ratio is about 0.4 to 0.5 irrespective of caching mechanism as stated in Waleed Ali et al. [2]. This is due to the browsing of new pages by end users.

So the hit ratio can be improved with the blend of web prefetching and web caching. Studies showed that this blend doubled the performance. As per recent study, web caching alone achieves the access latency by 26%, whereas the blend improves the latency by 60%. This phenomenon is also proved by the experiment carried in this paper.

Similar to caching, web prefetching can be implemented at client, proxy and server levels. The client side prefetching focuses on browsing patterns of single user across many webservers [5, 6]. The proxy side prefetching focuses on browsing patterns of group of users across many webservers. The server side prefetching focuses on browsing patterns of multiple users towards single web server. This paper implements the idea related to faculty browsing patterns that comes under the category of server side prefetching.

2 Related Works

Whenever a new page is requested by CPU and if the cache is full, the page replacement algorithm determines which pages needs to be replaced as mentioned in Tinghuai Ma et al. [1]. An ideal page replacement algorithm should have high hit ratio and thus low access latency. The page replacement algorithms that exist in literature are least frequently used algorithm, least recently used algorithm, size algorithm, function-based algorithm, randomised-based algorithm and weighting based algorithm. There is a significant value of research in web cache replacement [7–11].

(i) **First in First Out (FIFO)**:

The strategy in FIFO algorithm is that it replaces the oldest page in the cache memory. In other words, it replaces the page that came first into the cache memory.

(ii) **Least recently used algorithm (LRU)**:

The LRU algorithm mainly makes use of the property of algorithms, that is locality of reference which means that data or code requested recently will be requested again and again. In other words, it replaces the least recently used page inside the cache memory [4].

(iii) **Least frequently used algorithm (LFU):**

The LFU algorithm considers the frequency count of page, that is number of times page gets accessed. It assumes that the more the page's frequency, the more it gets accessed again. In other words, it replaces least frequently used page inside the cache memory if no space is there for newly requested page.

(iv) **Optimal replacement algorithm(OPT):**

The optimal page replacement algorithm replaces the page that will not be used for the longest period of time. This is an ideal algorithm which stands as bench mark for comparison.

3 Proposed Model

In the proposed model, we project our analysis on single affiliated college to the university. At first, we observe the gain in hit ratio by considering only single college, that is all 4 years, all braches and all sections, etc. Then we can easily scale up the gain for all affiliated engineering institutions of the university.

The number of sections that we assume for each branch in each year (around six branches are taken) of a single college are as follows (Table 1):

1. ECE-7, 2. CSE—5, 3. ME—3, 4. EEE—2, 5. CIV—1, 6. IT—5

Assuming that the faculty holding PhD may take two sections of a single subject, for seven sections of ECE, it requires four faculty (three Doctorates and one Assistant Professor) per subject. So, for six subjects 24 (6 × 4) faculty is required. Furthermore, assuming that faculty's one period covers four topics which is assumed to be spreaded over two unique pages (page size is at max 4 MB) from the syllabus of Unit 1 in a subject, it requires 8 (4 × 2) pages per subject. Hence, for all six subjects, it requires 48 (6 × 8) pages. So there are 48 references (page requests) from only ECE department. As there are two unique pages for subject and six subjects are assumed for each branch, we require 12 (6 × 2) unique pages for a branch and for six branches

Table 1 Time table III B.TECH CSE C SECTION

Day	I	II	III	IV		V	VI	VII
MON	OE-1		WT	TOC	LUNCH	PE-III	CN	MC
TUE	LR	TOC	CN	LUNCH	Computer Networks Lab			
WED	CN	TOC	PE-IV		LUNCH	MC		Coding Skill/NPTEL
THU	PE-III	TOC	PE-IV	LUNCH	Web Technologies Lab			
FRI	MC	WT	WT(T)	TOC(T)	LUNCH	LR		TECH SEM
SAT	PE-III	OE-I	WT	MC				

Table 2 Unit 1 Syllabus of python programming

Unit I	Introduction to Python: History, Features, Setting up path, working with Python Basic Syntax, Variable and Data Types, Operator. Conditional Statements (If, If-else, Nested if-else) Looping (for, While Nested loops) Control Statements (Break, Continue, Pass)

72 unique pages are required in a single year. So for all 4 years number of unique pages is around 288 per college.

Similarly for CSE, we have five sections, so total references will be $6 \times 3 \times 2 = 36$.

For ME, we have three sections, so total references will be $6 \times 2 \times 2 = 24$.

For IT, we have five sections, so total references will be $6 \times 3 \times 2 = 36$.

For EEE, we have three sections so total references will be $6 \times 2 \times 2 = 24$.

For CIV, we have one section so total references will be $6 \times 1 \times 2 = 12$.

So total references for all six branches will be total references $= 48 + 36 + 24 + 36 + 24 + 12 = 180$ references. Hence there are 180 page references per year which may not be unique. So for all 4 years there exists 720 (4×120) page references per college.

Furthermore, the case of faculty handling more than one subject is also covered implicitly in the above calculations (Table 2).

In the experimentation, we have taken 4 GB of RAM, and we have taken the percentage of frames used for paging versus percentage of other page references that are not browsed by faculty of the college.

We have taken 4 GB of RAM, that is 1024 frames as each frame size is of 4 MB.

We have implemented proposed algorithm with percentage of frames (10%, 20%, …., 100%) versus percentage of other references (10%, 20%, …, 100%).

The number of frames as per percentage of frames out of 1024 frames used is as follows:

nof is vector name where 0th element(103) of it is 10% of 1024 and 1st element(205) is 20% of 1024 frames and so on.

nof=[103 205 308 410 512 615 717 820 922 1024]

totreferences is the vector name where 0th element value is calculated as follows:

Let x be total number of references, and then 10% of it should be of other references.

Then

$$x = 720 + (10/100) * x$$
$$x(90/100) = 720$$
$$x = 720 * 100/90$$
$$x = 800$$

This 800 is the 0th element of totreferences vector.

Similarly, 1st, 2nd, 3rd position values can be calculated.

totreferences=[800 900 1028 1200 1440 1800 2400 3600 7200 8720];

Generating references:

The matlab code to generate references is as given below.

```
nof=[103 205 308 410 512 615 717 820 922 1024];
%No of unique pages is 288 for a college for all 4 years
pages=1:288;
%Generating references
refer=zeros(8720,1);
%No of references for one year is 180
%we have to generate references by branch wise and year wise
rhigh=0;
phigh=0;
pagefaultcount=zeros(10,10,100);
%The for loop iterates over 4 times for all 4 years of engineering course.
for i=drange(1:4)
%for CSE Branch 3 faculty 6 subjects
rlow=rhigh+1;
rhigh=rhigh+36;
plow=phigh+1;
phigh=phigh+12;
refer(rlow:rhigh)=repmat(pages(plow:phigh),1,3);
%for ECE branch 4 faculty 6 subjects
rlow=rhigh+1;
rhigh=rhigh+48;
plow=phigh+1;
phigh=phigh+12;
refer(rlow:rhigh)=repmat(pages(plow:phigh),1,4);
%for ME branch 2 faculty
rlow=rhigh+1;
rhigh=rhigh+24;
plow=phigh+1;
phigh=phigh+12;
refer(rlow:rhigh)=repmat(pages(plow:phigh),1,2);
%for IT branch 3 faculty
rlow=rhigh+1;
rhigh=rhigh+36;
plow=phigh+1;
phigh=phigh+12;
refer(rlow:rhigh)=repmat(pages(plow:phigh),1,3);
%for EEE branch 2 faculty
rlow=rhigh+1;
rhigh=rhigh+24;
plow=phigh+1;
phigh=phigh+12;
refer(rlow:rhigh)=repmat(pages(plow:phigh),1,2);
%for CE branch 1 faculty
```

```
44. rlow=rhigh+1;
45. rhigh=rhigh+12;
46. plow=phigh+1;
47. phigh=phigh+12;
48. refer(rlow:rhigh)=repmat(pages(plow:phigh),1,1);
49. end
50. %Declaring the Percentage of other references
51. others=zeros(10,1);
52. for i=drange(1:9)
53. others(i)=floor((720*10*i)/(100-10*i));
54. end
55. others(10)=8000;
56. %Generating referece string by including others
57. otherpages=289:8288;
```

The above instructions correspond to distribution of 288 unique pages over 720 (4 × 180, for 4 years of engineering course) references with repetition. The implementation of techniques FIFO, LRU, LFU and OPT without proposed idea proceeds in regular manner after 56th instruction in above code which are not covered here. We focus only on proposed versions of techniques. The main idea in proposed versions of four algorithms is as follows.

Strategy:

(i) The number of frames that correspond to prefetching of estimated number of pages are reserved (which is calculated as given below) inside the cache(RAM), and other frames are used for other references or prefetched references depending on the case whether number of frames is shorter than prefetched number of pages. Furthermore, we load the set of pages that are going to be referenced into the cache memory at every afternoon (12 PM).This is due to the reason that preparation of faculty for next day happens in between afternoons of 2 consecutive days as shown in Fig. 1.

```
noframesothers=floor((10*itero/100)*nof(iterf));
noframesreserved=nof(iterf)-noframesothers;
```

(ii) Furthermore, if number of frames reserved is less than number of unique pages, we bring only random number of unique pages into reserved frames. Otherwise, number of frames reserved will become fixed as in fourth line in the below code which is 288 and remaining frames are used for other pages which will be used

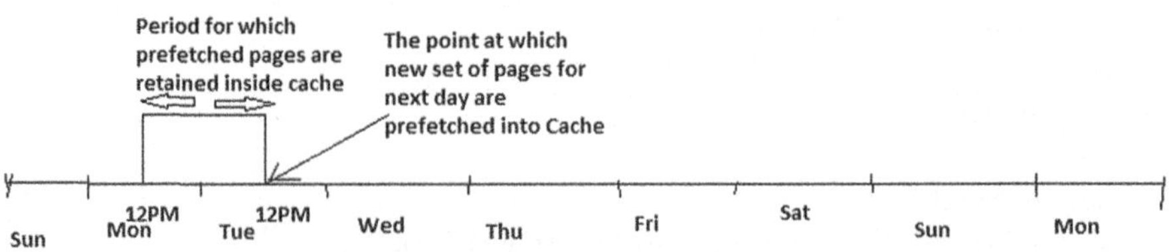

Fig. 1 The loading and clearing of cache during days of week

for replacement. But we never replace the reserved frames if any page fault occurs because they are reserved and we use other frames for replacement if that happens so. And we generate total references string, that is referred as in eight line given below and we generate random permutations of these references.

```
1. if(noframesreserved>=1 && noframesreserved<=288)
2. frames(1:noframesreserved)=randperm(288,noframesreserved);
3. elseif(noframesreserved~=0)
4. noframesreserved=288;
5. frames(1:noframesreserved)=randperm(288,noframesreserved);
6. noframesothers=nof(iterf)-noframesreserved;
7. end
8. refer(721:(720+others(itero)))=otherpages(1:others(itero));
9. lengthconsider=720+others(itero);
10. refernew=refer(randperm(lengthconsider));
```

3.1 Proposed FIFO/LRU/LFU/OPT Algorithm

1. Initialize nof vector.
2. Initialize totreferences array
3. Initialize array of references as shown in Generating references section.
4. Generate other pages references as given in 56th line above.
5. for each percentage number of frames
6. for each percentage number of other references
7. Calculate the no of frames reserved and other frames as given Strategy (ii).
8. Apply regular FIFO/LRU/LFU/OPT technique considering only other frames for replacement.
9. Count the total number of page faults in this case.
10. end for
11. end for
12. Print the number of page faults in each case and calculate Hit ratios.

4 Experimental Results and Analysis

This technique has been implemented in MATLAB (R2016b). Each technique FIFO/LRU/LFU/OPT is applied on 100 different permutations of input reference string and average of all page faults on 100 runs of input reference strings is calculated and from these values hit ratios are calculated. The results of FIFO and FIFO-PROPOSED versions are not given in this paper due to unimportance and same is the case with OPT and OPTPROPOSED. But graphs project the results of all eight techniques with and without the proposed techniques. The results indicate that there is double the performance of the technique without the proposed concept.

Results of LRU
See Tables 3, 4.

Results of LRU Proposed
See Tables 5, 6.

Results of LFU
See Tables 7, 8.

Results of LFU PROPOSED
See Tables 9, 10.

Table 3 LRU results

	10%	20%	30%	40%	50%
10%	0.536	0.476	0.417	0.357	0.297
20%	0.538	0.478	0.418	0.358	0.298
30%	0.539	0.479	0.419	0.358	0.299
40%	0.54	0.479	0.419	0.359	0.299
50%	0.54	0.48	0.419	0.359	0.299
60%	0.54	0.48	0.42	0.359	0.299
70%	0.54	0.48	0.42	0.36	0.299
80%	0.54	0.48	0.42	0.36	0.299
90%	0.54	0.48	0.42	0.36	0.3
100%	0.54	0.48	0.42	0.36	0.3

Table 4 LRU results

	60%	70%	80%	90%	100%
10%	0.23	0.151	0.078	0.023	0.016
20%	0.231	0.153	0.078	0.023	0.016
30%	0.231	0.152	0.079	0.023	0.016
40%	0.231	0.152	0.078	0.023	0.016
50%	0.231	0.153	0.078	0.023	0.016
60%	0.231	0.153	0.079	0.023	0.016
70%	0.232	0.153	0.079	0.023	0.016
80%	0.232	0.153	0.079	0.023	0.016
90%	0.232	0.153	0.079	0.023	0.016
100%	0.232	0.153	0.079	0.023	0.016

Table 5 LRU PROPOSED results

	10%	20%	30%	40%	50%
10%	0.309	0.26	0.211	0.161	0.119
20%	0.613	0.506	0.407	0.312	0.226
30%	0.881	0.723	0.577	0.443	0.323
40%	0.9	0.8	0.698	0.546	0.403
50%	0.9	0.8	0.7	0.6	0.467
60%	0.9	0.8	0.7	0.6	0.5
70%	0.9	0.8	0.7	0.6	0.5
80%	0.9	0.8	0.7	0.6	0.5
90%	0.9	0.8	0.7	0.6	0.5
100%	0.9	0.8	0.7	0.6	0.5

Table 6 LRU PROPOSED results

	60%	70%	80%	90%	100%
10%	0.08	0.048	0.022	0.006	0.002
20%	0.153	0.092	0.042	0.011	0.004
30%	0.218	0.131	0.063	0.017	0.005
40%	0.275	0.167	0.08	0.022	0.007
50%	0.324	0.199	0.097	0.027	0.008
60%	0.365	0.228	0.113	0.032	0.01
70%	0.399	0.253	0.127	0.037	0.012
80%	0.4	0.274	0.14	0.041	0.013
90%	0.4	0.294	0.153	0.045	0.014
100%	0.4	0.3	0.163	0.049	0.016

Table 7 LFU results

	10%	20%	30%	40%	50%
10%	0.194	0.161	0.128	0.1	0.074
20%	0.383	0.311	0.247	0.189	0.14
30%	0.515	0.421	0.338	0.261	0.192
40%	0.54	0.472	0.388	0.308	0.231
50%	0.54	0.48	0.413	0.335	0.258
60%	0.54	0.48	0.42	0.351	0.275
70%	0.54	0.48	0.42	0.358	0.288
80%	0.54	0.48	0.42	0.36	0.295
90%	0.54	0.48	0.42	0.36	0.299
100%	0.54	0.48	0.42	0.36	0.3

Table 8 LFU results

	60%	70%	80%	90%	100%
10%	0.049	0.03	0.014	0.004	0.003
20%	0.095	0.056	0.027	0.008	0.005
30%	0.132	0.08	0.039	0.011	0.008
40%	0.161	0.1	0.049	0.014	0.01
50%	0.183	0.117	0.059	0.017	0.012
60%	0.2	0.13	0.068	0.021	0.014
70%	0.213	0.141	0.075	0.023	0.016
80%	0.221	0.149	0.081	0.026	0.018
90%	0.228	0.156	0.086	0.028	0.02
100%	0.233	0.161	0.091	0.03	0.021

Table 9 LFU PROPOSED results

	10%	20%	30%	40%	50%
10%	0.309	0.261	0.214	0.167	0.124
20%	0.61	0.509	0.412	0.32	0.238
30%	0.883	0.729	0.587	0.453	0.334
40%	0.9	0.8	0.698	0.551	0.413
50%	0.9	0.8	0.7	0.6	0.47
60%	0.9	0.8	0.7	0.6	0.5
70%	0.9	0.8	0.7	0.6	0.5
80%	0.9	0.8	0.7	0.6	0.5
90%	0.9	0.8	0.7	0.6	0.5
100%	0.9	0.8	0.7	0.6	0.5

Table 10 LFU PROPOSED results

	60%	70%	80%	90%	100%
10%	0.086	0.051	0.026	0.007	0.003
20%	0.163	0.099	0.048	0.013	0.005
30%	0.232	0.141	0.07	0.02	0.008
40%	0.287	0.178	0.089	0.026	0.01
50%	0.332	0.209	0.106	0.032	0.012
60%	0.369	0.236	0.121	0.036	0.014
70%	0.399	0.258	0.135	0.042	0.017
80%	0.4	0.277	0.146	0.046	0.018
90%	0.4	0.294	0.158	0.05	0.02
100%	0.4	0.3	0.168	0.054	0.022

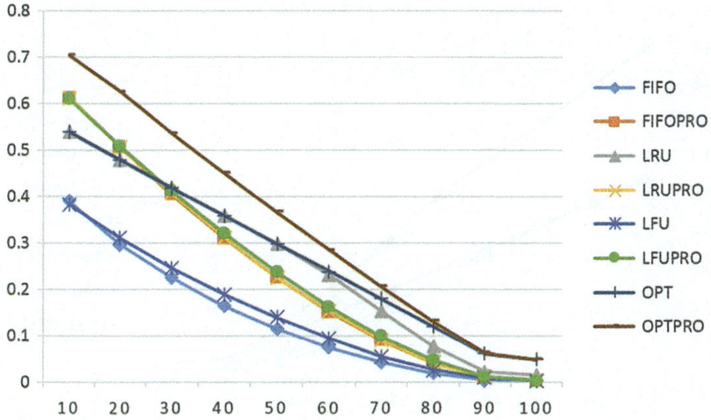

Fig. 2 The graphs of 20% percentages of frames

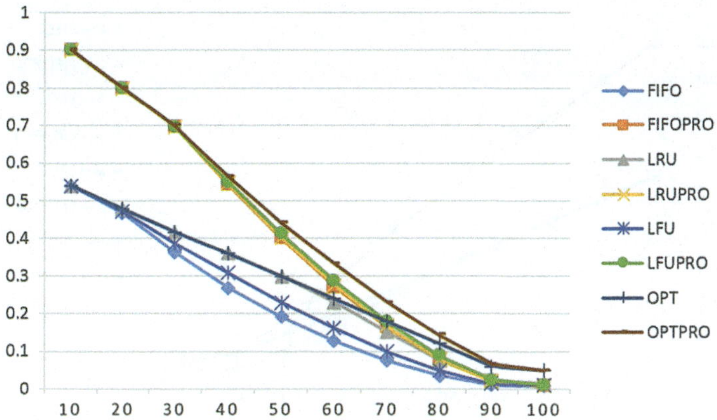

Fig. 3 The graphs of 40% percentages of frames

The graphs of all techniques with respect to 20, 40, 60, 80 percentage of frames for all percentages of other references are as given below (Figs. 2, 3, 4, 5).

5 Conclusion

This paper projects identification of faculty browsing patterns in affiliated engineering colleges. The results suggest a substantial improvement in hit ratio of websites like Wikipedia and tutorialspoint. This technique is not only applicable to affiliated engineering colleges but also to the autonomous engineering institutions where a significant overlap of subjects in semesters will derive substantial improvement in hit ratio. This is will be a future enhancement of this paper.

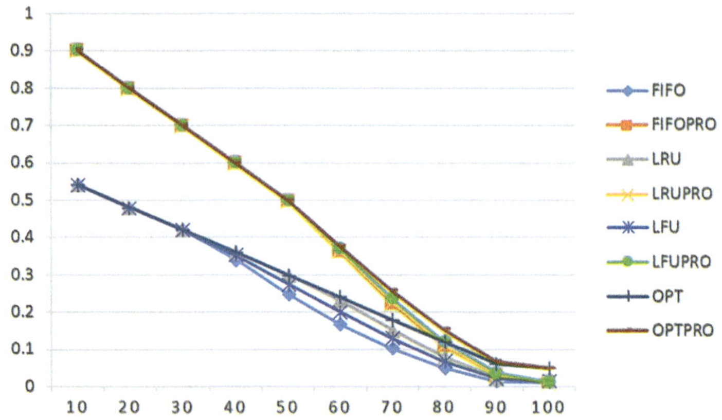

Fig. 4 The graphs of 60% percentages of frames

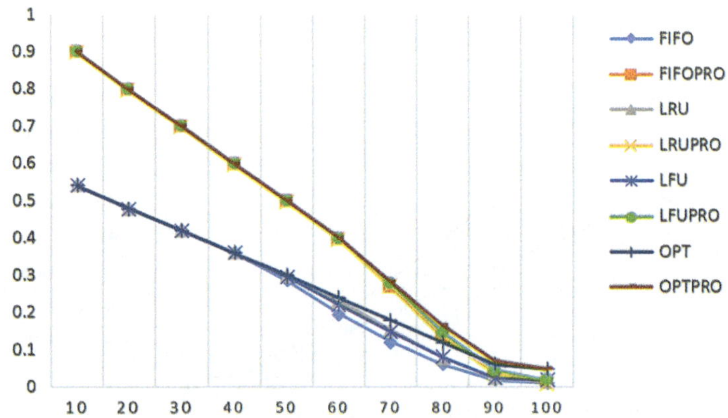

Fig. 5 The graphs of 80% percentages of frames

References

1. T. Ma, Y. Hao, Y. Tian, M. Al-Rodhaan, An improved web cache replacement algorithm based on weighting and cost, IEEE Access, vol 6, 2018
2. W. Ali, S. Mariyam Shamsuddin, A. Samad Ismail, A survey of web caching and Prefetching. Int. J. Advance. Soft Comput. Appl. **3**(1), March 2011 ISSN 2074–8523; Copyright © ICSRS Publication, 2011
3. D. Singh, S. Kumar, S. Kapoor, An explore view of Web caching techniques. Int. J. Adv. Eng. Sci. **1**(3), 38–43 (2011)
4. S.M. Shamsuddin, W.A. Ahmed, Integration of least recently used algorithm and neuro-fuzzy system into client-side Web caching. Int. J. Comput. Sci. Secur. **3**(1), 1–15 (2009)
5. W.-G. Teng, C.-Y. Chang, M.-S. Chen, Integrating Web caching and Web prefetching in client-side proxies. IEEE Trans. Parallel Distrib. Syst. **16**(5), 444–455 (May 2005)
6. K. Kim, D. Park, Reducing outgoing traffic of proxy cache by using client-cluster. J. Commun. Netw. **8**(8), 330–338 (2006)

7. X. Wu, H. Xu, X. Zhu, W. Li, *"Web cache replacement strategy based on reference degree',' in Proc* (Chengdu, China, Dec, IEEE Int. Conf. Smart City/SocialCom/SostainCom, 2015), pp. 209–212
8. S. Hiranpongsin, P. Bhattarakosol, Integration of recommender system for Web cache management. Maejo. Int. J. Sci. Technol. **7**(2), 232–247 (2013)
9. T. Ma et al., KDVEM: A k-degree anonymity with vertex and edge modificationalgorithm. Computing **70**(6), 1336–1344 (2015)
10. S. Jin, A. Bestavros, Popularity-aware greedy dual-size Web proxy caching algorithms. Proc. Int. Conf. Distrib. Comput. Syst. 254–261 (2000) https://doi.org/10.1109/icdcs.2000.840936
11. A.S.V. Praneel, T. Srinivasa Rao, M. Ramakrishna Murty, A survey on Accelerating the classifier Training Using various boosting schemes within casecades of bossted ensembles. Int. Conf. Spring. SIST Ser. **169**, 809–825 (2019)

Transient Response Improvement of Separately Excited D.C Motor Using Fuzzy Logic Controller

Venu Yarlagadda, V. Ramesh Babu, K. Veeresham, Kasala Rajesh, and S. N. V. Karthik Kumar

Abstract The fuzzy logic control draws the attention of Researchers to improve the system performance, when the control strategy involves ambiguity. Fuzzy means, things are not clear; hence the fuzzy logic control determines the proper control actions for non-deterministic systems. The article is aimed to determine the small signal model of a d.c shunt motor using laboratory based testing, and to design a PI controller for improving the system performance using Ziegler-Nichols tuning method. The performance of the machine can still need some improvement, which can be attained using fuzzy logic controller. The dynamic and steady state performance can be enhanced even for load disturbance. The fuzzy controller is designed with two input membership functions and one output function of triangle shape with twenty five fuzzy rules. The case study have been performed with simulation results. The machine is tested with two major cases, one is comparative analysis of ZN tuned PI controller with FLC in two cases one without load disturbance and the other is with load disturbance. The performance of the machine with fuzzy logic controller is robust and much superior to other method for both the cases. The Obtained simulation results are as follows −100% steady state error for the both without & with controller by consideration of without and with one percent load disturbance, while ZN tuned PI gives the settling time of 12 s with 3% steady state error and 12 s with 0.5% steady state error for without and with one percent load disturbance respectively FLC gives

V. Yarlagadda · V. Ramesh Babu · K. Veeresham · K. Rajesh (✉) · S. N. V. Karthik Kumar
Department of EEE, VNRVJIET, Hyderabad, India
e-mail: kasalarajesh566@gmail.com

V. Yarlagadda
e-mail: venu_y@vnrvjiet.in

V. Ramesh Babu
e-mail: rameshbabu_v@vnrvjiet.in

K. Veeresham
e-mail: veeresham_k@vnrvjiet.in

S. N. V. Karthik Kumar
e-mail: karthik.kumarsnv3297@gmail.com

C. Kiran Mai et al. (eds.), *Proceedings of International Conference on Advances in Computer Engineering and Communication Systems*, Learning and Analytics in Intelligent Systems 20,
https://doi.org/10.1007/978-981-15-9293-5_16

settling time of 0.82 s with zero steady state error and 0.9 s with zero steady state error for without and with one percent load disturbance respectively.

Keywords ZN tuned PID controller · Ziegler nichols tuning · Fuzzy logic controller · Dynamic performance of D.C motor · Root locus based design · FLC controlled D.C motor control

1 Introduction

In applications like paper mills, rolling mills, hoists, traction, printing press, excavators and mine winders etc. [1]. DC shunt motors are widely used, due to the advantages they offer like adjustable speed characteristics with good speed regulation and frequent starting, reversing and braking operations. Though there is improvement in technology of AC drives, they couldn't replace DC drives because of its low cost, simple control and reliability.

There are many methods available in literature to control the speed and position of the d.c shunt motor. The proportional-integral derivative controller has been widely used in the process industries. The output of a Proportional Integral and Derivative controller relies on the error between desired value and actual value [2]. So by using PID controller error can be corrected to control the position or speed of the motor to the desired point or speed. The tuning of PI and PID controllers can be obtained using ziegler-Nichols tuning method [3].

The working of PID controller involves computation of three separate parameters namely: Proportional, Derivative and ISntegral constants. The impact of proportional, derivative and integral controller on time domain specifications of the response are given in Table 1. Since most of the time the PID controllers are poorly tuned for the systems, for proper tuning a higher degree of technology and expertise is required,the transfer function PID controller [4] is given as:

$$G_c(s) = P + I + D = K_p + \frac{K_i}{s} + K_d s \tag{1}$$

Speed control of DC motor based on PID/FOPID controller is explained and recently, the inspiration is taken from nature to develop a host of nature-inspired algorithms namely genetic algorithms from evolution artificial neural networks, Fuzzy Logic Control, Ant Colony Optimization, Artificial Bee Colony optimization and particle swarm optimization [5–8]. Out of these techniques, stochastic search,

Table 1 PI controller values obtained by ZN Tuning

S.No	Type of controller	K_p	K_i	K_d
1	PI	0.1850805	0.15971223	–

Fuzzy Logic Control, particle swarm optimization and ant bee colony optimization techniques are widely used for optimal tuning of the parameters [9].

In this work, a good performance and robust speed control system for a DC motor are attained for all load scenarios using Fuzzy Logic Controller. The paper is organized and explained as follows. Section II presents small signal modelling of DC shunt motor drive. Section III deals with obtaining the small signal model parameters from the laboratory based testing and Section IV explains the design of PI controller using Ziegler-Nichols tuning method. Section V deals with design of fuzzy logic controller and the system still need some improvement, hence the fuzzy logic controller has been designed along with the root locus based designed PI controller at output terminals of the fuzzy logic controller and Section VI analyzes the simulation results with two cases, one is without any load, the machine is tested with ZN tuned PI controller and fuzzy controller, and load disturbance with both the controllers.

2 Small Signal Model of DC Shunt Motor

The basic block diagram of armature controlled DC shunt motor model with field circuit operation as in case of separately excited machine as

The following two equations:

$$\omega(s)/Va(s) = R_a I_a(s) + sL_a I_a(s) + K_b\omega(s) \tag{2}$$

$$K_t I_a(s) = sJ_m\omega(s) + B_m\omega(s) \tag{3}$$

Case (i) When $\mathbf{V_a = 0}$ and load disturbance alone is considered then the transfer function is

$$\frac{\omega 1(s)}{T_L(s)} = \frac{-(L_a s + R_a)}{L_a J s^2 + (RaJ_m + L_a B_m)s + (R_a B_m + K_b K_t)} \tag{4}$$

Case (ii) When $T_L = 0$ and armature voltage alone is considered then the transfer function is

$$\frac{\omega 2(s)}{Va(s)} = \frac{K_t}{L_a J s^2 + (RaJ_m + L_a B_m)s + (R_a B_m + K_b K_t)} \tag{5}$$

Case (iii) when both V_a and T_L are present in the circuit i.e. speed controls with load disturbances have been considered.

When $\mathbf{V_a = 1}$ and $T_L = 0.01, 0.05$ and 0.1 of three load disturbances from 1% to 10% load variations.

$$G(s) = \frac{\omega 1(s)}{T_L(s)} \frac{\omega 2(s)}{Va(s)} \tag{6}$$

3 Experimental Determination of Small Signal Model

The testing is performed on the dc shunt motor of ratings as 220 V, 5HP, 1500 rpm D.C Shunt Motor with 19A armature current and 0.8A Field current ratings as shown in Fig. 1. The retardation test is performed on the machine to obtain moment of inertia and friction of machine rotor. The armature resistance and impedance tests have been performed to compute armature resistance and inductance respectively. The transfer function model of armature controlled dc shunt motor have been obtained from these test results.

Retardation test results and computing the J and B values:

E_b = voltmeter reading = 206 V, I_a = Ammeter reading = 0.75A, t_1 = 51.3 s and t_2 = 23 s from the equations we can compute the J and B values as follows

$$P = \omega\left(\frac{Jd}{\omega}dt1 + B\omega\right) \tag{7}$$

$$P - W1 = \omega\left(\frac{Jd}{\omega}dt2 + B\omega\right) \tag{8}$$

By substituting the values, the computed rotor parameters as J = 0.002366 and B = 0.017

From the armature resistance and impedance tests, the computed values of armature resistance as 2 ohms and armature inductance of 0.0256 H calculated from the experimental test steup as shown in Fig. 2. The transfer function model of D.C motor determined by the laboratory based testing as given by Eq. (9).

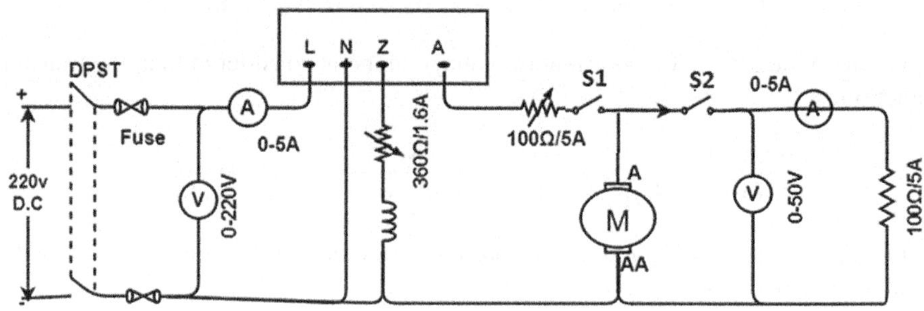

Fig. 1 Retardation test for finding J and B values

Fig. 2 Retardation test for finding J and B values

$$G(s)H(s) = \frac{0.124}{0.00512s^2 + 0.0547525s + 0.051076} \tag{9}$$

4 Ziegler Nichols Tuning of PI Controller

The Ziegler–Nichols tuning method is a heuristic method of tuning a PID controller [10–12]. It is used for the systems with first order type dynamic response with transport delay, the critical gain and time constant values can be computed from Routh Hurwith stability criterion. The proportional gain $K_{p,}$ the integral gain $K_{i,}$ and the derivative gain K_d are computed using critical gain and time constant values as illustrated in the below Table 1.

5 Fuzzy Logic Control

Input membership functions are taken as Triangular type which is varying from minimum value to a maximum value and mean is at its peak of the triangle as shown in Fig. 3, five input functions have been taken from −2.5 to +2.5 range as input 1 similarly input 2 and output variables, including fuzzy rule surface followed by rules shown in Table 2.

Strong rules has to be framed for better response of the system. It can be achived when the number of combinations between the inputs are considered more. Here we have considered two inputs as triangle membership functions,and number of rules can be formed as follows.

ENN, ENN1 and ENN2 are meant for the Extreme Negative Negative output, input1 and input2 respectively. MN, MN1 and MN2 are meant for the Medium

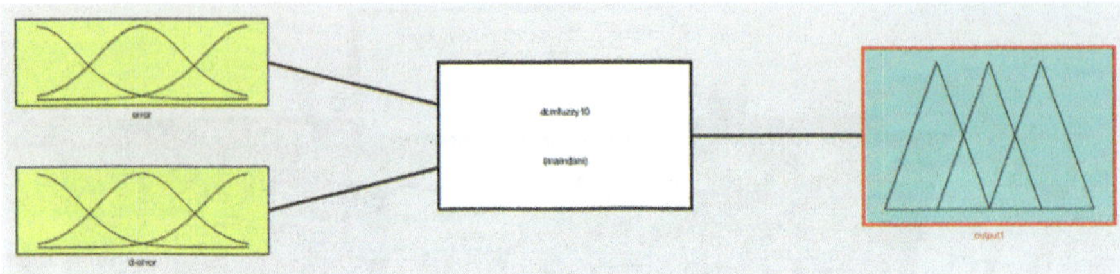

Fig. 3 Fuzzy controller two input and one output triangular membership association

Table 2 Membership function inputs and output relationship

de/dt e	ENN2	MN2	Z2	MP2	EPP2
ENN1	ENN	ENN	MN	Z	Z
MN1	ENN	MN	MN	Z	MP
Z1	MN	Z	Z	MP	MP
MP1	Z	Z	EPP	EPP	EPP
EPP1	Z	MP	MP	EPP	EPP

Negative output, input1 and input2 respectively. Z, Z1 and Z2 are meant for the zero output, input1 and input2 respectively. EPP, EPP1 and EPP2 are meant for the Extreme Positive Positive output, input1 and input2 respectively. MP, MP1 and MP2 are meant for the Medium Positive output, input1 and input2 respectively.

Fuzzy Rules are mentioned as below:

Rule1: If input1 is ENN1 and input2 is ENN2 then the output is ENN

Rule2: If input1 is ENN1 and input2 is MN2 then the output is ENN

Rule3: If input1 is ENN1 and input2 is Z2 then the output is MN

Rule4: If input1 is ENN1 and input2 is MP2 then the output is Z

And the remaining rules will be in similar manner

By using this rules we obtain the fuzzy rule surface output as shown in Fig. 4.

6 Case Study and Conclusions

The case study and result analysis have been presented in two major cases and major case is comprised of two sub cases of one is with out any load disturbance, and other is with load disturbance of one percent.

Case I: Comparitive analysis without any controller, with ZN tuned PI controller and Fuzzy Logic Controller with all cases as mentioned subsequently in case II.

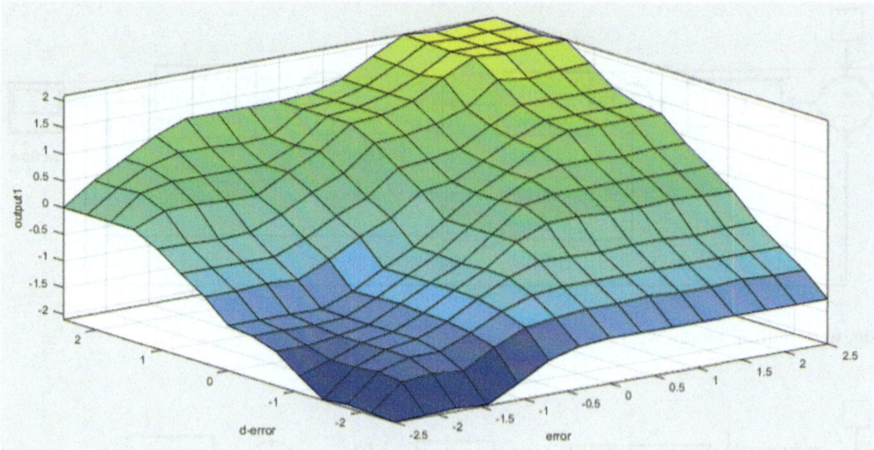

Fig. 4 Fuzzy Rule surface with variations from −2.5 to +2.5 in xyz planes

Case II: Comparitive analysis without any controller, with ZN tuned PI controller and Fuzzy Logic Controllers with two sub cases are:

Without any load disturbance

With load disturbance of one percent

Figures 5, 6 and 7 shows the simulink diagrams without any controller, with ZN tuned PI controllers and Fuzzy Logic Controller respectively.

Figures 8 and 9 shows the Dynamic responses for all three cases without Controller, with ZN tuned PI and Fuzzy Logic Controllers without any load disturbance, with one percent load disturbance.

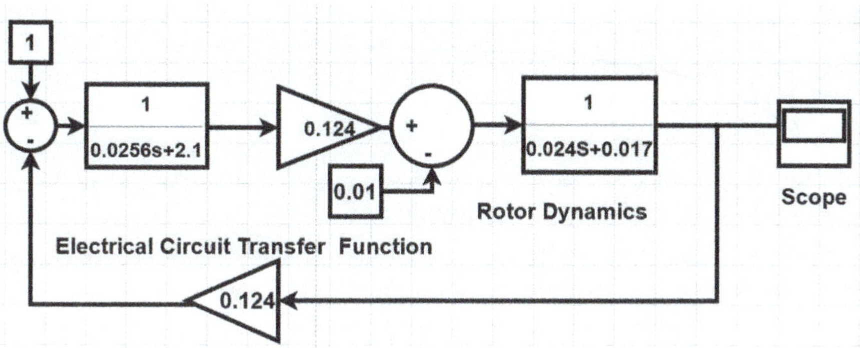

Fig. 5 Simulink model without any controller

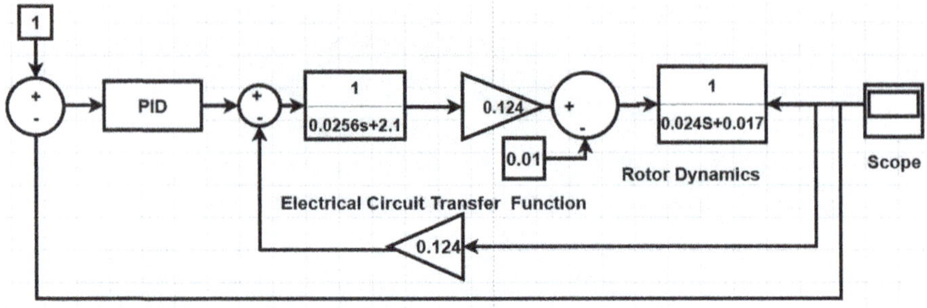

Fig. 6 Simulink model with ZN tuned controller

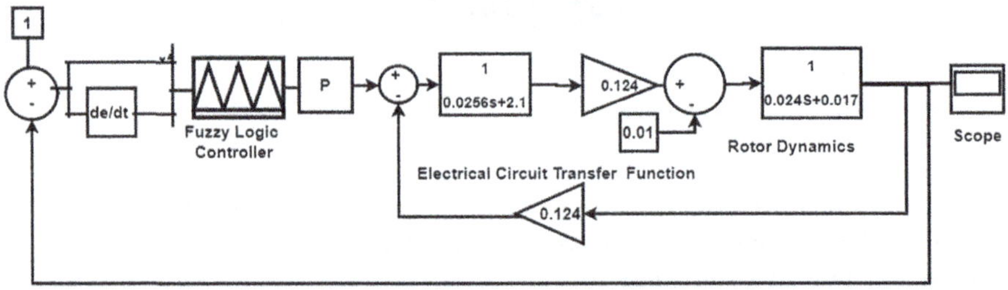

Fig. 7 Simulink model with fuzzy logic controller

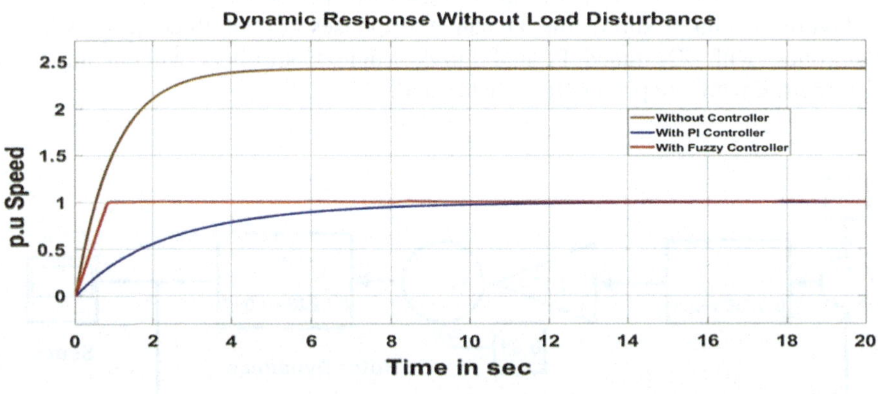

Fig. 8 Dynamic responses for all three cases without controller, with ZN tuned PI and fuzzy logic controllers without any load disturbance

Case I and II: Comparıtıve analysis without any controller, with ZN tuned PI controller and Fuzzy Logic Controller

Case I. Without any load disturbance

Simulation results without any load disturbance, the simulation results of uncontrolled machine is completely unstable without any external controller, the steady

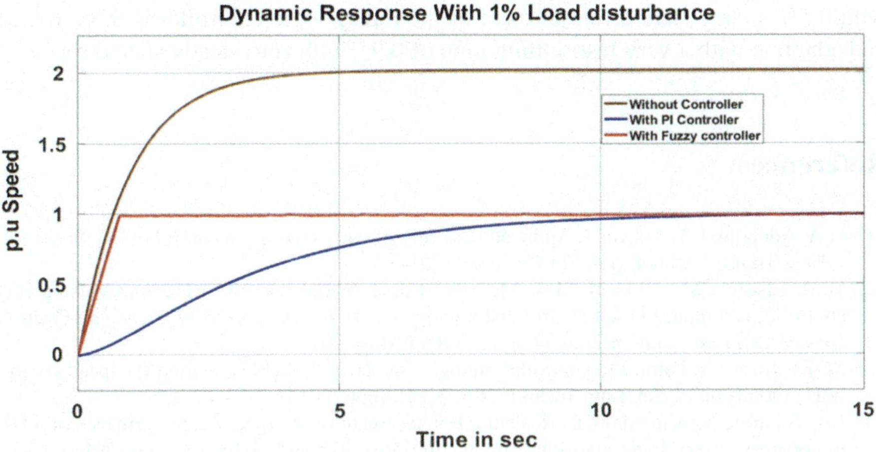

Fig. 9 Dynamic responses for all three cases without controller, with ZN tuned PI and fuzzy logic controllers with one percent load disturbance

state error is of -100%, Practically the machine is stabilized by including a four point starter with an additional resistance in series with the line. This starter acts as a proportional controller in the closed loop and the same system is tested with proportional controller. The simulation results of ZN tuned PI controller is also stable with settling time of 12 s with 1.3% steady state error, and that of the Fuzzy logic controller result shows that the controller is more robust and adaptive with a very fast settling time of 0.82 s with zero steady state error as shown in Fig. 8.

Case II. With load disturbance of one percent

Simulation results as shown in Fig. 9 with one percent load disturbance, which shows that the uncontrolled machine is completely unstable with the steady state error of −100%. The simulation results of ZN tuned PI controller is also stable with settling time of 12 s with 0.5% steady state error, and that of the Fuzzy logic controller result shows that the controller is more robust and adaptive with a very fast settling time of 0.9 s with zero error.

7 Conclusion

The simulation results reveal that the performance of Fuzzy Logic Controller is Adaptive and Robust Controller for the load disturbance. Fuzzy Logic Controller fed Machine is completely stable with very fast settling time and zero steady state error compared to Ziegler Nicholas tuned Proportional plus Integral Controller Case I result shows that Fuzzy logic controller is more robust and adaptive with a very fast settling time of 0.82 s with zero steady state error. Similarly the case II simulation results shows that the ZN tuned PI controller is also stable with settling time of 12 s with 0.5% steady state error, and that of the Fuzzy logic controller is more robust and adaptive with a very fast settling time of 0.9 s with zero steady state error.

References

1. G.A. Adepoju, I.A. Adeyemi, Application of fuzzy logic to the speed control of DC motor. Int. J. Eng. Trends Technol. (IJETT) **15**(5) Sep (2014)
2. W.M. Elsrogy, M.A. Fkirin, M.A. Moustafa Hassa, Speed control of DC motor using PID controller techniques based on artificial intelligence, in *International Conference on Control, Decision and Information Technologies*, CODIT May (2013)
3. M. Kushwah, A. Patra, Pıd controller tuning using Ziegler-Nichols method for speed control of Dc motor, Int. J. Sci. Eng. Technol. Res. **3**(13), June (2014)
4. I.A. Tohamy, A. Ahmed, D. El Koshairy, Enhancement of the dynamic performance of a DC motor using fuzzy logic algorithm. Int. J. Eng. Res. Technol. **8**(10) October (2019). (Ijert) 2278-0181 Ijertv8ıs100117 Published By: www.ijert.Org
5. D. Karaboga, B. Basturk, Artificial Bee Colony (ABC) optimization algorithm for solving constrained optimization problems IFSA 2007. LNAI **4529**, 789–798 (2007)
6. A. Rajasekhar, A. Abraham, *Fractional order speed control of DC motor using levy mutated artificial bee colony algorithm* (World Congress on Information and communication Technologies, IEEE, 2011), pp. 7–13
7. O. Ercin, R. Coban, Comparison of the artificial bee colony and the bees algorithm for PID controller tuning, in *Innovations in Intelligent Systems and Applications (INISTA) IEEE Conference*, pp. 595–598 (2011)
8. K. Sharma, D.K. Palwalia, A modified PID control with adaptive fuzzy controller applied to DC motor, in *IEEE International Conference on Information, Communication, Instrumentation and Control ICICIC*(2017)
9. M. Kushwah, A. Patra, Tuning PID controller for speed control of dc motor using soft computing techniques-a review. Adv. Electron. Electric Eng. **4**, 141–148 (2014)
10. K. Astrom, T, Hagglund, PID controllers: Theory, design, and tuning. 2nd Edition, Instrument Society of America (1995)
11. I.J. Nagrath, M. Gopal, Control systems engineering, New Age International Publishers Jan (2017)
12. P.S.Bhimbra, Electrical machinery. Khanna publications (1973)
13. K. Sundareswaran, V.T. Sreedevi, D.C. motor speed controller design through a colony of honey bees, TENCON (IEEE Region) conference, pp. 1–6 (2008)

Malware Detection Kit for Malware Analysis of Big Data

T. Sree Lakshmi, M. Govindarajan, and Asadi Sreenivasulu

Abstract Contemporaneously, Security is a major mission in technological world, irrespective of domains, fields and technologies. Like other areas, Big Data and IOT too experiencing security issues, threats and attacks in every single minute. These attacks can be on different components of Big Data and IOT, like data stored on various nodes, clusters, propagated through networks, and via various components of the system or sensors. Big data security deals with the measures, techniques and tools used to protect both the data and analytics methods from attacks, threats, or other unauthorized activities. In this paper, To apply Security measures to overcome vulnerability of infrastructure, proposed a method called Malware Detection Kit (MDK). MDK comprises of static and dynamic analysis. Malware detection kit identifies attacks and threats. To detect attacks analysis is performed in two iterations. Whenever a new data enters into system, data undergoes first iteration of MDK, by static analysis it identifies threats if any. In the second iteration, file is submitted to automated dynamic analysis tools. The tool reports whether the file is malware on benign. Static Analysis is done by using Random Forest classifier, it produced high accuracy and low error rate for Malgenome dataset. Dynamic analysis is done by using automated sandboxes. Finally, this method recognises whether the file is benign or malware.

Keywords Big data security · Malware · Vulnerability · Automated tools

T. Sree Lakshmi (✉)
Research Scholar, Department of CSE, Annamalai University, Chidambaram, Tamil Nadu, India
e-mail: lakshmi.nbvm@gmail.com

M. Govindarajan
Associate Professor, Department of CSE, Annamalai University, Chidambaram, Tamil Nadu, India

A. Sreenivasulu
Professor, Department of CSE, Sreevidyanikethan Engineering College, Tirupati, Andra Pradesh, India

C. Kiran Mai et al. (eds.), *Proceedings of International Conference on Advances in Computer Engineering and Communication Systems*, Learning and Analytics in Intelligent Systems 20,
https://doi.org/10.1007/978-981-15-9293-5_17

1 Introduction

Contemporaneously, Security is a major mission in technological world, irrespective of domains, fields and technologies. New Malwares are emerging within a short span of time. Like other areas, Big Data and IOT too experiencing security issues [1], Threats and attacks in every single minute. These attack can be on different components of Big Data and IOT [2], like data stored on various nodes, clusters, propagated through networks, and via various components of the system or sensors. Big data security deals with the measures, techniques and tools used to protect both the data and analytical methods from attacks, threats, or other unauthorized activities. Security measures are required as the data integrity [3] is compromised by malicious activities. Malicious activities are generated actively and passively by online and offline tasks. For data storage and processing, Big data utilizes distributed file system, popularly known as Hadoop distributed file system (HDFS) [4]. HDFS stores voluminous of data and processes multiple variants of data which are combined and stored in distributed data pool. Hence we can say Big data is suitable for voluminous storage of data, parallel speed of storage and processing operations is high. That is, data access rate is very high speed. Concept of big data became target for security attackers, started developing malicious programs. Goal of attackers [5, 6] is to make the system to work abnormally. To fight against threats, Security analysts developed various strategies like keeping out unauthorized users and intrusions with firewalls [7], making user authentication strong, giving training to end-user training, and many others. However, these strategies are suitable and good enough for conventional file systems. Today every organization is enjoying the potential of Big data. Conventional strategies [8, 9] are not sufficient to meet the technological advancements. Hence to secure data, distributed file systems requires the development of much more novel strategies.

This paper overcomes the above drawbacks by introducing the static and dynamic malware techniques combinedly in the system for improving security analytics in big data applications. Section 2 describes the malware and its analyses, Sect. 3 describes the existing malware techniques, Sect. 4 explains the proposed methodology and Sect. 5 explains the experiment results with dataset description.

2 Malware and Its Analysis

Malware is a computer program which is popularly known as malicious software [9]. It enters into the system and corrupts the files, hardware and so on. It creates a severe damage to the potentiality of business organisations. Malware is any program which has destructive functionality. Various malware are viruses, worms, rootkit, ransomware, trojan horse, spyware [10] and so on.

Malware analysis is required to assess the damage, detect vulnerability, and discover levels of compromise. From the perspective of big data in industries, it is needed to know does it spread, the competitors of organisation, who is stealing data or making the system to malfunction, how to prevent it so as to enable smooth-running of the system. Malware is analysed by considering network indicators, host based indicators [11, 12], date of compilation, date of installation and whether code is packed or unpacked and many more. Malware analysis is of two types

1. Static analysis.
2. Dynamic analysis.

2.1 *Static Analysis*

Static analysis, analyses the binary code of malware without actually running the code. It is a signature based approach. A regular pattern which is known as signature is stored in the database. If the data is matched with signature then it would be identified as malware, otherwise it is benign data. It performs dissection only [13]. However this approach is ineffective to big data applications as it is time consuming and hard. A single signature needs to be stored as multiple patterns so as to withstand multiple vulnerabilities. This approach [14] involves virus scanning, file obfuscations.

2.2 *Dynamic Analysis*

Dynamic Analysis, analyses malicious program in virtual environment. It involves API calls, registry changes, memory traces, process traces and many more. It is a behaviour based approach.

3 Existing Malware Detection Techniques

Malware detection techniques provides a platform to identify the occurrence of malicious code, which is popularly known as malwares. Various existing methods are explained as follows.

3.1 *Static Techniques*

It is suitable to detect only known attacks. Threats are detected either by patterns or rules. In signature based, signature is stored in database. It is compared with the data entering into the system through various means. Limitation of this approach is multiple signatures need to be stored for single vulnerability. Another approach in static analysis is rule based, which detects malware based on file types, weight based or a generic signature based. Static analysis is effective for known threats.

3.2 Dynamic Techniques

To deal with unknown threats, dynamic analysis techniques are preferable. The union of static and dynamic analysis is hybrid analysis. Dynamic analysis is carried out by sandbox, virtualization [5], and file type and built in tools. It is conducted by identifying and handling malware [6].

4 Proposed Methodology

In this paper, a method is proposed to detect malware by combining the basic analysis and automated tools. Proposed method is done in two steps, first step is to find known attacks and Second step is to find unknown attacks. As a primary step, in the first iteration, the file is examined for abnormalities, identify any embedded code, Extract malicious code from the file, if required De-obfuscate, debug malicious code. In the second iteration, dynamic analysis is performed by using available automated tool. The output generated by tool recognises whether the file entered to the environment is Benign or malware.

The flowchart in Fig. 1 describes about proposed methodology.

Whenever new file enters into HDFS, File manager [7] performs static analysis of the new file to detect known attacks. Static analysis is done to identify abnormalities, embedded code if any, or malicious code. In the next step dynamic analysis is carried out to identify unknown threats. It is done by using automated tools. Output generated by the tools is the status of file.

4.1 Static Analysis by Random Forest Algorithm

Static analysis is carried out by Random forest algorithm, As it is a strong classifier and performance is good. It performs row and column sampling with decision tree as base. During the classification number of base learners could be found using cross validation. Base learners (k) is inversely proportional to variance. Random forest classification is the sum of Decision tree (DT), row sampling with replacement (b), column sampling (fb),aggregation (agg). It is given as Eq. 1

$$\text{Random forest} = DT + b + fb + agg \tag{1}$$

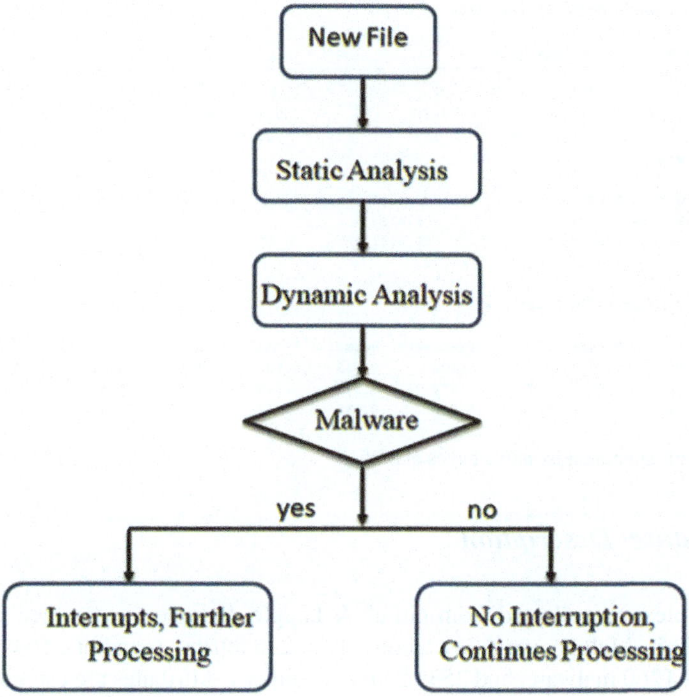

Fig. 1 Proposed methodology

4.2 Dynamic Analysis by Automated Tool

In the second phase, file is analysed by using automated tools to detect unknown threats. Some of the available tools are Cuckoo sandbox and Falcon sandbox.

Cuckoo Sandbox is the outstanding open source automated malware analysis approach. To get a detailed status of a file or event, user can upload any suspicious or benign file to it. File is analysed by executing it in a realistic isolated environment, once execution is completed, Cuckoo will generate a detailed analysis report outlining the behaviour of the file.

Falcon Sandbox performs analysis of deceptive, unknown threats. It generates actionable IOC (indicators of compromise), enabling security to better understand sophisticated vulnerabilities.

5 Experiment Result

In this section, description of Data set is given for comparison with five classifiers which is explained in the forthcoming descriptions.

```
Time taken to build model: 0.8 seconds

=== Stratified cross-validation ===
=== Summary ===

Correctly Classified Instances        3641              95.841 %
Incorrectly Classified Instances       158               4.159 %
Kappa statistic                          0.9069
Mean absolute error                      0.0419
Root mean squared error                  0.1852
Relative absolute error                  9.4596 %
Root relative squared error             39.3381 %
Total Number of Instances             3799

=== Detailed Accuracy By Class ===

                 TP Rate  FP Rate  Precision  Recall   F-Measure  MCC      ROC Area  PRC Area  Class
                 0.952    0.039    0.924      0.952    0.938      0.907    0.994     0.990     S
                 0.961    0.048    0.976      0.961    0.969      0.907    0.994     0.997     B
Weighted Avg.    0.958    0.045    0.959      0.958    0.959      0.907    0.994     0.994
```

Fig. 2 Result generated by naïve bayes classifier

5.1 Dataset Description

Data samples are collected from github & kaggle. Samples are merged from both websites [15]. Malgenome dataset, consists of 215 attributes and one class attribute. It consists 1260 malware, and 2539 benign instances. Attributes are categorised into API call signature, manifest permission, command signature, intent [8] and more. All attributes are Boolean. Class attribute gives output as Malware or benign.

5.2 Results and Discussion

Statistical analysis is done on Malgenome dataset, which consists of 215 attributes. It consists 1260 malware, and 2539 benign instances. Attributes are categorised into API call signature, Manifest permission, command signature, intent. Static Analysis is performed using naive Bayesian, Bayesnet, J48, Random tree, Random forest classifiers. Test mode is 10-fold cross validation. In 10-fold cross validation, the initial data are randomly partitioned into 10 mutually exclusive subsets, each of approximately equal size. Confusion matrix is a tool used for analysing how well classifier can recognize tuples of different classes. Compared with above mentioned classifiers, random forest is proven to be better in various aspects. Random tree performance is good because it performs row and column sampling with decision tree as base.

Dynamic analysis is performed by available automated tools. Cuckoo sandbox [15] and Falcon sandbox [16] is used for analysis. Report generated comprises of Name of the file, Analysis state, Threat Verdict, Score, AV Family name, Time of Analysis, File size and various other properties.

Table 1 Confusion matrix of naïve bayes classifier

	Predicted		
	Class	Malware	Benign
Actual	Malware	1200	60
	Benign	98	2441

```
=== Stratified cross-validation ===
=== Summary ===

Correctly Classified Instances         3694               97.2361 %
Incorrectly Classified Instances        105                2.7639 %
Kappa statistic                           0.9378
Mean absolute error                       0.0276
Root mean squared error                   0.1662
Relative absolute error                   6.2339 %
Root relative squared error              35.3112 %
Total Number of Instances              3799

=== Detailed Accuracy By Class ===

                 TP Rate  FP Rate  Precision  Recall   F-Measure  MCC      ROC Area  PRC Area  Class
                 0.962    0.022    0.955      0.962    0.958      0.938    0.970     0.931     S
                 0.978    0.038    0.981      0.978    0.979      0.938    0.970     0.974     B
Weighted Avg.    0.972    0.033    0.972      0.972    0.972      0.938    0.970     0.960
```

Fig. 3 Result generated by random tree classifier

Table 2 Confusion matrix of Random Tree classifier

	Predicted		
	Class	Malware	Benign
Actual	Malware	1236	24
	Benign	2	2537

Results generated by various classifier in static analysis phase is as follows Fig. 2 shows the output and Table 1 shows confusion matrix generated by naive Bayesian classifier. It classifies 95.841% instances as correctly. Run information generated by Random tree classifier is shown in Fig. 3 and Table 2. It classifies 97.05% instances as correctly. Run information generated by Bayes net classifier is shown in Fig. 4 and Table 3. It classifies 92.75% instances as correctly. Run information generated by J48 is shown in Fig. 5 and Table 4. It classifies 98.25% instances as correctly.

Confusion matrix of the above mentioned classifiers are shown. From the confusion matrix, it can be observed that accuracy is improved and error rate is minimised in random forest classifier. Various Measures are selected for assessing how "accurate "classifiers are at predicting the class label of tuples. The accuracy of each classifier is computed by using True Positive, True Negative, Positive and [10] Negative values. The equation to compute accuracy is

$$\text{Accuracy} = \frac{TP + TN}{P + N} \tag{2}$$

```
Time taken to build model: 1.97 seconds

=== Stratified cross-validation ===
=== Summary ===

Correctly Classified Instances        3523               92.7349 %
Incorrectly Classified Instances       276                7.2651 %
Kappa statistic                          0.8432
Mean absolute error                      0.0775
Root mean squared error                  0.2448
Relative absolute error                 17.4725 %
Root relative squared error             51.9874 %
Total Number of Instances             3799

=== Detailed Accuracy By Class ===

                 TP Rate  FP Rate  Precision  Recall   F-Measure  MCC      ROC Area  PRC Area  Class
                 0.980    0.099    0.831      0.980    0.899      0.850    0.992     0.987     S
                 0.901    0.020    0.989      0.901    0.943      0.850    0.992     0.996     B
Weighted Avg.    0.927    0.046    0.937      0.927    0.929      0.850    0.992     0.993
```

Fig. 4 Result generated by Bayesnet classifier

Table 3 Confusion matrix of Bayesnet classifier

	Predicted		
	Class	Malware	Benign
Actual	Malware	1235	25
	Benign	251	2288

```
=== Stratified cross-validation ===
=== Summary ===

Correctly Classified Instances        3731               98.2101 %
Incorrectly Classified Instances        68                1.7899 %
Kappa statistic                          0.9596
Mean absolute error                      0.0222
Root mean squared error                  0.1317
Relative absolute error                  5.0124 %
Root relative squared error             27.9784 %
Total Number of Instances             3799

=== Detailed Accuracy By Class ===

                 TP Rate  FP Rate  Precision  Recall   F-Measure  MCC      ROC Area  PRC Area  Class
                 0.974    0.014    0.972      0.974    0.973      0.960    0.983     0.967     S
                 0.986    0.026    0.987      0.986    0.987      0.960    0.983     0.986     B
Weighted Avg.    0.982    0.022    0.982      0.982    0.982      0.960    0.983     0.979
```

Fig. 5 Result generated by J48 tree classifier

Table 4 Confusion matrix of J48 classifier

	Predicted		
	Class	Malware	Benign
Actual	Malware	1235	25
	Benign	251	2288

Table 5 Confusion matrix of random forest classifier

	Predicted		
	Class	Malware	Benign
Actual	Malware	1236	24
	Benign	2	2537

Table 6 Various classifiers accuracy

S.No	Classifier	TPR (%)	FPR (%)	Precession (%)	F-Measure (%)	Accuracy (%)
1.	Naïve Bayesian	0.958	0.045	0.959	0.959	0.958
2.	BayesNet	0.927	0.046	0.937	0.929	0.927
3.	J48	0.982	0.022	0.982	0.982	0.982
4.	Random tree	0.972	0.033	0.972	0.972	0.9723
5.	Random forest	0.993	0.013	0.993	0.993	0.993

True positive rate is another metric, it is also referred as sensitivity. It is the proportion of positive tuples that are correctly identified. It is given as Table 5.

$$TPR = \frac{TP}{P} \tag{3}$$

True Negative rate is another metric, it is also referred as Specificity [11]. It is the proportion of negative tuples that are correctly identified. It is given as Tables 6 and 7.

$$TNR = \frac{TN}{N} \tag{4}$$

Tables 7 and 8 shows the performance of the used classifiers. Classifiers are evaluated with respect to the metrics such as TPR, FPR, Precission, F-Measure [12]. Figure 7 illustrates classifiers accuracy and Fig. 8 describes error rate.

In the first phase, static analysis is done by using the machine algorithm random forest classifier which is shown in Fig. 6. Figure 7 depicts the excellence of the

Table 7 Various classifiers error rate

S.No	Classifier	Root mean square error	Root relative squared error
1	Naïve Bayesian	0.1852	39.3381
2	BayesNet	0.2448	51.9874
3	J48	0.131	27.978
4	Random tree	0.1662	35.3112
5	Random forest	0.0913	19.4025

Table 8 Malgenome file by Falcon sandbox analysis

File Name	Malgenome.csv
Analysis state	SUCCESS
Threat verdict	No specific threat
Threat score	5/100
AV Detection ratio	n/a
AV Family Name	n/a
Time of analysis	29-02-2020 9.17
File size (bytes)	2965084
File type	ASCII text, with CRLF line terminators
Contacted domains	None
Contacted hosts	None
Environment	Windows 7.32 bit (ID: 100)

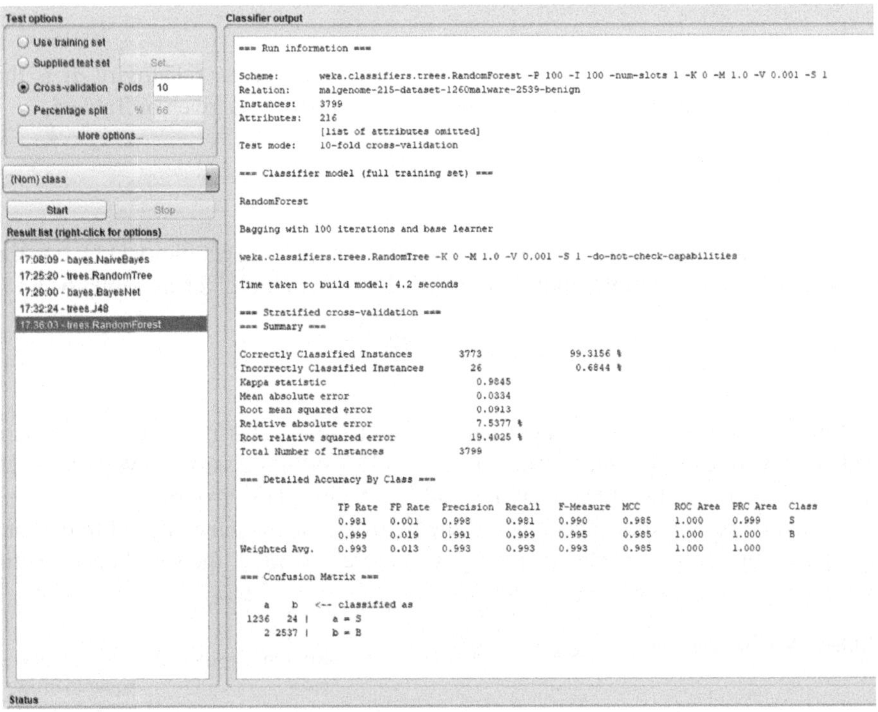

Fig. 6 Result generated by random forest classifier

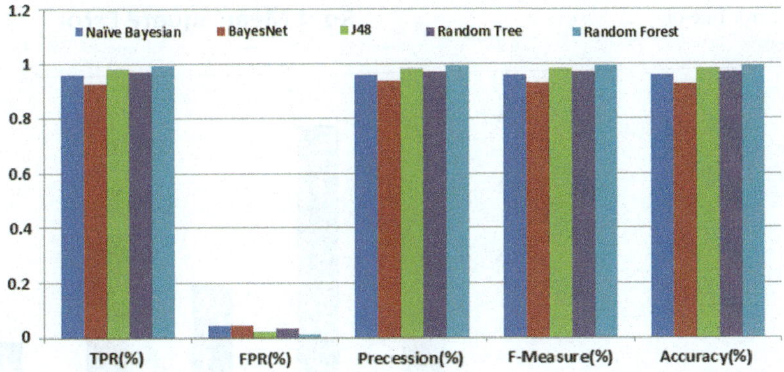

Fig. 7 Accuracy of classifier measures

Table 9 File detected as Malicious file by Falcon sandbox analysis

Input	097d327afe808550ca66af4dfda0dea9d794a4c96fadb352a9acbdd25ff38f6
	PF32 executable (DLL) (GUI) Intel 80386, for MS Windows
Threat level	Malicious
Summary	Threat score: 100/100
AV detection	68% Trojan.generic
	Matched 37 indicators
Environment	Windows 7 32 bit (HWP support)

proposed, Random forest classifiers decreased the error rate and increased the accuracy to 99.3%, which is 3.5% more accuracy than Naïve Bayes and 2.07% more than Random Tree classifier. In the second phase, to perform dynamic analysis, files are submitted to online automated tool. Two tools are used, one is Falcon Sand box and other is Cuckoo Sand box. Here, Results from falcon sandbox and Cuckoo Sand box are discussed in Tables 8, 9 and 10. In addition to malgenome dataset file, another file smart. Exe is also submitted for analysis. By Falcon analysis out of two files, one have been identified as malicious, and one as successful. Detailed summary report of malgenome file is shown in Fig. 8. The five reports generated by falcon sandbox are shown below. It gives the information about when file submitted and status of it. Summary report generated by Falcon Sandbox comprises of Name of the file, Analysis state, Threat Verdict, Score, AV Family name, Time of Analysis, File size and various other properties. Detailed summary report generated for file named malgenome.csv is as shown in Figs. 8 and 9.

Summary report generated by Cuckoo Sandbox comprises of Name of the file, Analysis state, Threat Verdict, Score, Av Family name, Time of Analysis, File size [13] and various other properties. Table 10 describes analysis report generated by Cuckoo sandbox the From the Score user can know whether the file is malware on

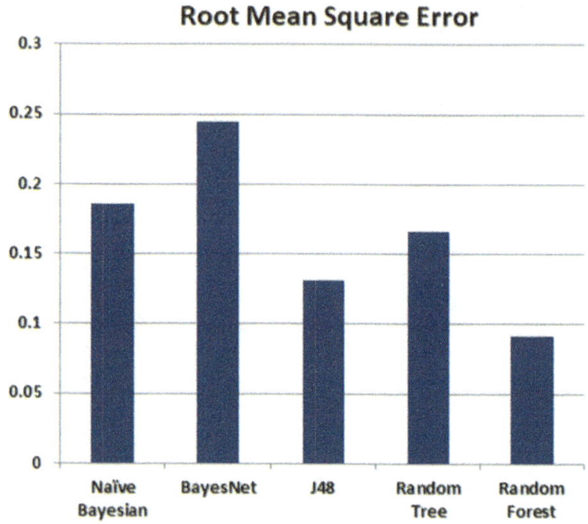

Fig. 8 Error rate of classifiers

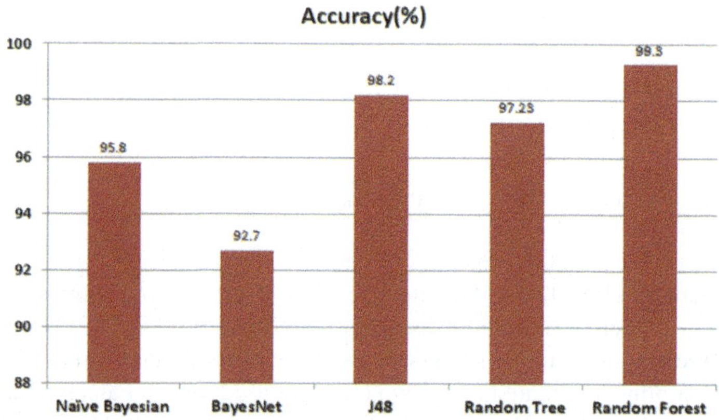

Fig. 9 Accuracy of classifiers

Table 10 Cuckoo sandbox analysis

File Id	Date and time of submission	Hash file	File name	Status	Score
1553864	2020-03-02 05:24	b64bd1c757eefc435569cd0a5839df5f	layer.arm6	Reporter	7
1553845	2020-03-02 04:58	a96ec877c036843c16d80afa439b1f33	index.html	Reported	4.4
1554064	2020-03-02 08:00	42c38f3432af9c0a7547ab28708f9051	sparc		10

benign. Sample output is shown below Score value ranges from 0 to 10. Score 10 indicates dangerous file shown in Table 10.

6 Conclusion

In this paper a new method malware detection kit (MDK) is proposed, through which data and infrastructure of the working environment could be protected against threats and attacks by unauthorised users by using static and dynamic analysis combinely for files. To safeguard data, static analysis is done by Random forest machine algorithm. It gives more accurate, effective and low error rate results. Random forest classifier is decreased the error rate and increased the accuracy to 99.3%, which is 3.5% more accuracy than Naïve Bayes and 2.07% more than Random Tree classifier. Root mean square error of random forest is 0.9. Dynamic analysis is done by using url hybrid analysis. Hybrid analysis performs the analysis by falcon sand box. Detailed Summary report is generated by automated tool. Hence proposed method is useful in protecting data against various threats and attacks.

7 Future Work

In future, system could be designed by incorporating the automated tools within data clusters. It enables and mandates the files to undergo the analysis process. The files which are free from attacks could be used further processing.

References

1. T.Y. Win, H. Tianfield, Q. Mair, Big data based security analytics for protecting virtualized infrastructures in cloud computing. IEEE Trans. Big Data **4**, 11–25 (2018)
2. W. Rankothge, J. Ma, F. Le, A. Russo, J. Lobo, Towards making network function virtualization a cloud computing service, in IM (2015), pp. 89–97
3. M. Tang, M. Alazab, and Y. Luo, Big data for cybersecurity: vulnerability disclosure trends and dependencies. IEEE Trans. Big Data
4. X. Wu, X. Zhu, G.-Q. Wu, W. Ding, Data mining with big data. IEEE Trans. Knowl. Data Eng. **26**(1), January (2014)
5. P.K. Chouhan, M. Hagan, G. McWilliams, S. Sezer, Network based malware detection within virtualised environments, in *Proceedings of Eur. Conf. Parallel Process* (2014), pp. 335–346
6. M. Watson, A. Marnerides, A. Mauthe, D. Hutchison, N.-ul-H. Shirazi, Malware detection in cloud computing infrastructures. IEEE Trans. Depend. Secure Comput. **13**(2), 192 205 Mar./Apr. (2016)
7. A. Fattori, A. Lanzi, D. Balzarotti, E. Kirda, Hypervisorbased malware protection with AccessMiner. Comput. Secur. **52**, 33–50 (2015)
8. E. Sayed Ali Ahmed, R.A. Saeed, A survey of big data cloud computing security. Int. J. Comput. Sci. Softw. Eng. (IJCSSE) **3**(1), 78–85 December (2014)

9. L.K. Breggin, et al. (editors). Big data, big challenges in evidence-based policy making 2014, West Academic Press
10. J. Kobielus, Deployment in big data analytics applications to the cloud: road map for success. Cloud Standards Customer Council, Technical report (2014)
11. "Impact of cloud computing on Healthcare, Version 2.0", Cloud Standards Customer Council, Technical Report, February 2017
12. A. Keeso, Big data and environmental sustainability: a conversation starter. Smith School Working Paper series, working paper 14–04 (2014)
13. M. Satish, M. RamakrishnaMurty, Document clustering with mapreduce using Hadoop Framework" published in International Journal on Recent and Innovation Trends in Computing and Communication **3**(1), 409–413 (2015). ISSN: 2321–8169
14. M. Alazab, S. Venkatraman, P. Watters, M. Alazab, Zero-day malware detection based on supervised learning algorithms of API call signatures, in *Proceedings of 9th Australas. Data Mining Conference*, vol. 121. Ballarat, Australia: Australian Computer Society, December (2011), pp. 171–182
15. T. Mahmood, U. Afzal, Security analytics: Big data analytics for cybersecurity: a review of trends, techniques and tools, in *Proceedings of. 2nd National Conference Information Assurance* (2013), pp. 129–134
16. www.cuckoo.com. February 2020. www.Hybridanalysis.com

Facial Emotion Recognition to Examine Human Face Using Hybridization Method

A. Sravanthi, B. Sindhuja, and D. Lakshmi Rohitha

Abstract Emotion recognition is a method used to inspect the ones internal feelings on a human face by using a technique Facial recognition technology can be used for authentification, recognition, and verification across various industries, including KYC, payment verification, and access control. Moreover, its aptitude is first to inspect facial expressions for emotions like happiness, sadness, surprise, anger, etc., Facial expression recognition is progressively receiving attention now a days. A person will express feelings through their emotions. The expressions are the exterior signals communicate the internal feelings of a person. Face appearances are divided into 7 feeling states (neutral, happy, sad, disgust, surprise, fear, and anger). The present paper is about hybrid method for face expression or emotion recognition. LBP and PCA are two methods individually used to recognize face. LBP method is used for local feature extraction. LBP method is incredibly sensitive to noise and can't distinguish between a powerful and a weak pattern. PCA used for global feature extraction. An Hybridization method can be used for face expression recognition to improve the authentication process.

Keywords Facial expression recognition · Local binary pattern · Principle component analysis

A. Sravanthi (✉) · B. Sindhuja
CSE Department, GRIET, Hyderabad, India
e-mail: sravanthi.dkk@gmail.com

B. Sindhuja
e-mail: bs50041@gmail.com

D. Lakshmi Rohitha
CSE Department, ACE, Hyderabad, India
e-mail: laksshmiroheta@gmail.com

C. Kiran Mai et al. (eds.), *Proceedings of International Conference on Advances in Computer Engineering and Communication Systems*, Learning and Analytics in Intelligent Systems 20,
https://doi.org/10.1007/978-981-15-9293-5_18

1 Introduction

There is no doubt that the emotions and feelings play a very vital role not just in the life of people but also in the computing domain as well. There can be hundreds of definitions for emotion but this paper defines emotion as "emotion is a reaction to stimuli that lasts for seconds or minutes" [1] The reason we are talking about emotions here is because the emotional state of a person can lead to influence the task given, decision making abilities. So if the computer is able to determine the minute emotion changes in the human then the computer can enhance itself and give a better productivity.

Facial emotion recognition is one of the active topic in image processing domain [2]. The facial expression recognition scheme is functional in diverse dimensions of life like security and surveillance [3]. Facial expression recognition Understand your audience enable real-time target marketing. Make knowledgeable selling decisions and optimize campaigns. Emotion recognition is valuable in many industry including marketplace, Online learning, Internet marketing and even in video Gaming. Emotion recognition is mostly used by many companies to know clients feeling about their products. The possibilities brought by this new-technology go further than research and Internet marketing. These newer technologies can also forecast the criminal actions by inspecting the images of their faces that are captured by the control-camcorder. The facial emotion recognition arrangement has been used in communication field to construct the machine more associated with people. It has very large number of application possibilities that are user-friendly or adaptable bonding man and machine, humanistic design of goods, and emotional robot etc. Recognizing facial features is done in two ways- global feature and local feature [4]. The global feature extraction process, focuses on entire image but it is less accurate, but local characteristic contemplates only the local area inside the image concentrates on nose, eyes and mouth regions which helps to identify a person uniquely, and it is more accurate. In the identification of global features of a face, Eigen vector, Principal Component Analysis [5], methods are used. For identification of local features Local Binary pattern (LBP) is used [6] but in reality there are many methods to extract features from face images. LBP is caption that describes texture of a rectangular block. LBP is one amongst the feature extraction methodology. Local Binary Pattern (LBP) method illuminates invariant narrative of look picture [7]. Anyhow, the current LBP method at rest undergoes much from non-monotonic enlightenment dissimilarity, arbitrary noise, and alters in cause, era, and expression. Even though, the comprehensive features are popular in face recognition but their performance declines in varying atmosphere. Now on the other hand Principle Component Analysis (PCA) is more often than not used as a tool in data analysis and prognostic or predictive models. PCA is the simplest of True Eigenvector-based multivariate analyses. PCA is an arithmetic course of action so as to use an orthogonal renovation to convert a set of values of possibly correlated 'M' variables into a set of values of 'K' uncorrelated variables called principle component. This paper presents a hybridization of face emotional recognition for providing better performance and improving efficiency. Therefore it is necessary to combine PCA and LBP smartly together for improving the authentication process [8]. Face expression recognition is done in three steps (1) Detecting face—detecting the face coordinates in any input image. The output is the bounding box of the detected faces. (2) Face Recognition—Comparison of multiple faces in order to identify which face belongs to the person. This is done by comparing face embedding vectors (3) Emotion Detection—Categorizing the emotion of the face into 7 has happy, angry, sad, neutral, surprise, disgust or fear.

2 Working of Local Binary Pattern Method

Local binary Pattern method is mostly accepted for face recognition [9]. In LBP Consider 3×3 matrix of pixels with nine elements {PIc, PI0, PI1 … PI7}, where PIc is the strength cost of the fundamental pixel and PIk ($0 \leq k \leq 7$) represent the intensity values of adjoining eight pixels. Each 3×3 figure chunk can be represented by {0, 1} values using, the equation

$$\begin{aligned} di = 1PI_k - PI_c > 0 \\ 0PI_k - PI_c < 0 \end{aligned} \tag{1}$$

Example:

LBP is an efficient and simple texture operator used for recognition.From the above figure it is well-known that an picture is separated into 3×3 matrix. P8 is the MSB (Most significant bit) and P1 (Least significant bit) (Fig. 1).

In the LBP method, the face image is divided into blocks. For each block we have to calculate LBP value. Likewise for the Total image we have to calculate LBP values mentioned in the procedure.

P1	**P2**	**P3**
P8	**pc**	**P4**
P7	**P6**	**P5**

MSB LSB

1	0	1	0	1	0	0	1

1*27	0*26	1*25	0*24	1*23	0*22	0*21	1*20

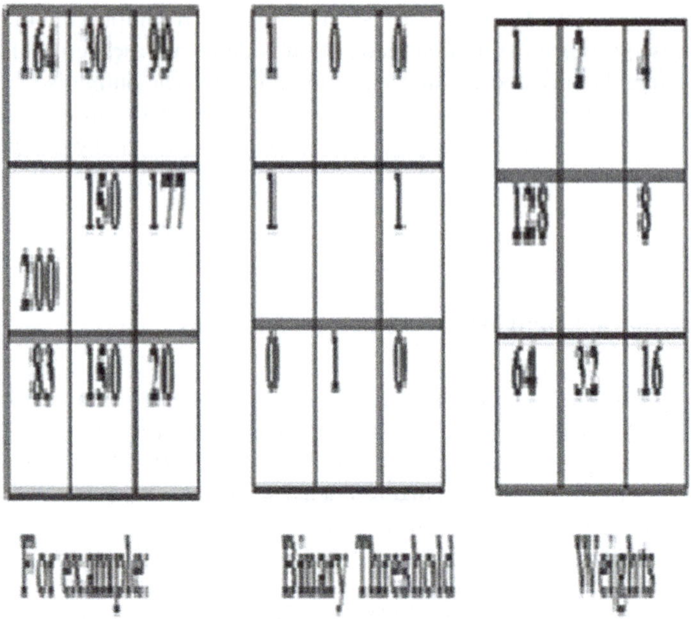

Fig. 1 Local binary pattern method calculated using binary threshold and evaluation of weights

In LBP after calculating Binary Threshold we have to multiply the binary values with $2i$ where $i = 0, 1 \ldots 7$.

In this paper JAFFE dataset is used for face expression recognition.An image is recognized by LBP using Haar-cascade method and performance is shown in Table 1. It is one of the deep learning object detection based approach for feature extraction. Some sample images presented in Figs. 2 and 3 are tested using LBP.

Table 1 Performance of LBP of face recognition

Method	Dataset	Accuracy (%)
LBP	JAFEE(64×64)	75

Fig. 2 Happy

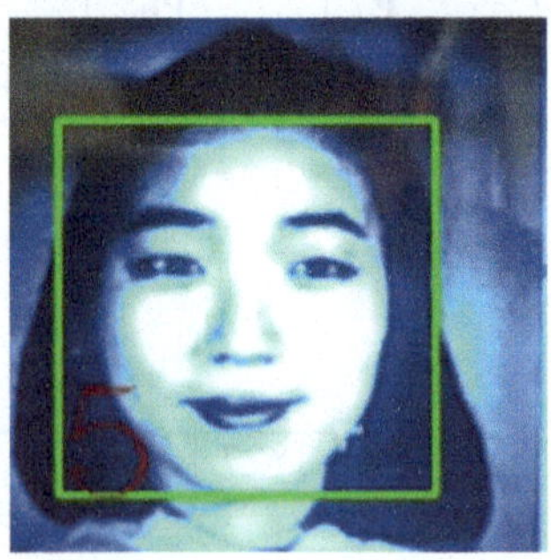

Fig. 3 Neutral

In testing phase the histogram is created for the test image and compares the histogram with training histograms and match give accuracy score. The accuracy varies between lighting conditions and quality of web cam. Haar cascade classifier is used. It is a deep and machine learning based approach used to detect objects. It helps in face detection and feature extraction.

3 Working of Principal Component Analysis

Principal Component Analysis, or PCA for short form, is a technique intended for tumbling the dimensionality of data. A picture might be spoken to as a vector of pixels where the estimation of every entry in the vector is the grayscale force estimation of the comparing pixel in picture.

For example, a 10 × 10 picture might be opened up and treated as a vector of length 100. The picture is said to be in N -dimensional space, where N is the quantity of pixels of picture (the length of the vector). This vector portrayal of the picture is considered to be the first space of the picture. This high dimensionality vector of pixels in unique space can be decreased by anticipating it into the lower dimensional space called subspace. A few subspace projection procedures are PCA (Eigen space), FDA (fisher spaces), and so forth.

The following are the steps

1. Taking the whole dataset
2. Computing the mean vector
3. Calculating the Covariance Matrix.
4. Calculating eigenvectors and corresponding Eigen values
5. Arranging the Eigen vectors in some predetermining order in increasing order
6. selecting *k* eigenvectors with the biggest eigen values
7. Converting the samples onto the new subspace.

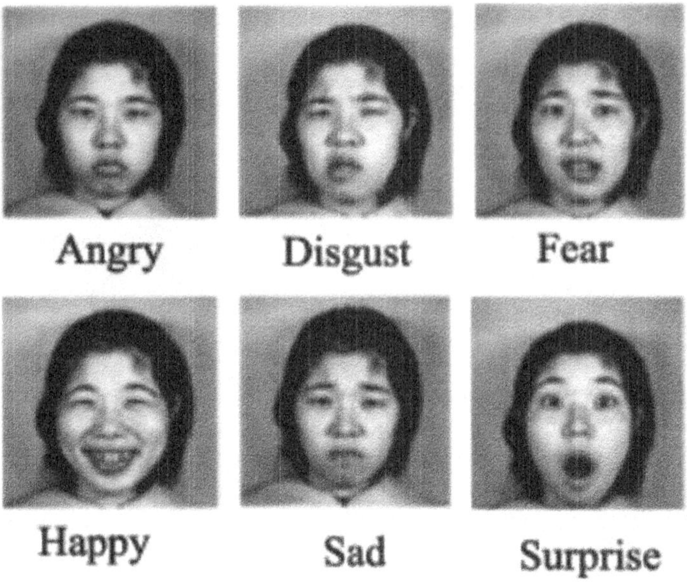

Fig. 4 Some of the sample images of JAFFE database

4 Proposed Method

LBP and PCA are two methods individually used in identification of face. PCA is used for universal feature extraction and LBP is used for local feature extraction. This method is incredibly sensitive to noise and can't distinguish between a powerful and a weak pattern. In PCA, standardization of data before implementation is must. Also, there is loss of information if number of principal components is not chosen wisely. The histograms which is produced in LBP are longer, as a result pace of recognition becomes less when a larger database is considered. The binary data i.e., 0 or 1 which is produced is sometimes noise sensitive and at some instants of time, local structure is missed out when centre pixel is ignored. Though LBP is a powerful technique for face recognition even though high memory and computational resources are used in LBP [10] (Fig. 4).

In this proposed method, to overcome all this disadvantages, An Hybridization method can be used for face expression recognition [11] can be used to improve the authentication process which is presented in Fig. 5.

5 Working of PCA and LBP Algorithm

Firstly the training images undergo face detection. In the next step, face parts are detected further. Next step involves extraction of features from specified regions. In

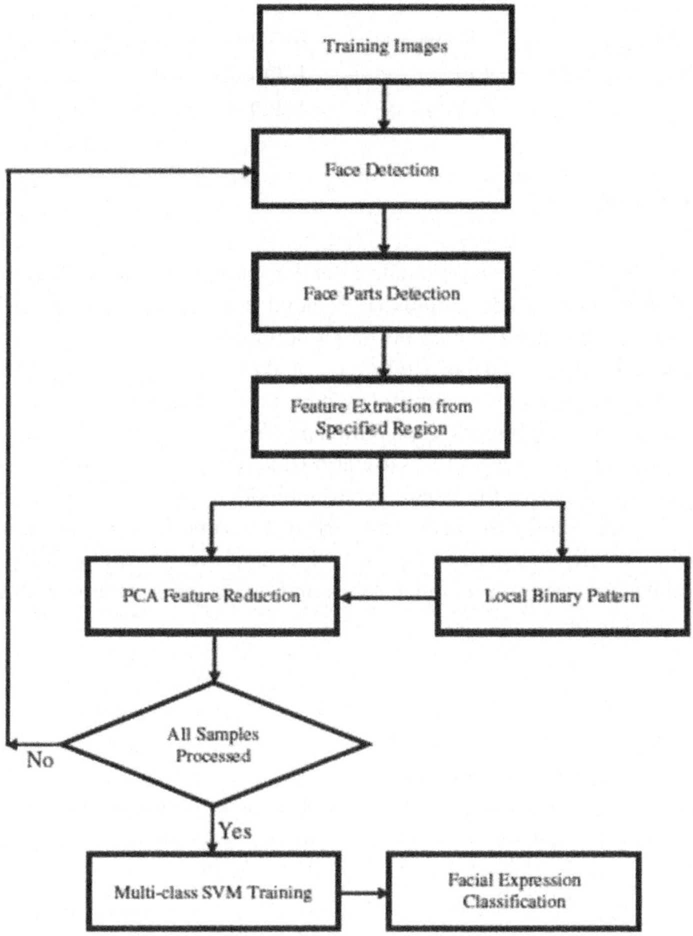

Fig. 5 Block diagram for proposed methodology to classify the facial expressions

the next step, hybrid model (PCA + LBP) is used. After this, if all the samples are processed, it undergoes multi class SVM training and finally expression is classified. Suppose if all the samples are not processed again the same process repeats from face detection as shown in Fig. 4.

To improve the authentication process of face recognition we have to combine PCA with LBP.

As shown in the Table 1 accuracy of 75% is obtained when only LBP is used. The accuracy obtained when this hybrid model is used is 90 percent.

As a result, the drawbacks of PCA and LBP when used individually is overcome and greater accuracy is achieved with hybrid model and results are presented in Table 2.

Table 2 Expected Performance of LBP and PCA for face recognition

Method	Dataset	Accuracy (%)
LBP	JAFEE(64×64)	75
PCA + LBP	JAFEE(64×64)	90

6 Conclusion

In this Paper JAFFE (Japanese Female Facial Expression) dataset is used. Which contains 10 Japanese female models of 7 facial expressions in total 213 images. Each image has been posed in 6 emotion +1 neutral.

In this paper, It is proposed that a hybrid method PCA and LBP is used to dig out the features of the complete expression image and the local texture features separately. PCA is used for dimension reduction. Which also reduces computational cost. Then, SVM used for feature expression classification and recognition. The process adopted in this paper will improve the accuracy in authentication process. This method is powerful to facial emotional recognition. Still the same kind expression, citizens possibly will have diverse shape. Our method, is used on static images. But in future this method can be used to recognize faces and emotions in videos with improved higher recognition rate.

References

1. A. Kołakowska, A. Landowska, M. Szwoch, W. Szwoch, M. R. Wróbel, Emotion Recognition and its Applications. Springer International Publishing Switzerland (2014)
2. D. voth, Face recognition technology 18(3), May-June (2003)
3. J. Kumaria, R. Rajesha, K.M. Pooja, Facial expression recognition: a survey, in *Second International Symposium on Computer Vision and the internet*, Vol. 58 (2015)
4. Anwar M. Mirza, *Gender Recognition Using Fusion of Local and Global Facial Features* (Springer-Verlag, Berlin Heidelberg, 2013)
5. L. Chandra Paul, Face recognition using principal component analysis method. IJARCET **1**(9), November (2012)
6. K. Lekdioui, Facial expression recognition using face-regions, in *3rd International Conference on Advanced Technologies for Signal and Image Processing - ATSIP'2017*, May 22–24 (2017)
7. Md. Abdur Rahim, Face recognition using local binary patterns (LBP) 13(4) Version 1.0, (2013)
8. O. Toygar, A. Acan, Face recognition using PCA, LDA and ICA approaches on colored images. J. Electr. Electron. Eng. **3**(1), 735–743 (2003)
9. T. Ahonen, Face recognition with local binary patterns. Springer-Verlag Berlin Heidelberg (2004)
10. M. Rao, Fast local binary patterns for efficient face recognition. Int. J. Sci. Eng. **2**(2), 22–26, Dec. (2011)
11. M. Revina, A survey on human face expression recognition techniques. J. King Saud Univ. Comput. Inf. Sci. (2018)

Emotion and Stress Recognition Through Speech Using Machine Learning Models

Druva Manasa and C. Kiran Mai

Abstract Emotion is a state, related with our sensory system brought out by the neurophysiological changes related with our thoughts like sad, happy, scared, excited and angry. These can be perceived through standard of conduct and outward appearances. Distinguishing discouragement through feelings is a troublesome errand. Being unhappy can't be identified with melancholy. There are explicit side effects that help recognizing a burdensome issue, through various researches and study. This paper centers around a feeling acknowledgment framework dependent on the discourse contribution of the person. The framework pre-forms the sound info and perceives the feeling utilizing the MLP classifier. The proposed model will be able to give the best prediction rate for the emotion recognition of the depressed.

Keywords Speech recognition · Emotion recognition · Pre-processing · MLP Classifier

1 Introduction

Emotions is specialized condition of consciousness where in different sensations are experienced inside. Emotions can be framed by memory, thought or inspirations made remotely and can change physical condition of an individual. Emotions are said to be long haul states. In organic setting pressure is a psychological, physical or passionate factor which causes mental or substantial strain. Stress can be interior or outer conditions like mental, condition, clinical methodology or disease that can impact numerous mental conditions like discouragement, uneasiness, hypertension, poor recuperating and other numerous body parts. The specialists utilize the mental tests to analyze the pressure of patient. In any case, on the off chance that

D. Manasa · C. Kiran Mai (✉)
Computer Science and Engineering, VNRVJIET, Hyderabad, India
e-mail: ckiranmai@vnrvjiet.in

D. Manasa
e-mail: dhruvamanasa1997@gmail.com

C. Kiran Mai et al. (eds.), *Proceedings of International Conference on Advances in Computer Engineering and Communication Systems*, Learning and Analytics in Intelligent Systems 20,
https://doi.org/10.1007/978-981-15-9293-5_19

the patient isn't perceived, at that point physical body parts may not react to the treatment. So right now, speech emotion recognition (SER) framework is structured in which voice input is given to framework and feeling of the individual is perceived inevitably and emotions of anxiety can be determined. A speech emotion recognition framework is broadly utilized in numerous applications like gaming, voice search, human-robot communication etc. In this investigation, MLP Classifier was utilized for characterization and acknowledgment of emotions was utilized for precision and manufacturing a superior speech recognition system.

2 Related Work

Huang et al. [1] in the traditional models of emotion recognition, extraction of passionate highlights are finished by following various techniques discussed and proposed a methodology for discovery of momentary temperament issue which depend on inspired speech reactions. Demonstrating of nearby variety of emotions were by using utilizing Convolution Neural Networks (CNN) to make emotions profile for every speech reaction. Long momentary memory (LSTM) - based consideration model was utilized to improve performance detection.

Yogesh et al. [2] suggested that discourse waveforms and the glottal waveforms are prepared through bispectral based highlights. The mean shift clustering was utilized for improving the segregation capacity of the removed bispectral features.

Hariharan et al. [3] builtup an improvement helped Biogeography-based calculation for selection of features. This is to beat the issues in reduction of dimensions and increased productivity in selecting the features.

Ngadiran et al. [4] tells that, motivation behind the examination is to choose a lot of higher order spectral features for emotion recognition system. Feature Subset Selection (FSS) was done here with the target of amplifying recognition pace of emotions. The FSS contains two phases: first stage is Multi-group highlight choice was adjusted to lessen include space and in the second stage Biogeography based optimization (BBO), Particle Swarm optimization (PSO) was utilized to diminish the dimensionality of feature space.

Kaya et al. [5] proposed the child emotion speech corpus in Russia called "EmoChildRu". This is intended to contemplate the passionate state in the qualities of voice and speech. They analyzed human and machine estimation on this corpus for the expectation of sex, age and solace state. The acoustics-based programmed framework was utilized here.

Karpov et al. [6] proposed a component standardization procedures forcross-corpus acoustic emotion recognition. For this they utilized standard arrangement of features that can be extricated by means of uninhibitedly accessible open SMILE tool.

Perez-Gaspar et al. [7] has set the principles for phonetic interpretation for Mexican speech. In addition, they proposed Hidden Markov Models and Artificial Neural Networks for classification of human-robot communications.

Jaywant et al. [8] describe whether passionate implications are handled from speech prosody and outward appearances. They tested by looking after the effects of vocal and outward appearances utilizing multi and cross model preparing of fundamental feelings.

Upon reviewing the current models, it can be comprehended that few tests were accomplished for the annoyance location, negative feeling recognition. Temperament issue like unipolar and bipolar was additionally identified. The emotion location and grouping didn't get the normal precision. This proposed method helps in improving the grouping and exactness of emotion and stress utilizing speech.

3 Methodology

The objective of the proposed system is to order the emotions and recognize the stress with greatest precision. Speech emotion recognition is associated with various modules as shown in Fig. 1, which includes preprocessing module, feature extraction module, feature selection module, classification and recognition module.

This work helps in expelling noise from the input audio signal. Librosa is utilized for sound preprocessing. For feature extraction Meel frequency cepstral coefficients (MFCC) is utilized which is included in Librosa library. To facilitate the way and fabricate a productive emotion classifier, MLP Classifier, have been discovered exceptionally precise.

In order to make a speech emotion system, firstly speech is recorded in audio format and then preprocessing on the audio dataset is performed where input is given in the form of signals. Librosa is the built-in library that is present in python. Both preprocessing, segmentation and feature extraction can be performed using this library. When we use this Librosa library on audio dataset it removes the missing values, and also does operations like display, feature extraction, Onset detection: which is used to calculate the energy function on multiple channels, beat and tempo: which is used for dynamic programming of beat tracker and estimation of beats per minute, spectrogram decomposition: for separating the closest-neighbours.

For feature extraction in Librosa there are numerous fragments that are performed inside. They are spectral features, rhythm features, feature inversion and feature manipulation. In spectral features, Meel- fequencycepstral coefficients (MFCCs)

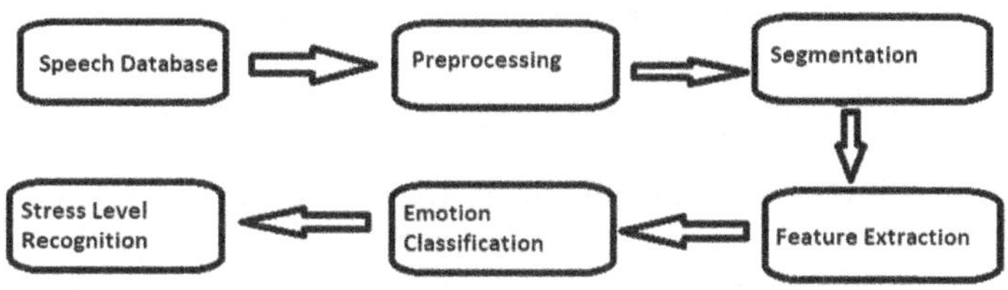

Fig. 1 Block diagram of speech emotion recognition working model

and root mean square is determined for estimation of each casing. In beat features, it figures the tempogram where nearby autocorrelation of the Onset and short-term Fourier change is determined. In succession, feature inversion and feature manipulation are performed to infer the nearby gauge of the information along with choosing pivot and connect information vectors with deferred duplicates of itself. Here MFCC is changed over to a period space sound signal in feature manipulation.

Along with this fleeting divisions: changing over information arrangement to a reference succession, changing over it again into lag-matrix for separating time slack and partitioning a division by feature clustering is made. Successive demonstrating: change of grids on various states is performed. Utilities and Filters were additionally performed on input information.

All this constitutes preprocessing, segmentation and feature extraction collectively in the Librosa library of python.After this preprocessing and feature extraction is to be done if essential feature selection is to be performed. Generally, the classification, feature selection is performed if needed. This is used along with scikit library in python.

MLP Classifier is utilized for the characterization of emotion and stress recognition from the features extricated. MLPC is a Multi-layer Perceptron classifier which enhances the order utilizing stochastic gradient descent by computing the loss function. This is an administered neural network model.

Multi-layer Perceptron is a machine learning algorithm which learns a function on preparing a dataset, where is the quantity of measurements for input and is the quantity of measurements for yield. Given a lot of features and an objective, it can get familiar with a non-linear function approximates for either regression or classification. It is not quite same as the logistic regression, as there can be at least one non-linear layers, called as hidden layers between the input and output layers. Figure 2 shows one input layer MLP with scalar yield.

The furthest left layer, known as the input layer, comprises of a lot of neurons representing to the input features. Every neuron in hidden layer changes the incentive from the past layer with a weighted linear summation, followed by a non-linear activation function—like the hyperbolic tan function. The yield (output) layer gets the qualities from the last hidden layer and changes them into yield esteems.

The module contains the public attributes coefs_ and intercepts_. coefs_ is a list of weight matrices, where weight matrix at index represents the weights between layer and layer. intercepts_ is a list of bias vectors, where the vector at index represents the bias values added to layer.

The benefits of Multi-layer Perceptron are:

- Ability to learn non-linear models.
- Ability to learn models progressively utilizing partial_fit.

MLP trains utilizing Stochastic Gradient Descent. Stochastic Gradient Descent (SGD) refreshes parameters utilizing the slope of the loss function as for a parameter that needs adjustment, for example

$$w \leftarrow w - \eta(\alpha\,(\partial R(w))/\partial w + \partial \mathrm{Loss}/\partial w).$$

Fig. 2 One hidden layer MLP

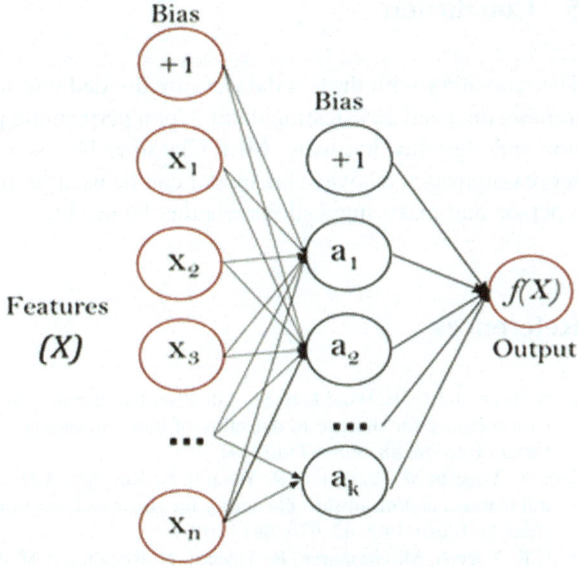

where the learning rate which controls the progression size in the parameter space search is the loss function utilized for the network. MLP Classifier trains the model iteratively since at every single step the partial derivatives of the loss function regarding the model parameters are figured to update the parameters. It can likewise have a regularization term added to loss function that therapists model parameters to reduce overfitting. This execution will work effectively with dataset represented to as dense numpy clusters or sparse SciPyof floating-pointesteems.

Utilizing MLP Classifier for emotion and stress recognition has acquired best outcomes. The best score acquired by utilizing MLP Classifier is 89.5%.

4 Results

In this paper, datasets used are RAVDESS(Ryson Audio-Visual Database of Emotional Speech Song) which consists of 24 actors (12 male, 12 female), TESS(Toronto Emotional Speech Set) a set of 200 target words were spoken by two actresses (aged 24 and 64 years), EMO_DB is a dataset recorded as part of funded research project, where in emotional statements spoken by actors are recorded. Customized dataset in which noisy data is located in data/train-custom for the training of data and data/test-custom for testing of the data in which you can add/remove recordings easily, by converting the audio to 16000 sample rate, mono channel and adding the emotion to the end of the file, name is separated with '_' (e.g. "20190616_125714_happy.wav" will be parsed automatically as happy). The best score obtained with this data is 89.5%.

5 Conclusion

The repository with these 4 datasets are divided into training and testing data (80% training data and 20% testing data). Upon performing preprocessing, feature extraction and classification using MLP Classifier best scores can be obtained. The best score obtained is 89.5%. This model can be used in finding the mental condition of a person and make suggestions whether he need to get diagnosed or not.

References

1. K.-Y. Huang, C.-H. Wu, M.-H. Su, Attention-based convolutional neural network and long short-term memory for short-term detection of mood disorders based on elicited speech responses. Pattern Recogn. **88**, 668–678 (2019)
2. C.K. Yogesh, M. Hariharan, R. Yuvaraj, N. Ruzelita, A.H. Adom, Y. Sazali, P. Kemal, Bispectral features and mean shift clustering for stress and emotion recognition from natural speech. Comput. Electr. Eng. **62**, 676–691 (2017)
3. C.K. Yogesh, M. Hariharan, R. Yuvaraj, N. Ruzelita, A.H. Adom, Y. Sazali, P. Kemal, A new hybrid PSO assisted biogeography- based optimization for emotion and stress recognition from speech signal. Expert Syst. Appl. **69**, 149–158 (2017)
4. C.K. Yogesh, M. Hariharan, N. Ruzelita, A.H. Adom, Y. Sazali, P. Kemal, Hybrid BBO_PSO and higher order septral features for emotion and stress recognition from natural speech. Appl. Soft Comput. **56**, 217–232 (2017)
5. H. Kaya, A. Ali salah, A. Karpov, O. Frolova, A. Grigorev, E. Lyakso, Emotion, age, and gender classification in childern's speech by humans and machines, Comput. Speech Lang. **46**, 268–283 (2017)
6. Heysem Kaya, Alexey A. Karpov, Efficient and effective strategies for cross- corpus acoustic emotion recognition. Neuro Comput. **000**, 1–7 (2017)
7. L.-A. Perez-Gaspar, S.-O. Caballero-Morales, F. Trujilo-Romero, Multimodal emotion recognition with evolutionary computation for human- robot interaction. Expert Syst. Appl. **66**, 42–61 (2016)
8. A. Jaywant, M.D. Pell, Categorial processing of negative emotions from speech prosody. Speech Commun. **54**, 1–10 (2012)

Dynamic Watermarking Using Python AST

V. Vivekananthan, K. Praveen, and M. Sethumadhavan

Abstract Watermarking refers to the process of embedding stealthy artifacts in a codebase, in order to determine and prove its origin and ownership. The artifacts should be stealthy, small, resilient but also efficiently detectable. In this paper, we will be exploring a novel way of automating watermark embedding, by manipulating the Abstract Syntax Trees (AST) of a given program. AST is an intermediary form obtained during the compilation process. AST's provide us with an accessible way to manipulate code entities while effectively preserving Semantic and Syntactic meanings of the original source code.

Keywords Watermark embedding · Python AST · Abstract syntax trees

1 Introduction

The software industry incurs heavy financial losses in the order of billions as a result of software privacy, each financial year, and is always looking forward to one-up its arsenal against privacy. One major technique to deter privacy and to establish ownership over software is Software Watermarking. Watermarking refers to the process of embedding stealthy software artifacts, in a software code base for the purpose of validating its origin and ownership [1–4]. Watermarking provides a means to discourage privacy by embedding hidden artifacts, which can be recovered at a later time to prove the legitimacy of given software or codebase. It is a concept closely related to steganography, in that they both hide a message inside a digital entity. However, in watermarking, there is a high degree of correlation between the watermark and

V. Vivekananthan (✉) · K. Praveen · M. Sethumadhavan
TIFAC-CORE in Cyber Security, Amrita Vishwa Vidyapeetham, Coimbatore, India
e-mail: vivekananthan979@gmail.com

K. Praveen
e-mail: kpraveen@cb.amrita.edu

M. Sethumadhavan
e-mail: msethu@cb.amrita.edu

C. Kiran Mai et al. (eds.), *Proceedings of International Conference on Advances in Computer Engineering and Communication Systems*, Learning and Analytics in Intelligent Systems 20,
https://doi.org/10.1007/978-981-15-9293-5_20

the actual content, whereas in steganography the entity is merely a superficial cover for the secret and has no dependency on it. While watermarking as a concept can be traced back several centuries, digital watermarking is relatively novel and is still in its incubation phase.

2 Literature Review

Software watermarking in its current state was formalized by Collberg et al. in the seminal paper [1], in which along with providing a formal background, they also analyzed the various existing models of watermarking, as well as the attacks, orchestrated on each of them. According to Collberg, software watermarking primarily has two approaches: Static and Dynamic.

2.1 Watermarking Approaches

In static watermarking, the data is embedded in the code or data section of the program's memory, whereas in dynamic watermarking the data gets embedded in the runtime behavior of the program.

- *Static Watermarks*: Static watermarking involves embedding the artifact in the code/data section of the program. A trivial case includes embedding a unique signature in a string, or a bit sequence in the binary of an executable. Static watermarks can also be embedded within the method bodies or in the variable pool of a program.
- *Dynamic Watermarking*: In dynamic watermarking, the artifact is encoded in the run-time behavior of the program, such as modifying the default looping behavior to echo a particular pattern or modifying the code to exhibit unique behavior for certain pre-defined inputs.

In addition to the embedding approaches, there are also marked differences in the functionality of the watermark themselves, in accordance with the various levels of security requirements.

2.2 Types of Watermark

There exist various types of watermarking schemes employed to achieve varying flavors of security. Nagra [5–7] and Thompson et al. [8] define four types of watermarks:

- *Authorship Mark*: Employed for recognizing the author of the software, these watermarks should be both visible and robust since they will be widely circulated.
- *Fingerprinting Mark*: Employed for identifying the distribution channel, each customer will be allotted a unique reference number, in order to trace back the source of leakage. These watermarks should be invisible and robust in order to prevent deletion.
- *Validation Mark*: Used for verifying the authenticity of a given software. These watermarks should be fragile, in order to detect tampering.
- *Licensing Mark*: Can enforce the expiry of a given software, the software can be designed to boot only if the licensing key is valid and unexpired.

In the following sections, some key constraints that must be considered while crafting a watermark are presented.

2.3 Watermarking Constraints

Every watermarking system has some important desirable properties. The watermarks usually have conflicting requirements, and one is forced to accept some trade-offs depending on the practical applications.

- *Robustness* The watermark should resist code preserving transformations, such as obfuscations/optimizations. On specific occasions, it is also desirable for the watermark to be fragile, such that any underlying modifications will render the watermark as invalid.
- *Code Fidelity* Watermarking alters a given program's functionality. Hence, ideally, the code's functionality should be preserved and should mirror the original, as closely as possible.
- *Artifact Size* Watermark sizes should be regulated as stringently as possible since an increased time/size in the artifact makes it susceptible to reverse engineer the location of the watermark.
- *Visibility* Watermarks can be designed to either be visible, or invisible. In visible watermarking schemes, the location or the retrieving scheme is public knowledge. Visible watermarking is widely employed in the digital media sector to establish ownership. In invisible watermarking, the retriever or some component of the watermarking scheme is not published, thereby hindering the removal of the watermark.

A comprehensive survey of watermarking concepts can be found in [9]. The various methodologies and techniques are also recorded in a comprehensive archive of literature such as [10–15]. In this paper, we will be discussing the process of embedding a dynamic watermark by employing Python AST.

2.4 Python AST

Abstract Syntax Tree is an intermediate data structure created during the compilation process. It is constructed by the compiler front end after scanning and parsing the source code. Let's have a brief walkthrough through the compilation process in order to better understand how AST fits into the overall procedure. The compilation process occurs as follows.

Source Code → Lexemes → Tokens → Parse Trees → Abstract Syntax Trees.

The compilation process begins with scanning the source text, an action performed by the scanner, to parse the given text into small logical units called lexemes. The lexemes are language agnostic and are nothing but a stripped-out version of the source text. Then they are passed to the tokenizer, which parses them into tokens. For example, a source line such as 2 + (30 * 4), will get parsed as [2, + , (, 30, *, 4,)]. Each individual language will have its own unique token set. The joint effort of the scanner and tokenizer make up the lexical analysis of compilation. Next up in the pipeline is the semantic analysis part which ensures the correct usage of the various code elements. It is at this junction that the compiler builds an AST and walks through it to ensure the correct usage of the elements. AST is built as a tree-like data structure having the previously processed tokens as its nodes. It is abstract since it captures only the structural information of the source and the non-trivial elements such as grouping parentheses are inferred from the structure of the AST, rather than being present as an explicit node. The AST aids in further stages of the compilation process such as generating symbol tables and ensures the syntactic validity of the given code. Python also facilitates the AST access by providing an in-built module out-of-the-box. A source line such as 'x = 2' finds its representation in Python AST as shown in Code Listing 1.1,

```
Module (
    body=[
        Assign (
            lineno =1 ,
            col_offset=0,
            targets=[Name (lineno=1, col_offset
               =0 , id=' x', ctx = Store ())],
            value=Num(lineno=1, col_offset=2, n
               =2) ,
        ),
    ],
)
```

Listing 1.1. A trivial AST representation

Now let's walk through the process of accessing AST's in Python. AST's in Python can be easily accessed since the language ports a standard library specifically for this purpose.

First, the AST library is imported. Then the line to be parsed is stored in a variable and then parsed into its AST form using the in-built ast.parse() function. The parsed AST can also be obtained using the in-built ast.dump() method as shown in Code Listing 1.2.

```
import ast
node = ast.parse( '''a = 2
a + 3 ''')
ast.dump(tree)
```

Listing 1.2. A basic code snippet demonstrating Python-3's AST interface

The plain ast appears as follows:

```
"Module(body=[Assign(targets=[Name(id='a', ctx=Store())
    ], value=Constant(value=2, kind=None), type_comment
    =None), Expr(value=BinOp(left=Name(id='a', ctx=Load
    ()), op=Add(), right=Constant(value=3, kind=None)))
    ], type_ignores=[])"
```

As one can see there is a distinctive tree-like structure and hierarchy in the AST dump, which becomes apparent when the dump is represented graphically as shown in Fig. 1.

The parsed AST can also be executed by the parser using the compile and eval commands.

3 Proposed System

Our proposed method can be utilized for automating any instances of source code modification. In this paper, we will be explaining the process of automating a specific watermarking scheme called Xmark [16]. Xmark is a dynamic watermarking scheme that embeds a stream of integers as the watermarked artifact into the target software. The integer sequence is generated by exploiting a well-known mathematical conjecture called the Collatz Conjecture. A brief introduction to the Collatz conjecture as well as the original Xmark scheme is due, before delving into the process of automating it via Python AST.

3.1 Collatz Conjecture

The Collatz conjecture is an unsolved mathematical problem concerning a sequence of integers nicknamed Hailstone sequences, which are obtained from a function called the Collatz function.

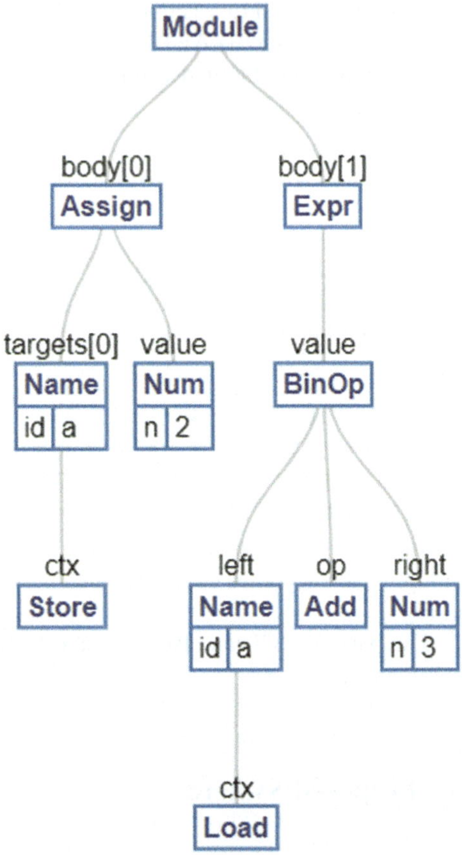

Fig. 1 A sample visualization of Python AST

$$\begin{aligned} a_n &= a_{n-1}/2 \rightarrow \text{if } a_{n-1} \text{ is even} \\ &\quad 3a_{n-1} + 1 \rightarrow \text{if } a_{n-1} \text{ is odd,} \end{aligned} \tag{1}$$

where $a0$ can be any positive integer. The Collatz conjecture asks if upon iterating through the Collatz function, does every positive integer $a0$, always terminate at 1. Let's, for illustration, take $a0 = 15$, the Hailstone sequence then obtained would be 15, 46, 23, 70, 35, 106, 53, 160, 80, 40, 20, 10, 5, 16, 8, 4, 2, 1.

3.2 *Internals of Xmark*

The original architecture involves a preliminary analysis of the code base for the identification of potential embedding points. After which the secret input to be embedded, as well as the earmarked embedding points are utilized for rewriting the original source code to inject the watermarked artifacts. Retrieval involves instrumenting the watermarked target code to track the control transfers. The data gained from which will reveal the embedded watermark. Below sections will give a brief summary of the embedding as well as retrieval procedures and delve into depth the proposed method for automation.

3.3 Watermark Embedding

The original Xmark scheme involves performing preliminary static code analysis to earmark potential embedding points, after which a series of secret inputs are generated to be embedded into the marked points. The inputs as well as the embedding points are then utilized to rewrite the original source to inject the watermark. The proposed preliminary analysis includes CFG (Control Flow Graph) to identify the potential embedding points. All basic branching structures which are reachable via multiple inputs are suitable candidates for embedding.

The next step involves the generation of the secret input lineup which when fed to the watermarked instance will reveal the embedded watermark.

Let's consider the example of **S**(μ), where μ is the input to the software and w the watermark to be embedded. Xmark selects from **S** a pair of conditional constructs (i.e., b1 and b2 in this example) that can be reached when running S with input cases $\mu = i1$ and $\mu = i2$, respectively. Both constructs are then transformed using a specialized Collatz conjecture-based obfuscation to produce the watermarked instance. Xmark also defines a pair of functions $\varphi 1$, $\varphi 2$ each of which takes μ the input to the target software and the original condition variable of $b1/b2$, i.e., (x1 or x2) as inputs, and outputs spurious variables $y1/y2$ for the Collatz conjecture routines to be used for obfuscation. These functions are also constructed to satisfy.

$$\varphi 1(x1,\ i1)\ =\ \varphi 2(x2,\ i2)\ =\ w \tag{2}$$

Care must be taken while designing the value of w since it directly impacts the runtime of the given software. Figure 2 represents the time plot of Collatz function. From the figure, it can be seen that the time taken to converge is significantly higher for a selected set of inputs. Hence, prithee or analysis should be performed in selecting a suitable input for embedding.

Then the original if-constructs containing the conditional logic are modified to infuse the Collatz conjecture-based signature into the target as shown in the Fig. 3. Such a construct can be automated via AST. As explained in the preceding section, everything in the source code is represented as a node in a tree-like structure during the compilation process in Python. Python also provides us with the ability to perform modifications on the nodes via the built-in module 'ast'. The ast-module provides facilities for editing, deleting or replacing a preexisting node with an entirely custom-made node. Such a facility is employed to automate the embedding process. First, a generic Collatz construct is created and converted into its AST form. Then the

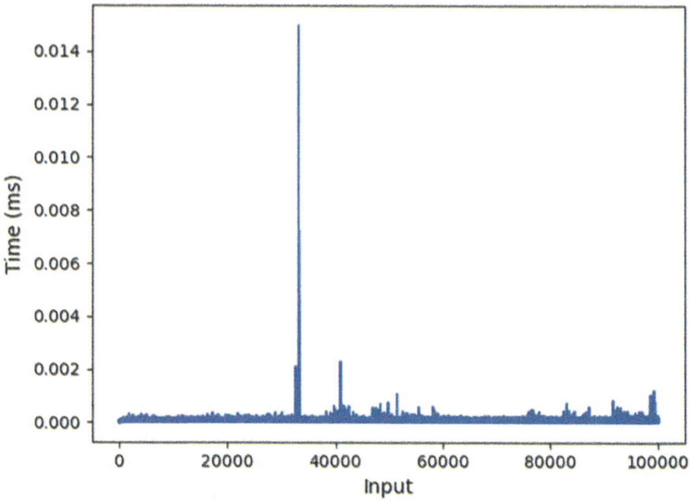

Fig. 2 Time plot for Collatz function

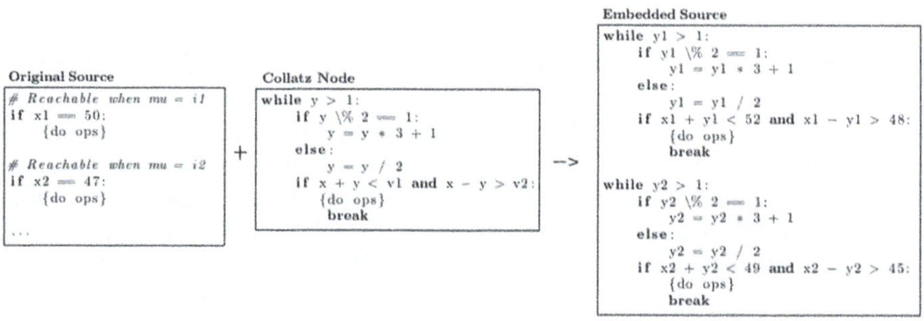

Fig. 3 An example of watermark embedding via node transformations

conditional variable from the original if the loop is parsed and utilized to modify the generic node, after which the modified Collatz node is used to replace the original if-conditional node, thus, concluding our embedding process. While the ast-module has some limitations with respect to the size of the programs which can be converted into its AST counterpart, it can be easily overcome with the help of some trivial textual parsing. Suppose consider a large-enough target source code, which when converted into AST raises a BufferOverflow exception, can be embedded by performing a pre-embedding static analysis that marks the boundaries of the if-conditionals to be modified. Then further steps can proceed by transforming only the source code within the marked boundary into AST and executing the necessary modifications. Thus, the ast-module can be used for manipulating very large codebases by transforming only the required context-specific targets.

3.4 Watermark Retrieval

After embedding, the retrieval part involves instrumenting the source code in order to harvest binary trials which will be walked to retrieve the embedded watermark.

In general, software instrumentation is mainly utilized for performance analysis, and for profiling source code in order to obtain insights such as code coverage, execution timings, functional dependencies, etc. Instrumentation can be achieved either via source-level modifications or binary-level modifications. For our retrieval procedure, we will be instrumenting the target source code, in order to harvest all branching trails, which are more inclined to carry the embedded watermark. The retrieval procedure mainly involves tracing the branching trails which arises due to the embedded Collatz payload.

The retriever is conceived with the simple observation that, the Collatz function is essentially an if–else branching structure that is nested within the body of a loop. Therefore, before actually running the watermarked software, our retriever examines statically to mark all loops containing at least one conditional branch and instruments them.

Intuitively, there should be many loops enmeshed within a software. Therefore, the recognizer creates a data structure called loop buffer arena (LBA) for all instrumented loops to track and record their internal branching. With the above mechanisms, the loop instrumentation phase allows us to specify the exact signatures to be traced while executing the embedded software in order to retrieve the watermark.

After done instrumenting, the recognizer runs the instrumented software to harvest the branching trails during the runtime.

The recognition proceeds as follows, the retriever instruments all the branching constructs in the original source code and harvests all the branching trails obtained when executing the given software with a chosen set of inputs. The branching trails are but a binary sequence constructed by appending a bit corresponding to the execution path undertaken. A trivial example is constructed by considering an if–else construct inside a while loop. For every iteration of the loop, the LBA records either a zero or one depending on the path which gets executed. The software is watermarked in such a manner that the branching trails will resonate only when the software is run with a specific input lineup.

The retriever recognizes the resonating branching trails and traverses backward to obtain the watermark as explained in the forthcoming chapter.

But before automating the retrieval procedure the Xmark requires the harvested branching trails to satisfy the following basic conditions.

- The trails must be respectively ported from two executions driven by contiguous cases of the secret input lineup.
- The trails must also be ported from different conditional branches of the software.
- Last but not least, the trails have to be either identical or bit-by-bit opposite to each other.

In our implementation, the LBA is constructed as a set of python list data structures that get appended with a bit whenever a particular execution path is undertaken. The python AST's capabilities become apparent when we are constructing such a retriever. The python AST along with its node editing capabilities also ports filtering capabilities, thus enabling us to filter all the nodes of a specific instance in a target source. An obvious and relevant use case will be filtering all the instances of if–else nodes since our Collatz signature is essentially an if–else node placed within a loop.

As shown in Fig. 4, our retriever implementation proceeds as follows, all the if–else instances which are nested within a while-node are identified and a separate list data structure is initialized for tracking the execution paths undertaken within a while–if–else construct, a fact worth reminding is that a single while-node may contain multiple if–else constructs, as well as, yet another while loop. Hence, the automation proceeds as follows, utilizing the ast-module's node filtering capabilities, which allows us to filter nodes with a particular signature, all the While-nodes with an if–else signature within the body are filtered. The query also conveniently handles nested while loops, after which a list structure is initialized for each individual pair of if–else nodes. The append logic is inserted by injecting custom nodes that append '1' or '0', for the if or else instances, respectively, into the conditional constructs. Since all these actions are handled at the node level, AST allows us to encode the instrumenting logic in a very high level, concise API calls to the library. From the context of Collatz Conjecture, the presence of a '1' indicates the odd path being traversed in that particular iteration, whereas a '0' indicates the even path being traversed.

Embedded Source

```
while y1 > 1:
    if y1 \% 2 == 1:
        y1 = y1 * 3 + 1
    else:
        y1 = y1 / 2
    if x1 + y1 < 52 and x1 - y1 > 48:
        {do ops}
        break

while y2 > 1:
    if y2 \% 2 == 1:
        y2 = y2 * 3 + 1
    else:
        y2 = y2 / 2
    if x2 + y2 < 49 and x2 - y2 > 45:
        {do ops}
        break
```

->

Instrumented Source

```
a1 = list()
a0 = list()

...

while y1 > 1:
    if y1 \% 2 == 1:
        a0.append(1)
        y1 = y1 * 3 + 1
    else:
        y1 = y1 / 2
        a0.append(0)
if x1 + y1 < 32 and x1 - y1 > 28:
    {do ops}

while y2 > 1:
    if y2 \% 2 == 1:
        a1.append(1)
        y2 = y2 * 3 + 1
    else:
        y2 = y2 / 2
        a1.append(0)
if x1 + y2 < 29 and x1 - y2 > 25:
    {do ops}
```

Fig. 4 Instrumentation via AST

Upon executing an instrumented instance, we will obtain a list of binary trails mirroring the paths undertaken within every while–if–else node instance.

3.5 Watermark Recognition

After extraction of the binary trails, the final phase involves recovering the originally embedded watermark. As one may recall, during the embedding phase, the software **S** is injected with Collatz constructs on two conditional branches b1 and b2, whose binary trails will resonate only upon executing **S** with the inputs i1 and i2, respectively.

Consequently, during retrieval, the software is executed with the input line-up [$i1$, $i2$] and all the resulting binary trails are harvested employing the instrumentation process outlined above. With high probability, only one resonant trail exists with the same signature across the input line-up, corresponding to the injected Collatz constructs.

Since the two paths of the Collatz function map, respectively, to 0 s and 1 s in the binary trials, the embedded watermark can be retrieved by traversing the resonant trail starting from its least significant bit and performing the inverse operation. Let the binary trail harvested for the embedded watermark $w = 6$ be $\{1,1,1,1,0,1,0,1\}$. Then the embedded watermark is retrieved as follows:

$$1 \rightarrow 1 * 2 = 2$$

$$1 \rightarrow 2 * 2 = 4$$

$$1 \rightarrow 1 * 2 = 2$$

$$1 \rightarrow 2 * 2 = 4$$

$$1 \rightarrow 4 * 2 = 8$$

$$1 \rightarrow 8 * 2 = 16$$

$$0 \rightarrow (16 - 1)/3 = 5$$

$$1 \rightarrow 5 * 2 = 10$$

$$0 \rightarrow (10 - 1)/3 = 3$$

$$1 \rightarrow 3 * 2 = 6$$

4 Advantages

AST's provide a very high-level handle to modify the code elements in a given source text. The modification also ensures that the code remains meaningful from a syntactic as well as semantic perspective. Since the AST's logic is recursive in nature, code elements which are nested multiple levels deep can be accessed and modified in a straight forward manner. Hence, the probability of introducing compile/run-time errors during modifications is significantly reduced. Also verbose modification procedures can be translated into concise and easy to interpret logic, which is not the case when employing more conventional compilers/lexical analyzers.

5 Conclusion

In this paper, a motivating example of utilizing Python AST for the purposes of automating dynamic watermarking has been explored. We have also covered the various advantages in adopting such an approach as well as the limitations encountered, with suitable workarounds in overcoming them. Since the AST provides granular access to the code elements, it makes a strong case for dynamic watermark automation.

6 Future Work

We have successfully embedded as well as retrieved the intended Collatz signature from the watermarked source, through AST manipulation, now going further, efficient recognition and retrieval techniques can be designed in order to optimize the watermarker's functionality. Also, a post-embedding analysis should be performed to quantitatively analyze the stealthiness of the embedded watermark along with various pre-embedding analyses, for the purpose of gathering enough information to effectively choose the embedding points to be watermarked.

References

1. C. Collberg, C. Thomborson, Software watermarking: models and dynamic embeddings, in *Proceedings of the 26th ACM SIGPLAN-SIGACT symposium on Principles of programming languages* (1999), pp. 311–324
2. C.S. Collberg, C. Thomborson, Watermarking, tamper-proofing, and obfuscation-tools for software protection. IEEE Trans. Softw. Eng. **28**(8), 735–746 (2002)
3. C. Collberg, C. Thomborson, D. Low, On the limits of software watermarking, Technical Report (1998)

4. W. Zhu, C. Thomborson, Recognition in software watermarking, in *Proceedings of the 4th ACM International Workshop on Contents Protection and Security*, pp. 29–36, ACM (2006)
5. J. Nagra, C. Thomborson, Threading software watermarks, in *International Workshop on Information Hiding* (Springer, 2004), pp. 208–223
6. J. Nagra, C. Thomborson, C. Collberg, A functional taxonomy for software watermarking, in *Australian computer science communications* (Australian Computer Society, Inc., 2002), vol. 24, pp. 177–186
7. J. Nagra, C. Thomborson, C. Collberg, Software watermarking: protective terminology, in *Proceedings of the ACSC 2002* (2002)
8. C. Thomborson, J. Nagra, R. Somaraju, C. He, Tamper-proofing software watermarks, in *Proceedings of the second workshop on Australasian information security, Data Mining and Web Intelligence, and Software Internationalisation* (Australian Computer Society, Inc., 2004), Vol. 32, pp. 27–36
9. W. Zhu, C. Thomborson, F.-Y. Wang, A survey of software watermarking, in *International Conference on Intelligence and Security Informatics* (Springer, 2005), pp. 454–458
10. D. Grover, The protection of computer software: Its technology and applications, *The British Computer Society Monographs in Informatics* (1992)
11. G. Hachez, A comparative study of software protection tools suited for e- commerce with contributions to software watermarking and smart cards, *Universite Catholique de Louvain* (2003)
12. J. Hamilton, S. Danicic, A survey of static software watermarking, in *2011 World Congress on Internet Security (WorldCIS-2011)* (2011), pp. 100–107. IEEE
13. A. Monden, H. Iida, K. Ichi Matsumoto, K. Inoue, K. Torii, Watermarking java programs, in *International Symposium on Future Software Technology* (1999), vol. 99, pp. 119–124
14. G. Myles, C. Collberg, Detecting software theft via whole program path birthmarks, in *International Conference on Information Security* (Springer, 2004), pp. 404–415
15. G. Myles , C. Collberg, Software watermarking through register allocation: Im- plementation, analysis, and attacks, in *International Conference on Information Security and Cryptology* (Springer, 2003), pp. 274–293
16. H. Ma, C. Jia, S. Li, W. Zheng, D. Wu, Xmark: Dynamic software water- marking using collatz conjecture. IEEE Trans. Inf. Forens. Secur. **14**(11), 2859–2874 (2019)

A Novel Framework for Video Retrieval Algorithm Evaluations and Methods for Effective Context-Aware Video Content Retrial Method on Cloud

T. Naga Raja, V. V. Venkata Ramana, and A. Damodaram

Abstract In recent time, multiple research outcomes have demonstrated diversified approaches for content management and showcased better outcomes. However, the parallel research outcomes are highly criticized for higher time complexity, reduction of the key contents, and finally, the lesser accurate indexing of the contents. Majority of the recent work outcomes have demonstrated the reduction of keywords based on language recommendations rather the context recommendations. This leads to non-context-aware reduction and further leads to incorrect extraction of keywords. Thus, the demand from recent research is to identify the keywords based on the context. Also, based on the previous claim, the proposed works must identify the actual frames if the keyword identification is based on the context. Thus, based on the recommendations by popular research outcomes, a framework is to be proposed to compare the existing video content retrieval methods and propose a novel process to identify the keyframes from the video contents using contextual mining and consider the optimal storage architecture for the proposed process input metadata and results for cloud-based storage service providers. The final outcome of this work is reduced complexity of the framework, compared with the parallel research outcomes, and higher accuracy for video content retrieval with reduction of the size for the searchable contents. The proposed work demonstrated nearly 15% improvements for content retrieval process and 86% improvements for time complexity over the parallel research outcomes for making the video content management and delivery mechanisms better and faster.

Keywords Segmented noise removal · Adaptive threshold · Video summarization · Cluster knowledge discovery · Quartic polynomial randomization · Similarity region extractor

T. N. Raja (✉)
NTUH, Hyderabad, India
e-mail: rajatnaga@yahoo.com

V. V. V. Ramana
National Informatics Centre, Amaravati, India

A. Damodaram
SIT, JNTUH, Hyderabad, India

C. Kiran Mai et al. (eds.), *Proceedings of International Conference on Advances in Computer Engineering and Communication Systems*, Learning and Analytics in Intelligent Systems 20,
https://doi.org/10.1007/978-981-15-9293-5_21

1 Introduction

The content management in recent years has gained a lot of attention due to the fact that the communication over the internet has increased to a greater extend. The recent works show for various domains multiple practices have started for making the content delivery online. Some of the business cases like for corporate practices, the meeting and the discussions are happening over the internet to increase the reachability and effective time utilization. Many of the medical practices are also using online content managements as the treatments over the internet via video conferences are getting high popularity for making knowledge sharing over various regions of the world. Also, in the field of education, many of the universities have showcased a higher rate of learning via online contents as the learners can learn at their own pace with greater engagements. So, the adaptability of the online contents, especially the video-based contents, is the demand of the recent research and practices.

Nonetheless, preserving the video quality is always a task for various reasons. In this work, the challenges are precisely identified based on the parallel research outcomes and further, the proposed solutions as a complete framework on cloud systems are also elaborated with significant improvements over recent and parallel research outcomes.

The later sections of this research are furnished as such a way that in Section 2, the parallel outcomes are discussed; in Section 3, the problems and research scope based on the earlier research outcomes are identified; in Section 4, the proposed frameworks and the sequence of the deployed algorithms are discussed; in Section 5 the results are analyzed; in Section 6, the real-time deployment of the proposed framework is analyzed on cloud; in Section 7, the comparative analysis is carried out; and, at the end, conclusion and future research directions of this research are discussed and on Section 8.

2 Literature Review

The parallel research outcomes will help in understanding the fundamental improvements in this domain of research and, at the same time, the gaps in the research can also be identified.

The work by Feng et al. [1] has recently demonstrated the applicability of the video data processing for a higher volume of the data. The work has clearly showcased the benefits of contextual video data processing for learning specific content retrieval. The personalized video searching, and retrieval process is definitely the demand for the current research.

The video content retrieval is not only the challenge of the present research context. Rather, the security or the sharing of the video contents are also to be considered with equal importance. The work by Yang et al. [2] has showcased the need for the time-dependent video content security approaches for a higher volume of the video data. This work also demonstrates the time-dependent query processing using the historical query results for making the query processing faster.

Further, the information, specifically the video information retrieval process is complex for the cloud- or cluster-based architectures. Thus, the information of the video content distribution is one of the keyfactors to be identified for context-based searching from distributed data sources. The work by Yang et al. [3] has listed the benefits and the challenges of the cloud-based content retrieval processes.

Many of the parallel research works have demonstrated similar outcomes focusing on the content retrieval processes. The majority of the works have showcased better outcomes for the static multimedia contents such as images. The work by Wang et al. [4] has established a process for managing, storing, and retrieval actions on the video data. The CHCF framework is widely accepted by the research community.

During the retrieval of the video or image or multimedia contents, the enhancements of the contents during the searching process is also important. Thus, the noise reduction processes come into the place for many of the parallel research works. The similar outcome by Song et al. [5] for content retrieval processes have confirmed this ideology.

The video content retrieval process is used for various purposes from security to education to entertainments. The distinguished work by Gao et al. [6] has demonstrated the use of video content retrieval methods for the identification of human or other objects from the video. The similar measures can be applied for identification or retrieval of the video contents, where the video contents are the search string or search input.

Yet another direction of the research community is extending their expertise for making the complete video content retrieval process with lower time complexity. The demand is eventually propagating throughout all the aspects of the content retrieval processes, where the searching device, the device generating the search string, is limited in terms of resources and processing capabilities. The work by Liu et al. [7] has showcased the process of video content retrieval from mobility devices with novel indexing methods. The identified problem from these works is discussed in the next section of the work.

3 Problem Identification

In this segment of the research, after a thorough analysis of the recent and parallel results are examined for identifying the bottlenecks of the parallel research outcomes and propose the research objectives as shown in Table 1.

Henceforth, the final research objectives are furnished here:

1. **Objective-1**: Apply a novel solution for the reduction of the keyframes with the reduction of the time complexity and improvement of the content retrieval accuracy.

Table 1 Identification of the research objectives

Research outcome	Proposed methodology	Identified bottlenecks	Research objectives
Kan Yang et al. [2]	Time slot-based content retrieval process	• The reduction of the informative content is grossly missing • Reduction of the keyframes or the important content can lead to the reduction in time for content retrieval	Objective-1
Xiaodan Song et al. [5]	Context-awareness during the video information retrieval	• The reduction in the frames can improve the complexity of the framework	Objective-1
Yinan Feng et al. [1]	A context-based online big data-oriented personalized video retrieval system	• Information about the cluster information is not considered • The lack of cluster information leads to the lack of information on data replication and contextual parameters	Objective-2
Hanli Wang et al. [4]	Distributed video information retrieval	• The video content is replicated for distributed systems. However, no approach is taken for generating cluster information	Objective-2
Guangyu Gao et al. [6]	Segmentation of the objects and detection for retrieval	• The threshold-based similarity measures can generate better performance measures	Objective-2
Wu Liu et al. [7]	Mobile instant content retrieval	• This direction of the search is also the need for further researches • The complexity of the framework or the retrieval time can be reduced further	Objective-2 & Further research directions

(continued)

Table 1 (continued)

Research outcome	Proposed methodology	Identified bottlenecks	Research objectives
Zhen Yang et al. [3]	The content retrieval process from cloud-based video library with data encryption	• The content retrieval process for video contents is significantly higher • During the encryption process, no optimization is proposed	Objective-3

2. **Objective-2**: Apply a novel solution for the improvement of content retrieval process on cloud-based or distributed online contents with the knowledge of data distributions over clusters.
3. **Objective-3**: Apply a novel solution for providing data encryption and decryption for data at rest and data on go for the complete framework with the reduction in time complexity and improved accuracy for content retrieval.

Based on the identified problems and the proposed solutions, the conceptual framework is formulated in the next portion of this work.

4 Proposed Framework

Based on the identification of the problems, research gaps and outcomes from the earlier research works, in this section, the proposed framework is elaborated (see Fig. 1).

The proposed algorithms are the outcomes from the works of the same authors and are summarized here.

Firstly, the Segmented Noise Removal Processing using Deep Segmentation Algorithm is displayed here [8].

Algorithm - 1: ***Segmented Noise Removal Processing using Deep Segmentation*** *Algorithm (SNRP-DS)*
Step - 1.*Accept all the video content as V[]* Step - 2.*For each V[i]* a. *Accept the video signal in analog form as f(t):V[i]* b. *For each time interval t/Δt,* i. *Calculate the signal variance as Amp(t/Δt) = (f(t/Δt):V[i]* ii. *Revise the variance difference, AmpDiff(t/Δt) = (|Amp(t/Δt) - Amp(t+1/Δt)|)/(t/Δt)* iii. *Report the overall signal variance, Amp[]* c. *Reduce the noise, V[i] - Amp[]* Step - 3.*Report the noise removed video content as V'[]*

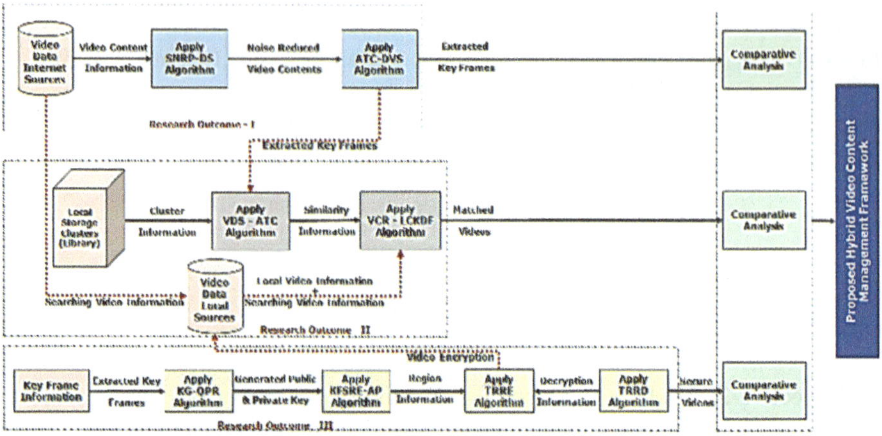

Fig. 1 Proposed framework for video content retrieval and management

Secondly, the Adaptive Threshold Calculation for Distributed Video Storage Algorithm is displayed here [8, 11].

Algorithm - 2: ***Adaptive Threshold Calculation for Distributed Video Storage*** *Algorithm (ATC-DVS)*

Step - 1. *Input the list of storage instances as S[]*
Step - 2. *At each level for every instance S[i]*
    a. *Accept the list of noise reduced video content set as V'[]*
        i. *For each V'[i]*
          1. *Extract the video meta data as*
            a. *Frames Per Second, FPS*
            b. *Scan System, SS*
            c. *Aspect Ratio, AR*
            d. *Augmented Aspect Ratio, AAR*
            e. *Channel information, C*
          2. *Calculate the key frame extraction threshold as f(FPS, SS, AR, AAR, C)*
          3. *Scan V'[i] for each frame V'F[t]*
          4. *If f(V'F[t])>=f(V'[i]) and V'F[t](AAR)> V'[i](AAR)*
          5. *Then,*
            a. *V'[i]_KeyFrame[j] = V'F[t]*
          6. *Else,*
            a. *Discard V'F[t]*
        ii. *Report the total key frames for V'[i]*
    b. *Calculate the overall threshold as, f(S[i]V'[i])*
    c. *If f(S[i]V'F[t]))< f(S[i]V'[i])*
    d. *Then,*
        i. *Discard the key frames in V'[i]_KeyFrame[j]*
    e. *Else,*
        i. *Report the final key frames for V'[i]*

Thirdly, the Video Data Summarization with Adjustable Threshold Correction is displayed here [9].

Algorithm - 3: ***Video Data Summarization with Adjustable Threshold Correction*** *Algorithm (VDS - ATC)*
Step - 1. Consider Set of Clusters as C[]
Step - 2. Consider set of Video data as V[]
Step - 3. For each V[i]
a. *Split video data into frames as F[]*
b. *For each C[j], Find similarity as S for V[i] and C[j]V[j]*
c. *Calculate F[k] = S*
d. *Update F[k](t+1) = F[k](t) ±(S/Δ S)*
e. *Find V[j+1] = C[j]V[j] Union C[j+1]V[j]*
f. *Update collaborative cluster as C'[{V[i]}] for V[i]*
Step - 4. Report C'[]

Fourthly, the Video Content Retrieval using Local Cluster Knowledge Discovery Factors is displayed here [9].

Algorithm - 4: ***Video Content Retrieval using Local Cluster Knowledge Discovery Factors*** *Algorithm (VCR - LCKDF)*
Step - 1. Consider Set of Clusters as C[]
Step - 2. Consider the searching video as V
Step - 3. For each C[i]
Calculate the KDF[i] = Selection of V[j] for V
Step - 4. Find the match based on KDF[] and C'[]
Step - 5. Report the final V[j] as match for V

Fifthly, the "Key Generation Based on Quartic Polynomial Randomization Algorithm" is displayed here [10].

*Algorithm - 5: **Key Generation Based on Quartic Polynomial Randomization** Algorithm (KG-QPR)*
Step - 1. Initialize the Quartic Polynomial function as f(X) with N order
Step - 2. Select two prime numbers based on f(X) as A and B with randomization of Order N
Step - 3. Calculate R as R = A product B
Step - 4. For each N as i
a. Calculate α(i) = [A{f(X^N)-1}] XOR [B{f(X^N)-1}] as polynomial component
b. Calculate β(i) = α(i) XNOR [A{f(X^N)-1}] XNOR [B{f(X^N)-1}] as public component
Step - 5. Generate the public key as PK = {α[] XOR β[]}
Step - 6. Generate the private key for similar regions as PSRK = {α[1..N/2] XNOR β[1..N/2]}
Step - 7. Generate the private key for dissimilar regions as PDRK = {α[N/2..N] XNOR β[N/2..N]}
Step - 8. Report final keys as PK, PSRK and PDRK

Sixthly, the Keyframe Similarity Region Extractor using Adaptive Progression Algorithm is displayed here [10].

*Algorithm - 6: **Key Frame Similarity Region Extractor using Adaptive Progression** Algorithm (KFSRE-AP)*
Step - 1. Accept the set of Key Frames as KF[]
Step - 2. For each KF[] as j
a. Calculate region similarity as KF[j], KF[j+1] as RS[i]
Step - 3. Calculate mean RS[] as RMS
Step - 4. For each KF[] as n
a. If region_similarity(KF[n], KF[n+1])> RMS
b. Then, Mark regions as SR[]
c. Else, Mark regions as DR[]
d. Report SR[] and DR[] for each KF[n]

Seventhly, the Time Restricted Region Encryption Algorithm is displayed here [10].

*Algorithm - 7: **Time Restricted Region Encryption** Algorithm (TRRE)*
Step - 1. Accept PK, PSRK and PDRK
Step - 2. Accept SR[] and DR[]
Step - 3. Accept KF[]
Step - 4. For each KF[] as m
a. Apply PK and PSRK for SR[m] for encryption
b. Apply PK and PDRK for DR[m] for encryption
Step - 5. Merge KF[] into video data V[]

Finally, the Time Restricted Region Decryption Algorithm is displayed here [10].

*Algorithm - 8: **Time Restricted Region Decryption** Algorithm (TRRD)*
Step - 1. Accept PK, PSRK and PDRK
Step - 2. Accept V[] and extract KF[]
Step - 3. Extract SR[] and DR[] from KF[]
Step - 4. For each KF[] as m
a. Apply PK and PSRK for SR[m]
b. Apply PK and PDRK for DR[m]
Step - 5. Merge KF[] into video data V[]

The proposed framework is deployed on the cloud infrastructure and the deployment environment is elaborated in the next section of this work.

5 Results and Discussions

After the identification of the problems and the detailed discussion on the framework, the results obtained from the proposed framework is discussed (Table 2).

The first outcome from the framework is nearly 96.83% accuracy for content retrieval process with mean of 95.71%, which also demonstrates the normal distribution of the accuracy for various datasets and various test runs (see Fig. 2).

The second outcome from the framework is nearly 97.93% accuracy for content retrieval process with mean of 97.15%, which also demonstrates the normal distribution of the accuracy for various datasets and various test runs (see Fig. 3).

The final outcome from the framework is nearly 99.93% accuracy for content retrieval process with mean of 99.41%, which also demonstrates the normal distribution of the accuracy for various datasets and various test runs (see Fig. 4).

Moreover, the incremental growth of the accuracy is also visible in the results and visualized graphically here (see Fig. 5).

Secondly, the time complexity analysis is carried out (Table 3).

It is natural to the observer that the time complexity is reducing progressively over the complete framework for three phases as nearly 9.54 s for the first phase, 7.93 s for the second phase, and, finally, 7.12 s in the final phase of the framework. The results are visualized graphically here (see Fig. 6).

Finally, with the detailed analysis of the obtained results, in the next section of the work, the comparative analysis is carried out.

Table 2 Accuracy analysis

Dataset	Test Run	Accuracy of SNRP-DS & ATC-DVS (%) [8]	Accuracy of VDS-ATC & VCR-LCKDF (%) [9]	Accuracy of KG-QPR & KFSRE-AP & TRRE & TRRD (%) [10]
BBC Motion Gallery Set-1	Test Run-1	94.68	96.98	99.12
BBC Motion Gallery Set-1	Test Run-2	96.55	96.64	99.18
BBC Motion Gallery Set-1	Test Run-3	94.49	97.72	99.30
BBC Motion Gallery Set-2	Test Run-4	96.11	97.38	99.21
BBC Motion Gallery Set-2	Test Run-5	94.04	97.93	99.70
BBC Motion Gallery Set-2	Test Run-6	95.65	97.78	99.05
BBC Motion Gallery Set-3	Test Run-7	95.98	96.80	99.74
BBC Motion Gallery Set-3	Test Run-8	96.83	96.63	99.64
BBC Motion Gallery Set-3	Test Run-9	95.83	96.62	99.93
TRECVID Set-1	Test Run-10	94.64	97.33	99.37
TRECVID Set-1	Test Run-11	96.15	96.61	99.26
TRECVID Set-1	Test Run-12	94.84	97.81	99.02
TRECVID Set-2	Test Run-13	96.80	96.91	99.78
TRECVID Set-2	Test Run-14	96.67	96.91	99.31
TRECVID Set-2	Test Run-15	96.42	97.18	99.52

6 Cloud-Based Deployment Models

One of the primary objectives of this research is to deploy the complete working framework on the cloud environment. Thus, this work selects the Amazon Web Services (AWS) cloud environment for the deployment of the work. The AWS framework provides multiple benefits for video content management and sharing. One the

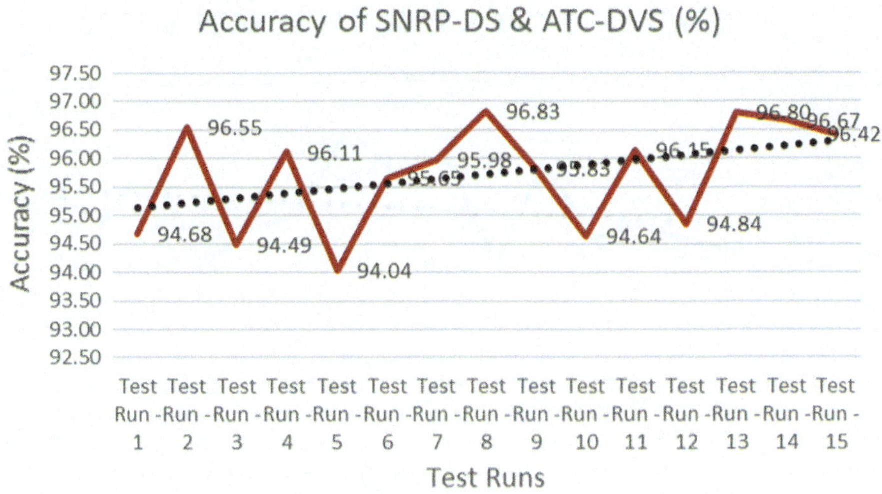

Fig. 2 Accuracy of SNRP-DS & ATC-DVS (%)

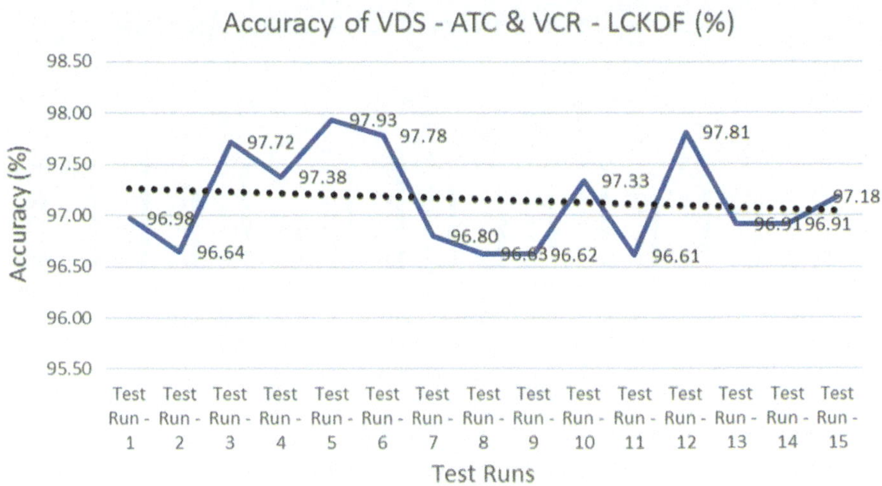

Fig. 3 Accuracy of VDS-ATC & VCR-LCKDF (%)

key components, which is used by the proposed framework, is the content management pipelines. The pipeline is one of the components which enables processing of the video data based on the sequences of the algorithms. The first two algorithms are deployed on the same component and the visualization is furnished here (see Fig. 7).

Further, the content management and retrieval algorithms can further be deployed on the jobs on AWS. The job component of the AWS enables multiple algorithms sequences to be applied on the video contents adaptively based on the pre-set

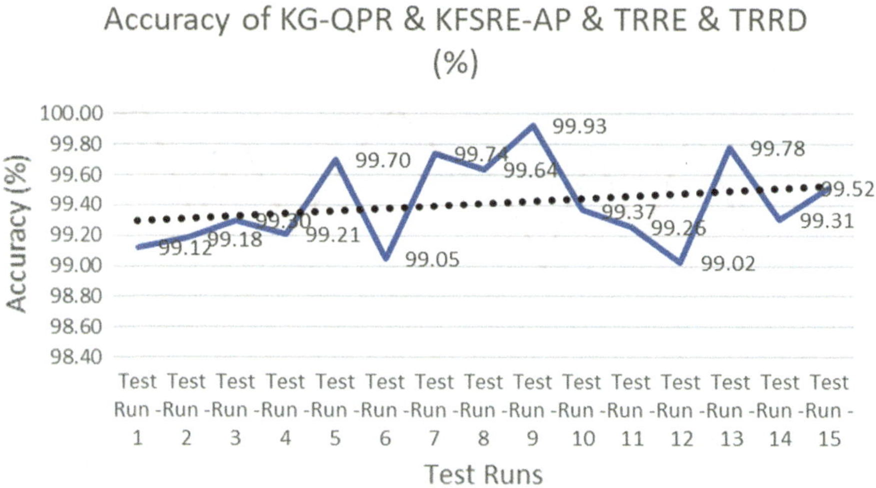

Fig. 4 Accuracy of KG-QPR & KFSRE-AP & TRRE & TRRD (%)

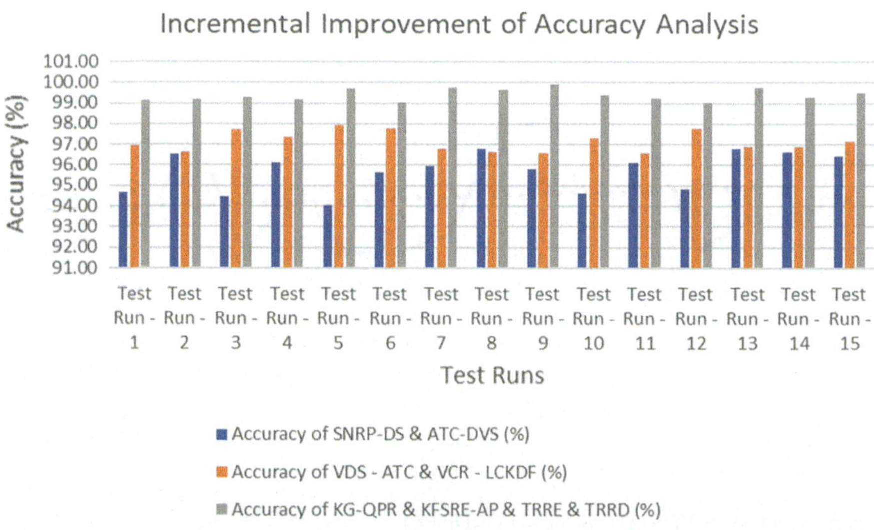

Fig. 5 Incremental improvement of accuracy analysis

sequences. The next two algorithms are deployed on the AWS and the deployment scenario is furnished here (see Fig. 8).

Further, the complete framework is again managed from the AWS EC2 instances and the deployment scenario is furnished here (see Fig. 9).

Furthermore, in the next section of this work, the comparative analysis is carried out.

Table 3 Time complexity analysis

Dataset	Test Run	Accuracy of SNRP-DS & ATC-DVS (Sec) [8]	Accuracy of VDS-ATC & VCR-LCKDF (Sec) [9]	Accuracy of KG-QPR & KFSRE-AP & TRRE & TRRD (Sec) [10]
BBC Motion Gallery Set-1	Test Run-1	9.07	7.80	7.44
BBC Motion Gallery Set-1	Test Run-2	9.80	8.24	7.70
BBC Motion Gallery Set-1	Test Run-3	9.16	7.44	6.71
BBC Motion Gallery Set-2	Test Run-4	9.56	7.49	6.03
BBC Motion Gallery Set-2	Test Run-5	9.61	8.07	7.94
BBC Motion Gallery Set-2	Test Run-6	9.10	7.99	6.45
BBC Motion Gallery Set-3	Test Run-7	9.79	8.28	7.92
BBC Motion Gallery Set-3	Test Run-8	9.58	7.16	7.15
BBC Motion Gallery Set-3	Test Run-9	9.87	8.30	7.04
TRECVID Set-1	Test Run-10	9.59	7.80	7.10
TRECVID Set-1	Test Run-11	9.94	8.89	7.58
TRECVID Set-1	Test Run-12	9.84	7.21	6.98
TRECVID Set-2	Test Run-13	9.13	7.92	6.28
TRECVID Set-2	Test Run-14	9.83	8.26	7.66
TRECVID Set-2	Test Run-15	9.22	8.13	6.77

7 Comparative Studies

For any research outcome, to be identified as the best, it is important to perform the comparative study of parallel research findings. Henceforth, the proposed framework is contrasted with the parallel major research findings in this portion of the study (Table 4).

In terms of retrieval accuracy, time complexity, overall framework complexity with additional features, it is observed that the proposed method outperforming. The improvements for content retrieval accuracy is nearly 15% and the improvement over time complexity is nearly 85%.

The accuracy improvements are visualized graphically here (see Fig. 10).

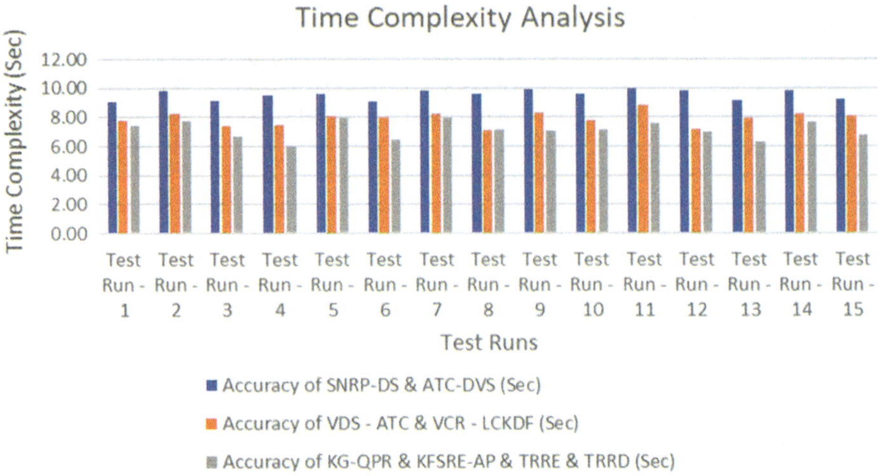

Fig. 6 Time complexity analysis

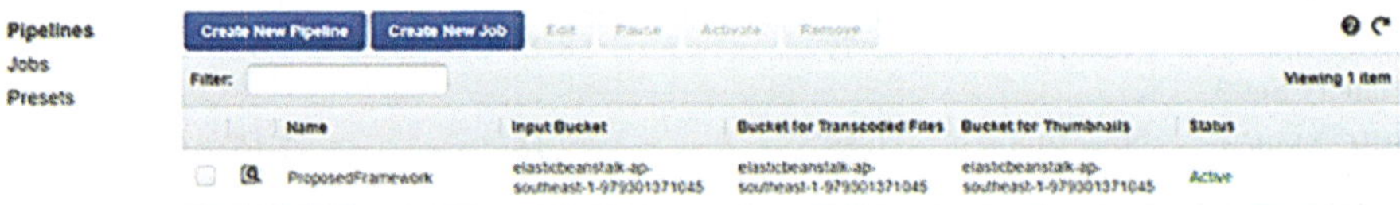

Fig. 7 Deployment on AWS pipelines

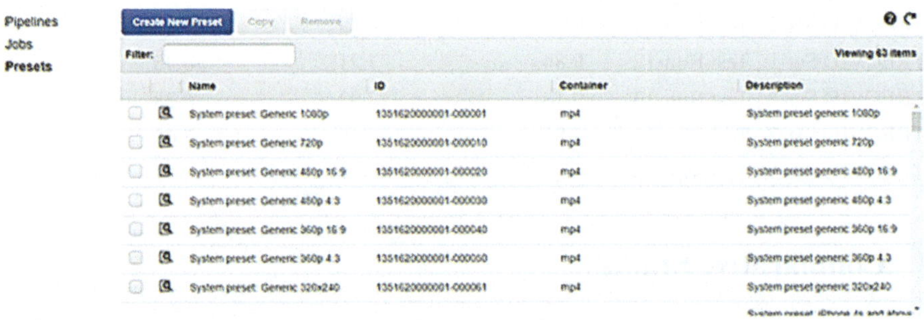

Fig. 8 Deployment on AWS jobs and pre-sets

The improvements over time complexity are also visualized graphically as (see Fig. 11).

Henceforth, in the next section of the work, the final research conclusion is presented.

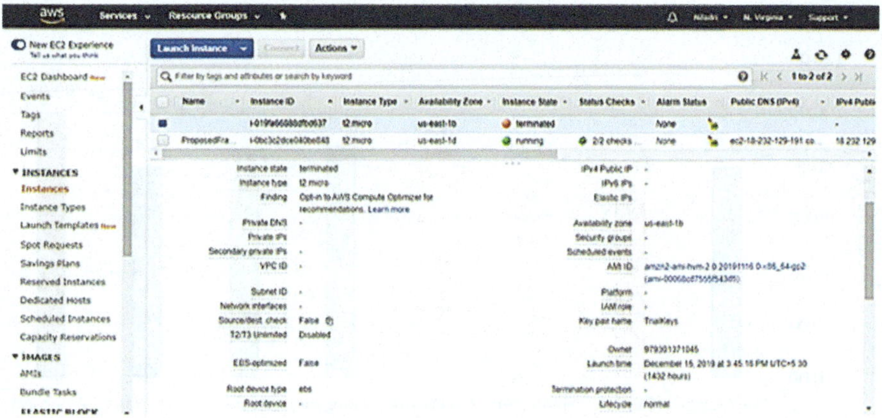

Fig. 9 Deployment on AWS EC2 instances

Table 4 Comparative analysis

Research outcome	Retrieval accuracy (%)	Time complexity (Sec)	Framework complexity	Applicable to distributed data
Kan Yang et al. [2]	79.98	51.36	$O(n * n^m)$	No
Xiaodan Song et al. [5]	71.96	86.31	$O(n^3)$	No
Yinan Feng et al. [1]	77.47	92.35	$O(n * m)$	Yes
Hanli Wang et al. [4]	85.49	42.62	$O(n^2)$	No
Guangyu Gao et al. [6]	87.86	80.59	$O(n * m)$	Yes
Wu Liu et al. [7]	75.82	78.10	$O(n^m)$	Yes
Zhen Yang et al. [3]	79.45	74.82	$O(n^2)$	No
Proposed Method, 2019	99.93	6.03	$O(\log n)$	Yes

8 Conclusion

The motivating factor for online video-based content management is wide adaptations and popularity. This catalyzed the growth in the recent research trends for making the video content management and revivals better and faster and more secure for cloud. However, the bottlenecks identified by many other researchers and this work as well are the setbacks for the growth of more adaptation of the video content delivery and management over cloud. Henceforth, this work designed a three-phase solution of the

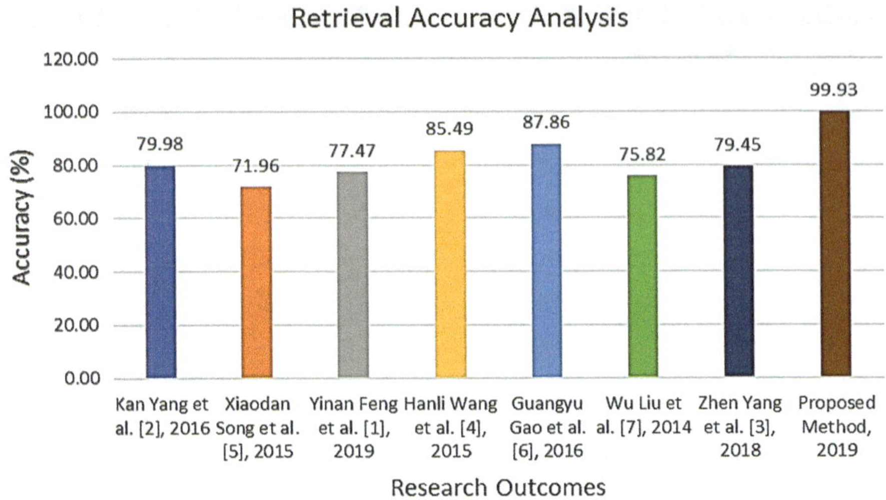

Fig. 10 Content retrieval accuracy analysis

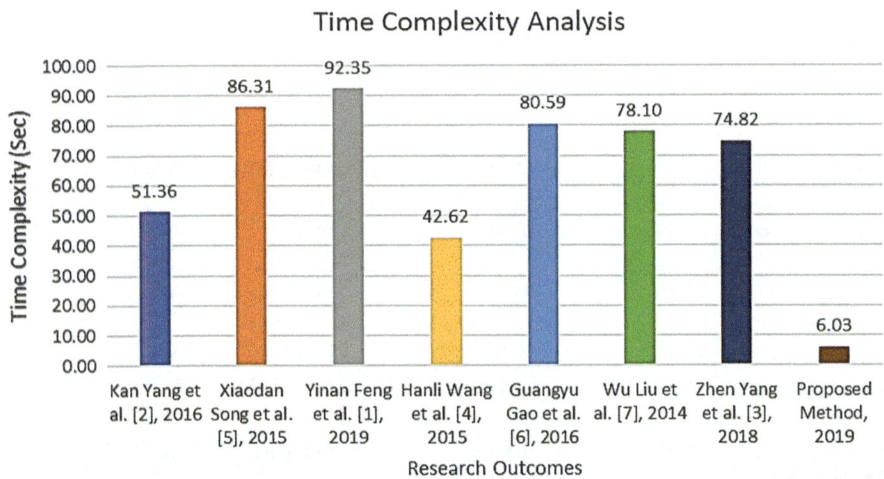

Fig. 11 Time complexity analysis

complete problem as in the first phase, designed a novel solution for the reduction of the keyframes with reduction of the time complexity and improvement of the content retrieval accuracy of 96.83%, in the second phase, designed a novel solution for improvement of content retrieval process on cloud-based or distributed online contents with the knowledge of data distributions over clusters with an accuracy of 97.83%, and, in the final phase, designed a novel solution for providing data encryption and decryption for data at rest and data on go for the complete framework with a reduction in time complexity and improved accuracy for content retrieval with an accuracy of retrieval 99.93%. Considering the final outcome and the comparative improvements over the other parallel research outcomes, this work is to be considered as one of the benchmarked work in this domain of research for making the video content management on cloud a faster, secure, and better area.

References

1. Y. Feng et al., Video big data retrieval over media cloud: a context-aware online learning approach. IEEE Trans. Multimed. **21**(7) July (2019)
2. K. Yang et al., Time-domain attribute-based access control for cloud-based video content sharing: a cryptographic approach. IEEE Trans. Multimed. **18**(5) May (2016)
3. Z. Yang et al., Cloud information retrieval: model description and scheme design. IEEE Access **6** (2018)
4. H. Wang et al., CHCF: a cloud-based heterogeneous computing framework for large-scale image retrieval. IEEE Trans. Circuit. Syst. Video Technol. **25**(12) Dec (2015)
5. X. Song et al., Cloud-based distributed image coding. IEEE Trans. Circuit. Syst. Video Technol. **25**(12) Dec (2015)
6. G. Gao et al., Cloud-based actor identification with batch-orthogonal local-sensitive hashing and sparse representation. IEEE Trans. Multimed. **18**(9) Sept. (2016)
7. W. Liu et al., Instant mobile video search with layered audio-video indexing and progressive transmission. IEEE Trans. Multimed. **16**(8) Dec. (2014)
8. T. Naga Raja et al., Video summarization using adaptive thresholding by machine learning for distributed cloud storage. Int. J. Eng. Adv. Technol. (IJEAT) (2019)
9. T. Naga Raja et al., A cloud based video content retrieval process by storage cluster aggregation using machine learning. J. Adv. Res. Dynam. Control Syst. (2019)
10. T. Naga Raja et al., Video content encryption as a service and performance implications of video data encryption on cloud. Int. J. Innov. Technol. Explor. Eng. (IJITEE) (2019)
11. Customized video processing modes For Hd-capable set-top decoders, Publication Number: WO/2006/112808, International Application No.: PCT/US2005/002750

Energy-Efficient Data Transmission in a Three-Hop Cooperative Cellular NB-IoT Network Using Double Auction

Srinivasa Rao Patri and L. Nithyanandan

Abstract A narrowband IoT (NB-IoT) device's battery life is vibrant for the future evolution of networks being wireless. The efficiency of energy for cooperative cellular NB-IoT networks is explored in this paper. In this context, the three-hop assignment problem is proposed; the theory of a double auction is used for its construction. The three-hop assignment aims to increase the battery life of a cell edge narrowband IoT users (CENUs) along with concentrating on energy-efficiency enhancement. In the proposed model, to get maximum battery life, the transmission power of NB-IoT user (NU) is taken into consideration and is decreased to the possible lowest magnitude. Also, to increase the battery life of CENU, an energy-efficient narrowband IoT user–maximum weight matching method (EENU–MWM) is recommended. Lastly, EENU–MWM performance is assessed in terms of capacity, EE, transmission time, which demonstrates that EENU–MWM can significantly enhance the efficiency of the cooperative NB-IoT cellular network.

Keywords NB-IoT · Cooperative communication · Double auction · Three-hop

1 Introduction

As of late, the world has seen a great deal of IoT based items in the market, for example, smart healthcare, smart agriculture, smart home, smart grid, etc. [1, 2]. These applications aid people in various aspects of their life. Narrowband Internet of Things (NB-IoT) unconfined by Third-Generation Partnership Project (3GPP), an important branch of IoT, is the emerging and sustainable 5G radio access technology that has the ability to support large number of devices with low power characteristics:

S. R. Patri (✉)
VNR Vignana Jyothi Institute of Engineering and Technology, Hyderabad, India
e-mail: srinivasarao.patri@gmail.com

L. Nithyanandan
Pondicherry Engineering College, Pudhucherry, India
e-mail: nithi@pec.edu

C. Kiran Mai et al. (eds.), *Proceedings of International Conference on Advances in Computer Engineering and Communication Systems*, Learning and Analytics in Intelligent Systems 20,
https://doi.org/10.1007/978-981-15-9293-5_22

lengthy battery life, wide coverage, and massive scale support. Technologies like Wi-Fi, ZigBee, Bluetooth, and 2G/3G/4G do not have the ability to support the above-mentioned requirements. In response, the applications are targeted by technologies of Low Power Wide Area (LPWA). The bandwidth of system utilized by NB-IoT by both uplink and downlink is set to be 180 kHz, so as to meet the requirements of LPWA. In LTE standard, the size of one physical resource block (PRB) is 180 kHz. According to the release 13, the NB-IoT device maintains a lifespan of 10 years and has a battery capacity of 5Wh within the site sector, supports at least 52,547 low-throughput devices for connection, where the traffic model assumes either 20 devices per person or 40 devices per household [3]. NB-IoT is designed in such a way that it can co-exist with Long-Term Evolution [4] and can be deployed in three operation modes which depend on LTE band used: in-band, guard-band, and stand-alone operation modes. NB-IoT requires a minimum bandwidth of 180 kHz for both downlink and uplink, i.e., one physical resource block (PRB) split into 12 subcarriers of 15 kHz each, irrespective of the operation mode. The energy-efficient broadcast problem in wireless networks is explored by Maric and Yates [5]; for multiuser cooperative communication the relay allocation problem is solved by Luo et al. [6]; the relay selection and power allocation scheme for decode and forward (DF) cooperative network based on energy pricing is proposed by Ke and Feng [7].

2 Relay Selection in a Three-Hop Cooperative Cellular NB-IoT Network

2.1 System Model

The battery life of an NB-IoT User Equipment (NU) depends on the transmission power of a NU. The transmission power of NU should be very less in order to get more battery life. The transmission power of NU also depends on the distance between NU and eNB. If the NU is in region r_1 shown in Fig. 1, as the distance between NU and eNB is very less, direct transmission is possible, if the NU is in region r_2, the distance between NU and eNB is more and we have to take the help of a relay as shown in Fig. 2. Based on the distance between NU and eNB, sometimes there is a need to take the help of a greater number of relays. If the NU is in region r_3 (CENU) we have to take the help of two relays (Three-Hop). In this paper, we have taken two relays (Three-Hop) as shown in Fig. 3. Presumed TDMA is made used for transmissions in two phases: its relay users and a source user. Obviously, orthogonal transmissions referring to NU and relay users can be guaranteed by the usage of other multiple access technologies.

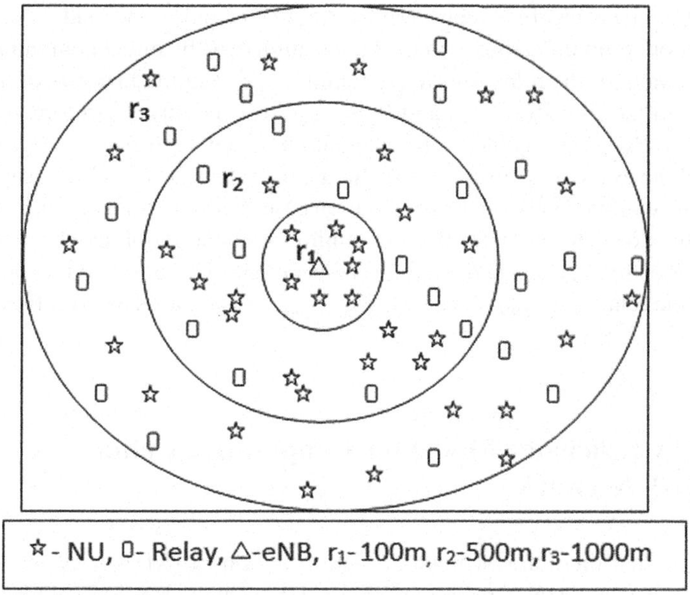

Fig. 1 NUs and relays random distribution

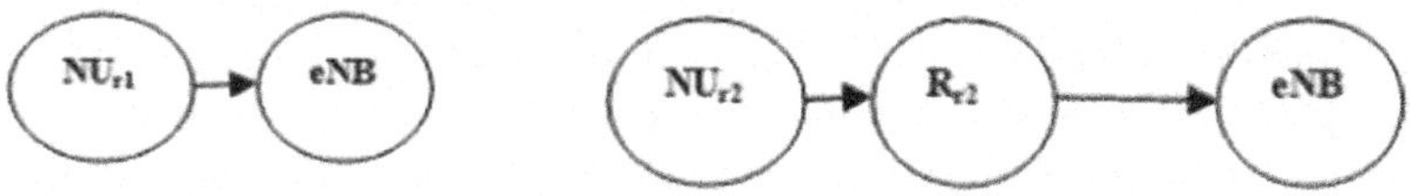

Fig. 2 Direct and two-hop transmission

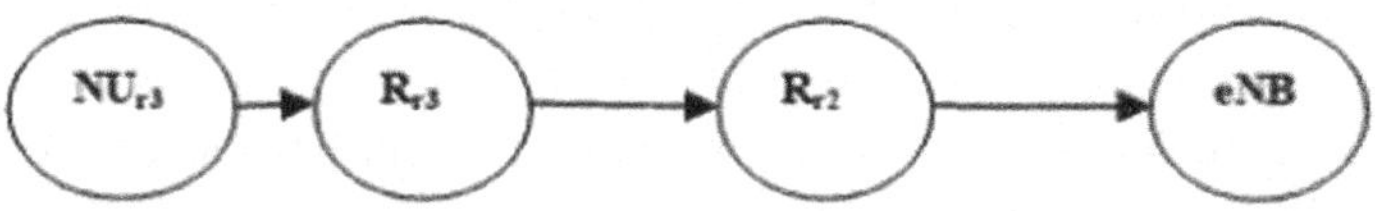

Fig. 3 Three-hop transmission

3 Double Auction Based Model for Three-Hop NB-IoT Cooperative Cellular Network

In this section, the double auction model [8] for negotiating among users at source (CENUs) and the relay cell users (CUs) is given, and in order to obtain high energy efficiency, a method is explored by assigning diverse relay users to CENUs, relay users to relay Users. In this double auction model, for obtaining services from the relay users CENUs play the role of buyers, and relay users behave as buyers and sellers for as long as relay services. Thus, users at source need to pay to first relay users in region r_3 and first relay users in region r_3 pay to second relay users in region r_2. For communicating with the base station eNB by direct communication or through the help of the relay user $R_{r3,}$ the minimal transmitted powers of the source user NU_{r3} utilized are $P_{NUr3-eNB}$ and $P_{NUr3-Rr3}$, respectively. Therefore, we can use $P_{NUr3-eNB} - P_{NUr3-Rr3}$ as the proper validation of the source user NU_{r3} when the situation chooses the relay user R_{r3} to estimate the relay service of the relay user $R_{r3.}$ In order to communicate with the base station eNB by direct transmission or through the assistance of the relay user $R_{r2,}$ the minimum transmission powers used by the relay user R_{r3} are $P_{Rr3-eNB}$ and $P_{Rr3-Rr2}$, respectively. Therefore, to reckon the relay service of the relay user R_{r2}, we can use $P_{Rr3-eNB} - P_{Rr3-Rr2}$ as the faithful validation of the relay user $R_{r3.}$

4 Energy Efficiency Model for Cooperative Cellular NB-IoT Network

We presume that there are M Source users in region r_3 (NU_{r3m}), N relay users in region r_3 (R_{r3n}) and Q relay users in region r_2 (R_{r2q}) in the NB-IoT cellular network. The main motivation behind the allocation of the relay users to users at the source is to maximize the capability of users at source. The CENUs information will be forwarded through these relay users when one NU_{r3m} is matched with one relay user R_{r3n} and this relay user R_{r3n} is matched to the relay user R_{r2q}. In the forward link, $NU_{r3m} \rightarrow R_{r3n} \rightarrow R_{r2q}$ obtains the capacity $C_{NUr3m,Rr3n,Rr2q}$ with the energy consumption of NU_{r3m}, $R_{r3n,}$ and $R_{r2q.}$ So, for each forward link, energy efficiency can be defined as follows:

$$EE_{m,n,q} = \frac{C_{NUr3m,Rr3n,Rr2q,eNB}}{P_{NUr3m} + P_{Rr3n} + P_{Rr2q}} \tag{1}$$

and the sum of efficiency i.e. independent energy of every link can be used to define EENU–MWM system energy efficiency (EE).

$$EE = \sum_{m=1}^{M}\sum_{n=1}^{N}\sum_{q=1}^{Q} \frac{C_{NUr3m,Rr3n,Rr2q,eNB}}{P_{NUr3m} + P_{Rr3n} + P_{r2q}} \times t_{m,n,p} \tag{2}$$

where Weight is,

$$W_{m,n,q} = EE_{m,n,q} = \frac{C_{NUr3m,Rr3n,Rr2q,eNB}}{P_{NUr3m} + P_{Rr3n} + P_{Rr2q}} \tag{3}$$

and $t_{m,n,p} = 1$, when the connection is established between NU_{r3m}, $R_{r3n,}$ and $R_{r2q.}$

The following example illustrates the suggested EENU–MWM. There are four users at source present in r_3, four relay users present in $r_{3,}$ and four relay users in r_2.

The bid matrix of the users at source B_1 and the ask matrix A_1 of the relay users R_{r3} are

$$B_1 = \begin{array}{c|cccc} & R_{r3,1} & R_{r3,2} & R_{r3,3} & R_{r4,4} \\ NU_{r3,1} & 4 & 3 & 5 & 1 \\ NU_{r3,2} & 3 & 5 & 2 & 5 \\ NU_{r3,3} & 5 & 2 & 1 & 3 \\ NU_{r3,4} & 1 & 2 & 3 & 4 \end{array}$$

$$\mathrm{A}_1 = \begin{array}{c|c} & \mathrm{NU}_{r3,1}/\mathrm{NU}_{r3,2}/\mathrm{NU}_{r3,3}/\mathrm{NU}_{r3,4} \\ \mathrm{R}_{r3,1} & 4 \\ \mathrm{R}_{r3,2} & 5 \\ \mathrm{R}_{r3,3} & 4 \\ \mathrm{R}_{r3,4} & 2 \end{array}$$

Based on bid values of the CENU and ask values of the relay in region r_3 and the matching condition, i.e., $B_{NUr3m,Rr3n} \geq A_{Rr3n}$, bipartite graph and the initial matching matrix T_1 are acquired as follows:

$$T_1 = \begin{array}{c|cccc} & R_{r3,1} & R_{r3,2} & R_{r3,3} & R_{r3,4} \\ NU_{r3,1} & 1 & 0 & 1 & 0 \\ NU_{r3,2} & 0 & 1 & 0 & 1 \\ NU_{r3,3} & 1 & 0 & 0 & 1 \\ NU_{r3,4} & 0 & 1 & 1 & 1 \end{array}$$

The corresponding matching matrix value is 1 if the bid value of CENU is greater than or equal to ask the value of relay in region r_3, otherwise 0. Data transmission path will be established based on the matching matrix as shown in Fig. 4.

Weigh of each data transmission path from CENU to relay will be calculated by using the Eq. (3).

The weight matrix of all 4 relay users and source users is

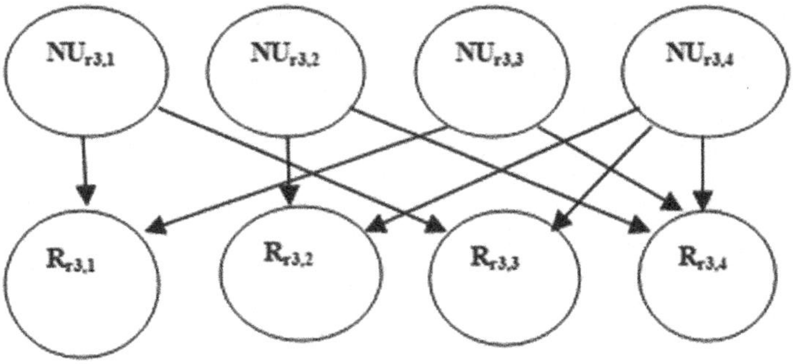

Fig. 4 Bipartite graph for matrix T_1

$$W_1 = \begin{array}{c|cccc} & R_{r3,1} & R_{r3,2} & R_{r3,3} & R_{r3,4} \\ NU_{r3,1} & 2 & 0 & 4 & 0 \\ NU_{r3,2} & 0 & 3 & 0 & 2 \\ NU_{r3,3} & 5 & 0 & 0 & 4 \\ NU_{r3,4} & 0 & 2 & 3 & 4 \end{array}$$

Based on the maximum weight of the data transmission path and matrix T_1 corresponding matching matrix as follows:

$$T_{w1} = \begin{array}{c|cccc} & R_{r3,1} & R_{r3,2} & R_{r3,3} & R_{r3,4} \\ NU_{r3,1} & 0 & 0 & 1 & 0 \\ NU_{r3,2} & 0 & 1 & 0 & 0 \\ NU_{r3,3} & 1 & 0 & 0 & 0 \\ NU_{r3,4} & 0 & 0 & 0 & 1 \end{array}$$

Final connection is established by using NUEE-MWM as shown in Fig. 5.

The bid matrix of the relay users (R_{r3}) $B2$, the ask matrix of the relay users (R_{r2}) $A2$, and the matching matrix T_2 and the final connection as shown in Fig. 6 are

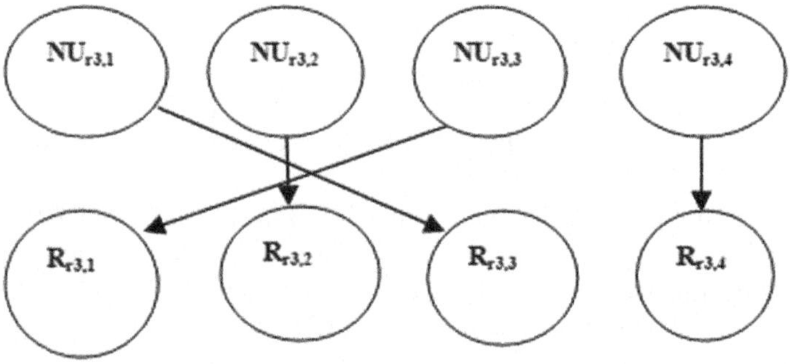

Fig. 5 Bipartite matrix for matrix T_{w1}

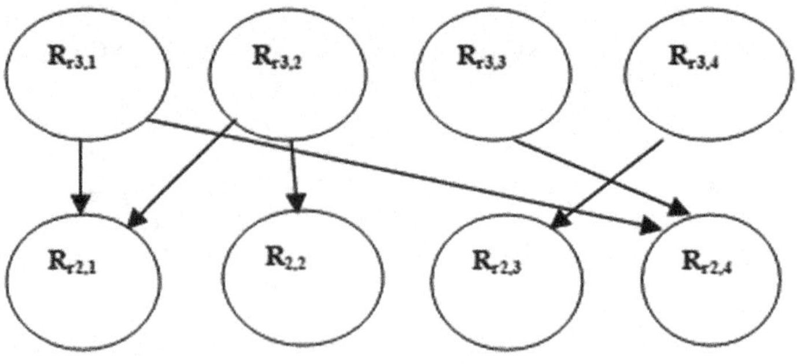

Fig. 6 Bipartite for matching matrix T_2

$$B_2 = \begin{array}{c|cccc} & R_{r2,1} & R_{r2,2} & R_{r2,3} & R_{r2,4} \\ R_{r3,1} & 3 & 1 & 2 & 4 \\ R_{r3,2} & 3 & 4 & 2 & 1 \\ R_{r3,3} & 1 & 3 & 2 & 5 \\ R_{r3,4} & 1 & 3 & 5 & 3 \end{array}$$

$$A_2 = \begin{array}{c|c} & R_{r3,1}/R_{r3,2}/R_{r3,3}/R_{r3,4} \\ R_{r2,1} & 2 \\ R_{r2,2} & 3 \\ R_{r2,3} & 5 \\ R_{r2,4} & 4 \end{array}$$

$$T_2 = \begin{array}{c|cccc} & R_{r2,1} & R_{r2,2} & R_{r2,3} & R_{r2,4} \\ R_{r3,1} & 1 & 0 & 0 & 1 \\ R_{r3,2} & 1 & 1 & 0 & 0 \\ R_{r3,3} & 0 & 0 & 0 & 1 \\ R_{r3,4} & 0 & 0 & 1 & 0 \end{array}$$

The weight matrix of all four relay users in region r_3, all four relay users in region $r_{2,}$ and the matching matrix and the connection as shown in Fig. 7 are

$$W_2 = \begin{array}{c|cccc} & R_{r2,1} & R_{r2,2} & R_{r2,3} & R_{r2,4} \\ R_{r3,1} & 4 & 0 & 0 & 3 \\ R_{r3,2} & 3 & 5 & 0 & 0 \\ R_{r3,3} & 0 & 0 & 0 & 2 \\ R_{r3,4} & 0 & 0 & 3 & 0 \end{array}$$

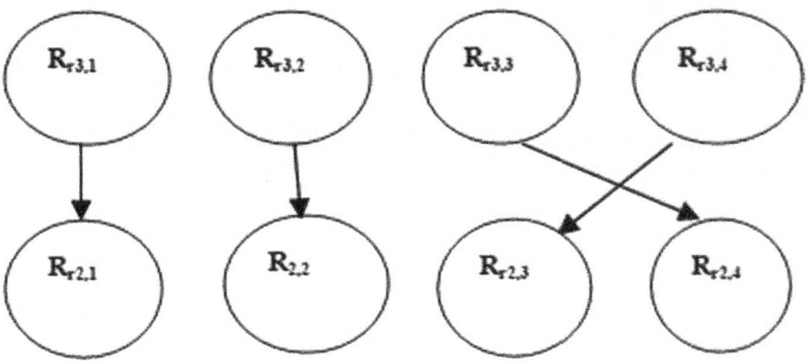

Fig. 7 Bipartite matrix for matching matrix T_{w2}

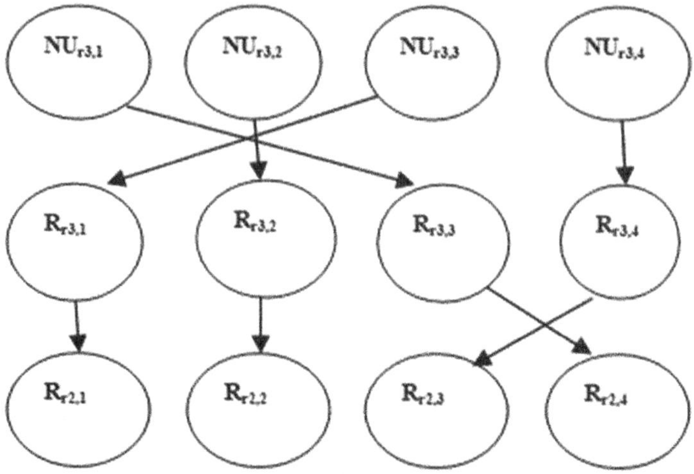

Fig. 8 EENU-MWM system bipartite graph

$$T_{w2} = \begin{array}{c|cccc} & R_{r2,1} & R_{r2,2} & R_{r2,3} & R_{r2,4} \\ R_{r3,1} & 1 & 0 & 0 & 0 \\ R_{r3,2} & 0 & 1 & 0 & 0 \\ R_{r3,3} & 0 & 0 & 0 & 1 \\ R_{r3,4} & 0 & 0 & 1 & 0 \end{array}$$

From matrices T_{w1} and $T_{w2,}$ we can get the EENU–MWM system bipartite graph as shown in Fig. 8.

NU$_s$ in region r_3 transmits the data to relays in region r_3, relays in region r_3 transmits the NU data to relays in $r_{2,}$ and relays in region r_2 transmits NU data to relays in $r_{1,}$ from that data will be sent to eNB to get maximum efficiency by using EENU–MWM method.

Total power saves in EENU-MWM method is

$$P_{NUr3-eNB} - P_{NUr3-Rr3} = \frac{SNR_{eNBTH}\sigma^2}{G_{NUr3-eNB}} - \frac{SNR_{Rr3TH}\sigma^2}{G_{NUr3-Rr3}} \tag{4}$$

The power saves of CENU device depends on relay selection distance. If the relay selection distance is less power save of CENU is more, if the distance is less power save is less. In this proposed relay, selection completely depends on EENU–MWM method.

5 Numerical Results

A cooperative cellular NB-IoT wireless network, where the source users and relay users are arbitrarily dispersed in a 1000-m radius globular area is considered. At the center of this area, the eNB is located. Region r_1 range is 100-m, range of region r_2 is 500 m and range of region r_3 is 500 m, i.e., r_1, r_2 and r_3 radii from the eNB are 100 m, 500 m, and 1000 m, respectively. The frequency bandwidth for each of these relay users (CU) is 25 kHz and the frequency bandwidth of each of the narrowband IoT users (CENU) is 15 kHz. The path loss component is set to 3.3. The noise power AWGN is 10^{-10} W.

5.1 *Energy Efficiency Versus SRP*

In this section, the energy efficiency in NB-IoT cooperative cellular networks evaluated. Energy efficiency in EENU–MWM is based on the number of source relay user pairs (SRP) and weight. The assessment is done based on the energy efficiency in the case of signal transmission from NU_{r2} to eNB and also from NU_{r3} to base station eNB (Fig. 9).

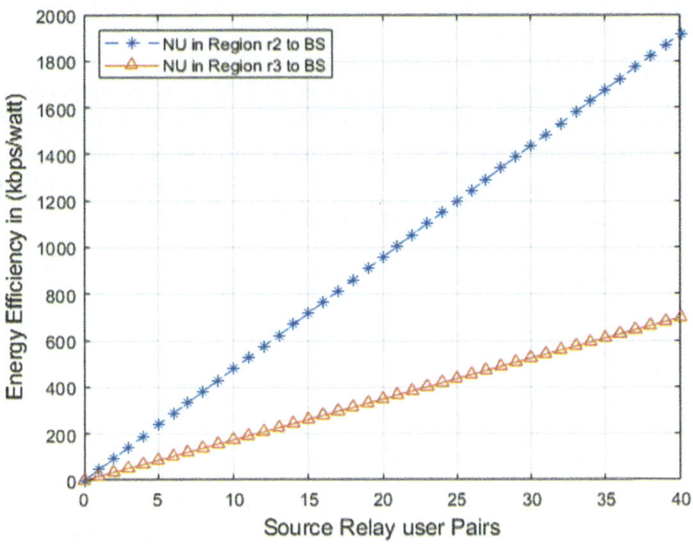

Fig. 9 Energy efficiency versus SRP

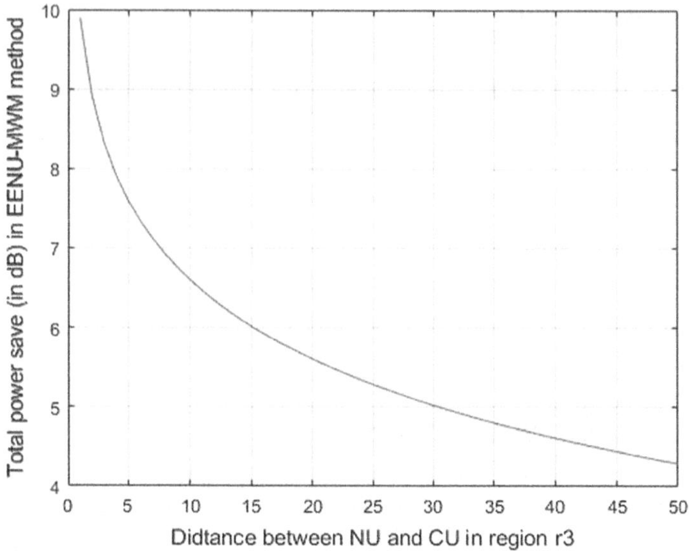

Fig. 10 Total power save versus distance

5.2 Total Power Save Versus Distance Between CENU and CU

In this section, the total power save in proposed EENU–MWM method is evaluated. In this proposed EENU–MWM method, total power save is evaluated for 5 dB SNRth at eNB and also CU in region $r3$. If the distance between NU and CU in region $r3$ is less power save is more. CU selection in region $r3$ is based on EENU-MWM algorithm to increase energy efficiency (Fig. 10).

6 Conclusion

In this model, the selfishness of the NU user and two relay users are considered. Moreover, based on offerings of NUs and demands of two relay users, an optimized NU–relay–relay matched scheme, called EENU–MWM, is suggested in order to increase the efficiency of systems energy. Limitation for further study is in the suggested double auction model, when CENUs and relay users shot their offerings and demands to the auctioneer (eNB), the auctioneer suits these diverse CENUs and diverse relay users agreeing to the testified offerings and demands, there will be no negotiating among CENUs and relay users. Dynamic negotiating mechanisms are taken into consideration for further improving the performance of the system.

References

1. I. Yaqoob et al., Internet of things architecture: Recent advances, taxonomy, requirements, and open challenges. IEEE Wirel. Commun. **24**(3), 10–16 (2017)
2. H. Sallouha, S. Rajendran, Localization in ultra narrow band IoT networks: design guidelines and tradeoffs. IEEE Internet of Things J. **6**(6), 2327–4662 (2019)
3. J. Xu et al., Narrow band internet of things: evaluations, technologies, and open issues. IEEE Internet of Things J. **5**(3), 2347–4662 (2018)
4. G. Tsoukaneri, M. Condolusi et al., Group communications in narrowband-IoT: architecture, procedures, and evaluation. IEEE Internet of Things J. **5**(3), 2327–4662 (2018)
5. I. Maric, R.D. Yates, Cooperative multihop broadcast for wireless networks. IEEE J. Sel. Areas Commun. **22**(6), 1080–1088 (2004)
6. J. Luo et al., New approaches for cooperative use of multiple antennas in ad hoc wireless networks, in Proceedings of 60th IEEE vehicular technology conference, September 2004, vol. 4, pp. 2769–2773 (2004)
7. F. Ke, S. Feng, Relay selection and power allocation for cooperative network based on energy pricing. IEEE Commun. Lett. **14**(5), 1089–7798 (2010)
8. Y. Li, C. Liao, Y. Wang, Energy-efficient optimal relay selection in cooperative cellular networks based on double auction. IEEE Trans. Wirel. Commun. **14**(8), 1536–1276 (2015)

Groundwater Level Analytics for Effective Water Level Predictions and Visualization of the Patterns

Myneni Madhu Bala, C. Keerthana, D. Pallavi Raj, and K. Prashanthi

Abstract In present days, everybody is depending on the borewells to get water for domestic as well as agriculture purposes. With submersible pumps set at few hundreds to thousand feet into the earth, there is no clue about the water availability levels and its recharge patterns. In existing systems, a sensor fixed to the motor tells us about the availability of water near the motor but is not a reliable indicator. Therefore, people are depending on a kind of guess and magnetic field techniques to assess the water level and its availability. In this paper, groundwater level analytics are performed on the data collected from Telangana region to visualize its patterns. This analysis helps to predict the water level and closely observe the recharge pattern by monitoring rainfall data.

Keywords Groundwater level analytics · Predictions · Visualizations · Linear regression

1 Introduction

Water is one of the essential substances for our survival as well as the earth's. Presently, we are facing a scarcity of the freshwater resources due to various reasons such as over exploitation, high degree of pollution, depletion of sources of recharge like glaciers and run off rainwater without percolating in ground. To avoid this

M. Madhu Bala · C. Keerthana (✉) · D. Pallavi Raj · K. Prashanthi
Computer Science and Engineering, Institute of Aeronautical Engineering, Hyderabad, Telangana, India
e-mail: ckeerthana53@gmail.com

M. Madhu Bala
e-mail: baladandamudi@gmail.com

D. Pallavi Raj
e-mail: pallavirajdamekar@gmail.com

K. Prashanthi
e-mail: koyyadaprashanthi@gmail.com

C. Kiran Mai et al. (eds.), *Proceedings of International Conference on Advances in Computer Engineering and Communication Systems*, Learning and Analytics in Intelligent Systems 20,
https://doi.org/10.1007/978-981-15-9293-5_23

problem, we can use other resources like groundwater. Groundwater is one among the country's precious resources. To execute any successful water management program efficiently, continuous monitoring of the groundwater level is necessary. In the continuous changing of environments, monitoring the levels of the groundwater is a critical issue. This monitoring of groundwater can also give us the information of the upcoming drought and other water supply conditions. In order to monitor the water levels, we will require more amount of data in order to assess and predict the groundwater conditions. This process can help us in situation like drought to, loss of pump age in various fields like agriculture, domestic, and industrial water supply. Using commercially available software, the data collected in PC or server or cloud can be automatically formatted as different types of graphs and tables. A rainfall gage installed can give the additional information on amount of rainfall in the region. In a case study of Andhra Pradesh, the data collected shows that in the last 2.5 years, the water level has risen by almost 2 m. After each monsoon, a good rise in the water table is seen which shows that the rainwater is recharging the subsoil aquifers, recovering the water table properly. This paper talks about the useful conclusions that can be drawn from the groundwater data, comparing water level of different regions and periods. To access the groundwater, we have installed hand pumps and wells. Presently, most of us drill borewell in our house to use groundwater. A borewell is a deep narrow well, typically equipped with pump to draw the water to the surface. The main source for borewell is groundwater. It is normally found in the underground between the cracks and spaces of soil. Rainfall is the important source which helps in recharge of the groundwater, during rainfall most of the rainwater flows toward lakes, rivers, and few percent of it goes inside the soil and there by recharging the groundwater. Groundwater represents 30% of world's fresh water. It is naturally recharged by rainwater, snowmelt, or water that leaks to the bottom of some lakes and rivers. The recharge patterns are directly proportional to the amount of rainfall. If you filter rainwater and let it into the well, it will recharge the well. In this present project, we are working on the analysis of the data of groundwater received from the government. We provide visualizations for various parameters which help us to analyze the groundwater, rain falls and borewells throughout the year in various districts of Telangana.

2 Literature Survey

Water is the primary source as it is available sufficiently and can be used for any requirement. Most percentage of the water we receive is form groundwater. It is a natural resource which supports daily routines, irrigation, and helps in the betterment of economy. Monitoring groundwater may help the country in various areas like industries, domestic and in economy. There are various methods available for us to work and get more information of groundwater and its existence in underground surface, from the traditional to the modern ways [1]. Groundwater System is heterogeneous in nature with dynamic patterns which makes it difficult to understand. Hence, reliable predictions are required to manage water resources in that region. We receive additional benefits when there is recharge of the groundwater. Study showed that river stream flow patterns had a great effect on natural recharge [2]. This analysis proves that floods and stream flow have a significant effect on recharge pattern of the groundwater [3]. India being a dependent country on groundwater, the users are given a direct access to use them. This led to the fall of wells mostly. The users started installing pumps at greater depths to pump the groundwater. This access led them to a continuous competition of installing high-capacity pumps, that left behind the poorer and marginal to the increasing burden of energy subsides per unit of pumped water to the exchequer. The digging and drilling are causing the exploitation of the groundwater [4]. To evaluate groundwater stability and flow, fluctuations in groundwater level can be used. The groundwater plays important role in agricultural areas.

The assessment of the groundwater by the agricultural water managers is done to provide the groundwater required during the sowing season. The government agencies collect the data very few times a year which leads to insufficient data. It is necessary to get an enough data to make accurate prediction of groundwater level and its variations. This groundwater level is influenced by various factors like urbanization, tidal effects, meteorological parameters, frost, precipitation, etc. When the physical model to stimulate groundwater is insufficient, the data-driven models like artificial neural network, support vector techniques machine are the alternatives. Even though support vector machine being the model that can overcome the drawbacks of artificial neural network, it consumes more time because of trial and error method [5]. Groundwater use is measured by factors such as its availability, accessibility, transportability, and cost-effectiveness. Due to some socio-economic conditions and eco-environment, the groundwater decreases [6]. For collecting the large amount of data, for example, data like air quality, health, groundwater level data we are using technology to capture data from sources like satellites and ground sensors. Their applications and theories of probabilistic have been limited for real problems and they have not become the regular tools for hydrological modeling [7]. The prediction model by using linear regression is giving high appropriateness on time series data [8].

3 Problem Definition

The prediction of water level and monitoring of borewell health will give a solution to a social problem. Through this setup, indicator can develop the water levels data in borewell which leads to appropriate decisions in making prediction. This led to use of sensors to detect water and these sensors push the data to a server which captures the data into a file system. In this project, to analyze the given file data and give graphical output representing various parameters. Using this approach, one will have clear understanding about the existing borewell, groundwater, rainfall reports.

New borewell positions can be well identified by looking at these statistics from surrounding ones in various districts in Telangana.

The data taken is used to implement the following in a graphical format:

- Rise of borewells in 1 year
- Fall of borewells in 1 year
- Month wise average rainfall
- Month wise average water level depth.

4 Methodology

The methodology adopted in groundwater level analytics and predictions are based on sensor data from borewells located at various locations in Telangana region. It includes data collection from borewells, transformation, and cleaning of data, filteration of useful data for predictions and visualizations of various patterns included and prediction of groundwater levels (Fig. 1).

In this paper, the data is collected from the Groundwater Department of Telangana. The Government of Telangana is monitoring groundwater levels at 1278 stations using 699 piezometers and 253 dug wells. Most percentage of groundwater recharge is contributed by rainfall. The behavior of groundwater table is essentially governed by rainfall, its quantity, intensity, and frequency. The state received 971 mm rainfall (till January 2020 end) against a normal rainfall of 852 mm, which is ~14% more than the normal annual rainfall during in 2019–20.

Using this data, the analysis was performed and visualization of groundwater level, rainfall and borewell is provided in various districts of Telangana. This can help us to predict future possibilities and unknown patterns between rainfall and

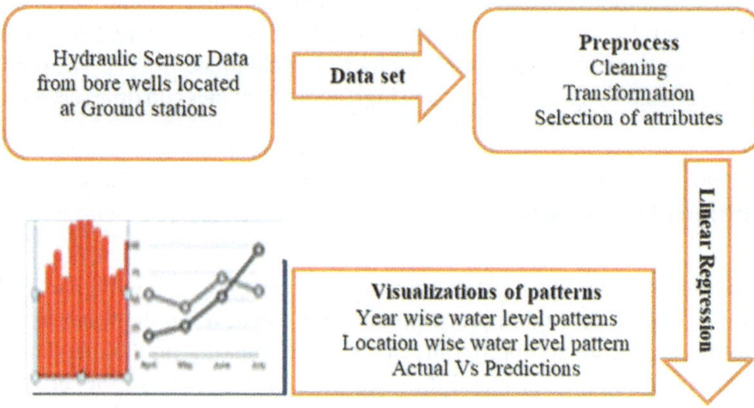

Fig. 1 Block diagram of proposed groundwater analytics

groundwater. We used a linear regression method to make predictions. We can see the change in groundwater level based on rainfall in that area.

The groundwater mainly depends on rainwater to recharge. Working on both groundwater data and rainfall helps us in many ways. Analysis of this data may help us predict and forecast the future pattern. In this project, we use linear regression in order to predict the data. Regression technique is mainly used for predictions and to forecast the data. Here, in linear regression, we followed $Y = mX + c$, where X is the rainfall and Y is groundwater level depth. Based on this, we can know the relationship between rainfall and groundwater level and find out how groundwater level is depended on rainfall. Using the equation, we substitute $x =$ rainfall values and $y =$ water level depth. By this, we get

$$Y = mX + c \tag{1}$$

$$X = mean(x) \tag{2}$$

$$Y = mean(y) \tag{3}$$

$$m = \frac{Sum\big((x - X)^*(y - Y)\big)}{Sum\big((x - X)^*(x - X)\big)} \tag{4}$$

By using these values, we find the value of c and hence we get an equation of regression line. Using this regression line, we can predict the future values which help us in forecasting.

5 Results and Discussion

Based on the data which is taken from the Government of Telangana [Department of Groundwater], we can visualize the average rainfall and average water level depth from February 2019 to January 2020.

In Fig. 2, we observe the data of month-wise average rainfall, rainfall deviation, and average water level depth.

In the above Fig. 3, we can observe how there is a fluctuation in water level depth. If we investigate the above graph of rainfall, we can see that when there is scarcity in rainfall the water level depth is increased.

In this above Fig. 4, we can observe the rainfall pattern throughout the year from February 2019 to January 2020. This data is taken from various districts of Telangana.

Figure 5 shows the average water level rise in borewell. It is observed that the percentage of rise in borewell in various districts of Telangana from January 2019 to January 2020.

	month	avg_rainfall	rain_dev	avg_waterlevel_depth
0	feb_19	722.00	-16	12.53
1	mar_19	723.00	-16	13.40
2	april_19	723.00	-16	14.14
3	may_19	738.00	-18	14.56
4	june_19	87.10	-33	14.40
5	july_19	302.30	-19	14.12
6	aug_19	545.90	-8	11.15
7	sep_19	794.84	10	9.85
8	oct_19	953.40	17	7.92
9	nov_19	961.10	14	7.53
10	dec_19	964.00	14	8.12
11	jan_2020	971.00	14	8.88

Fig. 2 Month-wise data of Telangana from February 2019 to January 2020

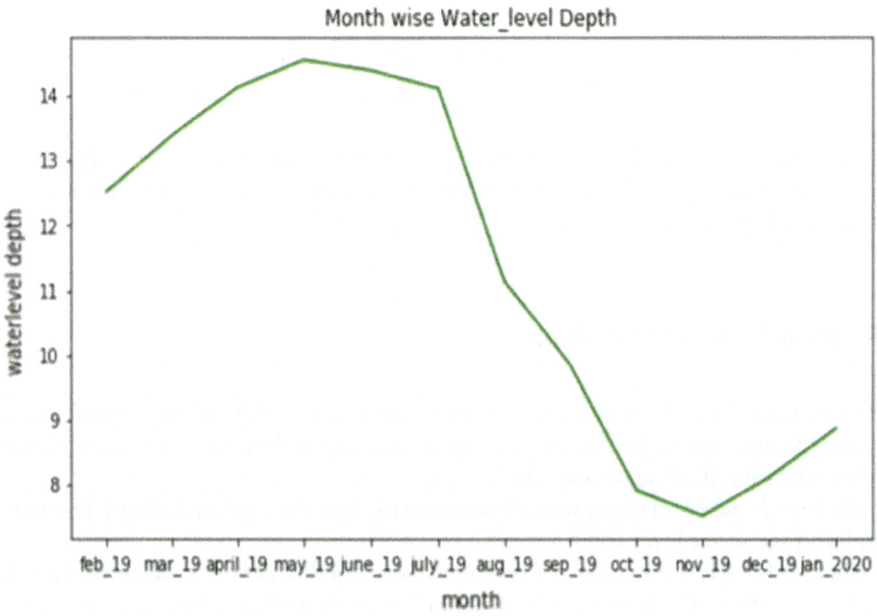

Fig. 3 Line plot of water level depth observed from February 2019 to January 2020

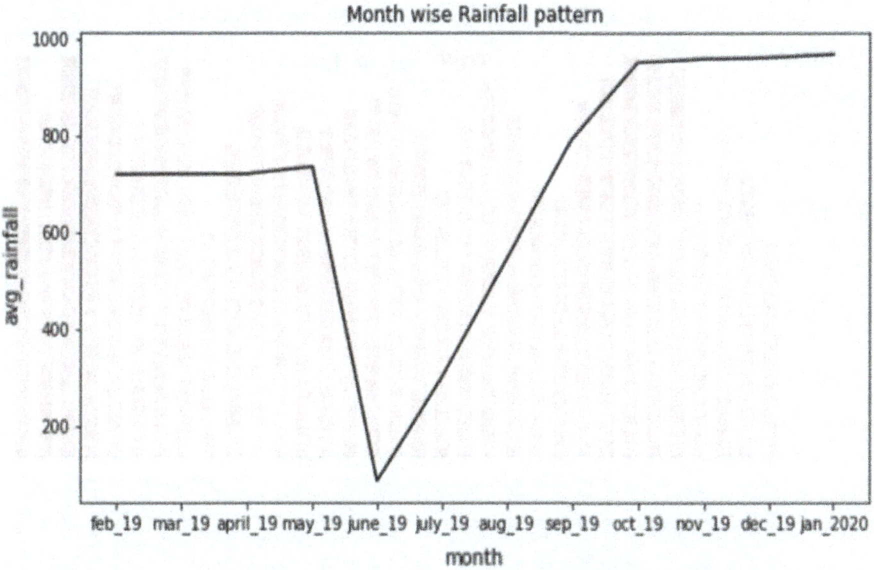

Fig. 4 Line plot of rainfall pattern observed from February 2019 to January 2020

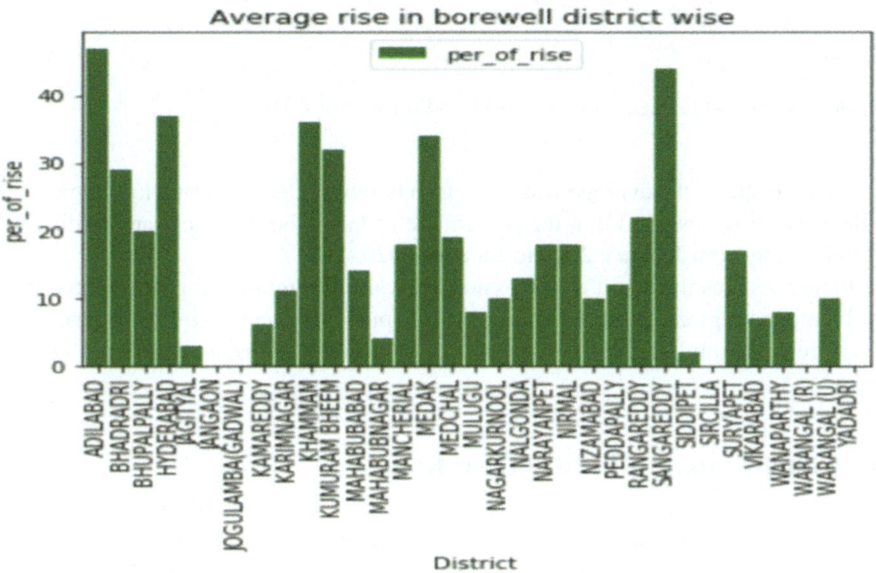

Fig. 5 Bar graph of average rise of percentage in Borewells of Telangana

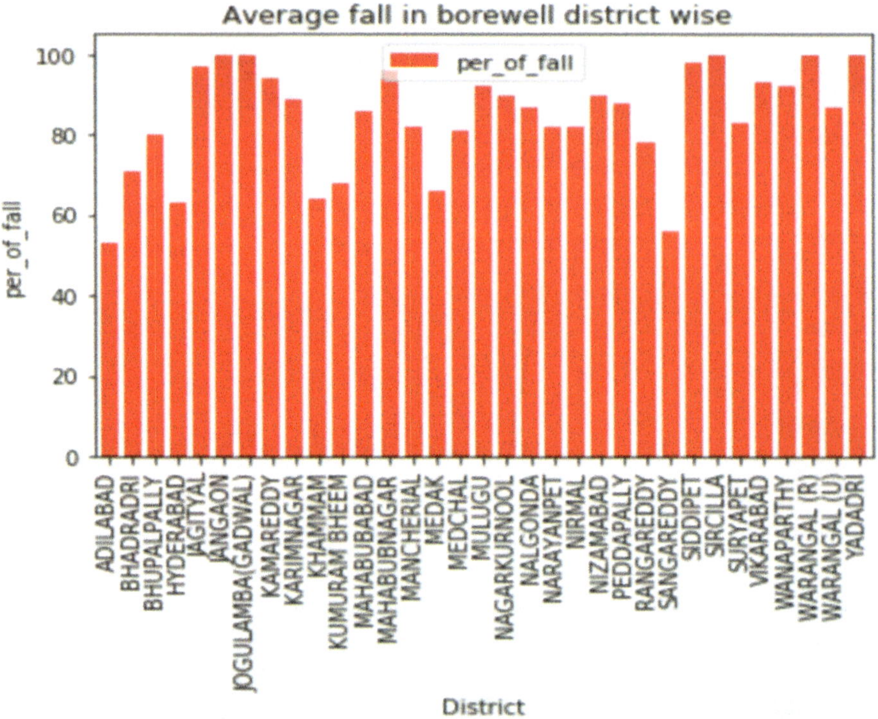

Fig. 6 Bar graph of average fall of percentage in Borewells of Telangana

Figure 6 shows the average water levels in borewells located at various regions of Telangana. It is observed that the percentage of fall in borewell in various districts of Telangana from January 2019 to January 2020.

Figure 7 shoes the result of regression analysis on rainfall and groundwater level of Telangana region. The prediction model depicts the increase in water level with the increase in rainfall. This prediction model is based on linear regression.

6 Conclusion and Future Directions

In this paper, we have seen the patterns and visualization of various parameters like rainfall, groundwater level depth, and borewells in various districts of Telangana. This helped us to analyze and visualize the relationship between rainfall and groundwater. We can also customize this to analyze the water patterns at our own houses.

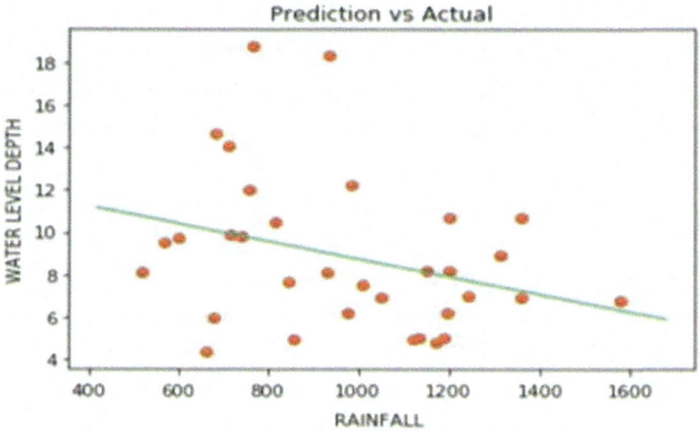

Fig. 7 Prediction of water level versus actual graph

Acknowledgements Authors gratefully acknowledge the computational facility created in the college under DST's FIST Programme (SR/FST/College-2017/28(C)) which helped them to carry out the work. Authors thank the management of IARE for their support and kind encouragement.

References

1. K.G. Berhanu, S.D. Hatiye, Identification of groundwater potential zones using proxy data: case study of Megech Watershed, Ethiopia. J. Hydrol.: Reg. Stud. **28**, 100676 (2020). https://doi.org/10.1016/j.ejrh.2020.100676
2. F.-J. Chang, L.-C. Chang, C.-W. Huang, I.-F. Kao, Prediction of monthly regional groundwater levels through hybrid soft-computing techniques. J. Hydrol. **541**, 965–976 (2016). https://doi.org/10.1016/j.jhydrol.2016.08.006
3. I. Fuentes, R. Willem Vervoort, Site suitability and water availability for a managed aquifer recharge project in the Namoi basin, Australia. J. Hydrol.: Reg. Stud. **27**, 100657 (2020). https://doi.org/10.1016/j.ejrh.2019.100657
4. H. Kulkarni, M. Shah, P.S. Vijay Shankar, Shaping the contours of groundwater governance in India. J. Hydrol.: Reg. Stud. **4**, 172–192 (2015). https://doi.org/10.1016/j.ejrh.2014.11.004
5. X. Wang, T. Liu, X. Zheng et al., Short-term prediction of groundwater level using im- proved random forest regression with a combination of random features. Appl. Water Sci. **8**, 125 (2018). https://doi.org/10.1007/s13201-018-0742-6
6. S. Lamichhane, N. Man Shakya, Alteration of groundwater recharge areas due to land use/cover change in Kathmandu Valley, Nepal. J. Hydrol.: Reg. Stud. **26**, 100635 (2019). https://doi.org/10.1016/j.ejrh.2019.100635
7. E.H. de Moraes Takafuji, M.M. Rocha, R.L. Manzione, Groundwater level prediction/forecasting and assessment of uncertainty using SGS and ARIMA models: a case study in the bauru aquifer system (Brazil). Nat. Resour. Res. **28**, 487–503 (2019). https://doi.org/10.1007/s11053-018-9403-6
8. M. Myneni, L.V. Narasimha Prasad, G. Geetha Reddy, Automatic assessment of floods impact using twitter data. Int. J. Civ. Eng. Technol. **8**(5), 1228–1238 (2017)

Identification of Security Threats Using Honeypots

Rajitha Bhargavi Movva and N. Sandeep Chaitanya

Abstract The number of devices connected to computer networks is growing rapidly and also the network attacks are growing. Attacks on the internet keep on increasing and it causes harm to our security system. In the field of network security, a honeypot is a computer system or application created to attract attackers, who are trying to attack computer networks through the use of spam, phishing, DDoS (Distributed Denial of service), or other nefarious methods. This paper proposes that, once an attacker falls into this trap, the honeypot allows administrators to obtain valuable data about the type of attacker, the activity he was attempting, and helps in identifying the attacker. Even though if an attacker using encryption to steal the data, honeypot can capture the malicious activities and send the alert messages to the admin. Through this resultant data, we can analyze hacker's intensions and new hacking techniques by using visualization tool which helps preventing from future attacks.

Keywords Network attacks · Network security · Honeypot · Attacker · Hacking techniques

1 Introduction

Honeypot is a well arrange that attracts hackers. By luring the hacker into the system, the procedures are begun and run by the programmer on the system that is checked. It's a snare machine that resembles a genuine program to draw the interloper. Honeypots might be utilized for assembling purposes to keep away from, track, or react to assaults. Honeypots can likewise be utilized for examination, gathering information about dangers, so we can all the more likely to comprehend and ensure ourselves

R. Bhargavi Movva (✉) · N. Sandeep Chaitanya
Department of CSE, VNR Vignana Jyothi Institute of Engineering and Technology, Hyderabad, India
e-mail: rajithabhargavi222@gmail.com

N. Sandeep Chaitanya
e-mail: n.sandeepchaitanya@gmail.com

C. Kiran Mai et al. (eds.), *Proceedings of International Conference on Advances in Computer Engineering and Communication Systems*, Learning and Analytics in Intelligent Systems 20,
https://doi.org/10.1007/978-981-15-9293-5_24

against them. Attributable to its arrangement and upkeep costs, little scope organizations despite everything don't pick interruption location frameworks. For these associations, honeypots and honeynets are significant other options. A honeypot can be just a machine that can go about as a hotspot for assaults. It welcomes the programmers to attempt to hack it, which can as a result record the hacker's systems. This log is helpful for shielding the legitimate system from these assaults. Honeypot machine commonly doesn't need to secure any significant information or data. Just to draw the aggressors, it has counterfeit administrations running on its ports. Next are a couple of attacks and a few kinds of honeypots.

1.1 Types of Attacks

1. **Distributed Denial Of Service (DDOS)**:
   A Distributed Denial-of-Service (DDOS) attack will occur when several systems overload a targeted system's bandwidth or resources, typically one or more web servers. This attack is often the result of several compromised systems overwhelming the targeted network with traffic (for example, a botnet).
2. **Phishing**:
   Phishing is the type of social assault frequently used to take client's significant information, including passwords for login and numbers for credit cards. This happens when an injured individual is tricked into opening an email, text, or instant message by an aggressor taking on the appearance of a confided face to face. The client is then tricked into clicking a pernicious association which can result in malware download, gadget freezing as a component of a payment product assault, or private data uncovering.
3. **SQL injection**:
   SQL injection is an imperfection in the Web security system that permits an attacker to connect with the questions that an application makes to its database. This generally causes an aggressor to get to the information they commonly can't get. It might incorporate information from different clients, or whatever other information that could be accessible to the application itself. An attacker can much of the time to adjust or erase this information, making lasting changes the substance or activities of the application.
4. **Spamming**:
   Email spam will be spam mail's online form. This implies conveying superfluous messages to a wide number of clients, frequently spontaneous advertisements. Spam is a significant worry for a guard, as it tends to be utilized to convey Trojan ponies, malware, worms, spyware, and focused on phishing assaults.
5. **Trojan attack**:
   A Trojan horse is a sort of malware that is regularly camouflaged as programming that is real. Digital cheats and programmers will utilize Trojan endeavoring to access client frameworks. Clients are typically fooled into stacking and

executing Trojans on their frameworks by some type of social designing. Whenever empowered, Trojans will keep an eye on you, take your private information, and increase secondary passage access to your gadget from digital crooks. These exercises may include information erasure, information blocking, information control, information duplicating, gadget, or PC arrange yield disturbance.

6. **Spoofing attack**:
   A spoofing attack is a point at which another PC or client is imitated on a system by a malevolent gathering to start assaults against arrange has, take malware spreading information or sidestep get to controls. There are different kinds of mocking assaults that can be utilized by malignant gatherings to accomplish this. The absolute most regular strategies incorporate mocking assaults on the IP address, ARP caricaturing assaults and satirizing assaults on DNS servers.

1.2 Types of Honeypots

Honeypots can be classified based on their deployment (use/action) and based on their level of involvement. Based on deployment, honeypots may be classified as follows:

- production honeypots
- research honeypots

Production honeypots are anything but difficult to utilize, just gather little data and are fundamentally utilized by organizations. Inside the creation arrange, information honeypots are introduced by an organization with other information servers to help their general wellbeing status. Creation honeypots are generally low-cooperation honeypots, which are all the more immediately sent. We are giving less data about the assaults or assailants than honeypots from the investigation.

Research honeypots are racing to gather information about black hat network's inspirations and techniques focusing on different systems. Such honeypots don't carry direct profit to a solitary association; rather, they are utilized to break down the dangers looked by associations, and to figure out how to best shield against such dangers. Research honeypots are intricate to introduce and keep up, gather huge data, and are for the most part utilized by scholastic, military, or legislative associations.

Based on design criteria, honeypots can be classified as follows:

- pure honeypots
- high-interaction honeypots
- low-interaction honeypots

Pure honeypots are finished manufacture frameworks. The attacker's activities are followed by the utilization of a bug tap mounted on the honeypot's system association. Any need to introduce some other applications. Regardless of whether an unimportant honeypot is helpful, an increasingly controlled framework will ensure stealthiest of the security components.

High-interaction honeypots imitate the practices of the creation arranges that have various administrations, such huge numbers of administrations can be empowered to burn through their time by an interloper. Various honeypots might be facilitated on a solitary physical PC, utilizing virtual machines. At that point, it tends to be re-established quicker regardless of whether the honeypot is undermined. High-interaction honeypots with high connections have more insurance as they are hard to distinguish yet they are exorbitant to keep up as shown in Fig. 1. At the point, when virtual PCs are not usable, one physical gadget for every honeypot must be kept up which can be extremely exorbitant. Example: Honeynet.

Low-connection honeypots speak to just certain administrations that assailants frequently demand. Since they expend moderately hardly any assets, it is easy to have different virtual machines on one physical gadget, the virtual frameworks have a short reaction time, and less code is required, in this manner decreasing the multifaceted nature of security for the virtual gadget as shown in Fig. 2. Example: Honeyd.

Some systems engineers tend to classify honeypots based on the targeted software they're trying to protect or expose. So while a list of honeypots could be extensive, we've listed some of the most popular ones here:

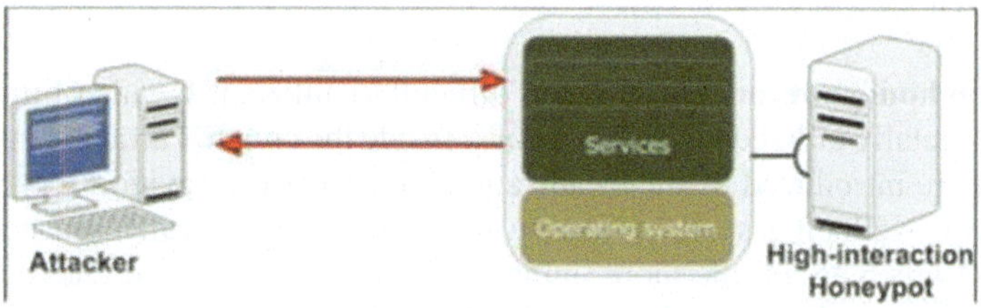

Fig. 1 High-interaction honeypot

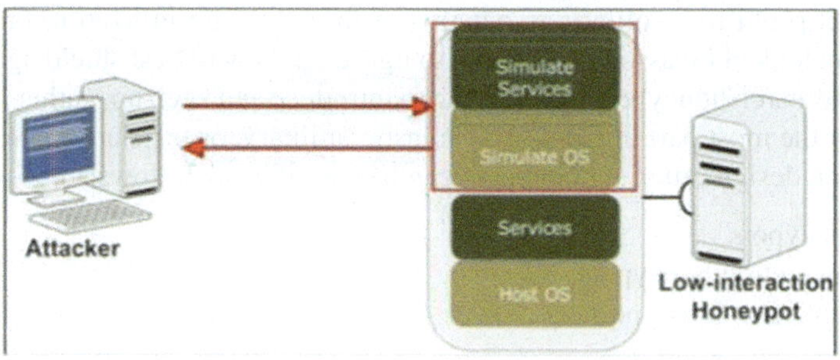

Fig. 2 Low-interaction honeypot

- **Spam honeypot**: Also known as spam trap, this honeypot is planned explicitly for getting spammers before they arrive at authentic email boxes. To get focused on, these likewise have open transfers and work intimately with RBL records to square malignant traffic.
- **Malware honeypot**: This type of honeypot is intended to reproduce powerless applications, APIs, and frameworks for malware assaulting purposes. The information that is then gotten will be utilized later for recognizable proof of malware patterns, to help set up effective malware indicators.
- **Database honeypot**: Databases are a well-known objective for web assailants, and you can watch and learn different assault procedures, for example, SQL infusion, benefit abuse, control of SQL administrations, and substantially more by setting up a honeypot database.
- **Spider honeypot**: This type of honeypot works by making bogus pages and connections which are open just by web crawlers, not by individuals. At the point, when the crawler gets to the honeypot, it is distinguished for later survey alongside its headers, normally to help square malevolent bots and promotion arrange crawlers.

2 Related Work

Presented a new efficient honeypot model to solve all existing problems by opening a virtual port of correspondence for any one of a kind correspondence between an endorsed customer and server, and by giving office to any honeypot to work as an Active Server (AS) [1].

One point of this paper is to show the potential outcomes of honeypots and their utilization in both research and profitable atmosphere. Contrasted with an interruption location framework, honeypots have the significant advantage of not creating bogus cautions in light of the fact that each traffic recognized is suspicious, there are no productive segments running in the system. So right now, the technique for giving system insurance and quickly about firewalls is about honeypots [2].

Right now present the consequences of an action of an exploration honeypot, which attempted the capacity of a web snare for aggressors focusing on the SSH administration to get unapproved access to servers. Throughout a few sequential months, the phoney gadget has stayed on the web and totally operational, recording assaults, and logging every vindictive action. It has appeared during this survey that honeypots stay compelling apparatuses for gathering data on SSH assaults. They additionally noticed that aggressors consistently target servers in the wild utilizing prepared to utilize programming and word references. At long last, they present a perception strategy proposed to help security specialists during the periods of study and drawing conclusions [3].

Talk about enemy of phishing arrangements dependent on the honeypots being used right now. What's more, propose limiting or conquering these restrictions/issues by transforming genuine internet banking framework into a major honeypot stacked with nectar tokens. Some extra honeypots should assist this huge honeypot, with making it progressively powerful. A phishing endeavor identification calculation, called PhishDetekt, is utilized to naturally detect the suspicious ideas by phishers to take cash from the financial balances of unfortunate casualties [4].

Have the discoveries of two particular honeypots from the investigation. The first filled in as a gatherer of malware, an apparatus regularly sent to get self-engendering malware and track its activity. The second went about as a bait server, which dropped however logged any endeavor at pernicious correspondence. Both of these projects have, since a long time ago, stayed online to screen the previously mentioned malevolent conduct. It was likewise demonstrated that the utilization of honeypots for the observing and logging of malware assaults can be exceptionally fruitful and give valuable information. At long last, present an open-source perception device that was created to help security experts and specialists in the examination and finish of periods of drawing [5].

The objective of this examination was to look at developing examples in built-up honeypot inquire about with the aim of adding to the holes in data in the honeypot atmosphere. The related writing has been gathered from an assortment of outlets, for example, books, diary articles, contemplate, and so on. The outcomes show that honeypots draw in specialists' consideration as an important wellbeing procedure that can be applied to moderate system assaults and give a chance to become familiar with the substance of those assaults. In this way, a honeypot can be utilized as a testing strategy to assemble information on organizing assaults [6].

This paper utilizes the honeypots standard to give assurance to business arranges that might not have custom interruption recognition frameworks or firewalls. The proposed model gathers the different strategies programmers use and produces a rundown of all programmer exercises. Hence, utilizing this log, aggressors can be kept from utilizing the creation organize system [7].

This paper proposes an interruption recognition framework (IDS)-based honeypot model to get the most helpful information about the interloper. Honeypots' usefulness and inadequacies have been checked and parts of it have been seen that need as improved. Later on, we expect to utilize this pattern for early anticipation so that pre-emptive activity [8].

This paper comprises of the engineering of honeypot that incorporates nectar net. This additionally incorporates how honeypot capacities for security purposes in the system, and honeypot points of interest, downsides. This paper additionally contains all honeypot subtleties and types of honeypot on cooperation level in the system [9].

Talk about another SDN-based engineering applied to the cross-breed honeypot system for reproducing system topology and relocating traffic assaults. It reinforces the inadequacies of the current honeynet organize satirizing innovation and stream control technology [10].

Examine an asset upgraded fuzzy strategy for the identification and security of a low-communication honeypot satirizing assault. It proposes, principally, a satirizing assault location instrument dependent on the investigation of test information got from the honeypot and its interior system. The paper at that point recommends a fluffy way to deal with envision and caution the parodying assault to low-collaboration honeypots in an opportune way so as to keep away from the assault. At last, test recreation is utilized to show that by utilizing the proposed fuzzy strategy, each low-collaboration honeypot can be rendered a satirizing assault honeypot [11].

3 Proposed Detection Mechanism for the Security Threats

A honeypot is a PC system or application made to pull in attackers, who are endeavoring to attack PC sorts out utilizing spam, phishing, etc. At the point when an attacker falls into this catch, the honeypot licenses chiefs to procure significant data about the kind of assailant, the activity he was charming, and helps in perceiving the assailant. Regardless of the way that if an attacker using encryption to took the data, honeypot can get the noxious activities and send the alert messages to the overseer. Through this resultant data, we can separate software engineer's objectives and new hacking strategies which helps from thwarting future assaults. Not just takes the assailant's exercises, it watches each action from the aggressor and sends the alarm messages to the administrator. From this, administrator can be aware of secure his information. Rather than producing nonstop alarms, better to send when the aggressor attempting to assault the system, so we can spare assets. Utilizing scripting language (like Python) in the execution of honeypot can give effective code.

3.1 Architecture of Proposed System

See Fig. 3.

3.2 Honeypot Deployment

There are various ways to deploy a Honeypot:

- In front of the firewall (over the Internet).
- DMZ can add an extra security layer to the neighborhood of an undertaking. A DMZ or peaceful area in PC security is a physical or coherent sub-arrange containing and uncovering the outside administrations of an association to a bigger un-confided in organizing, as a rule, the Internet. The term is normally alluded to by IT experts as a DMZ. It is some of the time alluded to as a system edge. A DMZ has the capacity of including an extra layer of security to the neighborhood (LAN) of an association; an unapproved gate crasher just approaches gear in the DMZ, as opposed to some other piece of the system.
- Behind the firewall.

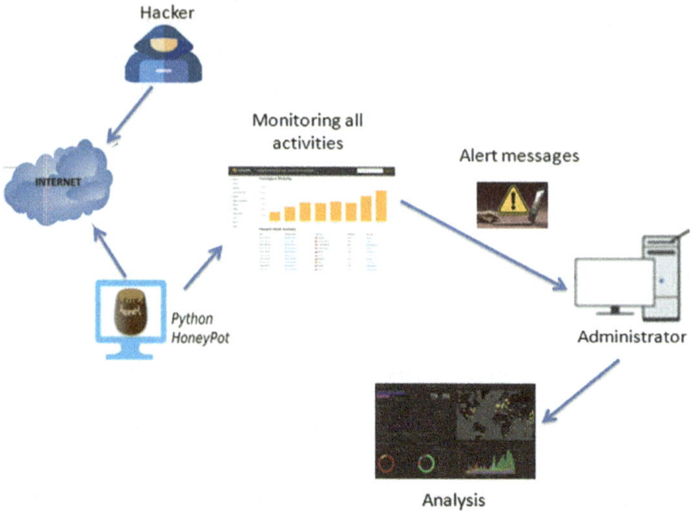

Fig. 3 Architecture of proposed system

4 Workflow of Proposed Work

4.1 Procedure

The process below outlines the workflow for the system proposed. The honeypot is initially built using Python scripting language. After installing honeypot in the device, it begins to track any operation that is carried out over the network, locate the malicious activities, and then send the administrator warning messages. Through this administrator, the intruder can save his valuable information. Furthermore, the honeypot can catch the actions of the intruder and collect their information. By analyzing this resulting knowledge, we can forecast future attacks and collect trends of the hacker and new techniques of hacking (Fig. 4).

4.2 Analyzing the Resultant Data

Deploying the honeypot systems is just the start. Valuable information can be obtained by diligent data analysis that is gathered by the honeypots deployed. From this point, the real work with those systems begins. Apparently, however, most efforts about the development of honeypot software are concerned with developing new forms of honeypots, or hiding existing ones. Sadly, the production of honeypots analytical program is ignored. Huge amounts of reported data require additional analytical functionalities. The functionalities of visualization and filtration, initially

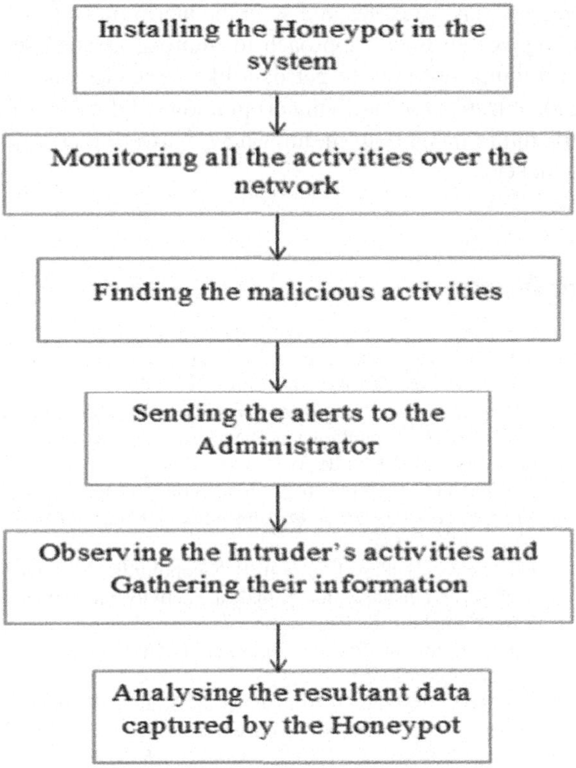

Fig. 4 Flowchart of proposed work

introduced as support devices, soon verified their usefulness. It can be very useful for people doing honeypot systems data analysis. The visualization helps evaluate and describe the behavior observed in the specified time period by honeypot systems. Through this, we will discover the intention of the hacker, and new methods of hacking that can deter potential attacks.

5 Conclusion

An ever-increasing number of individuals utilize the web each day. Today, a web day is only a piece of the life of everybody. Individuals check their messages, purchase items, peruse the web, play web-based games, and take care of online tabs (advanced exchanges) and so on. To request to determine every one of these dangers, assurance is an essential factor. Honeypot is an all-around structured system that attracts attackers. This paper recommends that, when an assailant falls into this snare, the honeypot permits executives to get significant information about the kind attacker, the action he was endeavoring and helps in distinguishing the assailant. Despite the fact that if an aggressor utilizing encryption to take the information, honeypot can catch the noxious exercises and send the alarm messages to the administrator. Honeypot is an extraordinary approach to improve arrange security executive's information and figure out how to get data like strategies and exercises from an injured individual framework utilizing criminological devices. Honeypot is additionally helpful for future dangers to monitor new innovation assaults which are performed by the attackers.

References

1. V.V. Das, Honeypot Scheme for distributed denial-of-service attack, in *International Conference on Advanced Computer Control* (2008)
2. A. Chandra, K. Lalitha, Honey pots: a new mechanism for network security, in *IJPaper* (2012)
3. I. Koniaris, G. Papadimitriou, P. Nicopolitidis, Analysis and visualization of SSH attacks using honeypots, in *IEEE* (2013)
4. S. Chauhan, S. Shiwani, A honeypots based anti-phishing framework, in *International Conference on Control, Instrumentation, Communication and Computational Technologies (ICCICCT)* (2014)
5. I. Koniaris, G. Papadimitriou, P. Nicopolitidis, M. Obaidat, Honeypots deployment for the analysis and visualization of malware activity and malicious connections, in *IEEE ICC* (2014)
6. R.M. Campbell, K. Padayachee, T. Masombuka, A survey of honeypot research: trends and opportunities, in *The 10th International Conference for Internet Technology and Secured Transactions* (2015)
7. S. Deepa Lakshmi, G. Arunkumar, V. Madhu Viswanatham, Network security enhancement through honeypot based systems. Int. J. Eng. Technol. (IJET) (2015)
8. J. Reddy Kondra, S.K. Bharti, S.K. Mishra, K. Sathya Babu, Honeypot-based intrusion detection system: a performance analysis, in *International Conference on "Computing for Sustainable Global Development"*, 16th–18th March (2016) (2016), https://www.researchgate.net/publication/298263572
9. K. Ashwini, G. Pratiksha, K. Anuja, S. Varsharani, S. Gayatri, Secure network system using honeypot. Int. J. Adv. Res. Comput. Commun. Eng. (ISO 3297:2007) Certified 6(2) (2017)
10. H. Wang, B. Wu, SDN-based hybrid honeypot for attack capture, in *IEEE 3rd Information Technology, Networking, Electronic and Automation Control Conference* (ITNEC 2019) (2019)
11. A fuzzy approach for detecting and defending against spoofing attacks on low interaction honeypots, in *21st International Conference on Information Fusion (FUSION)* (2018)

A TDMA Scheduling for Particle Swarm Optimization-Based Unequal Clustering in WSN: PSO-UFC-TDMA

A. Manisha and S. Renu Deepti

Abstract The lifetime of a network is maximized in wireless sensor networks (WSNs) by an effective energy-saving technique named as clustering. The energy of cluster heads (CHs) that are adjacent to the base station (BS) gets depleted because of traffic load at high inter-cluster relay in the multi-hop approach and the issue of hot spot arises. Therefore, a clustering protocol must have effective energy and should be fault tolerant. By using the process of TDMA which is stated as PSO-UFC-TDMA, a particle swarm optimization (PSO) is proposed in this paper on the basis of unequal as well as fault-tolerant clustering. Additional CH, also known as Surrogate Cluster Head (SCH), is selected because of unexpected MCH failure in order to restore connectivity of network in PSO-UFC protocol. Energy is saved using TDMA by allotting timeslots in the transmission of the data. The lifespan of the network is extended by PSO-UFC-TDMA technique compared to PSO-UFC, EBUC, PSO-C as well as LEACH-C protocols which are presented in the simulation results.

Keywords Particle swarm optimization · Unequal clustering · Fault tolerance · MAC protocol · TDMA · Packet delivery ratio · Network lifetime

1 Introduction

Recently, Wireless sensor networks (WSNs) became one of the developing technologies [1, 2]. In order to monitor the applications, investigations were initially performed on WSN. However, WSNs were widely applied in various situations and objectives such as Healthcare, Military Surveillance, Smart Grid as well as Industrial Automation using strong proliferation in microelectromechanical systems (MEMS)

A. Manisha (✉)
Department of Computer Networks & Information Security, VNR VJIET, Hyderabad, Telangana, India
e-mail: nisha661997@gmail.com

S. Renu Deepti
Department of Information Technology, VNR VJIET, Hyderabad, Telangana, India
e-mail: renudeepti_s@vnrvjiet.in

C. Kiran Mai et al. (eds.), *Proceedings of International Conference on Advances in Computer Engineering and Communication Systems*, Learning and Analytics in Intelligent Systems 20,
https://doi.org/10.1007/978-981-15-9293-5_25

[3]. Automated sensing, embedded processing as well as wireless transmission are incorporated within very small embedded devices named sensor nodes. Because of the reduced as well as non-rechargeable battery sources, the energy supply is limited by the respective sensor node. However, limited onboard processing power and storage capabilities are exhibited by their processors.

The performance of the WSNs is enhanced by limiting the energy resources [4]. In order to conserve energy in WSNs, the study of clustering has been widely performed recently. The whole network is divided within small clusters and individual cluster is allotted with a node named as a Cluster Head (CH) and participant nodes. When the clusters are formed, the communication within the nodes is performed in two kinds, namely intra-cluster and inter-cluster communication. The data is transmitted toward the CH from the remaining nodes in network and the data which is collected is forwarded toward the base station (BS) directly or by multi-hop routing [5]. Nevertheless, the energy of the CHs which are located adjacent to the BS gets depleted rapidly than other CH nodes due to the involvement of the CH in high inter-cluster relay traffic load in multi-hop routing. Generally, this problem is referred to as the hot spot problem [6] in literature. Because of the faulty performance of the hardware components and limited battery power, sensor nodes are susceptible to failure. The communication in network between the nodes and the neighboring CHs [7] is interrupted due to this failure. Hence, the issues of hot spot, imbalanced clustering as well as fault tolerance are discussed collectively in this paper.

Due to the selection of m optimal CHs in n number of sensor nodes with the possibilities as n m C, the selection of a CH becomes a Non-Deterministic Polynomial (NP)-hard optimization problem. For solving such various issues, the method Swarm intelligence is implemented. A stochastic optimization technique which is formed on the basis of is named as Particle Swarm Optimization (PSO) [8].

And it is given by the performance of bird flocking and fish schooling. A series of iterations are implemented for enhancing the candidate solution with respect to the application's quality and to optimize an issue. Due to its easy hardware or software implementation and the capability of rapidly concentrating an optimal solution, it might be an enhanced option in performing optimal CH selection. The optimization technique becomes simple and the efficiency of the network can be improved since the clustering is a repeated process. The selection of the cluster has been optimized in a wider range by implementing PSO in various clustering protocols. With respect to the PSO procedure, the issue of hot spot is discussed using Energy-balanced unequal clustering (EBUC) protocol [9]. Traffic load at high inter-cluster relay is supported by forming clusters with different sizes whereby the nodes adjacent to the BS consists of a smaller amount of participant nodes. The energy is distributed excessively between the CHs in considering the degree of the node and residual energy of CHs by EBUC during the process of clustering. In addition, the problem of fault tolerance is unconsidered by the EBUC protocol. On the basis of imbalanced and fault-tolerant clustering protocol named as PSOUFC, a new particle swarm optimization method is proposed in order to reduce the fault tolerance and clustering imbalance.

A new TDMA scheduling method is proposed for intra-cluster communication by implementing an energy-effective TDMA protocol on behalf of clustered WSNs (ECTDMA) in this paper. With respect to the sum of sensor nodes and also traffic load at sensor node in a cluster, TDMA length is changed by EC-TDMA. In addition to that, the idle listening is reduced, the efficiency of the channel is enhanced and therefore the lifespan of the network is extended.

2 Related Work

This unit mentions various existing clustering protocols.

2.1 LEACH-C

Heinzelman et al. [10] A lifespan of the network is extended where the overall process of CH selection is governed by the BS in a centralized clustering protocol called LEACH-C. The maximum numbers of CHs are selected all over the network by employing a simulated annealing technique by the BS. Single-hop routing in inter-cluster communication is a major disadvantage in LEACH-C, where the distribution of the energy is done in an unbalanced way whenever the distance between various CHs and the BS is so far. Therefore, during the communication with the BS, the energy is consumed highly by the CHs.

2.2 PSO-C

Latiff et al. [11] The network duration is extended by monitoring the completed procedure of CH selection as well as the formation of cluster by BS in a centralized clustering protocol named as an energy-awake clustering with the help of PSO procedure (PSO-C). In order to select an efficient CH, the intra-cluster distances among the nodes along with existing energy of entire CH nodes are considered by the BS. However, it doesn't consider the distance from the nodes which in turn enhances the energy to be consumed in the communication between the CHs and BS.

2.3 EBUC

Latiff et al. [11] The network duration is extended by monitoring the completed procedure of CH selection as well as formation of cluster by BS in a centralized clustering protocol named as an energy-awake clustering with the help of PSO procedure (PSO-C). In order to select an efficient CH, the intra-cluster distances among the nodes along with existing energy of entire CH nodes are considered by the BS. However, it doesn't consider the distance from the nodes which in turn enhances the energy to be consumed in the communication between the CHs and BS.

2.4 *IPSO*

Jiang et al. [9] Duration of the network is expanded by considering the improved form Particle Swarm Optimization (IPSO) in current EBUC protocol. Investigation regions of search particles are extended in order to overcome the issues of standard PSO using IPSO. With the implementation of the perturbations in a direct way toward the global best as well as personal best, it can be achieved. However, the EBUC protocols problems still remain to be solved.

2.5 *PSO-ECHS*

Rao et al. [13] The selection of CH and formation of cluster are optimized to run the PSO using the BS in an energy-efficient clustering algorithm called as PSO-ECHS. On the basis of intra-cluster and BS distances and sensor node's residual energy, a new fitness function is introduced by the protocol in order to lessen the linear combination of objective functions. Distance, energy as well as node degree of CHs are various constraints that are employed to connect the non-cluster head nodes toward the CHs for the balanced cluster formation. However, the performance of the network decreases even though the BS position is changed from center to corner.

Kaur and Kumar [14] Issue of hot spot is resolved using PSO-based clustering method in WSN in an effective clustering method called as PSO-UFC. The energy consumption of the intra-cluster and inter-cluster is balance is deriving the cost functions of imbalanced clustering method. The connection of networks in MCHs is ensured by constructing a multi-hop routing tree. The problem of fault tolerance is discussed by selecting a surrogate cluster head in every cluster. This is splitted within rounds and set-up phase as well as steady-state phase is included in every round. The network is divided into imbalanced clusters in set-up phase by the BS and an MCH and an SCH are included in every cluster. In addition to that, a multi-hop routing tree is constructed between MCHs that are selected. However, through the intra-cluster and inter-cluster communications, data that is sensed is transmitted toward the BS in the steady-state phase. This phase is implemented for reducing the consumption of energy and the calculation period of set-up phase. This phase is implemented only when the role of its MCH is taken by the SCH if not existing MCHs are employed in the succeeding rounds.

3 Proposed Framework

The cluster set-up in addition to a steady-state phase is formed by partitioning the EC-TDMA into rounds. Set-up phase forms the clusters & the steady-state phase transmits the data to the CH. n number of frames is formed by splitting EC-TDMA in steady-state phase. Every frame doesn't exhibit a fixed duration because every source node may not have data at all times for the transmission. A schedule period as well as a data transmission period is there for every frame. In order to achieve a collision-free intra-cluster communication, a TDMA schedule is established by CH in the respective schedule period. Similar time slots were assigned to the cluster members by the CH in the initial scheduling of every round. Various time slots were assigned to the source nodes by the CH in the additional scheduling period. The number of time slots and the source nodes are equally divided at each data transmission period in a cluster. Through the allocated transmission time slot, the data, as well as estimated traffic load of next frame, is transmitted by respective source node towards the CH by keeping its radio off in the remaining time. In the period of data transmission, the entire non-source node keeps their radios off. Moreover, data that resembles the leave message is transmitted by the source node to CH. Thus, with respect to the number of sensor nodes as well as traffic load at sensor node, TDMA frame length can be varied by the CH. Using a scheduled period, the succeeding frame is started with similar process whenever a frame is completed. The data from the entire sources is collected and transmits the data toward the BS. The succeeding round is started and the entire process is repeated after some time.

3.1 *Protocol Definition*

The EC-TDMA process is partitioned into rounds identical with the LEACH as well as BMA protocols. Cluster set-up as well as a steady-state phase is included in every round. The EC-TDMA frame structure is shown in Fig. 1. The set-up phase forms the clusters, as well as the steady-state phase, transmits data to CH. n number of frames

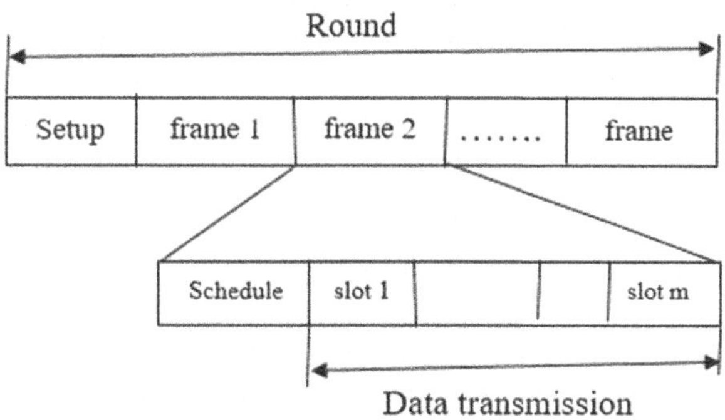

Fig. 1 EC-TDMA frame structure illustration

is formed by splitting EC-TDMA in steady-state phase in contrast with LEACH as well as BMA protocols. Every frame doesn't exhibit a fixed duration because every source node may not have data at all times for the transmission. A schedule period as well as a data transmission period is there for every frame. In order to achieve a collision-free intra-cluster communication, a TDMA schedule is established by CH in respective schedule period. Similar time slots were assigned to the cluster members by the CH in the initial scheduling of each round. Various time slots were assigned to the source nodes by the CH in the additional scheduling period.

The number of time slots and source nodes are equally divided at each data transmission period in a cluster. Through the allocated transmission time slot, the data, as well as estimated traffic load of succeeding frame, is transmitted by respective source node toward the CH by keeping its radio off in the remaining time. In the period of data transmission, the entire non-source node keeps their radios off. Moreover, data that resembles the leave message is transmitted by source node to CH. Thus, with respect to the number of sensor nodes as well as traffic load at the sensor node, TDMA frame length can be varied by the CH. Using a scheduled period, the succeeding frame is started with similar process whenever a frame is completed. The data from the entire sources is collected and transmits data to BS. The succeeding round started and the entire process is repeated after some time.

3.2 Size of Time Slot

Intended for intra-cluster transmission, an efficient slot is assigned by EC-TDMA protocol in WSNs. With respect to the topology of network and traffic load at the cluster members, the time slot size may be adjusted. A frame, time slot size of ith source node is given below:

$$ti = \min\left\{\frac{li}{CCi}, \frac{li}{\sum_{j=1}^{m} lj}(Tn_{frame} - Tn_{schedule})\right\} \quad (1)$$

where the traffic load of succeeding frame that is expected by the source node is denoted by *li*, the source node in the cluster is characterized by m, channel capacity is denoted by *CCi*, the maximum frame length is given by *Ti* frame, the time of schedule period in a frame is denoted by *Ti* schedule. As given in (1), the time slots are allocated to source nodes schedule period by CH.

4 Experimental Results

This paper proposes EC-TDMA method by implementing network simulator (NS-2). Over a 1000 × 1000 m^2 field, 21 sensor nodes are assumed which are distributed randomly including four obstacles. This paper assumes that no gap is present in

Table 1 System parameters

Parameter	Value
Application traffic	CBR
Area	1000×1000
Transmission rate	512 bytes/0.5 s
Routing protocol	AODV
Time of simulation	1200 ms
Node count	21
Radio range	250 m

detecting field and identical capabilities are exhibited by the static sensors. However, the mobile sink whose location is on the top-left corner of the 2D region is assumed along with its coordinates (250, 250 m). From the initial position, the periodical obstacle-avoiding movement is initiated by mobile sink returns in the end. The system parameters utilized in the simulations are presented in Table 1 and the simulation time is assumed as 1200 ms. The deferral tolerant data that is collected by sensor nodes is considered in this paper for the simplification of the mobile sink that makes a delay for the mobile sink to motivate them.

4.1 Performance of Proposed System

In regard to PDR (packet delivery ratio), E2E (end to end) delay, throughput as well as consumption of total energy consumption in the network, evaluation of PSO-UFC-TDMA protocol performance can be done. With respect to the consumption of energy, the amount of communication rounds whenever the first node dies (FND), or the expiration of a determined part of nodes defines the lifespan of the network. Within the sparsely deployed WSNs, the usage of the FND parameter is done. On the other hand, the connectivity of the network, as well as the process of sensing, is affected by the expiry of a single node in densely deployed WSN. However, the deployment of the WSN is degraded if a part of the nodes expires.

In above Fig. 2, shows all nodes placed in the network and deployment of the nodes in the network. Here, every node displayed on the basis of topology values and all properties of NAM window it should be mentioned. In Fig. 3, broadcasting occurs throughout network. Here, broadcasting occurs for the purpose of communication. Every node should involve in this process. In Fig. 4, cluster formation started based on network area. In this cluster, members should transmit the data to cluster head (MCH or SCH). The selection of cluster head based on the distance parameter. In Fig. 5, cluster member to cluster head data communication, this data should be reached to cluster head properly. Here, we use CBR protocol as traffic purpose and how many bytes of data sent and when it sending is shown in Fig. 5. In Fig. 6, it shows the cluster formation file. In this file cluster division and cluster selection based on

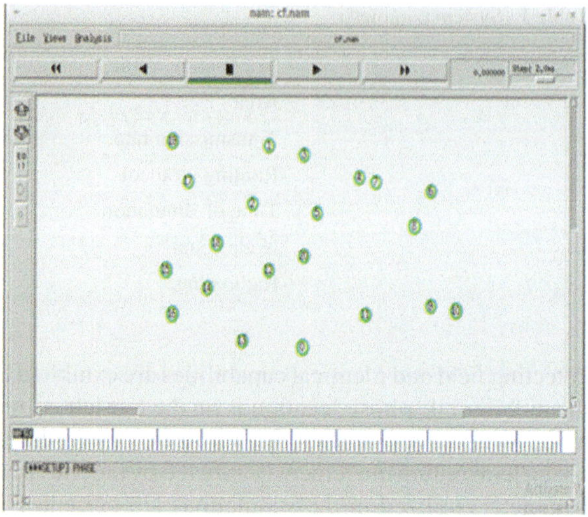

Fig. 2 Network deployment

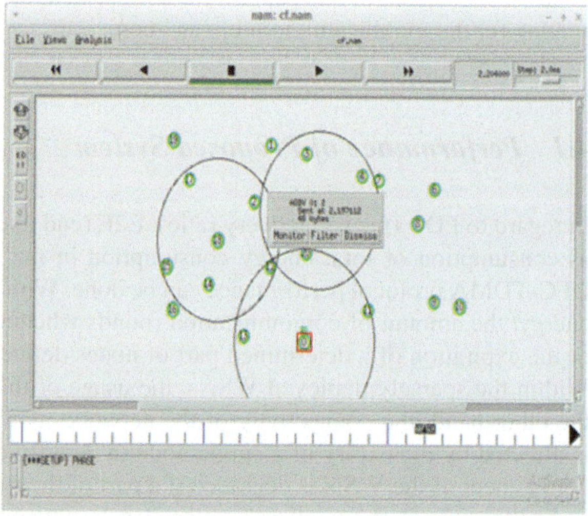

Fig. 3 Broadcasting process in network

parameters. Figure 7 shows and represents the trace file of the network. It shows all the attributes, variables, energy levels for individual nodes, and data transmission process. Figure 8 shows the graphs and representing the delay. It shows the delay in the network. Here, in EBUC, PSO-UFC, and PSO-UFC-TDMA, the verification of this says that PSO-UFC-TDMA shows the reduction of delay in network than the previous methods. Figure 9 shows the graphs and representing energy consumption. It shows the energy consumption in network. Here, in EBUC, PSO-UFC, and PSO-UFC-TDMA, the verification of this says that PSO-UFC-TDMA shows reducing

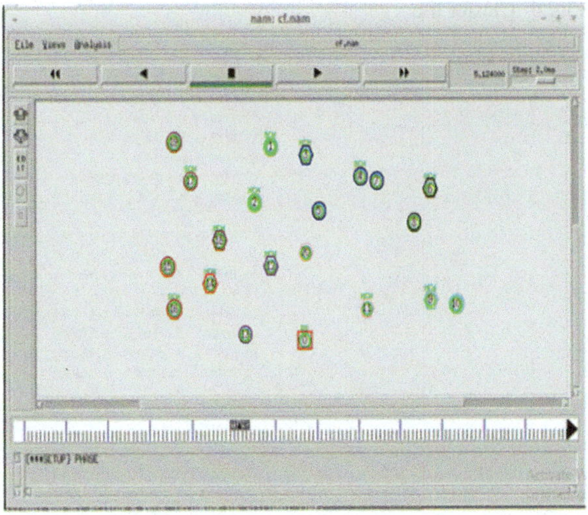

Fig. 4 Cluster formation and selection of MCH and SCH

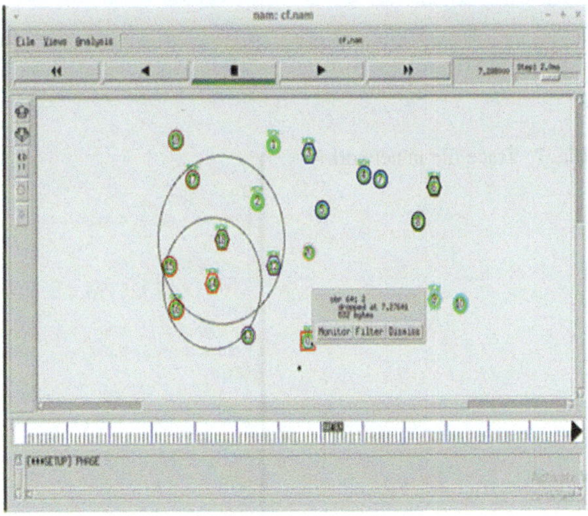

Fig. 5 Data process in between cluster members and cluster heads

energy consumption in the network than the previous method. Figure 10 shows the graphs and representing the throughput. It shows the throughput in the network. Here, in EBUC, PSO-UFC, and PSO-UFC-TDMA, the verification of this says that PSO-UFC-TDMA shows that maximizing the throughput in the network than the previous method. Figure 11 shows the graphs and representing the Packet Delivery Ratio. It shows the Packet Delivery Ratio in the network. Here, in EBUC, PSO-UFC,

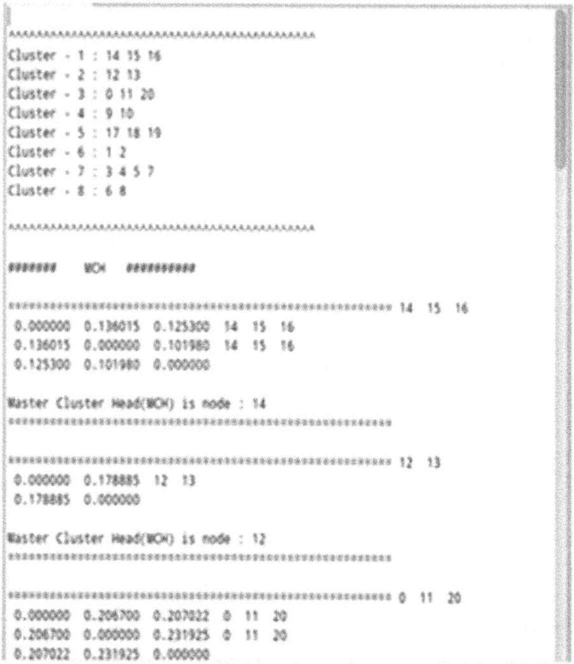

Fig. 6 Cluster file

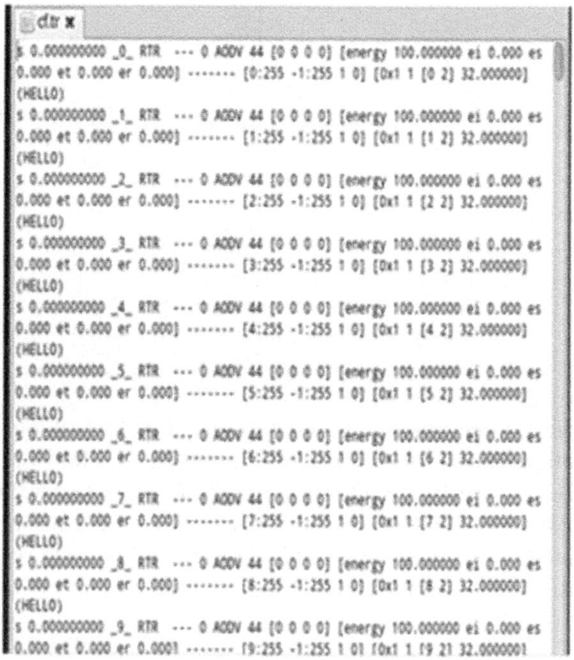

Fig. 7 Trace file in network

Fig. 8 End-to-End delay

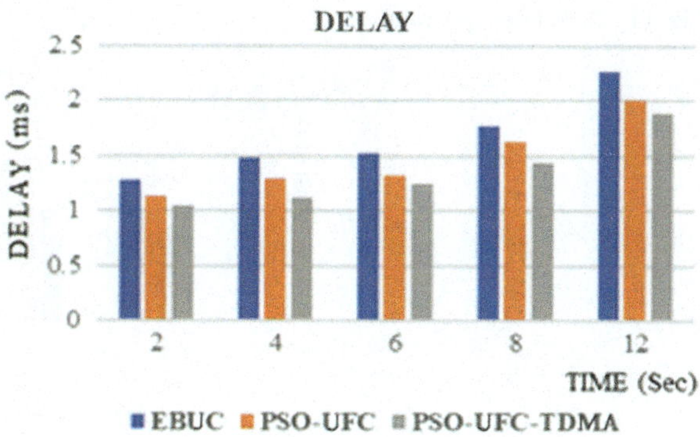

Fig. 9 Energy consumption

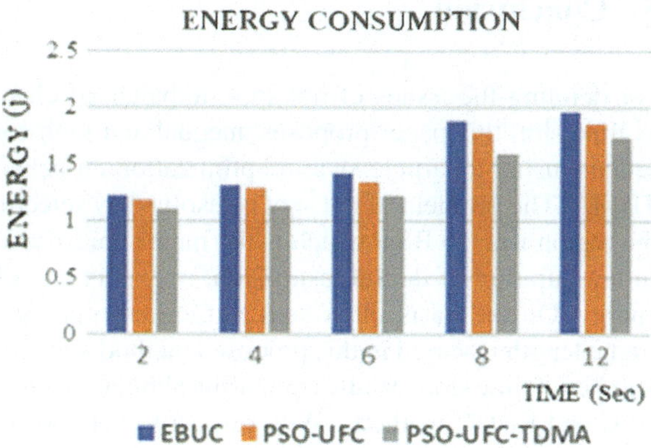

Fig. 10 Throughput

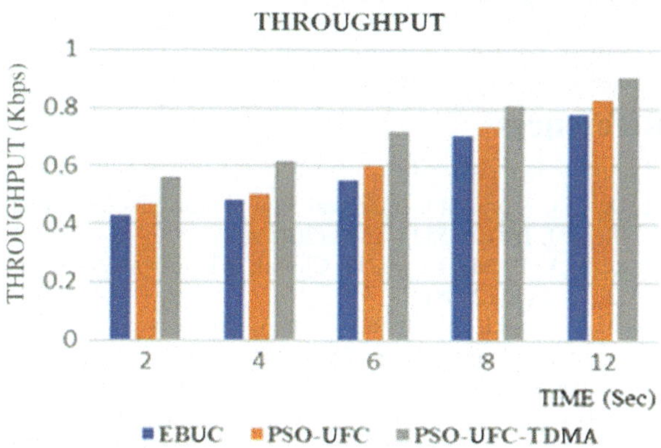

and PSO-UFC-TDMA, the verification of this says that PSO-UFC-TDMA show maximizing the Packet Delivery Ratio in the network, than the previous method.

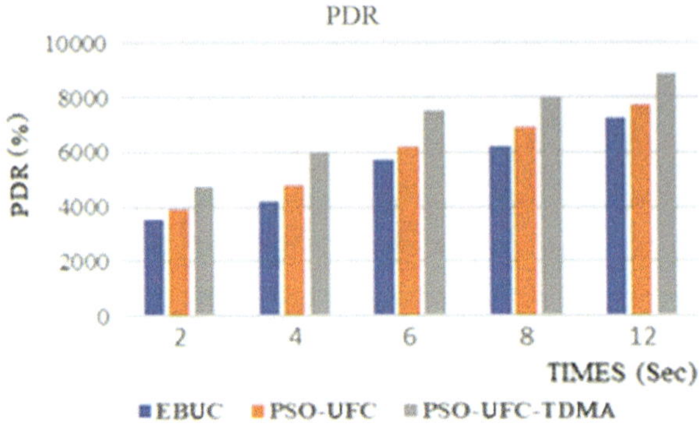

Fig. 11 Packet delivery ratio

5 Conclusion

For defining the issues of hot spot, imbalanced clustering, fault tolerance as well as time slot, this paper proposes unequal and fault-tolerant clustering with TDMA scheme using a particle swarm optimization on behalf of WSNs called PSO-UFC-TDMA. The problem of hot spot is resolved by selecting a large number of MCHs in the region near to BSs, which is the major objective of PSO-UFC-TDMA method. In the duration of data transmission, timeslots are allocated by TDMA for saving energy. On the basis of present traffic and topology of the network, the TDMA frame length is varied in this proposed method and the energy is saved by decreasing number of time slots. While comparing simulation results is performed with the PSO-UFC and EBUC methods. With respect to network lifetime, PDR (packet delivery ratio), end-to-end delay and throughput, it is obvious that superior energy efficiency comparatives are ensured by PSO-UFC-TDMA method.

References

1. X. Liu, A typical hierarchical routing protocols for wireless sensor networks: a review. IEEE Sens. J. **15**(10), 5372–5383 (2015)
2. J. Yick, B. Mukherjee, D. Ghosal, Wireless sensor network survey. Comput. Netw. **52**, 12, 2292–2330 (2008)
3. M.A. Yigitel, O.D. Incel, C. Ersoy, QoS-aware MAC protocols for wireless sensor networks: a survey. Comput. Netw. **55**(8), 1982–2004 (2011)
4. G. Anastasi, M. Conti, M. Di Francesco, A. Passarella, Energy conservation in wireless sensor networks: a survey. Ad Hoc Netw. **7**(3), 537–568 (2009)
5. W.R. Heinzelman, A. Chandrakasan, H. Balakrishnan, Energy-efficient communication protocol for wireless microsensor networks, in *Proceedings of the 33rd Annual IEEE Hawaii International Conference on System Sciences*, August 2000, pp. 10–16 (2000)
6. A.F. Liu, X.Y. Wu, Z.G. Chen, W.H. Gui, Research on the energy hole problem based on unequal cluster-radius for wireless sensor networks. Comput. Commun. **33**(3), 302–321 (2010)
7. G. Gupta, M. Younis, Fault-tolerant clustering of wireless sensor networks, in *Proceedings of IEEE Conference on Wireless Communications and Networking*, March 2003, vol. 3, pp. 1579–1584 (2003)

8. R.V. Kulkarni, G.K. Venayagamoorthy, Particle swarm optimization in wireless-sensor networks: a brief survey. IEEE Trans. Syst. Man Cybern. **41**(2), 262–267 (2011)
9. C.J. Jiang, W.R. Shi, X.L. Tang, Energy-balanced unequal clustering protocol for wireless sensor networks. J. China Univ. Posts Telecommun. **17**(4), 94–99 (2010)
10. W.R. Heinzelman, A. Chandrakasan, H. Balakrishnan, An application-specific protocol architecture for wireless microsensor networks. IEEE Trans. Wirel. Commun. **1**(4), 660–670 (2002)
11. N.A. Latiff, C. Tsimenidis, B.S. Sharif, Energy-aware clustering for wireless sensor networks using particle swarm optimization, in *Proceedings of IEEE 18th international symposium on personal, indoor and mobile radio communications (PIMRC)*, December 2007, pp. 1–5 (2007)
12. S. Salehian, S.K. Subraminiam, Unequal clustering by improved particle swarm optimization in wireless sensor network. Proc. Comput. Sci. **62**, 403–409 (2015)
13. P.S. Rao, P.K. Jana, H. Banka, A particle swarm optimization based energy efficient cluster head selection algorithm for wireless sensor networks. Wirel. Netw. **23**(7), 2005–2020 (2017)
14. T. Kaur, D. Kumar, Particle swarm optimization-based unequal and fault tolerant clustering protocol for wireless sensor networks. IEEE Sens. J. **18**(11), 4614–4622 (2018)

Construction of Phylogenetic Tree for Cyclooxygenase Dataset by Clustering

Abdul Khadeer, Chukka Santhaiah, Perepa Koushik, G. Manohar, Y. Deepthi, and N. V. Krishna Rao

Abstract Phylogenetic trees represent headway relationship among characteristic collection, residing topics. The tree finds the likeliness or divergence of received skills. I construct phylogenetic tree for developments coming across comparison among wonderful species developments using institution exam, as an example, Hierarchical Bunching, KMEANS and SOM with PCA. As in line with author species, tendencies can be comparative at the off threat that they have got comparative examples or conduct contrast can be removed in conduct, examples. For actualizing the understanding, I applied COX dataset from NCBI website online for manufacturing the phylogenetic tree.

Keywords Phylogenetic tree · Cyclooxygenase · Dendogram

1 Introduction

The method for the significance has been changed considering the formation of giant proportions of characteristic data.

High-throughput sequencing made gene progressions in different animals. It is regarded that the courses of action are inconceivably basically proportionate living things, anyway not tantamount of formative distant creatures. Thus, clustering is thoroughly used to construct phylogenetic trees [1].

Comparingly, COX data are furthermore delivered past decade. Examination by using of COX data in natural assessments in any case is called Cyclooxygenase. From animals, it is instructive [2].

Speculatively, formative far away species should have unmistakable aggravation activities and models, while solidly related species should have practically identical models. Along these lines, it is appealing for using Cyclooxygenase data for examination. COX is used looking at amazing characteristics that was fruitful to standard scientists. Various tasks can be done by utilizing group examination [3].

A. Khadeer (✉) · C. Santhaiah · P. Koushik · G. Manohar · Y. Deepthi · N. V. Krishna Rao
Department of CSE, Institute of Aeronautical Engineering, Hyderabad, Telangana, India
e-mail: abdulkhadeer1998@gmail.com

C. Kiran Mai et al. (eds.), *Proceedings of International Conference on Advances in Computer Engineering and Communication Systems*, Learning and Analytics in Intelligent Systems 20,
https://doi.org/10.1007/978-981-15-9293-5_26

Genes are never-ending. Billions of genes are evolving day by day. It is a bit difficult to find the similar genes from that huge data. There are many methods available to find the similar genes. Many researches and studies have been done by our scientists since decades. Here, in this paper, we have studied the previous models and created a very effective and easy model to find the similar genes [4].

We have applied different clustering techniques to find the similarity between genes. As clustering is the best way to group the similar genes.

We have different clustering algorithms like Hierarchical Clustering, K-Means Clustering, PCA with SOM.

2 Literature Survey

The author has explained that MUSCLE exhibits enhancements in exactness and decreases in computational unpredictability by abusing a scope of existing and new algorithmic strategies. While the structure commonly for viable different arrangement apparatuses seemingly needs tastefulness and hypothetical soundness, valuable enhancements Ire accomplished through various components [5]. Generally, significant of these are choice of heuristics, close thoughtfulness regarding subtleties of the usage, and cautious assessment of the effect of various components of the calculation on speed and exactness. MUSCLE empoIrs high-throughput applications to accomplish normal exactness similar to the most precise apparatuses already accessible, which I hope to be progressively significant taking into account the proceeding with quick development in grouping information [6]. Study of the Cyclooxygenase 1 and Cyclooxygenase 2 datasets. The structures of prostaglandin endoprise and other lipids are deeply studied here. The various algorithms for clustering the data can be used [7].

3 Problem Statement

The key incentive here is to build phylogenetic tree and to develop the dendrogram for Cyclooxygenase (COX) with the help of different clustering methods [8]. Probably, the most popular genetic sequence alignment algorithm is Clustal W. Although successful in its domain, Clustal W is very sensitive to highly divergent sequences. Clustal W has become the dominant part of appreciated calculation for various grouping arrangements. It applies a dynamic strategy for different arrangements. As a dynamic calculation, Clustal W adds successions individually to the current arrangement to assemble another arrangement [9]. The guide tree is built utilizing the closeness of every single imaginable pair of successions. We construct phylogenetic tree for cyclooxygenase by overcoming the defects of Clustal W approach. We develop a Phylogenetic tree that has high accuracy than the previous method.

We collect the COX dataset from NCBI, process the data, convert to excel and apply Hierarchical Clustering, K-Means Clustering, and Principal Component Analysis and Self-Organizing Maps. When we collectively apply these together, we get output as clusters. The similar genes are grouped in to one single cluster [10].

4 Methodology

The methodology in our paper is: we constructed a phylogenetic tree for the COX datasets. We are applying various clustering methods; hierarchical clustering, K-Means, PCA along with SOM. The process is clearly explained in Fig. 1.

4.1 COX Dataset

The experimental values of cyclooxygenase are collected. Cyclooxygenase is an enzyme that is present in every living organism. It clearly portrays the internal mechanisms of any organism, so we have used this in our project [11–13].

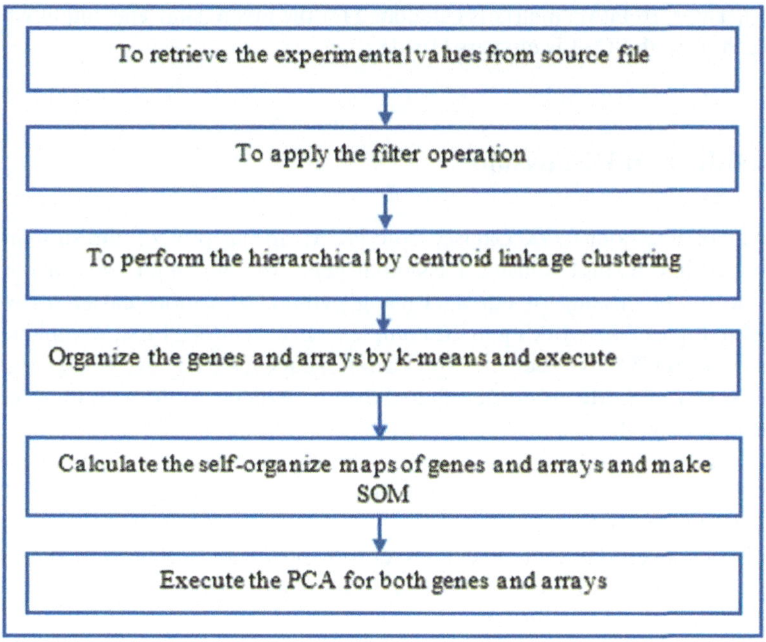

Fig. 1 Methodology

4.2 Performing Hierarchical Clustering, K-MEANS Clustering

Hierarchical Clustering bundles relative things to packs known as bundles. Great deal of gatherings is bundle specific to each other gathering, and in every pack are completely similar [14].

The output of this clustering is dendogram. The length of each branch in dendogram means the distance between species. Shorter the distance between species, more closer and similar they are.

K-MEANS calculates the attempts, which piece dataset into K preportrayed explicit things where every point includes a single party. We give a value for K, each cluster has its mean (i.e., centroid). We keep on assigning each object to group that has the closest centroid till the centroids are fixed and no longer more [15].

4.3 Performing PCA and SOM

PCA and SOM are executed together to cluster the similar genes. In Artificial Neural Network, there are many types, SOM is one of them. We train it by unsupervised learning to get a map, which is a low-dimensional representation. PCA is used in data analysis. The correlation matrix is calculated for the given data. The output when we execute PCA with SOM is clusters [16–18].

5 Results and Discussion

We have collected our COX Dataset from NCBI. In our project, the similar genes are grouped into a single cluster. Dissimilar genes remain single. We have created a console for processing of our data using python. It is more easier to work on new technologies than working in the complex software tools. First, we upload COX dataset, convert CSV to Excel, then run Hierarchical Clustering, we get dendogramas output. Next, run K-Means clustering and then we execute SOM with PCA, clusters are obtained (Fig. 2).

This is the console for our data processing. This is the console we have created with python. We first upload our genes dataset. Here we are uploading BG086892 gene dataset (Fig. 3).

Click on convert dataset to excel. Excel format of data is obtained from its CSV format. The Dataset size and number of records are displayed. Now, Hierarchical Clustering is performed (Fig. 4).

Hierarchy of clusters is formed. The data from excel are converted into distance matrix. Closest clusters are identified and similar clusters are merged here. The output of this is dendogram (Fig. 5).

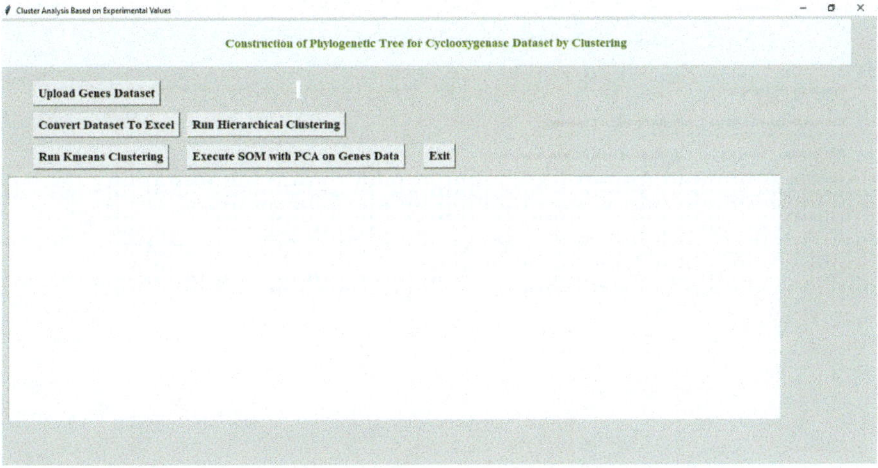

Fig. 2 Console

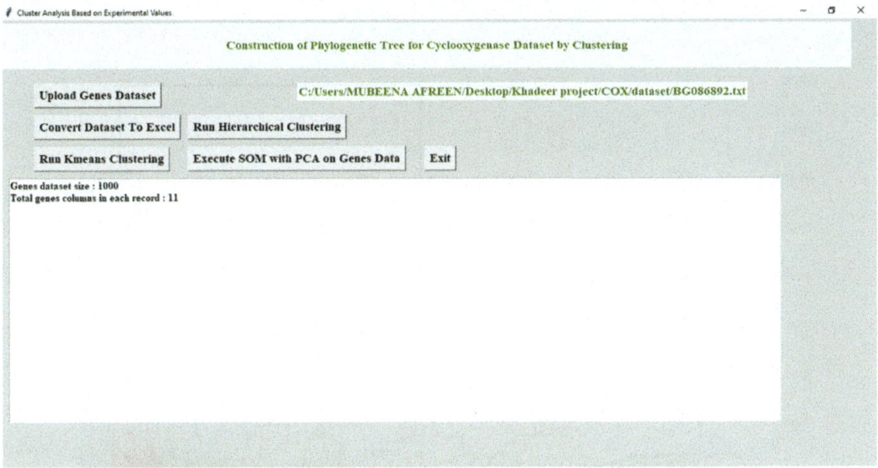

Fig. 3 Data conversion

Dendogram is the phylogenetic tree. In this dendogram, the length of the branch represents the distance between species. Similar species are closest. Next, we run K-Means clustering (Fig. 6).

In this, for each object, we have assigned a particular group that has the closest centroid. Our first record belongs to cluster 0 here. Next, PCA along with SOM is performed (Fig. 7).

For 1000 records, 1000 circles are formed. Here, similar genes are in the same cluster. Dissimilar genes remain single without forming clusters.

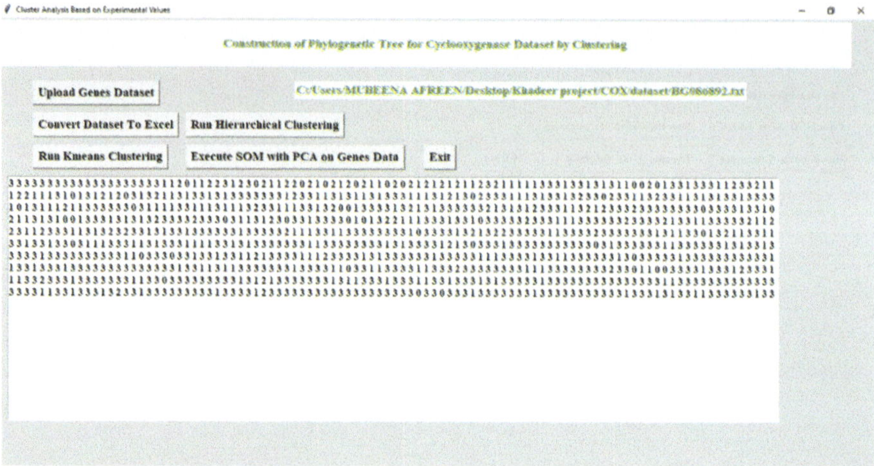

Fig. 4 Hierarchical clustering

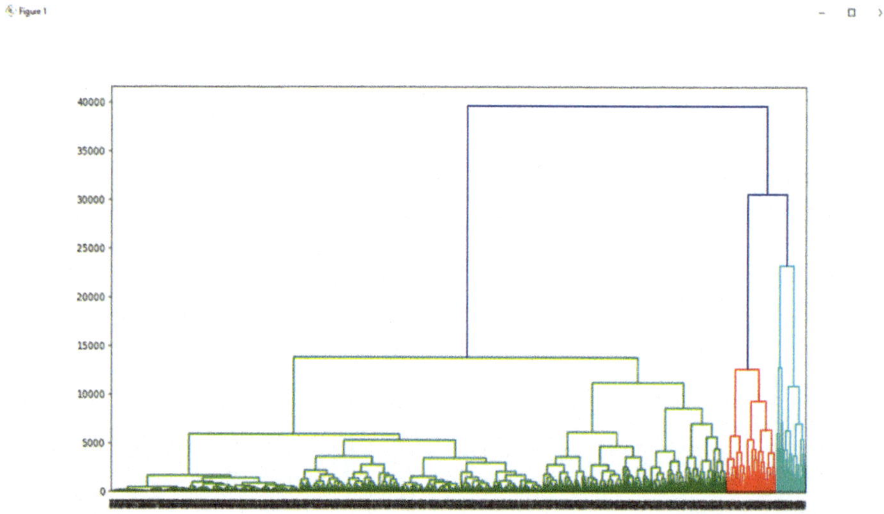

Fig. 5 Dendogram

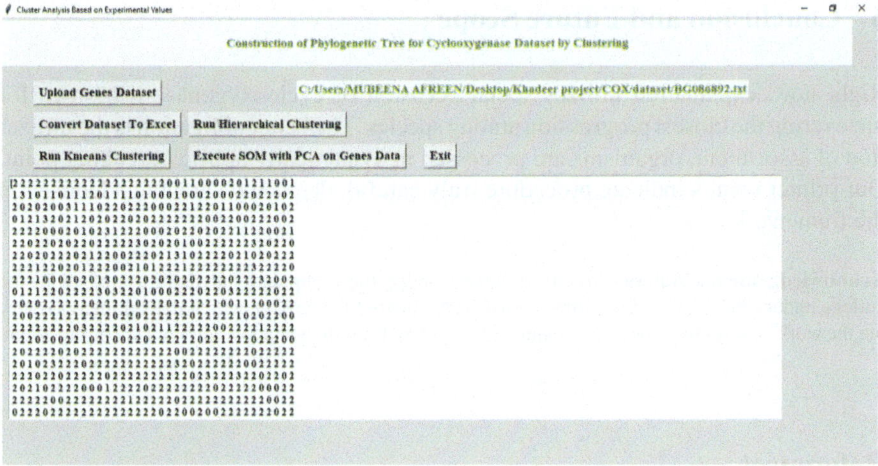

Fig. 6 K-means clustering

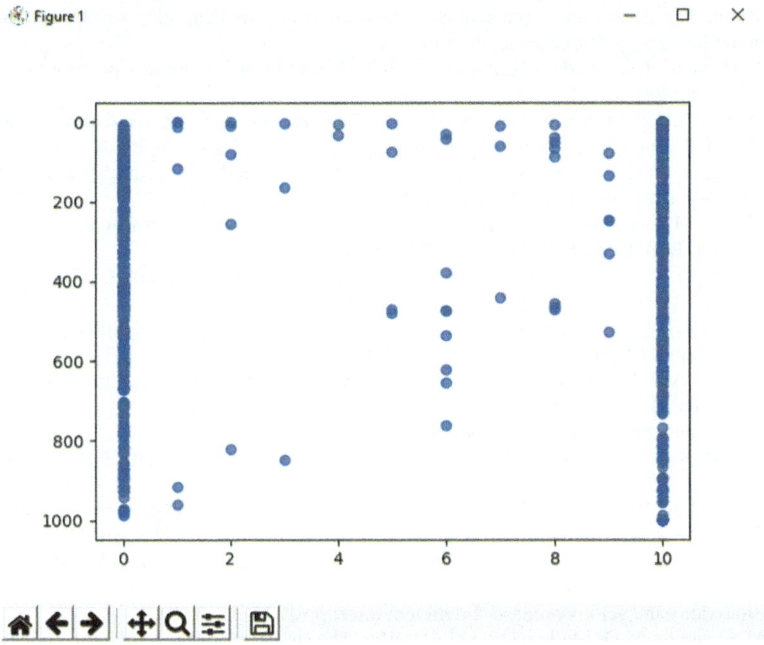

Fig. 7 PCA with SOM

6 Conclusion and Future Scope

Right now, explanation utilizes social occasion of cyclooxygenase responses for uncovering the closest progression among species. The better obligation is the utilization of assortment, organisms are processed in dynamically uplifted breaking point. Our primer results indicate procedure truly careful, decidedly showing adequacy in the framework.

Acknowledgements Authors gratefully acknowledge the computational facility created in the college under DST's FIST Programme (SR/FST/College-2017/28(C)), which helped them to carry out the work. The authors thank the management of IARE for their support and kind encouragement.

References

1. B. Snel, Genome phylogeny based on gene content. Bioinformatics 108 (1999)
2. L. Ming, An information based sequence distance and its application to whole mitochondrial genome phylogeny. Bioinformatics, 149 (2001)
3. E.A. Herniou, Use of whole genome sequence data to infer baculovirus phylogeny, J. Virol. 8117–8126 (2001)
4. H.W. Ma, A.P. Zeng, Reconstruction of metabolic networks from genome data and analysis of their global structure for various organisms. Bioinformatics 270–277 (2003)
5. R.C. Edgar, Muscle: a multiple sequence alignment method with reduced time and space complexity. BMC Bioinform **5**, 113 (2004)
6. N. Saitou, M. Nei, The neighbor-joining method: a new method for reconstructing phylogenetic trees. Mol. Biol. Evol. **4**(4), 406–425 (1987)
7. J.R. Vane, Y.S. Bakhle1, R.M. Botting, Annu. Rev. "Pharmacol. Toxicol, Cyclooxygenases 1 and 2", 97–120 (1998)
8. W.L. Smith, D.L. DeWitt, R.M. Garavito, Cyclooxygenases: structural, cellular, and molecular biology. Annu. Rev. Biochem. 145–182 (2000)
9. P.C. Kam, A.U. See, Cyclo-oxygenase isoenzymes: physiological and pharmacological role. Anaesthesia 442–449 (2000)
10. https://genomebiology.com/2007/8/1/R14/table/T1.
11. A.K. Jain, R.C. Dubes, *Algorithms for Clustering Data* (Englewood Cliffs, N.J., Prentice Hall, 1988)
12. Z. Feng, D. Li, P.C. Fung, Z. Pei, D.B. Ramsden, S.L. Ho, *COX-2-deficient mice are less prone to MPTP-neurotoxicity than wild-type mice*, Neuro Report, pp 1927–1929 (2003)
13. C. Iadecola, K. Niwa, S. Nogawa, X. Zhao, M. Nagayama, E. Araki, S. Morham, M.E. Ross, Reduced susceptibility to ischemic brain injury and N-methyl-D-aspartate-mediated neurotoxicity in cyclooxygenase-2-deficient mice, pp 1294–1299 (2001)
14. R.M. Garavito, M.G. Malkowski, D.L. DeWitt, The structures of prostaglandin endoperoxide H synthases-1 and -2. Prostaglandins other lipid, pp. 68–69 (2002)
15. Mag.rer.nat.Kerstin Kitz, Transcriptional regulation of Cox-2 expression in human osteosarcoma cells (2011)
16. R. Madhuri, M. Ramakrishna Murty, J.V.R. Murthy, P.V.G.D. Prasad Reddy et al. Cluster analysis on different data sets using K-modes and K-prototype algorithms, in *International conference and published the proceeding in AISC and Computing* (indexed by SCOPUS, ISI proceeding DBLP etc), vol. 249 (Springer, 2014). ISBN 978-3-319-03094-4
17. P. Pecina, H. Houstková, H. Hansíková, J. Zeman, J. Houstek, Genetic defects of cytochrome coxidase assembly. Physiol. Res. **53** (2004)

18. J.M. Zee, D.M. Glerum, Defects in cytochrome oxidase assembly in humans: lessons from yeast. Biochem. Cell Biol. 859–869 (2006)

Efficient MAC Protocol Analysis Systems for Wireless Sensor Network

Palamakula Ramesh Babu, A. Rajasekhar Reddy, and A. Narayana Rao

Abstract Energy preservation for effective WSN applications is one of the major challenges, as the small sensor nodes have very limited power, storage, communication and computing resources. The sensors remain unattended, and after deployment, rechargeable batteries are almost difficult. The algorithm proposed to pick a nearest hop node based on the token-ring algorithm, compared with the other current S-MAC, T-MAC and HEED algorithms, which pick the next hop node based on parameters such as shortest-path, maximum residual power or probability. More specifically, it is easier to reduce the total power consumption on a multihop route when driving by carefully choosing suitable intermediate nodes. The combination of cluster-based routing and containerized WSN routing algorithms is proposed to be used as an Advanced Energy Efficient Sensor-based MAC (AES-MAC) protocol. The routing process involves the maintenance of routing and routing. The illustrations show a simple, distributed, localized routing algorithm, which can be easily implemented for various practical applications.

Keywords MAC · Hybrid protocols · AES-MAC protocol

1 Introduction

1.1 Overview of WSN

Thousands of low-cost small nodes consist of wireless sensor networks with limited capacity for communication. There are thousands of inexpensive, low-power, low-cost digital signal processors.

P. Ramesh Babu (✉)
Department of CSIT, Siddharth Institute of Engineering & Technology, Puttur, Chittoor, AP, India
e-mail: palamakularamesh@gmail.com

A. Rajasekhar Reddy · A. Narayana Rao
Department of CSE, N.B.K.R Institute of Science & Technology, Vidyanagar, SPSR Nellore, AP, India

C. Kiran Mai et al. (eds.), *Proceedings of International Conference on Advances in Computer Engineering and Communication Systems*, Learning and Analytics in Intelligent Systems 20,
https://doi.org/10.1007/978-981-15-9293-5_27

Thousands of low-cost small nodes consist of wireless sensor networks with limited capacity for communication. There are thousands of inexpensive, low-power, low-cost digital signal processors [1, 2], including WSNs. The potential for these sensing networks has been improved by MEMS [3, 4] and radiofrequency applications. The microcontroller consists of the actual transceiver and the sensor. The sensing circuit is used to test and convert the sensor in the environment to an electrical signal. The data collected are sent, as shown in Fig. 1, to the central plate directly or through the gateway via a radio transmitter.

The main challenge in sensor networks is to increase life span of sensor nodes that involve the energy-efficient computing of sensor nodes and protocols. Other priorities of the WSNs are equity, accuracy, cost-quality, efficiency and health [5]. Our mobile contact network focuses on the default mobile wireless networking infrastructure. The rapidly evolving and more complex communication, embedded and sensor technology worldwide develops micro sensors with a view to digitalization and communication.

Wireless sensor networks are able to track and collect information from multiple environments in network areas or track and process artifacts in order to track data and delivery in real time, receive precision and reliable information and then send it to the user who needs it. Since its processing power, storage, communication and transmission energy are limited to sensor nodes, each node can only obtain general

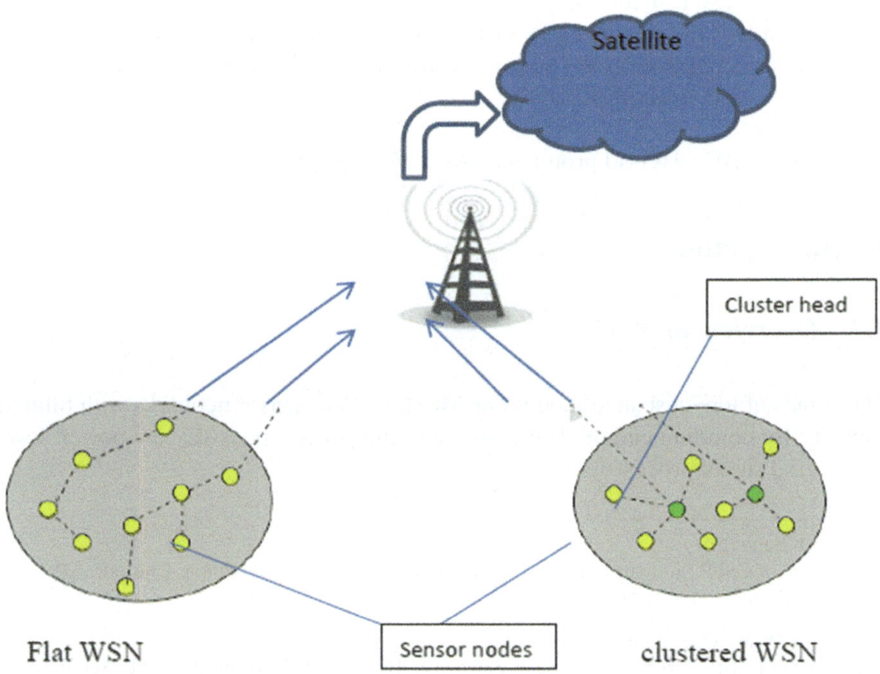

Fig. 1 Wireless sensor networks

information on the topology of the local network and also the network protocol that nodes cannot use. The restriction of energy supplies at the sensor nodes has stimulated a significant number of researches into sensor networks. The research can be split into two general classes, the processing of signals and radio communication that discuss the main causes of energy use. The first class of works is planned to extend the network life by choosing a subset of environmental surveillance sensors. In low power sleep mode, the sensor cannot be picked. The goal of this project is to maintain as many live sensors as possible in order to protect the zone of interest also known as a network organization. The sensor organization often needs signaling technique and regional mapping to ensure sufficient coverage for the least number of deployed sensors.

1.2 Problems in the Existing Protocols, Motivation to Proposed Protocol

- Increased silent listening is the first drawback of existing (S-MAC, T-MAC, TRAMA) protocols.
- The second restriction is overhearing. This is because the sensor nodes listen to signals that have not been expected.
- Collisions are the third issue. When refilling collided packets, the sensors waste energy.

To solve these questions, but only at a certain level, the proposals (R-MAC, OB-MAC and SIFT) are designed.

TRAMA is a timeline-based protocol that uses the distributed algorithm to resolve these problems. It uses the NP to collect information from adjacent nodes. It uses the NP. The reason is to develop a new protocol: the energy-efficient Token Bus MAC Protocol (ETB-MAC) that can deal with such problems. This is increasingly overhearing. The ES-MAC protocol for collision-free environment is developed for wireless networks. A new protocol idea inspired by an Advanced Energy Efficient MAC Protocol (AES-MAC) is to reduce energy waste.

2 Related Works

Sensor-MAC (S-MAC), specifically, designed for WSNs, is a fundamental and essential MAC protocol. Nodes are normally idle in WSNs. The continuous listening of sensor nodes contributes to energy loss as the key source. SMAC reduces idle listening time by requiring the sensor node to enter slept mode periodically on the basis of the duty cycle to address this problem. The SMAC is also aimed at the energy loss caused by crashes, overhears and overhead checks. Nevertheless, the data quality is in line with the S-MAC delay.

The enhanced S-MAC variant Timeout MAC (T- MAC) incorporates a duty cycle adaptive to accommodate changes in loads and time and ends the active section dynamically. This reduces the amount of energy waste when the sensor nodes wait for messages while keeping equal performance. T-MAC shows comparable results in uniform traffic conditions and outperforms S-MAC in a scenario with varying transport loads. T-MAC has an early sleep problem, which can result in delays in the transfer of data [6].

The T-MAC, which includes a future FRTS and the maximum buffer priority, is the solution to this problem. Although these solutions can solve the problem of early sleep, the delay is reduced by one hop. For continuous data transmission, the data collection MAC (D-MAC) calls a service loop hop chain environment. D-MAC will not have to provide RTS or CTS procedures as the schedule is set up. However, it uses the ACK platform to relay data packets effectively.

However, in the specific networking environment in which data are collected on a sick note, D-MAC does not support WSN with mobile Sensor Nodes. The MAC (E2-MAC) energy efficiency suggests that the sensor node buffers have a threshold value that further increases the sensor node power consumption. Information can be transmitted to other knots only if the buffer occupancy reaches the default threshold value under this scheme. The No Activation-Event Timer in sleep mode is marginally lower than in T-MAC.

The S-MAC achieved energy efficiency by setting the duty cycle and thus raising traffic volatility resources in low-traffic conditions. If no contact occurs on the channel, all nodes will follow the set schedule. The timing of the network load during traffic fluctuations is not adjusted by the system. T-MAC therefore based its attention on management of fixed active time, allowing for low-traffic and busy traffic (thereby moving from active to sleep without losing more energy efficiency). The idea that shortened listening time is silent in case of low traffic or canals.

Thus, T-MAC implements adaptive listening and helps the node to go back to sleep mode after a synchronization. No-touch reasons cannot be due to the events mentioned below and suggest the time for the radio to close. The above may be: if there is time for the given frame length, if the data have received properly or if the received data have occurred with the acknowledgment or where a positive data exchange in the neighborhood has occurred as a result of the other transfer.

Medium access control is a broad area of research, and many researchers work in the new field of WLAN networks and low performance. Current MAC designs for wireless sensor networks can be broadly divided into TDMA and contention-based protocols. The Structured IEEE 802.11 Distributed Coordination Function (DCF), based principally on the MACAW research protocol, is one example of this controversial protocol. It is widely used in adhoc wireless networks to solve the hidden end problem due to its simplicity and robustness. Nevertheless, recent work has shown that this MAC's energy usage is exceptionally high while nodes are inactive, especially, due to the idle audition. To prevent neighboring nodes from being overheard, PAMAS is modified. Our paper also uses the same form of saving resources. Our research with PAMAS is largely different because we do not use outdoor signals. In PAMAS, two separate radio channels are needed, which in most cases mean two distinct radio systems per node. PAMAS does not answer the issue of rising idle listening.

For other MAC protocol classes, the reservation and timing are dependent, e.g., Protocols for TDMA. TDMA protocols have an intrinsic advantage of saving time in contrast to the contending protocols as the radio transmission duration is minimized and no overheads and collisions are threatened. However, TDMA typically includes specific communication clusters for nodes such as Bluetooth and LEACH. The control of interconnection and conflict is not an easy job. Furthermore, changing its frame length and allocating time slots through dynamic changes in the number of nodes within the cluster is not easy for a TDMA protocol. It also does not depend on disagreement as flexibly as a protocol. For example, in a cluster, Bluetooth can have up to eight nodes.

Sohrabi and Pottie proposed a protocol for self-organization of the wireless sensor network. Each node has a TDMA-like structure, named an excellent framework in which different time slots can interact with your friendly neighbors. Only a neighbor is spoken to at any location. In order to prevent the intervention of the neighboring links, the protocol assigns various channels (FDMA) and code delivery (CDMA). The structure of the superframe is identical to the TDMA frame, but it does not at the same time limit access to the media through two nodes. The actual multiple accesses are completed by FDMA or CDMA. Low bandwidth use of the system is an inconvenience. For instance, if a node already has packets to be sent to a neighbor, it cannot reuse time slots for other neighbors. Piconet is an ad hoc wireless network with low power. It is also an interesting feature that the Piconet put nodes in daily sleep to save resources. A node will broadcast its address before listening, the program used by Piconet to synchronize neighbors' nodes. Should a node want to talk to a nearby node, it must wait for a broadcast by its neighbor to be broadcast.

Woo and Culler reviewed several Carrier Sense (CSMA) configurations and suggested an adaptive rate control mechanism, with the main objective of ensuring that all nodes on the multihop network are fairly assigned bandwidth. They have tested and measured different MAC systems using motes and the TinyOS Platform. In comparative terms, our approach does not promote equity per node and even compensates for further energy savings.

3 Medium Access Control Protocols

The wireless sensor network provides a wide range of low-cost and powered nodes. Sensor networks consisting of small instruments that track conditions such as pressure, temperature and noise, physical or environmental conditions. Sensor system applications are used in military, commerce, civil, vehicle tracking, intelligence and environmental monitoring. Note, the capacities for data processing and communication are a network sensor. The data routing protocols aim to avoid replication of data and thus increasing the lifetime of storage. A special feature of WSN is the linked data problem. There are very similar collected data from the surrounding nodes and shifts in environmental amounts are very slow. These features can be used to establish effective techniques for data collection. The existence and stability of the device are central considerations throughout the sensor network. The Medium Access Control protocols play an important role in accessing wireless channels. The design issues in MAC protocols include the cached terminal issue, collision, overhearing, QoS, etc. There are a lot of difficulties. Research projects were focused on tackling these problems. Energy is highly critical for sensor nodes, as all battery-powered sensors.

They are normally mounted and not recharged in a wild environment. Therefore, energy efficiency is the most important requirement for MAC protocol design. Crash, data replication, or idle audition include some energy-consuming issues.

3.1 CSMA-Based Protocols

The sensor notes wake up periodically, hear the channel and sleep again in the CSMA-based protocol. The dispute happens during the listening times of multiple transmitters as they want to send packets. The downside of CSMA is that the lower traffic levels are more costly and promising. Nevertheless, to deal with collisions, more collision avoidance methods are needed.

3.2 TDMA-Based Protocols

Slots are reserved for TDMA-based protocol sensor nodes and these nodes are only accessible in these slots and they listen to the channel. The great advantage is that, through collision-free media access, it improves the capacity of high traffic loads. The problem is that the low traffic load is reduced because of idle spaces. In decentralized world, it is not easy to modify the slot assignment on TDMA-based protocols. A further downside is that all nodes are difficult to synchronize and adapt to topology change in order to prevent wandering clocks. TRAMA is a TDMA protocol used to avoid waste of space by using traffic planning. The E-MAC does not require a central manager or a base station. The nodes are going to select their spaces. The nodes may decide to become active or passive based on local information.

3.3 Hybrid Protocols

Although TDMA is better suited to that traffic loads, CSMA makes it simpler, flexible and needs no up-to—date global details. The network topology can change constantly. CSMA has some inconvenience, however, such as hidden terminal problems and reduced traffic performance. This led to some researchers proposing a combination of CSMA and TDMA.

4 MAC Routing Mechanisms

The WSN communication routing architecture is shown in Fig. 2. Sensor nodes can gather environmental and route data information on the discharge by means of adjacent sensors. The gateways can be fixed or mobilized with externally established communication structures to connect sensor networks such as the internet and satellite networks.

The network layer determines how data from wireless nodes can be passed to the sink node as soon as possible. Features of the key network layer:

- As a result of very high overhead sensor implementation, the Internet Addressing System (IP) does not apply [1] to the wireless network of sensors.
- Send data via multihop routing to the sink via sensor nodes in networks with wireless sensors.
- Data traffic created by the sensors is highly redundant as the sensor nodes generate the same data at a distance equal to the phenomenon. This problem is eliminated by the appropriate routing protocols.
- Stack resource management protocols need to be taken into account in terms of their transmission, processing and storage capacity.
- Several specific algorithms have been suggested to solve these problems in wireless sensor networks. Depending on network structure and applications, these Routing Structures are classified into data-centered, hierarchical and placed.

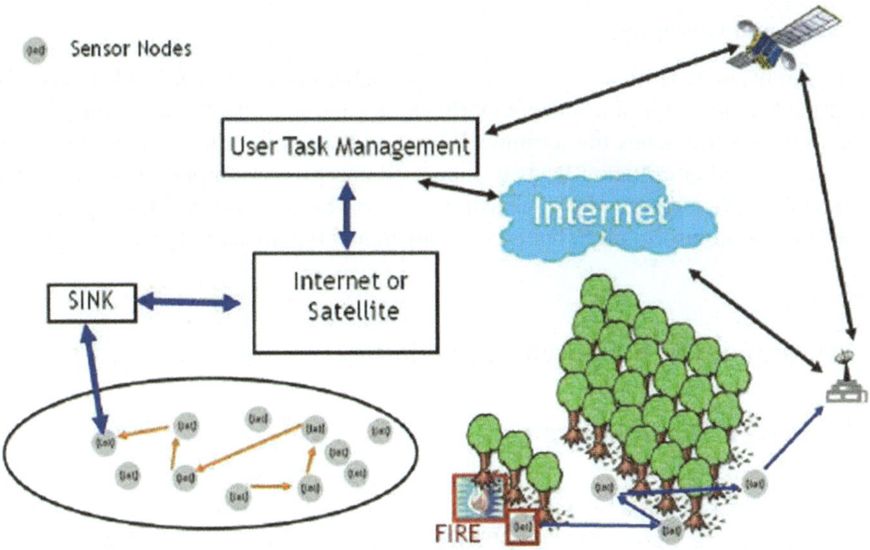

Fig. 2 MAC routing

4.1 Date-Centric Protocols

Date-centered protocols can select a number of sensor nodes and aggregate data based on queries. Connections are built on request and node information is required on the sink waiting for the sensors to respond in this area.

SPIN

Metadata [5] are much more efficient and relatively simple to use by the SPIN Information Sensor Protocols through negotiation. It means that neighboring nodes have the same form of data. Data collected through nodes are exchanged between sensors through a data publicity mechanism allowing data sharing between nodes. SP IN ensures a network-based transmission of low redundancy information to address bandwidth and energy loss problems and flood diffusion mechanisms [1]. The sensor node can be stronger and more energetic in SPIN. The limitation of SPIN is that the data collection process cannot guarantee data and therefore needs more time.

Directed Diffusion

Frequently built on a data-centric routing framework, the aim is to use naming data for all communications to distribute data through sensor nodes. Data are collected at intermediate nodes for aggregation, and distributed through unicast, broadcast and multicast from the sink to other nodes. When sensor nodes collect data from the other nodes, they compare it against data already stored in the cache. Path repairs can also be carried out with numerous guided diffusion paths in advance.

Flooding and Gossiping

Both systems include flooding and gossiping. Each node, when it is flooded, receives a data packet then sends it to all neighboring nodes and stops the broadcasting process when the packet reaches the terminus. Flooding is easy to enforce, but with many problems. Another collision leads to increased energy consumption with the same node with duplicated messages. Gossiping solves the collapse problem by sending a random friend information, but it is a delay that is the main challenge.

5 Design of the Proposed Methodology

5.1 MAC Protocol Design

The MAC Protocol has a big impact both on the probability that the radio is efficient and, on the energy, consumed by the radio. In particular, a WSN MAC protocol can monitor the operational process and perform carrier detection. For some of the experiments, a perfect communication model was required. The simulator is to provide every message that requires no other communication with a single network-wide communication interface. A node should therefore not be transmitted to protect the packet from collision. A radio is used to listen if you are listening for inbound signals. It is known to be sleeping when the radio is not broadcast, received or heard. To calculate the impact of different routing protocols better, sleeping energy was reduced to 0 mw. In any simulation, several resources are used in sleep. Even if the radio transmission consumes about 44 times the sleep consumption, it only lasts for a short period. For example, 250 kbps of data is transmitted by CC2420.

5.2 Handling Multiple Networks

It should be tested for overall capacities for a routing protocol for the conservation of (source and sink) diversity in a variety of networks with specific parameters. Moreover, as some parameters are random, such as those seeds regulating random radio communication or node positions (which could be positioned on the random levels), it is necessary to repeat the experiment several times and change the chosen random parameter. If the experiment is not replicated by various random parameters, the only one where a single routing protocol functions successfully can be the network generated randomly. It also appears to be good if the average protocol is bad.

5.3 Design of AES-MAC Protocol

The following is the AES-MAC communication method: when the node transmits the information to the recipient nodes, the improved performance is that after every data packet has been sent it cancels the ACK packet and only leaves the final ACK packet as confirmation of approval. The recipient records the time of the conflicting nodes and notifies the node sent in the data packet phase. If the neighbors of the recipient listen to the radio, the recipient will send the transmitter a short message and can alert the neighbors in the sleep. When all data packets are received, the recipient sends an ACK packet to inform you that all data packets have been sent. Under the specified time span, the recipient can enter the serial number of packets not received by recording the number of packets when not all packets are received from the recipient. The recipient sends these numbers to the sender, and after receiving those numbers, the transmitter is able to return the data packet. All node 1 nodes are distributed and send node 2, receiver node 3 and recipient, node 4 and receiver are neighbors. Ha, Fig. 3 indicates the ES-MAC and AES-MAC protocol communication process.

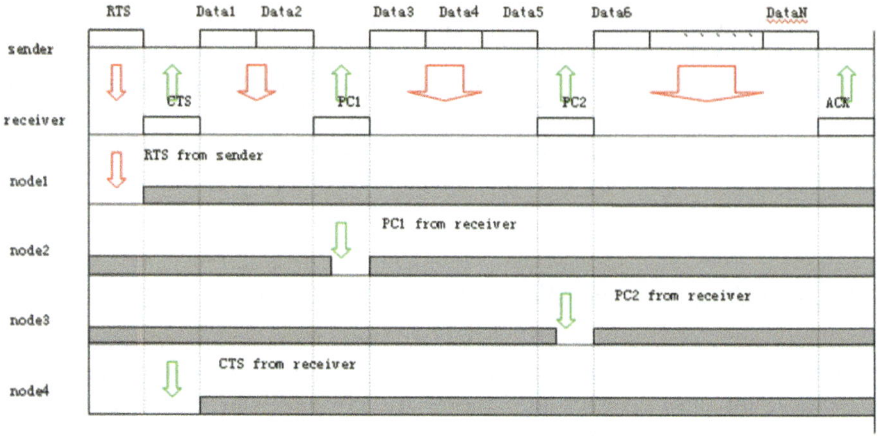

Fig. 3 The transmission of ES-MAC protocol and AES-MAC protocol (shaded part represents the sleep state)

5.4 The Theoretical Analysis of Advanced ES-MAC Protocol

In this region, we equate the more energizing and time-consuming AES-MAC, ES-MAC and S-MAC. The Advanced energy consumption compares to the SMAC when the energy consumption and the sending time are specified, CTS, ACK, PC and DATA and the total calculation is carried out during the communication stage while the SMAC compares the time consumed by the ES-MAC protocol with the S-MAC. Energy consumption of the S-MAC protocol is the energy consumption of the STR, CTS, N data and NACK packets, meaning that the N packets are sent to the receiver. The Advanced ES-MAC Protocol is the energy consumption for RTS, CTS, N and M ACK packets sent via the ES-MAC Protocol. M is higher or lower than $N - 1$, so $M + 1$ is higher or lower than N. Nonetheless, as N increases, M is significantly less than -1 and $M + 1$ is marginally smaller than N, and thus the energy of the transmitted M PC and ACK is significantly smaller than N. The Advanced ES-MAC works better on ES-MAC, S-MAC and uses less power. Advanced ES-MAC saves more time, in other words.

6 Implementation and Simulation

6.1 Simulation Environment

See Figs. 4 and 5.

This paper uses experimental data simulation to compare ES-MAC, S-MAC protocol with Advanced ES-MAC protocol. Receiving, sending and sleeping

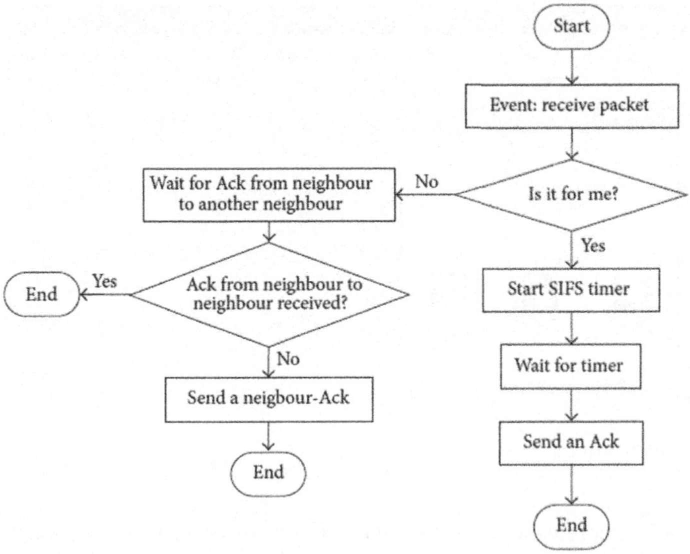

Fig. 4 Flowchart for receiving sensor node

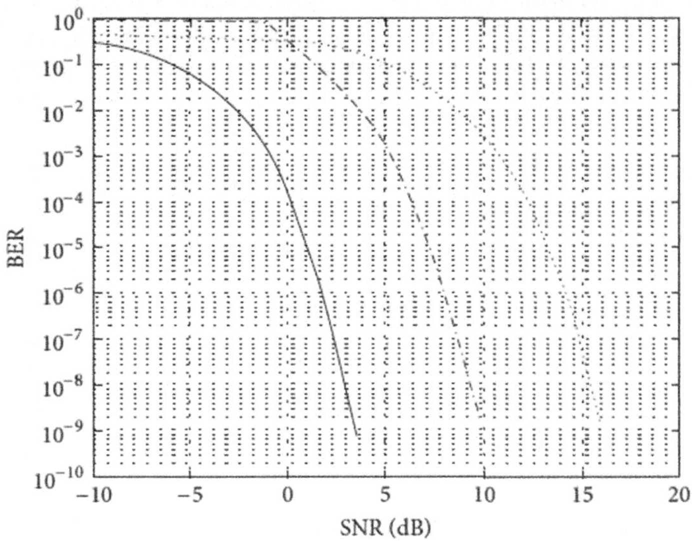

Fig. 5 Theoretical BER to SNR for different communication protocols

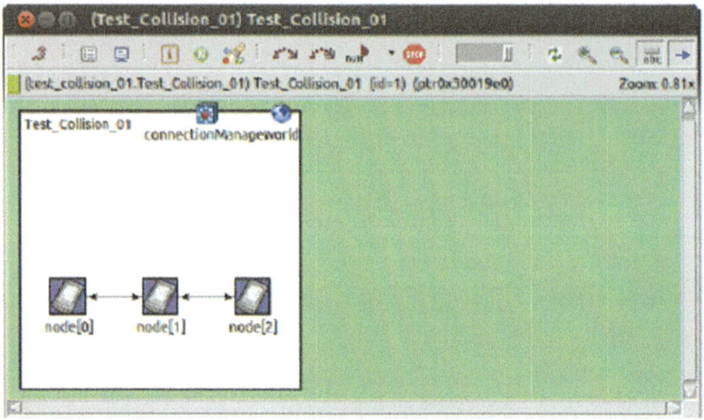

Fig. 6 Collision scenario

consumption is increased. The data packet transmission time is less than 5 ms. The test was carried out on more than 50 sensor nodes, the node distribution and the communication range are shown in Fig. 3 and the data packets are transferred between the simulation throughputs and back off shown in Figs. 7, 8. Their scenario for collision of S-MAC and Advanced MAC protocol throughput and back off was also presented in Fig. 6. The theoretical diagram and flowchart were provided in the fourth and fifth categories.

Fig. 7 Comparison of simulation results of with throughput and error rate

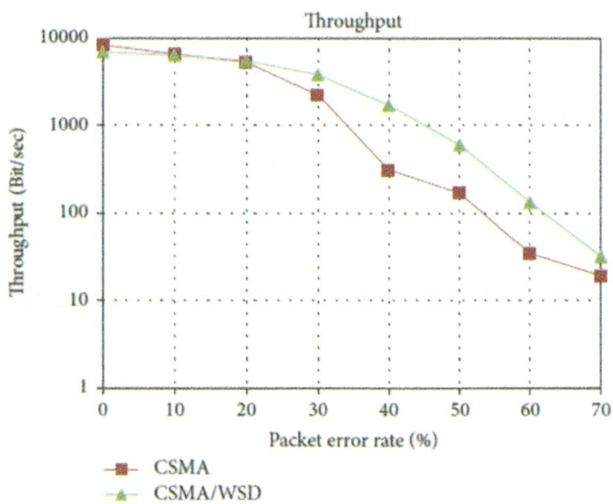

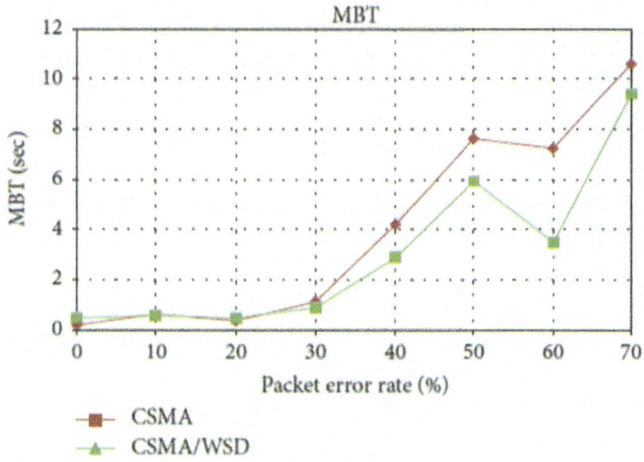

Fig. 8 Performance comparison with mean back off time and packet error rate

7 Conclusion

Wireless sensor networks have become more interested in recent years. In such systems, sensors are usually redundant and will last until they are depleted. For a certain mission, energy must therefore be handled wisely to extend the service life of the sensors. The energy conservation is normally targeted at all contact stack layers. This research proposes the design and analysis of the Advanced ES-MAC, which improves the fundamental ability of SMAC. The implementation of the schemes suggested allows for improved energy efficiency for medium access wireless sensor networks. This Advanced MAC saves energy well and also enhances the performance and reduces the packet collision. Simulation checks the effectiveness of this protocol.

References

1. I.F. Akyildiz, W. Su, Y. Sankarasubramaniam, E. Cayirci, Wireless sensor networks: a survey. Comput. Netw. **38**(4), 393–422 (2002)
2. S. Tilak, N.B. Abu-Ghazaleh, W. Heinzelman, A taxonomy of wireless micro sensor network models. ACM Mob. Comput. Commun. Rev. (MC2R) **6**(2), 1–8 (2002)
3. G.J. Pottie, W.J. Kaiser, Wireless integrated network sensors. Commun. ACM **43**(5), 51–58 (2000)
4. J. Rabaey, M. Ammer, J. Silva jr., D. Patel, and S. Roundy, Pico Radio supports ad hoc ultra-low power wireless networking, IEEE Computer, **33**(7), 42–48 (2000)
5. R. Min, M. Bhardwaj, S. Cho, E. Shih, A. Sinha, A. Wang, A. Chandrakasan, Low power wireless sensor networks, in *Proceedings of International Conference on VLSI Design*, Bangalore, India, January 2001 (2001)
6. D. Estrin, R. Govindan, J. Heidemann, S. Kumar, Next century challenges: Scalable coordination in sensor networks, in *Proceedings of the 5th IEEE/ACM Annual Conference on Mobile Computing and Networks (MobiCOM'99)*, Seattle, WA, August 1999 (1999)

Analysis of Wireless Mesh Networks in Machine Learning Approaches

A. Narayana Rao, Palamakula Ramesh Babu, and A. Rajasekhar Reddy

Abstract Now days many people utilizing the internet is to get the Wireless Mesh Network (WMN). Naively, all nodes on the network are not malicious underlying routing protocol. WMN's open architecture, multishop communication nature, different styles of management and wireless communication pave the way for malicious attackers. In the multiway mesh routing protocol, the attackers may exploit hidden loopholes to carry out a suction attack called the blackhole attack. Wireless Mesh network efficiently improves with ping mesh nodes equipped with multiradios tuned to nonoverlapping channels. Thus, there are multiple connection ranges for data transfer between the two nodes and the bandwidth between the node pair is dynamically different. In this case, a mesh node uses machine learning to select the best possible data bandwidth. A new heterogeneity key management system was created, which combined logical key hierarchy with localized threshold technology. The heterogeneity of wireless mesh networks is being developed. In order to recognize profile profiles and intrusion, we have a cross-layer detection model using machine learning algorithms to exploit linked routing features. With a generic response model to define system and resource dependency services, we address the wireless network automatic intrusion response issue. In this article, we present a dispersion calculation dependent on machine allowing transmission node to select the next step dynamically with the maximum bandwidth possible to resume communication, based upon the study algorithm.

Keywords Machine learning · Wireless mesh networks · And deep learning

A. Narayana Rao (✉) · A. Rajasekhar Reddy
Department of CSE, N.B.K.R Institute of Science & Technology, Vidyanagar, SPSR Nellore, AP, India
e-mail: narayanaraoappini@gmail.com

A. Rajasekhar Reddy
e-mail: arsreddy521@yahoo.com

P. Ramesh Babu
Department of CSIT, Siddharth Institute of Engineering & Technology, Puttur, Chittoor, AP, India
e-mail: palamakularamesh@gmail.com

C. Kiran Mai et al. (eds.), *Proceedings of International Conference on Advances in Computer Engineering and Communication Systems*, Learning and Analytics in Intelligent Systems 20,
https://doi.org/10.1007/978-981-15-9293-5_28

1 Introduction

The mesh routers and customer devices are separate and autonomous and form wireless networks [1]. A device named a node in WMNs. Node has multiple radios that enhance WMNs total capacity [2]. Radiographies in WMNs are cognitive radios that allow radios to learn and adapt from their environment [3]. Radio coding [4], learning, and reason [3], which is also referred to as programmable radio, can be programmed. This research includes a comprehensive safety framework that includes protection of WMNs against intrusion, intrusion detection and intrusion reaction system. Intrusion prevention combines the protection feature with a specific function to prevent possible intrusions. In this context, we are investigating principal group management in WMNs. In combination with the localized threshold-based technology, we recognized the architectural heterogeneity of WMNs [2, 5, 6] and introduced a new heterogenous community-based key system of management.

The multicast WMN session particularly includes both mobile and static backbone nodes: for spine hubs that are topologically steady yet can be spread over an expansive range, the found limit-based key administration method was implemented because this method needs reliability and the delay of rekeeping is independent of network size. Interference avoidance is the first component of the protection program. However, the first intrusion detection device is required for attacks and alarms to be identified if the first security line is breached. Intrusion prevention for the network also depends on an analysis of network traffic. New routing measures for WMNs have been developed to support and deliver high-performance multiradio and multichannel features, integrating with the routing option for the efficiency of wireless communication.

In this case, cross-layer architecture is necessary for intrusion detection. The machine training algorithms are used for the trainings and the recognition of intruders in our cross-Canada anomaly detection model. For both the network layer and the data link layer, statistical characteristics are obtained and analyzed. A WMN test bed has been used to implement the anomaly detection concept. Experimental studies have shown that the cross-layering approach has a better detection rate and a better false alarm rate in comparison to one-layer intrusion detection (e.g. network layer). Detection of intrusion is also associated with behavioral responses to mitigate damage caused by intrusion. Current responses are limited to manual responses that require the network manager to manually detect intrusions. A standardized reaction model is necessary between automated intrusion reaction steps to describe system service and resource dependence and determine the harm to intrusion and response costs to select a correct answer.

1.1 Applications

The most important applications drive research and development of wireless mesh networking. Many applications have clearly shown their promising market value. For instance, home internet, business and government networks, corporate networks, metropolitan networks, transit infrastructure and telecoms and healthcare networks [7]. A single access point is connected to the internet in the current broadband home network. There are often dead areas in the building, so adding more access points to boost coverage is costly. However, it is not effective to communicate through various access points. The WMNs will fix both of these problems. The data traffic can be routed through the wireless network with a good coverage and a powerful home wireless network if the access point is replaced with a variety of mesh routers.

A single access point is connected to the Internet in today's broadband home networks [8]. There are often dead areas in the building, so adding more access points to boost coverage is costly. However, it is not effective to communicate through various access points. The WMNs will fix both of these problems. The data traffic can be routed through the wireless network with a good coverage and a powerful home wireless network if the access point is replaced with a variety of mesh routers. A WMN can be created for a group or between areas for data sharing and communication improvement. WMNs can also be used for wireless communications via corporate networking between buildings and offices. WMNs are cost-efficient options for broadband networking in vast areas such as towns in particular in underdeveloped regions. WMN also needs instant traffic information for passengers on trains, buses and ferries via infrastructure [9, 10]. This is also necessary that WMNs also support the medical and health systems with high bandwidth and fast access.

1.2 Challenges and Issues Wireless Mesh Networks

Wireless mesh networks are inefficient and are intended to provide a secure communication system through efficient mechanisms. WMN's work can address intrusion prevention, intrusion detection and intrusion response problems in a comprehensive protection architecture. WMN is an important function of group communication. The key management teams, which include key team development, delivery, cancelation and update, play a major role in coordinating the group members. Wireless routers are relatively stationary in WMNs and mobile customers are often subject to community modifications. The key must be changed if the members of the group leave and leave the party. Traditional solutions for key updates using a single application repository are not successful as any key update needs to spread across the entire network. Different radio technologies are widely used to enhance wireless network performance. However, the channel must be carefully planned for the radio range and negotiations. Cross-layer architecture has become a need for WMNs, multiradio and multichannel communication. For example, a route with many channels can have a higher efficiency than a road with the shortest number of hops. As data can be transmitted simultaneously on consecutive links when any link on different channels is correctly configured. Inspired by this, we are looking at cross-layer architecture for intrusion detection.

2 Literature Review

Shams and Rizaner [2] will ensure stability and short-term opportunity efficiency in the long term for the wireless mesh routing system. Second, wireless routing must maintain its intensity in order to avoid a range of hard and soft defects, including passing channel interruptions, transient failure rates, frequent channel disconnections, DoS attacks and failed nodes. The state of the art approaches do not address both of these problems.

In Narayana Rao [7], an ad hoc network could be considered as a particular type of network. The network is structured and can be used in a region for complex and economic connectivity. On the other hand, when wireless devices get into contact with one another, an ad hoc network is created. The mesh routers can be moved and moved according to various networking requirements. Not all network nodes must reach the Access Point with WMN. The connection points transmit the packets from adjacent nodes to the entry point.

In Gonzalez et al. [10], traffic may occur in this situation also proposed a fluid logic algorithm for the estimation of mistakes and for efficient network routing to overcome these disadvantages. The revised NEURO fuzzy is based on the shortest possible route for the WLN in order to avoid the fail-out node and to calculate network performance.

Narayana Rao [9] suggested routing protocol is used between source and target node for one route, i.e., only one route for meshing and ad hoc networks. Multi-tracking's main goal is to ensure that many good tracks, not just the best, are used. It can be done in maintaining these routes without excessive overhead costs. This traffic-conscious routing of fluid logic, packet transmission and distribution is more efficient than their current network routing.

Maccari et al. [11] and many investigators investigated traffic problems in the wireless network, which is the subject of the representative network for the overall congestion and delay problem. After a thorough overview of current research, these algorithms need high overall APs. The Neuro-Fuzzy algorithm was developed to improve fluidity. A new logic routing is recommended to ensure effective CSPR-based wireless network routing.

3 Cross-Layer Design of Intrusion Detection in Wireless Mesh Networks

Figure 1 reveals our cross-cutting intrusion detection architecture. This model consists of four components: an input module, a training profile module, an anomaly detection module and an alert generation. The four modules are located in each mesh node and cooperate in detecting anomalies in the wireless network backbone.

According to the defined functionality, the data collection module collects audits of data from local activities and network transport. The data recovery module collected multilayer data sets through our cross-cutting design, including channel assignments and physical layer changing channel frequencies, links data link layer and network layer routing information and transmission of network layer data. The standard data set collected is inserted into the profile training module into normal profiles during the data training process; data collection module pipes are inserted into the anomaly detection module. Apply machine training algorithms to profile a default profile on each mesh node. The Bayesian network, the decision tree and the Vector Assistant Machine (Support Vector Machine Algorithm) apply three algorithms of machine learning. These three algorithms are the standard algorithms for machine learning and prevent biases leading to a type of detection. A regularly profile-based intrusion is detected by the anomaly detection module. Any crossing traffic different from the usual profile is marked as an alert and additional identification or reaction intrusion. Any conduct substantially different from the profile is an anomaly in the detection of anomalies.

Fig. 1 Detection model

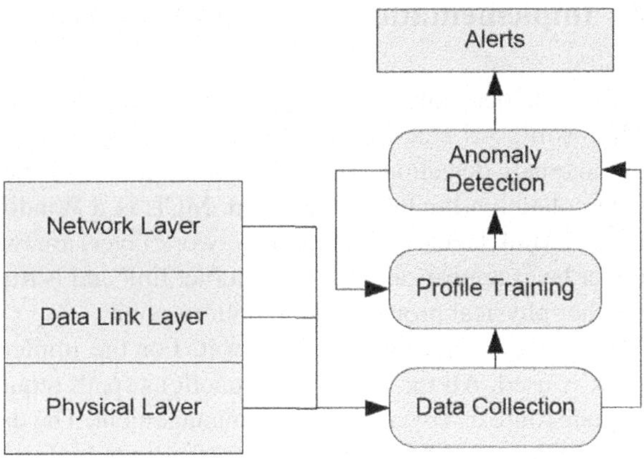

3.1 Anomaly Detection

This segment deals with the anomaly detection module. A variety of elements from this module are also used, apart from data collections and training kits. The study of phenomena is based on the presumption that there is an inherent and measurable distinction between normal and abnormal behavior. Applications with a high volume of information are often preferred to identify normally and abnormally. In this section, first threat models for WMNs are examined and functions collected based on model threats are addressed. The description of the classifier is followed.

3.2 Feature Selection

In the anomaly detection process, we would like to collect trace data to show normal or abnormal results. The layer of the MAC is defined by who accesses the medium. The medium can be used only for a malicious node and cannot use any legitimate nodes. A broken connection can be detected by a valid node. An attacker may be targeted to two distinct problems at a network layer: routing disruption and data transmission disruption. The characteristics that can accommodate such anomalies are therefore selected. The first features of the MAC layer and network-level features are defined in this section and then settings are made for every layer.

4 Implementation of System Prototype

This test bed integrates our cross-cutting intrusion detection. In particular, Microsoft's Ad Hoc Routing System-Mesh Communication Layer is integrated in our anomaly detection data collection module. For the implementation of MCL, our data collection module is included. MCL is a Windows driver that can be loaded. Layer 2 (link layer) and Layer 3 (network layer) are two of these driver systems. The upper layer appears as another Ethernet link and is translucent. It seems that MCl is another physical protocol at the bottom point.

The MCL routes the QSR packet. For the implementation of LQSR in MCL, DSR is used. All the basic DSR functions (path requests and path reply messages) include route discovery and route management. The data collection module contains all LQSR routing functionalities and the linked information can be registered by the connection cache. The cache database link functions are then saved.

When implementing LQSR, WEST is measured using the associated data. All our MAC layer functionality is compiled by the WCETT implementations. In order to achieve statistical accuracy, our experiments are loaded into the pre-processing module. We use a sliding window system for average and standard feature set deviations. The windows have been set to k sample time and the average and standard difference from the k sample is the last data trace, which can be loaded for classification into the training module. Then you go through windows with a slot to get the next record.

4.1 Attacks Scenarios in Wireless Mesh Networks

We performed three separate attacks for specific network protocol stack layer, a probe flood attack, a gray-hole stack attack and a black hole string, in order to test the effectiveness of the cross-layer detection. These three attacks are chosen because they are common attacks that a wireless attacker can launch, with substantial damage to the wireless network. In order to calculate reception rate and rate on every outbound connection, each node regularly sends test packets to LQSR. This helps a mesh knot measure the probability of loss to calculate the ETT value of that relationship. As our test disables RTS and CTS, an intruder could launch an attack on the medium by a sample flood. The malicious node gains additional access to the medium by its sample size. It is a form of DoS attack that is in violation of the law of MAC justice.

Blackhole attack illustrated. In this section, we define the operation by AODV protocol analysis of a blackhole attack. We look at the AODV protocol [6] road discovery stage and then determine vulnerabilities that the attacker could exploit in AODV protocol.

Vulnerabilities of AODV. The AODV convention is a directing convention on request [12] which possibly begins to discover courses when the starting MR wishes to advance some traffic to an obscure area. A bundle with an obscure sequence number is communicated from the starting MR on a course demand (RREQ). Next, if there is not sufficient course in the neighbors, RREQ parcels are recommunicated just (for example, grouping numbers are more prominent than those of promoted RREQ packets). This procedure proceeds until the RREQ arrives at the goal MR or another MR. In any case, if the system contains a malevolent blackhole hub, it delivers a bogus RREP for all the RREQ bundles it gets. The malignant blackhole hub makes a phony RREP parcel in any case in the event that it has a course to a specific goal. While acquiring a RREQ, MR first checks for the goal in quite a while course table during the typical procedure of the course goals. It creates a RREP to the source in the event that you know about the way to the goal. If not, the NULL worth is returned. A vindictive blackhole hub anyway goes around this pursuit procedure and consistently makes a RREP. It publicizes to be like the rot (signs with lower hopcount esteem in RREP) and furthermore distorts the succession number to be a subjective high incentive to guarantee that the source favors that RREP. The first MR at that point sends information bundles to the blackhole hub that conscientiously drops all information traffic. The vindictive blackhole hub foundationally traps all its neighboring MRs by sucking information traffic. This assault prompts an extreme corruption of the exhibition of WMNs, especially if the blackhole malignant hub is situated close to the IGW. The hub likewise de-empowers arrange execution, bringing about system apportioning, the postponement from end up and most genuinely adding to forswearing of administration for customers utilizing the User Datagram convention (UDP) sort of traffic (e.g., VoIP, FTP). Thus, it is essential to forestall a blackhole assault in a WMN invasion.

Vulnerabilities in WCETT. The WCETT is an additional metric of routing, which takes the intraflow interaction between multiplenum channels into account. The interference occurs if the same flow is transmitted by different nodes [2]. This is a description of the pause from one end to another. In the routing algorithm, the track with the lowest WCETT is decided. The vulnerability attackers will exploit the least end-to-end delay reported. The blackhole attacker played the same role by advertising himself to attract all the traffic and then dropped all the networking traffic. This allows the assailant to draw all traffic and then drop all traffic in the network. It is therefore essential that a blackhole assault in a WMN does not infiltrate.

Impact of black hole attack. In this article, simulations in ns-2 demonstrate the effect of a malicious blackhole node on WMNs. In the following two topologies, we demonstrate the impact of a blackhole attack:

1. Grid Topology and
2. Random Topology.

A simple grid-based network of IEEE 802.11 with 49 MRs (7–7) in the size 1500–1500 m is considered. The MRs communicate with one another using a wireless backbone based on the IEEE802.11 interface. We have two or three mesh clients. We assume that there is no interference in communication between an MR and an MC between two MRs.

We start streams from the MCs provided by the MRs. At the point when we state from here on that a MR stream is being started, we mean the MC has started the stream. We have 20 UDP streams, which send traffic to 200 kbps at a steady rate. We are utilizing a steady 512 byte parcel size. The transmission range and channel limit of 250 m and 11 Mbps separately are utilized in IEEE 802.11 channel intervention. The protocol that is at the root is AODV. The total time for the simulation is 500 s. The 10 traffic profiles with randomly selected sources were replicated in each simulation. The destination is modified periodically in order to achieve different flows and to evaluate the impact of the blackhole nodes on the network.

Blackhole node, the closest route through advertisements (higher sequence and shortest hop count) for all network traffic. For a selected random traffic profile, Fig. 2 demonstrates the effects on instantaneous output of the blackhole nodes of three of the affected IGW flows. When selecting the blackhole MR, we consider that it is positioned alone to initiate the flow of one MR into the other MR. The flows in which there are blackhole MRs on the net are very low in comparison to other flows in which nodes of blackhole do not exist. The number of network blackhole nodes increases and performance falls. Consequently, the number of blackhole nodes determines instant flow efficiency. Ten such flows were triggered randomly and flows tracked. We check the profiles in three such streams. An Attack Flow was implemented with several blackhole nodes were observed along the chosen path during simulation. These blackhole nodes give the destination the RREQ packets and do not permit other RREQ packages from neighboring innocent MRs. Thus, all malicious packets

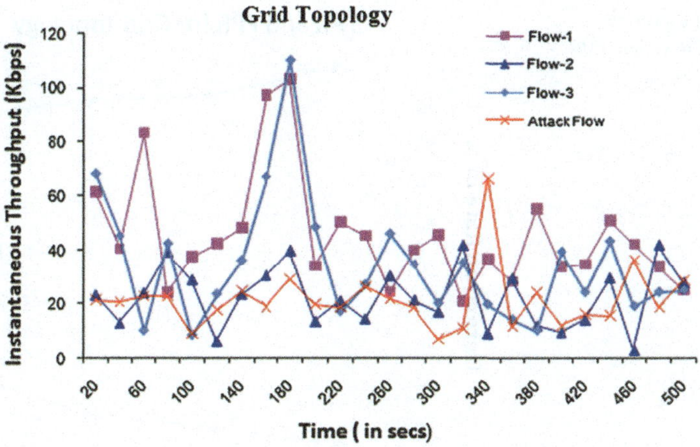

Fig. 2 Instant flow rates for grid topology malicious under the blackhole attack

are flooded in the flow, which does not allow the transmission to destination of other innocent MRs.

5 Experimental Results

Figure 3 specifies that the irregular topology aggregate throughput ranges from 85 to 96% as the blackhole MRs rate is expanded from 5 to 20%. The total system creation in the network topology ranges from 100 to 90% as the blackhole MR rate increases somewhere in the range of 5–20%.

Figure 4 shows that the TPR decreases from 100 to approximately 87% at the point when the quantity of system assailants has been expanded from 5% to 20%.

Fig. 3 Aggregate throughput random and grid topology

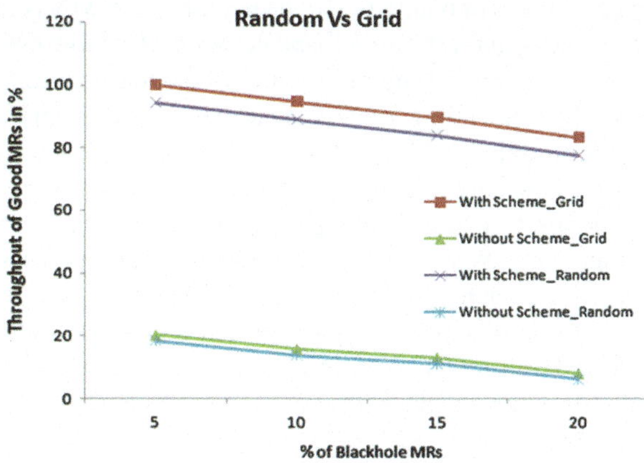

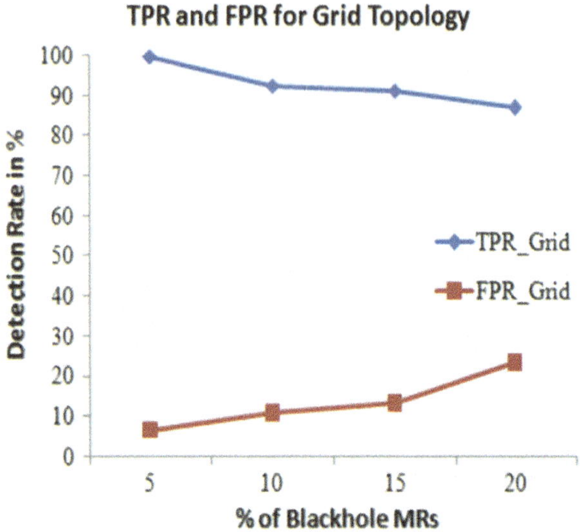

Fig. 4 Grid topology variation TPR versus FPR

This shows, regardless of whether the quantity of aggressors has been expanded, the system detects almost all network attackers.

When there are not many aggressors, the quantity of false alarms was very small. However, FPR is increasing as number of attackers is increasing. Figure 4 proofs that a high TPR (100%) and low FPR (23%) for grid topologies are possible in the proposed honeypot-based detection method.

6 Conclusion

A machine-based communication algorithm in wireless mesh networks and cognitive radios is introduced in this paper. Under this algorithm, any moving system has three changes of bandwidth to its neighbors. WMN is inexpensive, widely advertised access system for the broadband network. Both low network installation costs administrators and reliable internet connectivity end-users enjoy this consistently. However, as WMNs expanding, issues concerning safety and privacy are becoming increasingly apparent and should therefore be resolved in connection with this communication model. A security scheme is not easy since a variety of considerations need to be considered: intrusion anticipation in which security structures are remembered for components; WMNs defenseless against malignant assaults; interruption counteraction is inadequate to secure the network. Intrusion detection can also be used as a second line of security to prevent attacks against illegal activities. If an intrusion is detected and understood, it is important to mitigate attack reactions and the harm caused by the attack. This is considered as a method of intrusion response.

References

1. F. Aghoubi, M. Furdek, A. Rostami, P. Öhlén, L. Wosinska, Consistency aware weather disruption tolerant routing in SDN based wireless mesh networks. IEEE Trans Netw Serv Manag. **15**(2), 11 (2018)
2. E.A. Shams, A. Rizaner, A novel support vector machine based intrusion detection system for mobile ad hoc networks. Wirel. Netw. **24**(5), 1821–1829 (2018)
3. X. Cheng, L. Fang, L. Yang, S. Cui, Mobile big data: the fuel for data driven wireless. IEEE Internet Things J. **4**(5), 1489–1516 (2017)
4. Cisco Aironet 1500 series, http://www.cisco.com/en/US/products/ps6548/index.html
5. W. Cheng, J. Ni, K. Zhang, Ranking causal anomalies for system fault diagnosis via temporal and dynamical analysis on vanishing correlations. ACM Trans. Knowl. Discov. Data **11**(4), 1–28 (2017)
6. Y. Ghasempour, C.R.C.M. da Silva, C. Cordeiro, Knightly EW: IEEE 802.11ay: next generation 60 GHz communication for 100 Gb/s Wi-Fi. IEEE Commun. Mag. **55**(12), 186–192 (2017)
7. A. Narayana Rao, Subba Rao Ch D V: way-point multicast routing framework for improving QoS in hybrid wireless mesh networks. Wirel. Netw. **22**(8), 2681–2694 (2016)
8. A. Narayana Rao, Subba Rao Ch D V: QoS-aware service selection and multicast framework for wireless mesh networks. IJBDCN **12**(1), 1–15 (2016)
9. A. Narayana Rao, ChDV Subba Rao, Dynamic mobility management scheme for inter handoff in hybrid wireless mesh networks. IJMNDI **7**(2), 118–125 (2017)
10. J.M.N. Gonzalez, J.A. Jimenez, J.C.D. Lopez, G.H.A. Parada, Root cause analysis of network failures using machine learning and summarization techniques. IEEE Commun. Mag. **55**(9), 126–131 (2017)
11. L. Maccari, M. Maischberger, R.L. Cigno, On the optimal selection of multi point relays. Ad Hoc Netw. **77**, 69–83 (2018)
12. S. Kobayashi, K. Otomo, K. Fukuda, H. Esaki, Mining causality of network events in log data. IEEE Trans. Netw. Serv. Manag. **15**(1), 53–67 (2018)

Analysis of Detecting White Blood Cells by Computer Vision Methods

A. Rajasekhar Reddy, A. Narayana Rao, and Palamakula Ramesh Babu

Abstract Experts in both medicine and computer vision were engaged in the study of WBC images. Since WBC can be approximated by ellipsoid shape, an algorithm for the ellipse detector can be implemented to recognize them successfully. This paper introduces a complicated and unpacked novel algorithm for scanning WBC automatically, which sees the whole process as a problem for the sensing of multi-ellipse. The process converts the detection function into an optimization problem based on the novel differential evolution algorithm (DE), which is used to mimic ellipses of individuals. An objective function assesses whether these ellipses really exist in the edging image of the smear. The encoded candidate group creates ellipses (electors) using the novel DE algorithm to fit into the WBC, based on the values of a single edge map image.

Keywords Blood · Image segmentation · Image edge detection · Image color analysis · And diseases

A. R. Reddy (✉) · A. N. Rao
Department of CSE, N.B.K.R Institute of Science and Technology, SPSR Nellore, Vidyanagar, A.P, India
e-mail: arsreddy521@yahoo.com

A. N. Rao
e-mail: narayanaraoappini@gmail.com

P. R. Babu
Department of CSIT, Siddharth Institute of Engineering and Technology, Puttur, Chittoor, A.P, India
e-mail: palamakularamesh@gmail.com

C. Kiran Mai et al. (eds.), *Proceedings of International Conference on Advances in Computer Engineering and Communication Systems*, Learning and Analytics in Intelligent Systems 20,
https://doi.org/10.1007/978-981-15-9293-5_29

1 Introduction

1.1 White Blood Cells

White blood cells (WBCs) are also known as leukocytes. They are immune cells that battle infectious diseases and foreign materials inside the body. These cells help fight infections by attacking bacteria, viruses, and germs that invade the body. All leukocytes are produced and derived from multipotent cells [1]. This is often referred to as a blood-thirsty stem cell. They stay in the human body for around 3–4 days. The segmentation has to be done to extract the nucleus of the white blood cell image. A scanning electron microscope image of a human blood is shown in Fig. 1.

The main task of the segmentation is to extract the objects from an image. During segmentation, the size and shape of the nucleus should be maintained. Many segmentation algorithms have been developed for various applications [2–5]. So it may use the modified versions of moving K-Means algorithm called adaptive fuzzy moving new K-Means and algorithm of Standard Fuzzy C-Means and Standard Fuzzy Possibility C-Means may be used. To overcome from the problem of image shaping, it may use the classification of Relevance Vector Machine, RVM. This study offers a WBC-based Relevance Vector (RVM) detection technique. This proposed method works efficiently for the WBC identification, reducing computational time and preserving the image shapes and borders. The key goal is to segment the WBC to identify an immature cell from the blood smoking picture.

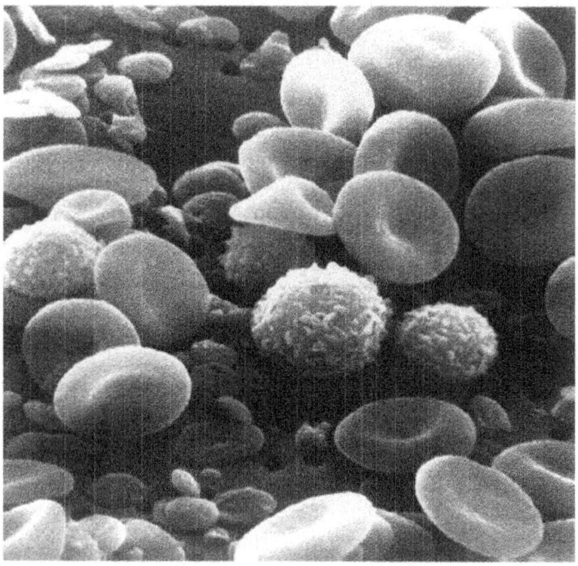

Fig. 1 A scanning electron microscope image of a human blood

1.2 Image Segmentation

Various segmentation methods are depending upon the nature of segments, diversified backgrounds, contrast, etc. The most popular segmentation procedures are edge and border detection, area growth, filtering, mathematical morphology, clustering of the watersheds, thresholds, neighborhood diagrams, mean shift and deformable models, color division. In this work, they have used the HSV segmentation method. The primitive step of segmentation is to convert the RGB image to an HSV image. To extract the cell portion, the hue plane alone is considered [6]. Based on observation, the hue value of a white blood cell was found to lie between 0.7 and 0.98. The image is thresholded based on the observed value, which yields the result as shown. Then, dispensable regions such as platelets are eliminated by area thresholding. The area threshold value is fixed on an experimental basis to obtain the entire region of the cell. The next step is to extract the nucleus portion. For this, the saturation plane alone is considered and the nucleus is found to occupy a saturation level above 0.40. The chain of steps results in the binarized images of the nucleus and cell region. Subsequently [7], the binary image of the nucleus is subtracted from the cell image to extract the cytoplasmic region.

The extraction of features is a specific form of dimension reduction in image segmentation. When an algorithm processes the input information too that, the input information is converted into a reduced set of features. The transformation of the data into a set of functions is called function removal. For image segmentation, the standard Extreme learning machine technique, ELM classification is used. The ELM classifier is, however, little changed in its training level. A fast RVM computer provides a much better test time than the standard ELM classification. For this, the fast RVM system is a good example. Quick RVM is more suited for applications that are likely to require less complexity and less real-time classification. The following paper is structured as follows: "Literature Survey" section summarizes principles and literature surveys. The proposed approach section is discussed and the experimental and the outcomes obtained by the proposed techniques are given in the Experimental results section. The end of the thesis can eventually be found in the section "Conclusion."

2 Literature Review

The algorithm proposed is an extension to the RVM standard algorithm, which only spars the sample domain. In the section of the nucleus and cytoplasms of WBCs, Wang et al. [3] address new procedures To ensure the precision of two well-known image segmentation techniques, namely watershed transformation and levels processes, image preprocessing by SMMT has been demonstrated in the segmentation of the nucleus.

Wu et al. [4] segment medical pictures according to an Active Learning approach that permits fast interactive segmentation, reducing human operator interaction requirements. Automatic leukocyte segmentation may allow drug makers to make their decisions about finding medicines and promotes the development of an automated recognition system for leukocytes and propose to segment the leucocytes of mice's skin parts stained with H&E stain and obtained at an increment of 40 mn based on a new technique, based on a differential development algorithm (DE).

Wang et al. [5] have the goal to reduce the watershed algorithm's repeatedly segmented error by removing 1% of the local minimum. The main aim of the proposed work is to provide a similar analysis of techniques used for the identification of trends in state-of-the-art technology, which have artificially trained and heuristically focused regulatory techniques. The learned RSFM models are scarce in either samples or functional fields with a Bayesian approach and Gaussian priors.

In this research, Muammar et al. [6] explored the use of the French–American-English classification for subtyping leukemia for image morphometry and patent recognition techniques. It offers an automated software method for classifying and segmenting medical 2-D/3-D images. The powerful segmentation algorithm, which can securely separate touching cells, note that the field of application of imagery algorithms in medical imagery. The study of WBC images was maintained by researchers from the fields of medicine and computer vision.

3 The White Blood Cell Detector

3.1 Image Processing

The proposed sensors need to be preprepared to get two new pictures: the portioned picture and the relating edge outline. To utilize the locator proposed by the division procedure delivers a sectioned picture, while an outskirt extractor calculation produces the edge map. The target capacity to decide the closeness of an applicant circle to a current WBC is considered for this limit map. The objective of this paper is to isolate white platelets (WBCs) from different structures, including red platelets and foundation pixels. Shading, brilliance and inclination information are broadly utilized for creating labels for classifying each pixel within a thresholding scheme. Even if the WBC can be segmented with a simple histogram thresholding, the diffuse anticipation maximization (DEM), during this study, was used to provide better performance. DEM is an algorithm based on expectation maximization (EM) used for the segmentation of complex medical images [6]. DEM considers spatial similarities between pixels as part of the minimum requirements, in contrast to traditional EM algorithms. Despite noisy and complex environments, this adaptation helps the segment of objects. The method models an image as a finite mixture to use the expectation–maximization (EM) algorithm, together with the class diffusion

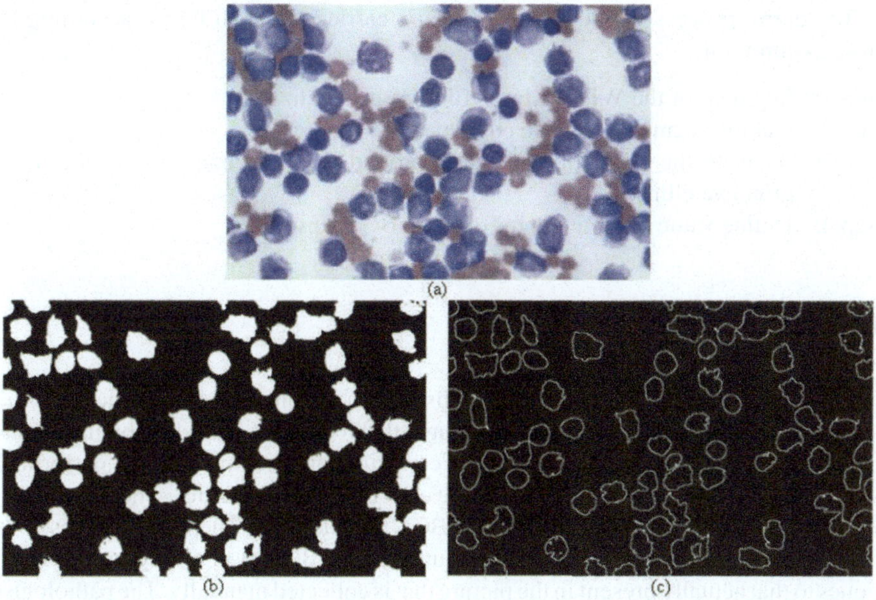

Fig. 2 The cycle of pretreatment. **a** Original image, **b** Segmented image from DEM and **c** Edge map from the morphologic edge detection process

anisotropy, to evaluate the parameters of each category, to take account of spatial dependencies between pixels (Fig. 2).

The implementation of the DEM given in [8] was used in the WBC segmentation. Since the implementation allows gray-level images and color images to be segmented, it can be used for the process of all straight pictures regardless of their acquisitions.

3.2 Ellipse Detection Approach

The map for the ellipse detector is used as the input image. Table 1 displays, after many calibration trials, the parameters used for the DE algorithm in this work. The final configuration coincides with the optimal calibration suggested in [5] where the effect of DE-parameter adjustment on several generic optimization problems has been studied. The parameter of population size (m = 20) was chosen to ensure the best possible balance between overload and convergence. Once specified for every test image used in the experimental trial, this configuration has been preserved.

Table 1 DE criteria for scientific photographic leukocyte detection

m	F	CR	NI
20	0.25	0.80	200

The entire process for detecting WBCs is carried out as follows, according to those assumptions:

Step 1: Segment of the WBC using DEM (described in 3.1).
Step 2: Get the segmented image edge map.
Step 3: Activate the DE ellipse detector around the rim chart to save the most effective ellipses.
Step 4: Define value parameters for every WBC ellipse.

4 Experimental Results

The tests are conducted using photographs from a hospital. The implementation is carried out using MATLAB. For each image, the WBCs are identified by the hematology expert and registered for database construction. The procedure is performed on various images of blood cells to test the effects of the techniques. The representation of the blood cells contains RBCs, WBCs, and platelets. The WBCs themselves are separated from these, comparing the number of WBCs observed by various techniques to that actually present in the picture that is collected manually. The pathologist gathers the fundamental truth for the whole dataset (Figs. 3, 4 and 5).

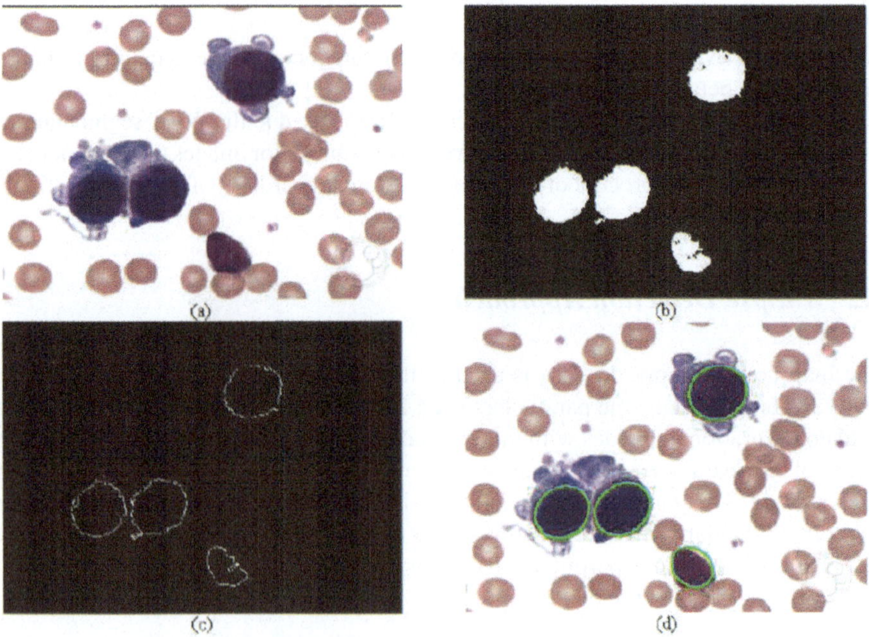

Fig. 3 Images of first test proceeded with the use of a WBC detector: **a** a source image; **b** a DEM divided image; **c** an edge map; and **d** a detected white blood cell

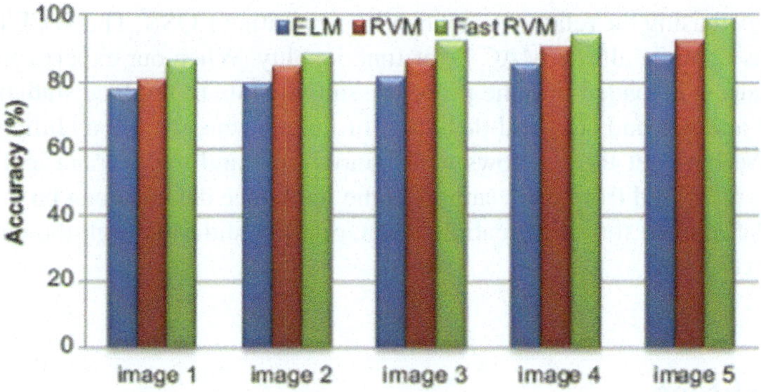

Fig. 4 Precision of possible techniques for WBC classification

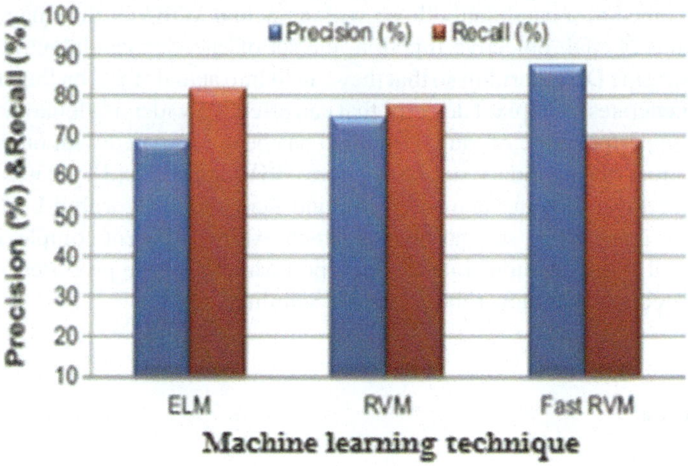

Fig. 5 Precision and reminder of the existing WBC system of classification

Cross-cutting intrusion detection has been incorporated in this test bed. In particular, Microsoft ad-hoc routing system-Mesh Communication Layer (MCL) is included in our anomaly detection data collection module. With the introduction of MCL, our data collection module is incorporated. MCL is a Windows driver that can be loaded. Layer 2 (link layer) and Layer 3 (network layer) are located between these two driver systems. The upper layer is translucent because it appears like another Ethernet link. At the bottom layer, it appears that MCl is another physical protocol. The QSR packet routes the MCL. DSR is used to implement LQSR in MCL. This includes all basic DSR features including route discovery and route management (path requests and path reply messages). All LQSR routing-related functions are logged in the data collection module and a link cache is used to record all linked information. The link-related functions of the database cache are then stored. WEST is determined using the related data when implementing LQSR. The WCETT implementations compile all our MAC layer functionality. When our experiments obtain the raw data, it is loaded into the preprocessing module to produce statistical functions. For average and standard deviation of feature sets, we use a sliding window system. We have set the windows to k sample time and the average and standard difference measured from the k sample is the final trace data that can be loaded into the training module for classification. Then, one slot slides through the windows to get the next record.

5 Conclusion

The proposed method uses the encoding of five edge points as candidate ellipses in the edge map of the smear. An objective function allows to accurately measure the resemblance of a candidate ellipse with an actual WBC on the image. Guided by the values of such objective function, the set of encoded candidate ellipses are evolved using the DE algorithm so that they can fit into actual WBC on the image. The approach generates a subpixel detector that can effectively identify leukocytes in real images. The performance of the DE method has been compared with other existing WBC detectors (the boundary support vectors (BSV) approach [3], the iterative Otsu (IO) method [4], the Wang algorithm [5], and the genetic algorithm-based (GAB) detector [2]) considering several images which exhibit different complexity levels. Experimental results demonstrate the high performance of the proposed method in terms of detection accuracy, robustness and stability.

References

1. X. Zhuang, Q. Meng, Local fuzzy fractal dimension and its application in medical image processing. Artif. Intell. Med. **32**, 29–36 (2004)
2. I. Scholl, T. Aach, T.M. Deserno, T. Kuhlen, Challenges of medical image processing. Comput Sci Res Dev **26**, 5–13 (2011)
3. M. Wang, R. Chu, A novel white blood cell detection method based on boundary support vectors, in *Proceedings of the 2009 IEEE International Conference on Systems, Man, and Cybernetics* (San Antonio, TX, USA, 2009)
4. J. Wu, P. Zeng, Y. Zhou, C. Oliver, A Novel Color Image Segmentation Method and Its Application to White Blood Cell Image Analysis, in *8th International Conference on Signal Processing* (2006)
5. S. Wang, F.L. Korris, D. Fu, Applying the improved fuzzy cellular neural network IFCNN to white blood cell detection. Neurocomputing **70**, 1348–1359 (2007)
6. H. Muammar, M. Nixon, Approaches to extending the Hough transform. Proc. Int. Conf. on Acoustics, Speech, and Signal Processing ICASSP-89, 3, (1989), 1556–1559.
7. M. Fischer, R. Bolles, Random sample consensus: A paradigm to model fitting with applications to image analysis and automated cartography. CACM **24**(6), 381–395 (1981)
8. T. Atherton, D. Kerbyson, *Using Phase to Represent Radius in the Coherent Circle Hough Transform* (IEEE, IEEE Colloquium on the Hough Transform, 1993), pp. 1–4

Smart Healthcare Monitoring for Drivers

K. Prasanna and L. Jai Vinita

Abstract Internet of Things (IoT) is utilized to monitor electronic devices by everyone at every place. There are so many applications like medical care, home appliances, vehicles, industries where sensors operated. To avoid health issues, this smarter health care system is introduced. Most of the expensive cars are getting the inbuilt sensors. If the health monitoring kit is available at an affordable price, then it is easy to avoid accidents for people who drive the vehicle. The sensors are used to monitor temperature sensor LM35, Arduino UNO, GSM module sim800L, finger clip pulse sensor, buzzer, and LCD. Here the health is recorded continuously with the help of sensors and is fixed some threshold value for temperature and pulse sensor so that if the readings crossed the threshold value, then the message immediately goes to the healthcare provider and also the transport office. Some parameters are used to monitor the pulse rate, body temperature by using the sensors and that data transferred to the mobile with the help of Bluetooth.

Keywords GSM · GPS · Internet of things

1 Introduction

Health is a principal factor in one's life. We need to maintain healthy conditions at our best. Some of the parameters of a human are glucose levels, the temperature of a person, heart rate, here we use healthcare monitoring systems. Currently, these types of systems are used in medical fields but also used in all kinds of fields like industries, home needs, and electronic appliances. There are two types of monitoring of the healthcare system and they are wired and wireless connections. In a wireless connection, there are again classified into two types and they are smart wearable devices or

K. Prasanna (✉) · L. J. Vinita
Department of CSE, VNRVJIET, Secunderabad, India
e-mail: prasanna.kch31@gmail.com

L. J. Vinita
e-mail: jaivinita_l@vnrvjiet.in

C. Kiran Mai et al. (eds.), *Proceedings of International Conference on Advances in Computer Engineering and Communication Systems*, Learning and Analytics in Intelligent Systems 20,
https://doi.org/10.1007/978-981-15-9293-5_30

implantation in the body itself. In a wired connection, the sensors are having connections with the help of wires and a person cannot involve in all activities. To avoid that problem, smart wireless devices are used but they are more expensive. The sensors like heart rate monitors, pulse oximeters, body temperature are connected for the human with the help of jumpers. In India, there are so many accidents happening with the driver's rash driving, and also health problems. While driving they do not take much care about their health because they just want to earn money. Some drivers are very old and weak heart and for them, it is very important to keep their health properly. We have used the sensor for the drivers and those sensors are temperature sensors and pulse sensors along with GPS trackers to track the vehicle if anything goes wrong for the driver. By that, we can save the driver as well as passengers in that vehicle.

2 Existing System

People are not taking health seriously and driving continuously without any rest for the body. Mainly the drivers do not concentrate on their health and many health issues were raised, to avoid that some people gave a solution for health. Durai Raj Vincent and Soundarajan [1] used IoT and cloud technology, and also RFID technology is used as a security measurement for protecting the patient's life. They used multiple sensors to collect different data from different users. Priyanka and Kaur [2] have used the parameters for path identification using ACO and they are path length, path identification duration, and efficient threshold, energy, time, and latency. These parameters used to identify the shortest path with less time are possible, which is more energy proficient, and this optimization is used in healthcare services. Jaiswal et al. [3] introduced how data are integrated using Rasberry pi and Docker container which stores the data on the server and sends the information for the user. Here Raspberry Pi collects and stores the medical data through the sensors attached. The received data transferred to the user through mobile apps. The information provided through apps improves the health of the patients. Lavanya et al. [4] introduced an IoT-based intelligent home-centric platform that connects smart sensors attached to the human body for physiological monitoring for daily medication management and also offered different types of medical health applications with small changes.

Mumtaj and Umamakeswari [5] proposed an intelligent real-time patient monitoring system that can be used in hospitals and homes and parameters such as temperature and ECG as well as to detect any abnormality accurately. The system sends an alert message to the caretakers and doctors in case of any abnormal values. It also enables the clinicians to optimize the usage of available medical resources and minimize the costs in monitoring the patients. The adaptive learning ability of the developed system improves the forecasting accuracy and results in enhanced decisions. The system remains more reliable and robust. There are some more people who found the smart healthcare and they are Subasi et al. [6] introduced smartphones, health checking gadgets, and individual computerized collaborations in the health sector. They offer quick information at a cheaper cost that enables them to keep if for the users and methods used in human activity recognition by using data mining techniques that are necessary for the health applications. Dejana and Gordana [7] proposed the Contiki Operating System that acts as a simulator and cooja for analyzing the health status that indicates the patient's health by using the Constraint Application protocol. They used the web browser to get details like a heartbeat, blood pressure, and also the level of glucose in the body. Their main intention is to collect the user data and communicate through the gateway where details can be seen over the world. Chatterjee et al. [8] used the Decision support system where data are analyzed and predicted. They have taken the sample set of people who are suffering from cardiovascular diseases and applied k-means clustering by differentiating as red and blue color where red indicates high risk, blue indicates normal as per the reports of the patient. Shaikh and Vidya [9] have used a decision tree algorithm along with raspberry pi B + that is written in python where data are stored on cloud that reduces the hospital visit for heart patients but the disadvantage is, they have taken the old records where real-time data is not detected.

Banka et al. [10] proposed the healthcare by using raspberry that reads the values, and those values sent to the database that is MySql at last, the result is displayed in the web interface. Here the emergency alert system is present with the help of some threshold values for blood pressure and temperature of the body. Carnaz and Vitor [11] have given the relation for IoT and the Medical field where they mentioned different technologies like cloud computing, big data, AR, wearables, etc. There are also some healthcare securities where data are stored securely. Zakaria et al. [12] have reviewed all the risks that are facing by the people till now by collecting the data from one Government hospital in Malaysia, they have used apriori algorithm to predict the result. They also mentioned the categories of risk like security, third party suppliers, infrastructure, etc. The advantages of predicting risks are reducing cost, the time it gives a better performance that will be effective for further steps. Sravanti, Lakshmi, Srivalli, Sneha proposed an android application for healthcare and used a heartbeat sensor, temperature sensor, which activates when a person requires an emergency, the alert message is sent to the medical staff and doctors with the help of medical graph. Here doctor can chat with the person who needs medical advice using Wi-Fi. Tastan [13] used Arduino and via Bluetooth for communication and pulse sensor, temperature sensor activates, and passes the data to doctor and family members using Wi-Fi where the data stored in the Blynk server. There are different kinds of approaches to health monitoring; one of the approaches is that they have used the data of a person with the help of a network by using the optimization technique that is an ant colony.

3 Proposed System

In this system, we are using different types of sensors and they are temperature sensor LM35, Arduino UNO, GSM module Sim800l, finger clip pulse sensor, mini piezo buzzer, LCD 16 × 2, and GPS which tells the location to know the illness of the

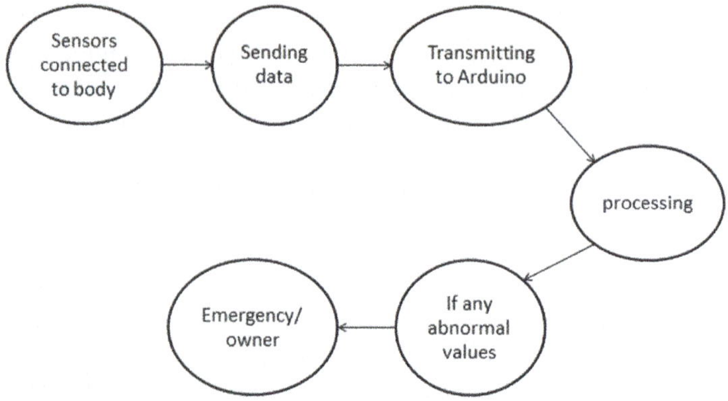

Fig. 1 Data flow diagram

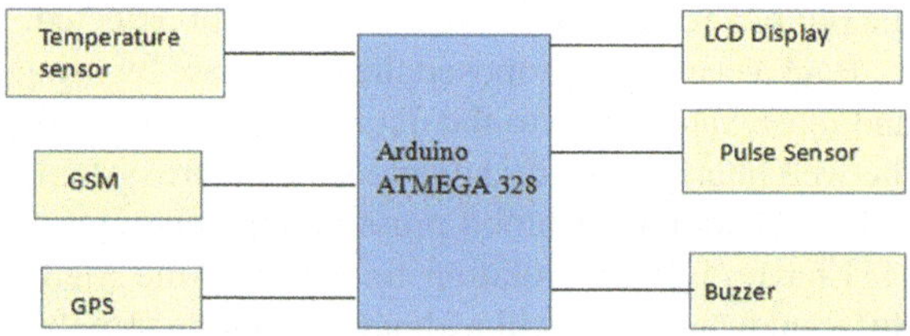

Fig. 2 System block diagram

driver with the heartbeat to the particular owner of that car and also in the hospital. The message will be sent to both people with some parameters like temperature, heartbeat and also the exact location of that person. So, they can be alerted with the message and save the person from the risk. In this, we have included a pulse sensor that monitors the heartbeat of a person continuously. The temperature sensor gives the driver temperature values and also displays on the LCD screen, the GPS tracker is to track the location of that particular vehicle and here we get the location through the link in the form of a message so that the ambulance can get the exact location and the customer get another vehicle without any problem with the help of transport owner (Figs. 1 and 2).

4 System Architecture and Analysis

This block diagram contains an Arduino board that is a type ATMEGA 328, Temperature sensor, GSM module, GPS module, LCD, Pulse sensor, and Buzzer. Each contains its functionalities where the Arduino board is connected to all other sensors and other hardware things. The temperature sensor gives the temperature of the surrounding vehicle that is attached to the Arduino kit. GSM module contains the sim slot that is the micro size and it is connected to GPS and LCD screen with the help of Arduino. It also holds some ports like mic and speaker where we can connect to the external sources, but the drawback is we cannot keep the volte sim cards in this slot. GPS gives the exact location of the kit and sends the message to the relative or family member and the ambulance when he/she is ill while driving the vehicle. The buzzer used when the kit is ready with a power supply. The LCD screen is 16 × 2 that gives the message and it should be connected to the Arduino kit with the help of jumper wires. Here pulse sensor works by checking the pulse of a person and gives the response to those people, also there are some conditions. If conditions are exceeded then they receive the message from the kit along with the location.

5 Implementation

There are many open-source platforms and one of them is Arduino which is easy for the beginners. Arduino takes the input data from the sensors and reads the functionalities of the code that is written by the programmer and gives the output in the form of LCD Displays, through mobiles and other parameters. It contains both the Analog pins and Digital pins that are for the sensors. Here we can modify or update the changes in the code whenever needed with the help of USB AB cable that is attached to the Arduino and system that is an advantage of Arduino. The Arduino and GSM connected for ground, transceiver, and receiver with the help of jumping wires. GPS connected to the Arduino power supply where location and be detected.

Fig. 3 After connecting all the sensors

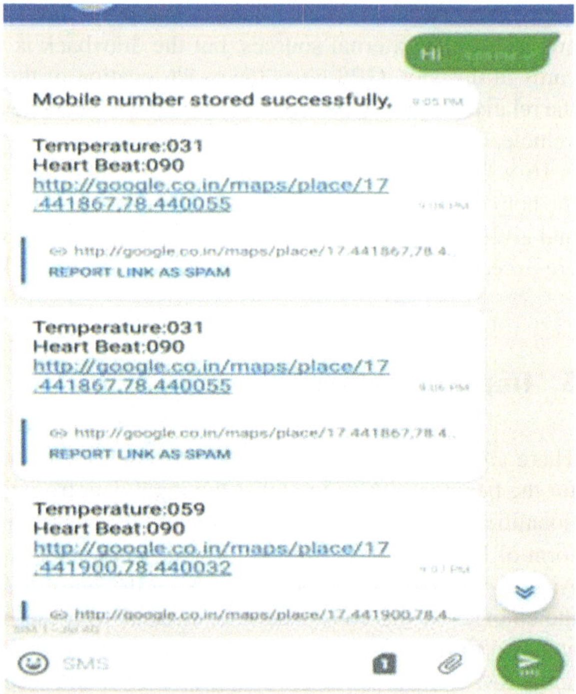

Fig. 4 Message is sent to receiver

6 Result Analysis

Figure 3 implies that the LCD, Buzzer, GPS, GSM are connected through the power supply. GPS sends the exact location of the driver who is at risk. The buzzer also activates when the values of temperature and pulse are abnormal.

The message is sent to the receiver where the heartbeat and temperature are shown and by clicking the link, we can find the person who is having the health problem and the emergency vehicle will be sent (Figs. 4 and 5).

7 Conclusion

There are many approaches to monitor healthcare in India and so many accidents happening with driver's rash driving, and also heart-related problems. Some people who are very old mainly for heart patients, it is difficult to find if there is no location. So, for that, the GPS location is present to the driver and if his/her condition is very poor, the message will be sent to the transport owner and friend or relative. We have used the sensor for the drivers and those are temperature sensor and pulse

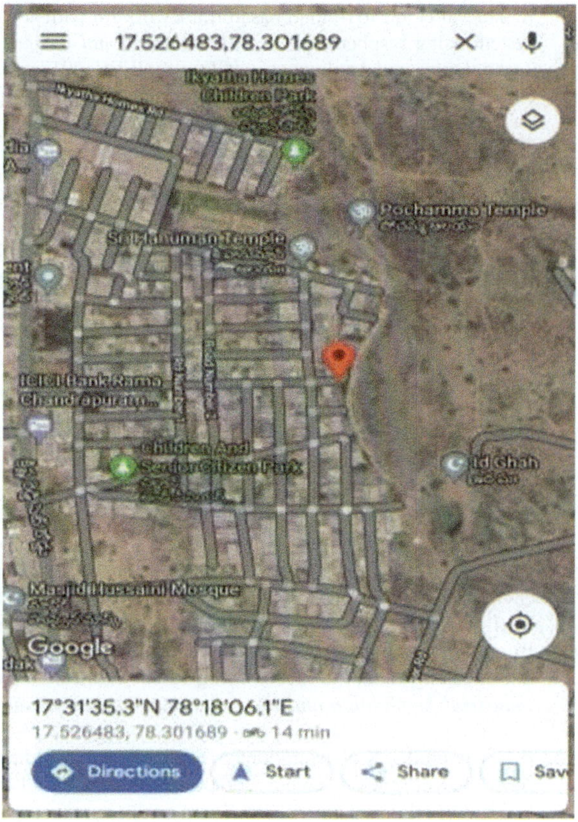

Fig. 5 Location from the sender

sensor along with the GPS tracker to track. By that, we can save the driver as well as passengers in that vehicle.

Future Scope

In the future, we can add a camera in the kit so that we can see the driver's condition and passenger count in the vehicle. If the driver is sleeping then we can alert him/her by some buzzer sounds.

References

1. J. Soundarajan, P.M. Durai Raj Vincent, An IOT based Human healthcare system using Arduino Uno board, in *2017 International Conference on Intelligent Computing, Instru- mentation and Control Technologies (ICICICT)* (IEEE, 2017)
2. P. Priyanka, J. Kaur, Ant colony optimization based routing in IoT for healthcare services, in *2018 Second International Conference on Intelligent Computing and Control Systems (ICICCS)* (IEEE, 2018)

3. K. Jaiswal et al., IoT-cloud based framework for patient's data collection in smart healthcare system using raspberry-pi, in *2017 International Conference on Electrical and Computing Technologies and Applications (ICECTA)* (IEEE, 2017)
4. S. Lavanya, G. Lavanya, J. Divyabharathi, Remote prescription and I-Home healthcare based on IoT, in *2017 International Conference on Innovations in Green Energy and Healthcare Technologies (IGEHT)* (IEEE, 2017)
5. S.Y. Mumtaj, A. Umamakeswari, Neuro fuzzy based healthcare system using iot, in *2017 International Conference on Energy, Communication, Data Analytics and Soft Com- puting (ICECDS)* (IEEE, 2017)
6. A. Subasi et al, IoT based mobile healthcare system for human activity recog- nition. in *2018 15th Learning and Technology Conference (L&T)* (IEEE, 2018)
7. U. Dejana, G. Gordana, CoAP protocol for Web-based monitoring in IoT healthcare applications, in *2015 23rd Telecommunications Forum Telfor (TELFOR)* (IEEE, 2015), pp. 79–82
8. P. Chatterjee, L.J. Cymberknop, R.L. Armentano, IoT-based decision support system for intelligent healthcare—applied to cardiovascular diseases, in *2017 7th International Conference on Communication Systems and Network Technologies (CSNT)* (IEEE, 2017)
9. S. Shaikh, V. Chitre, Healthcare monitoring system using IoT, in *2017 International Conference on Trends in Electronics and Informatics (ICEI)* (IEEE, 2017)
10. S. Banka, I. Madan, S.S. Saranya, Smart healthcare monitoring using IoT. Int. J. Appl. Eng. Res **13**(15), 11984–11989 (2018)
11. G. Carnaz, V.B. Nogueira, An overview of iot and healthcare (2016)
12. H. Zakaria et al., IoT security risk management model for secured practice in healthcare environment. Procedia Comput. Sci. **161**, 1241–1248 (2019)
13. M. Taştan, IoT based wearable smart health monitoring system. Celal Bayar Üniversitesi Fen Bilimleri Dergisi **14**(3), 343–350 (2018)
14. S. Chaudhury et al., Internet of Thing based healthcare monitoring system, in *2017 8th Annual Industrial Automation and Electromechanical Engineering Conference (IEMECON)* (IEEE, 2017)
15. B.K. Bhoomika, K.N. Muralidhara, Secured smart healthcare monitoring system based on Iot. Int. J. Recent Innovat. Trends Comput. Commun. **3**(7), 4958–4961 (2015)
16. M.M. Baig, H. Gholamhosseini, Smart health monitoring systems: an overview of design and modeling. J. Med. Syst. **37**(2), 9898 (2013)
17. H.M. Chang, L. Ji, B.S. Renger, Wearable health monitoring system. U.S. Patent No. 9,357,921 (2016)
18. A. Pantelopoulos, N.G. Bourbakis, Prognosis—a wearable healthmonitoring system for people at risk: Methodology and modeling. IEEE Trans. Inf. Technol. Biomed. **14**(3), 613–621 (2010)

Analyzing Vocal Patterns to Determine Gender, Age and Emotion

N. Sandeep Chaitanya, Pilli Shivani, Nallapuneni Sahithi, Muttevi Sravanthi, and Juvas Aditya

Abstract The speech represents a significant role in human communication as we can convey our feelings through it. Emotion is a strong feeling that is derived from one's circumstance or surroundings. Speech analysis is important to have natural interplay between human beings and machines and also to reduce the alienation and isolation in human beings. In speech emotion detection, the emotional state of an individual is extracted from their speech. Our project aims to design and develop a hybrid system to detect emotion, gender and age by analyzing vocal patterns (or speech) in which the current system fails to do since it uses separate systems for the detection of gender, age and emotion. We will be taking speech signals as an input, which will be converted into NumPy array and later classified using the SVM algorithm.

Keywords Gender classification · Age classification · Emotion classification · SVM (Support Vector Machine) · Machine learning

1 Introduction

Speech-based communication contains both linguistic and paralinguistic information. The latter is predominantly significant in stipulating factors of behavioral and

---

N. S. Chaitanya (✉) · P. Shivani · N. Sahithi · M. Sravanthi · J. Aditya
Department of CSE, VNRVJIET, Secunderabad, India
e-mail: sandeepchaitanya_n@vnrvjiet.in

P. Shivani
e-mail: shivani27.p@gmail.com

N. Sahithi
e-mail: sahithi.nallapuneni04@gmail.com

M. Sravanthi
e-mail: sravanthimuttevi1998@gmail.com

J. Aditya
e-mail: juvasaditya123@gmail.com

C. Kiran Mai et al. (eds.), *Proceedings of International Conference on Advances in Computer Engineering and Communication Systems*, Learning and Analytics in Intelligent Systems 20,
https://doi.org/10.1007/978-981-15-9293-5_31

functional status, and particularly emotional states. There are two main approaches to the study of emotional speech. One approach focuses on the psychological and physiological aspects of emotions, which include studies of brain activity and processed data and cross-cultural investigation of emotional states in speech. The second approach is stressed on the physical speech signal and its analysis. Hence, it is geared toward software applications for human–computer interaction, such as automatic speech recognition and speech synthesis [1–3].

Automatic voice classification based on gender, age and emotions is a growing area of research with numerous real-world applications. While there is a large body of research in related audio classification fields such as gender and age, work on the classification of emotions is comparatively scarce. Similarly, observing the current progressions in the field of image classification where convolutional neural networks are being used to classify images with high precision and at scale, it begs the question of the applicability of these procedures in other fields, namely voice classification. From a machine learning aspect, voice recognition can be perceived as one label multiclass text classification problem where the role of classes is played by contestant authors [4, 5].

With the advancement of technology, human–machine interaction is broadly used in many applications. Humans find speech the most natural way of communication. Therefore, speech can be an effective method of interaction between humans and machines. For effective communication to happen, humans sense the emotional state of their communication partner. Emotion plays a significant role in human to human communications. When two individuals interact with each other they identify the emotion in the speech of another person. To make the interaction between humans and machine natural, the machine needs to have the ability to respond to the emotions of humans in the same way as humans do. The emotion detection is a very difficult task for the machine. The main objective of the speech analysis is to improve human–machine interaction [5].

In applications such as e-commerce, the system can be used to identify the user by age and gender to deliver product recommendations. In social networking applications, identifying the gender, age and emotion of an individual have massive potential business value. The age and gender of an individual play an important role in the area of speech recognition. The age identification can be considered as a classification problem as there is a prediction problem with different age groups and continuous age values [1].

To overcome this obstacle, we want to build a system that can identify the emotions and also classify the age and gender of a person efficiently and effectively. This project uses two labeled datasets to train and test our models. Several most advanced deep learning models are utilized at different levels to evaluate the accuracy of voice. We built a program with different modules that take real-time speech data as input and determine the age, gender and emotion. This can be done by extracting the features from the speech and training is done for the speech database. The further steps include taking the speech data as an input which is converted into NumPy array and specific features are extracted, this is later classified using the SVM algorithm to produce the required results [6, 7].

2 Literature Survey

The methods that are already in existence are used for gathering effective learner data, such as physiological sensors or validated questionnaires, are more limited, and they automatically disrupt the process of learning. Most of the research is dealing by using emotions for adapting learning concepts and content or learning tasks. This insight has led to the research and development of effective tutoring systems.

Zhao et al. [1] studied the speech-based classification of age and gender and discuss the existing key features and different classification models. The methodology follows first by detecting the gender and then proceeds to recognize the age based on the gender from the given input speech. This process uses MFCC for Feature Extraction and for the classifier, they used classifiers like Random Forests and Support Vector Machine.

Zheng et al. [2] in this methodology, the convolution neural networks algorithm is used for extracting the features from the standardized spectrogram, and random forest is used for the classification of the extracted features from the speech to recognize the emotion. This system detects emotion at 84% of accuracy because of the RF classifier used.

Basu et al. [3] described the usage of the MFCC algorithm. The database used for their methodology was the Berlin Emotional Speech dataset (EmoDB). Here they have using CNN algorithm as a classifier with an LSTM approach. Around 80% of accuracy for the test data was noted.

In Venkataraman et al. [4], this project was done for building a program input using a webcam and detects age, gender and emotion from facial features. Furthermore, our goal was to create a universal solution to this problem. This system used CNN to detect age emotion and gender through facial expressions.

In Rawat et al. [5], this model uses MFCC for feature extraction. First, then inputs the speech signal, which is to be validated. The input given and the data model must be of the similar period and occurrence. Next step involves the removal of noise. The above step helps us to smoothen the signal. There is no need of manipulation of the noted or input model data. In the following step, features are extracted using the MFCC algorithm and at the last fuzzy theory, the approach is applied and then matched with the trained data.

Vinay et al. [8] explain the methodology of two subsystems, i.e., Gender Recognition (GR) system and Emotion Recognition (ER) system. A fitting emotion detection technique is applied after the extraction of features like pitch, energy, etc. This model uses MFCC, Naive Bayes and gives an approximate of 70% accuracy of emotion detection based on voice.

3 Implementation

3.1 Proposed Predicted Process

In order to solve the limitations of the existing systems, we hereby introduce a novel system that uses machine learning algorithms on preprocessed voice samples to determine gender, age group and emotions of the speaker. The input to the system is an audio file from the database, which contains audio files in.wav (waveform audio file) format. The following move is to extract the features from the audio records using the MFCC algorithm. After the feature extraction, we send the extracted features for training where we use the SVM classifier to train the data and generate a training model. When an audio file is given as an input for testing, the system extracts the features and compares them against the training model to predict the age, gender and emotion of an individual [9, 10] (Fig. 1).

3.2 Implementation

Data Collection

We created our own datasets by recording the voices of our families, relatives and friends when we are having conversations with them by seeking their permission and

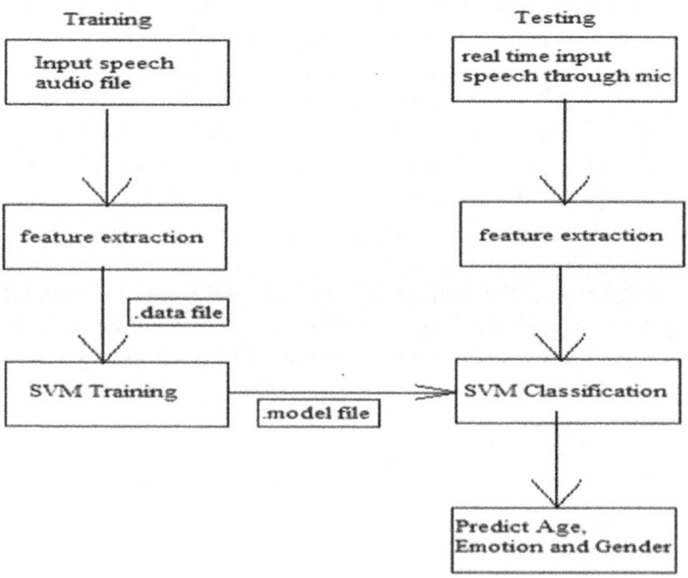

Fig. 1 System architecture of emotion, age and gender recognition system

also used the data set "The Ryerson Audio-Visual Database of Emotional Speech and Song (RAVDESS)." This database contains 24 professional actors. The actors participating are 12 females and 12 males, vocalizing or enunciating two lexically matched statements. This can be passed in a neutral North American accent. Speech includes sad, angry, happy expression with an additional neutral expression.

Data Preprocessing

There are some audio properties to make certain or to make sure for the consistency that needs the preprocessing across the whole dataset that are Audio Channels, Sample Rate, Bit-depth. These converts the sampling rate to 16 kHz, normalizes the data so the bit-depth values range between -1 and 1 and flattens the audio channels into mono. We are also using a low pass filter for noise cancellation. IIR and FIR filters are being used in this project using SciPy.

FIR: There are two primary types of digital filters, and FIR filter is one of the primary types. It is used in Digital Signal Processing (DSP) applications. And the other type of primary filter is IIR. "FIR" stands for "Finite Impulse Response." If we place an impulse, that is, if we keep the sample of "1" at primary and later this sample is followed by many of the "0" samples, all the zeroes will come out after the "1" sample has made its way through, which means all the samples of "1" should pass through the delay line of the filter. In the frequent case, there will be no feedback arising in this FIR filter as the impulse response is limited or finite. This lacking in feedback guarantees or ensures that the impulse response will be limited or finite. Therefore, the term "finite impulse response" or "limited impulse response" is nearly synonymous or identical with "no feedback."

IIR: IIR refers to Infinite Impulse Response. An infinite impulse response (IIR) filter is a digital filter that is contingent on or depends linearly on a finite or limited number of input samples or any tryouts and also depends on a finite number of previous or predecessor filter outputs. In other words, it combines or amalgamates a FIR filter with feedback from previous or predecessor filter outputs. DSP filters can also be involved as an IIR (Infinite Impulse Response). IIR filters use feedback, so when you give impulse as an input then the output conceptually or theoretically rings indefinitely.

Feature Extraction

Since a Machine Learning algorithm cannot understand voice data, we need to convert it into numeric data. This is essentially Feature Extraction. We extracted 34 features using a two-stage approach. Following features are extracted from the sample using FFT (Fast Fourier Transform). Speech contains very important frequency information about human beings. The features or characteristics of Audio, especially or exceptionally a speech signal may be extracted or selected out using FFT (Fast Fourier Transform) to detect to perceive the frequency information of the signal. 1. Spectral Centroid, 2. Spectral Spread, 3. Spectral Entropy, 4. Spectral Flux, 5. Spectral Roll-off. MFCC features are extracted from audio file using DCT (Discrete cosine transform) [11, 12].

MFCC: the Mel-frequency cepstrum (MFC) is a portrayal of the sound which in turn represents the short-term power spectrum that is based on the log power spectrum which is in turn based on an additive cosine transform on a sporadic Mel-scale of frequency. Mel Frequency Cepstral Coefficients (MFCC) form a cepstral portrayal where the frequency bands are not linear or not additive or sporadic but distributed according to the Mel-scale. The coefficients that accumulatively make up an MFC are called as Mel-frequency cepstral coefficients (MFCCs). They are derived or acquired from a type of cepstral representation or delineation of the audio clip. The differentiation or dissimilarity between the cepstrum and the Mel-frequency cepstrum, which is MFC is that in the MFC, on the Mel-scale, the frequency bands are uniformly or evenly spaced, which comes closely or approximates or approaches to the human auditory system's response or reaction or retaliation more closely or nearly than the linearly spaced or additively spaced frequency bands used in the normal cepstrum. This frequency warping or twisting or distorting can allow for better representation or portrayal of sound, for an instance, in audio compression.

Chroma Vector: In music, the term Chroma gram or the chroma feature or its characteristic is nearly consisting of or relates to the 12 various or different pitch classes. Chroma-based features, which are also called as "pitch class profiles," are a powerful or a sturdy tool for scrutinizing music whose pitches can be meaningfully or relevantly categorized or classified (often into 12 categories) and whose tuning approximates or approaches to the equal-tempered scale. One main property of chroma features is that they capture or catch hold of sympathetic or harmonic and melodic characteristics or features of music, while being robust or vigorous to changes or responding very fast to changes in timbre and instrumentation. The bins portray or represent the 12 equal-tempered pitch classes, which are of western-type music or also called as semitone spacing is a 12-element representation of the spectral energy. The following images show chroma representation of an audio sample.

Training and Testing the Model

The whole data are split into training set for algorithm learning and validation set for testing the model's performance. Validation set size is chosen as 10% of whole data. After the splitting of the data, we follow the following process for training and testing of the model.

Feature Extraction: Any machine learning model can understand only numeric data. Since the data set contains text data, it needs to be converted into numeric data. This is feature extraction and can be done using several techniques. The audio feature extraction methodology involves two stages for the extraction:

1. Short-term feature extraction: This is implemented in the function feature_extraction () of the ShortTermFeatures.py file. It splits or cleaves the input signal into short-term windows, which is also known as frames and computes or calculates a number of features for each and every frame. This process or procedure leads or approaches to a sequence of short-term feature or characteristic vectors for the whole signal.

2. Mid-term feature extraction: In most of the cases, statistics are the one which are responsible for representing the signal based on the extracted short-term feature sequences described above. Toward this end, the function mid_feature_extraction () from the MidTermFeatures.py file extracts or draw out a number of statistics. These statistics involve mean and standard deviation over each and every short-term feature sequence. The total number of short-term features implemented or developed is 34 in pyAudio Analysis.

Training the Model: The process of or procedure for training a machine learning model necessitates furnishing or supplying an ML algorithm with training data to learn from. The creation of the model artifact which is created by the training process is referred as ML model. The training data must contain the labels, which is known as a target or target attribute. We give the specific algorithm of our choice for training the model. In this project, we have used SVM algorithm among Support Vector Machine, Random Forest and KNN. The learning algorithm discovers patterns or design in the training data that map or plots the input data attributes to the target or to the answer that you want to predict or forecast, and it outputs an ML model that captures or catch hold these patterns or the designs that are obtained. We are considering our training model as ML model.

Testing the model: We give the input through mic for real-time testing or audio files, which are containing the real-time data. For this data, the system extracts the features in the same way that is explained in the above module. Then these features are stored as vectors, which are tested against the training model. We use the ML model to get predictions on new data for which you know the labels. This process is called testing or evaluating the model.

4 Result

In this section, experimentation or the trial and error results are presented and are in turn discussed. We report the recognition accuracy of using SVM classifiers. Experimental evaluation is performed on databases. All classification results are obtained under tenfold cross-validation. Cross-validation is a most widely used or frequent practice used in performance analysis or scrutiny that randomly or haphazardly partitions or divides the data into N complementary subsets, in which $N-1$ of them used for training in each validation and the remaining all are used for testing.

Before applying classifiers, data are scaled to $[-1, 1]$ from the features. Scaling features before recognition or identification are very important or crucial, because when a learning phase is fit or pertinent on unscaled data, it is possible or having chance for large inputs to slow down or leisured down the learning and convergence and in some cases prevent or avert the used classifier from effectively learning for the systematization or classification problem. These experiments use feature set without feature selection. SVM classifier yields better results with feature combination of MFCC and other features in the database. Our results have improved compared to previous results because we changed the SVM parameters for each type of features to develop a good model. In our data exploration, we have noted that the difference between some of the classes is very difficult to visualize. We will examine if the final model also felt difficult or struggled to differentiate between these classes by using the confusion matrix. The Confusion Matrix tells a different story. This gives us some in-sights into the features by showing us that the problem is more nuanced or refined than our initial assessment so that the algorithm is extracting to make its classifications.

5 Conclusion

In this paper, the real-time data that are collected by our own and RAVDEES emotion database are used for training the model. MFCC features are extracted from the speech files, which are in.wav format. For the testing purpose, the input is given in real time through mic to detect the Emotion and Gender and Age is detected through the audio files given as an input.

In a Word, this paper studies the audio-based age classification, gender classification and emotion classification and discusses the current main features and classification models. It provides the reference for the research based on audio processing, such as speech recognition, speaker attributes recognition, etc (Figs. 1, 2 and 3).

Fig. 2 Emotion and gender classification

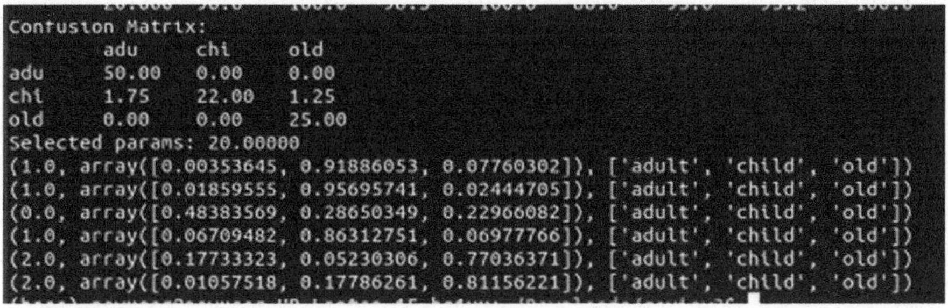

Fig. 3 Age classification

References

1. H. Zhao, P. Wang, A short review of age and gender recognition based on Speech (2019)
2. L. Zheng, Q. Li, H. Ban, S. Liu, Speech emotion recognition based on convolution neural network combined with random forest (2018)
3. S. Basu, J. Chakraborty, Md. Aftabuddin, Emotion recognition from speech using convolutional neural network with recurrent neural network architecture (2017)
4. R. Venkataraman, D. Rochiramani, M. Falak Bhardwaj, Real time age, gender and emotion detection of facial images using deep learning (2015)
5. A. Rawat, P. Kumar Mishra, Emotion recognition through speech using neural network (2015)
6. N. Sandeep Chaitanya, Implementation of DHS for effective usage of resources and providing security using ECC in multi cloud environments. Int. J. Eng. Technol. (UAE)(IJET) (2018)
7. M. Navya Sri, M. Ramakrishna Murty, et al., Robust features for emotion recognition from speech by using gaussian mixture model classification, In International conference and published proceeding in SIST series, vol. 2 (Springer, Berlin, 2017), pp. 437–444
8. Vinay, S. Gupta, A. Mehra, Gender specific emotion recognition through speech signals (2014)
9. I.R. Murray, J.L. Arnott, Toward the simulation of emotion in synthetic speech: A review of the literature on human vocal emotion. J. Acoust. Soc. Am. **93**, 1097–1108 (1993)
10. Y. Chavhan, M.L. Dhore, P. Yesaware, Speech emotion recognition using support vector machine (2010)
11. N. Sandeep Chaitanya, Implementation of security and bandwidth reduction in multi cloud environment published, in IEEE Digital Explore IEEE ISBN: 978–1–5090–5256–1/16/$31.00_c 2016 page no 758–763
12. N. Sandeep Chaitanya, Usage of DHS and de-duplicating encrypted data using ABE and ECC for secured cloud environment, in IEEE Digital Explore IEEE ISBN: 978–1–5386–1442–6/18/$31.00 ©2018 IEEE page no 614–619

A Robust and Secured Mechanism for Sharing Encrypted Data in Cloud Systems

Jangili Narendra

Abstract Citizens are embracing the major power of cloud computing, make sure that cloud providers can not truly believe in hosting information that is vulnerable to privacy, as access to cloud control is lacking. Due to this, the data owners will share the encrypted data instead of plaintexts to ensure confidentiality. By using the Ciphertext Policy-Attribute based Encryption (CP-ABE), we can share the encrypted files with the other users, for solidified and owner-significant monitoring of access. Through this, we do not get safe enough by means of some unknown attacks. Many existing methods will not allow the cloud provider to check to see if the downloader is able to decrypt. Such files will also be open to anyone usable for storage in the cloud. A malevolent attacker can access hundreds of files to conduct Economic Denial of Sustainability attacks (EDoS) that use the cloud resource to a large extent. For this purpose, the responsibility rests with cloud service payer and the payer has to bear the expenses. The cloud provider simply acts as both the assessor and the technology use tax payee, minimizing data owners' responsibility. In real-public cloud storage, these issues should be addressed. In this paper, we suggest a solution for protecting Encrypted cloud storage from Economic Denial of Sustainability assaults and providing resource usage transparency. In a black-box approach, it uses CP-ABE methods along with complying with CP-ABE's arbitrary access policy. We have two protocols proposed for a particular environment, performance and safety review are followed.

Keywords Ciphertext-Policy Attribute-Based Encryption (CP-ABE) · Managed access · Storing the public cloud · Accounting · Safeguarding privacy

1 Introduction

Cloud storage [1, 2], spaces have much compensation. Throughout the time, additional information is redistributed toward open cloud designed for industrious limit,

J. Narendra (✉)
VNR Vignana Jyothi Institute of Engineering and Technology, Hyderabad, India
e-mail: jangilinarendra@gmail.com

C. Kiran Mai et al. (eds.), *Proceedings of International Conference on Advances in Computer Engineering and Communication Systems*, Learning and Analytics in Intelligent Systems 20,
https://doi.org/10.1007/978-981-15-9293-5_32

together with individual as well as business reports. It brings a protection stress [3] toward information proprietor: the way to the cloud supplier without the approval as of information proprietor. Various limit organizations utilize server-administered get to manage, similar to mystery word based and underwriting based approval. They too much faith the cloud supplier toward makes sure about their delicate information. The cloud supplier plus their agents are able to examine some record paying little notice to data proprietors' as entrance course of action. Likewise, the cloud provider [4] be able to embroider the advantage procedure about details storing along with accuse the clients need structure meant in favor of certain figure about benefit exercise. Dependent ahead their server-ruled obtain toward management is not checked. Information proprietors who store report resting on cloud servers despite everything require toward manage the passage independently along with pernicious clients. Encryption [5] is not sufficient: toward integrate the protection assurance, data manager be able to encode the documents along with locate to find a workable pace simply qualified customers can interpret the report. Among Ciphertext Policy—Attribute based [6, 7] Encryption (CP-ABE) we able toward contain together fine-grained find a workable pace strong protection. In any case, this finds a good pace available for data proprietors, which ends up being inadequate. In case, the cloud provider cannot approve customers previous to downloading, similar to numerous presented CP-ABE [6] conveyed capacity system the cloud wants toward permit each one download toward make sure accessibility. This creates boundary arrangement feeble against benefit weakness ambush. Within casing, we settle this problem through have information proprietors approves the downloader's formerly permitting them toward download, we are unable to find the versatility about administration manage as of CP-ABE. Now proceedings two issues are supposed toward be tended to within our effort.

Problem 1 (Resource-Exhaustion assault):

In event that the cloud cannot do cloud-side administration [8] manage, it needs toward permit everyone, tallying noxious assailants, toward energetically download, regardless of the way that only a couple of customers can decipher. The server is feeble against resource weariness attacks. Right when malevolent consumers dispatch the DoS/DDoS attacks [9] toward the conveyed stockpiling, the advantage use might augment. Payers (in pay-all the more just as expenses emerge model) need toward disburse in favor of the extensive practice contributed through those assaults, which is broad and stunning cash. This process is displayed as Economic Denial of Sustainability (EDoS) [10], that show payers be monetarily assaulted at last. In adding, still reviews be combined unapproved downloads can decrease defense through sporting comfort toward detached examination and spilling statistics similar to record period before inform repeat.

Problem 2 (Source Expenditure responsibility):

Within compensation all more just as expenses emerge model, customers disburse cost toward the cloud supplier meant for restriction businesses. The price is picked way of aid use. Nevertheless, CP-ABE [6] primarily based designs for conveyed capability discover a true tempo make online an assertion to the information proprietors previous to downloads. It is required about cloud pro affiliation toward reveal to the real customers about the resource utilization. Something exclusive, the cloud issuer able to price more without life form found.

1.1 Summary of Challenges and Approaches

Many existing CP ABE [6, 11] based plans method the cloud suppliers as semitrusted or else latent aggressors. Be that as it may, such a definition is confined and it avoids some potential assaults in the genuine world, for example, misrepresented asset use. To demonstrate such assaults, we think a fewer confined productivity method, clandestine enemy, for the cloud supplier [3]. Practically speaking, the cloud administrations are typically given by a few huge IT ventures like Google, Amazon, Microsoft. They need to keep up great notoriety and guarantee secure distributed storage [2] administrations to their clients. In the event that any endeavor the cloud supplier strays from the convention should be gotten with a plausibility (for example, $p = 0.001$), the cloud supplier challenges do not lie. Since being gotten would not just disregard the administration contracts, yet in addition lead to media introduction in addition to pulverizes notoriety. Mindful of consequence, the cloud supplier has toward avoided assaulting, like cheating is able to recognize. This method, incognito Security, has been utilized in many comfortable frameworks. Such gathering may not deceive through meaning, regardless of whether different gatherings be able to distinguish its dishonest. The incognito version, which dwells amongst "malignant" and "semi-legit", models this gathering in an unexpected way. It would not execute an inappropriate program in particular if there is a system to distinguish its cheating. In the event that no location exists in the framework, the gathering may even trade off the information, designed for instance. Subsequently, it is progressively useful on behalf of open cloud capacity. Approach: model cloud suppliers as undercover enemies, and structure conventions flexible to a secret foe.

There are various advancements and varieties for CP-ABE [7]. We do not structure another variety about CP-ABE [6] toward decide primary test, as it is far difficult to reap all the functionalities in those systems and besides it is extra. Other than the functionalities, a couple of sorts provide additional protection and guarantee assure. For example, the composed works cover the passageway plan. If the cloud-facet right of entry manage make the cloud issuer understanding the passage plan, it is not visible as cozy plus awesome. It requires the cloud-side [4] get admission toward manipulate to be zero-data meant for self-self-confident CP-ABE plans. Approach: utilize CP-ABE [6] within phonetic along with revelation method what is more, make sure improvement not spilling arrangement plus traits. The framework just realizes whether the client is genuine or not, what is more, nothing else.

To guarantee the circulated stockpiling efficiently toward the gain tiredness assault, the cloud-side get entry to manage have toward worthwhile what is more and more, lightweight, regardless get right of entry to control, it will end up being a computational resource weak spot ambushes, which may be utilized by poisonous aggressors for DDoS [10] as well as EDoS. The display overhead, being little in like manner, benefits the records, customers who download the facts from the circulated stockpiling, making the estimation not beautiful to resource incomplete devices. Approach: plan a talented admission manipulate for the cloud dealer which ought no longer to include an excess of in the clouds.

1.2 Our Work and Contribution

For the purpose of cloud-side admission controlling, we utilize CP-ABE [6, 7] decoding/encryption as a challenge reaction. Although transfer encoded record, information owner right off the bat creates some irregular plaintexts plus relating ciphertexts. The ciphertexts [5] be identified with a similar access arrangement with the explicit record. For an approaching information client, the cloud server inquires him/her toward decode arbitrarily chosen dispute ciphertext. In the event that the client shows a right outcome, which implies he/she be approved within CPABE [6], Cloud-side access control [12] allows the document to be downloaded. For making our solution, confidential plus effective [13, 14] within genuine global applications, we propose two of the conventions about cloud-side plus information owner side joined admittance manages.

2 Literature Survey

Bethencourt et al. [6] Among circulated frameworks, the client should possibly have the option for getting information so that client gangs a particular arrangement of certification or properties. From nowadays, the key techniques that preserving these arrangements should be used computing believed for storing information plus intercede for control. In such case, if any computing that puts the matter away is undermined, at that point the privacy of the information will be undermined. By using our tactics, confused information may classify irrespective whether there is capability server is distrusting; moreover, In the structure, our techniques are secured against attacks. Past property—mechanisms used to encrypt [15] ascribe the scrambled information and integrate approaches with client keys. Though credits are used in our system to represent the credentials of a client and the set of encoding details determines by who it is it can be unscrambled. By this way, our strategies are similarly near to traditional control techniques, for example, job-based access control (RBAC). Furthermore, we can furnish usage by our structure and forecasts for execution.

Ren et al. [2] Talking of distributed computing of today's most energizing shift in technology creation perspective, however, defense and safety seen to be as important obstacles for its wide-ranging appropriation. Here the developers diagram some specific security challenges and convince for a stable open cloud condition to further analyze security responses.

Zhang et al. [16] From the late days, distributed computing has evolved as another paradigm to promote and transfer Web Administrations. The Distributed computing is appealing to entrepreneurs because it illustrates for the need of consumers for conceiving about supply, it enables undertakings that starts small by which raise assets when there is a rise in the request for administration. Notwithstanding the way in which distributed computing provides enormous opportunities for IT business magnets, the work under distributed computing innovation is still in its starting stages, with many problems still have to be addressed. Right now, present a study of distributed computing, fostering their screw opinion, building standards, using cutting corners, just as research difficulties. The aim of this paper is to provide a clearer understanding of the plan's difficulties with distributed computing and to identify important research features in the current territory.

Harn and Ren [17] Open key advanced testament has been broadly utilized mainly in an openkey framework (PKI) for giving client openkey validation. In any case, the open key advanced declaration cannot utilize as a security factor among verifying client. Right now, recommended an idea of estimated with advanced testament (GDC) which can be applied for giving up client verification and key understanding. A GDC contains client's open data, for example, the data of client's computerized driver's permit, the data of an advanced birth authentication, and so forth., and an advanced mark of the open data marked by a confided in endorsement authority (CA). Be that as it may, the GDC does not contain any client's open key. Since the client does not have that personality and open set of keys, keys administration for utilizes of GDC a lot small difficult that of utilizing unlocked keys computerization authentication. The GDC computerized the markings are used for mystery token for any customer that is never discovered by any checker. Sooner, the owner demonstrates to verify that he had that information on the mark by reacts for verifier test. Among the light idea, we propose both discrete logarithm (DL) and number calculating (IF)-based conventions can accomplish client confirmation and mystery key foundation.

3 Methodology

Cloud Computing allows users to store or access data from anywhere and anytime with cheap cost. All data storage at cloud side will be at security risk due to unavailable control of data owner on store data. To provide security to data, many data security algorithms are introduced and the most famous one is CP-ABE (Cipher Policy Attribute Based Encryption). In this algorithm, data owner can encrypt data by specifying attributes of those users who can access data and the CP-ABE will generate encryption public and private keys by using those attributes and then encrypt and upload data to cloud. Any user with access control can request file from the cloud and then download that file and if user has permission in his attributes then file will be decrypted otherwise file will not be decrypted. With this algorithm, access control and data security can be achieved but the drawback of CP-ABE is first it allows user to download file whether it has permission or not and after download he can decrypt the file if he has permission. Due to file downloading, prior decryption can raise Economical Denial of Sustainability (EDoS) attack. In this attack, malicious users will download files (attackers know they cannot decrypt file but still to raise problems they will download files to make cloud busy and put charges on customers) and consume cloud resources and these consumption charges will be applied on customers.

To avoid author has introduce concept called Robust and Secured mechanism for sharing encrypted data in cloud systems. In this technique, while uploading file, user will generate secret data [17] and encrypt that secret data with bloom filter algorithm and then encrypt file data with CP-ABE and then upload encrypted file with secret data and bloom filter data to cloud for storage. If any user wants to download file then cloud will ask secret data from user and then encrypt that data with bloom filter and check the existing data owner bloom filter with user bloom filter and if match found then only cloud send download file to user. By applying secret data, bloom filter match author has prevented EDoS attack, author introduces two techniques two avoid EDoS attack:

(1) POP (Partially Outsource Protocol): using this technique cloud allow data owner to generate secret data for user verification before file download.
(2) FOP (Fully Outsource Protocol): using this technique instead of user cloud will generate secret data with bloom filter for user verification before download file.

Right now, first, portray the three-party method in favor of distributed storage space. Furthermore, the protection from vindictive in order customers plus clandestine cloud provider is characterized. The distributed storage framework comprises of three elements: Information Owners, Information Clients, and the Cloud Supplier. Information proprietor be the founders and others distributer about documents plus compensate [4] intended for asset utilization lying on document distribution. Because the payers meant for data benefits, the information proprietor [8] need straightforwardness about the usage of properties toward guarantee reasonable was charging. The information proprietors need the cloud supplier toward legitimize [18] property utilization. In this framework, the information proprietor is not continuously lying on the network [1].

Information clients need toward obtaining a small amount of records from the cloud supplier place away on top of the disseminated storage space [5]. They should be confirmed through the cloud dealer previous to the download (toward impede EDoS attack) [9]. The approved clients at that point affirm and sign for the asset utilization, this update was intended for the cloud dealer. Provider in the cloud [3] has the prearranged stockpiling be consistently resting on the network. Its proceedings the positive feature utilization plus charge in order owner dependent on document [2]. Now in order clients fulfilling the entry agreement are able to download the relating documents. The cloud dealer additionally gathers the proof [16] about benefit utilization toward legitimize the charge [18]. We have three manages along with three substances within our framework (Fig. 1).

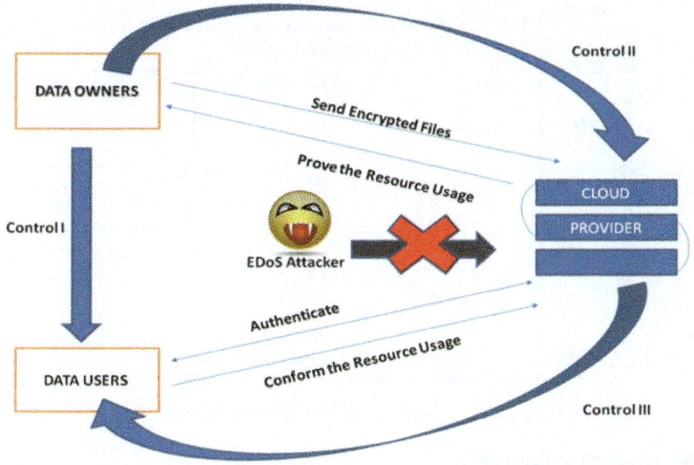

Fig. 1 Encrypted cloud storage system model with prevention of EDoS attacks and resource use accounting transparency

Control I. Data proprietors/owners allocate the entry arrangement within the archive that inspects agreement about facts clients who have benefits toward decode the material.

Control II. Information owner checks asset utilization as of the cloud trader [4], which manage the cloud supplier not toward misrepresent use of quality.

Control III. The cloud trader checks whether the client is able to decode previous to download, which controls the capability about pernicious customer who transmit DDoS/EDoS [10, 9] assaults. Additionally, our structures vary as of went before distributed storage [14] developments, since we think about advantage utilization [13]. Practically speaking, the cloud administrations are normally charged as per the consumption [12], which incorporates the positive feature, exhausted lying on aggressor. The DDoS/EDoS physical attacks [10] determination constantly achieves, which be controlled in our structure as of the presentation about cloud-side administration control.

4 Results

The results obtained [19] in this paper have paved a path to the decreasing number of malicious accesses with cloud-side access control [19]. The EDoS attacks have been evaded by using POP and FOP protocols, respectively. Figurer 2 depicts the communication under attacks (Fig. 2).

The CP-ABE [6, 7, 20] and bloom filter [15, 21, 22] algorithms are applied to encrypt [23] the uploaded file in the cloud and secret data, respectively. The receiving

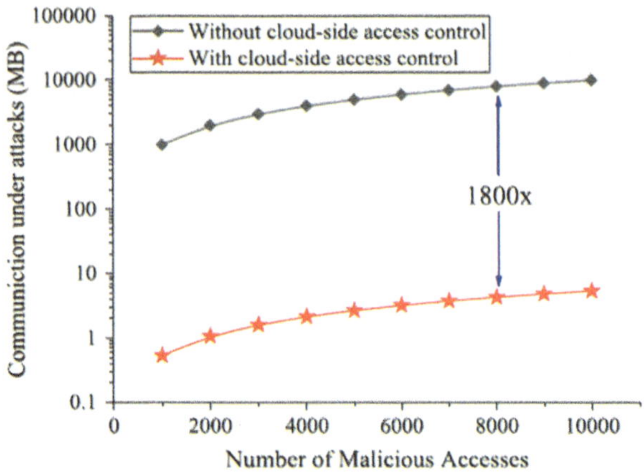

Fig. 2 Communication under attacks

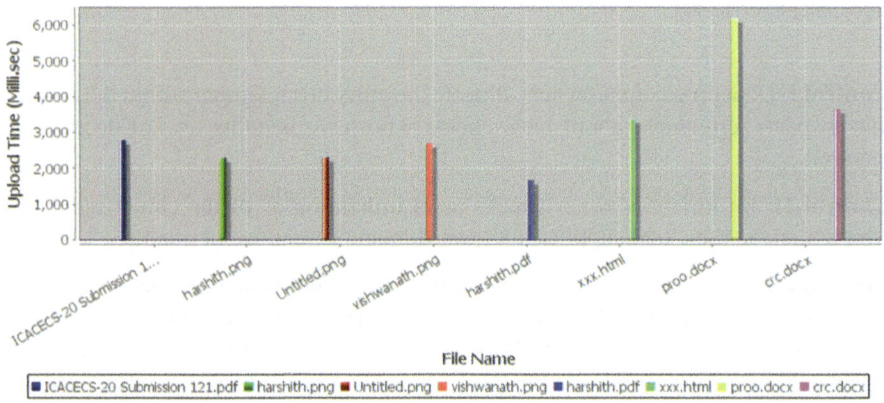

Fig. 3 Computation Graph

user can access the uploaded file only if he can match his bloom filter with the owner side bloom filter. Figure 3 shows the improved computation time, where the encryption and uploading time of a file have been represented.

5 Conclusion

In this paper, we suggest a hybrid access control on the Server-side and Data proprietor/owner-side in encrypted server storage [23] that is immune to DDoS/EDoS [10, 9] attacks and offers resource accounting. This program allows random constructions of the CP-ABE [6, 7, 11]. This design is safe from malevolent data users and secret computing service. They waive the cloud provider's security [2] provision for secret opponents, which are the most realistic and comfortable belief than that of outfit sincere opponents. We use bloom filters and probabilistic control in the accounting of resource use to minimize up above to allow the use of the covert protection. Analyzing the output shows that over current systems [19], the overhead of our construction is minimal.

References

1. W. Li, K. Xue, Y. Xue, J. Hong, TMACS: A robust and verifiable threshold multi-authority access control system in public cloud storage. IEEE Trans. Parallel Distrib. Syst. **27**(5), 1484–1496 (2016)
2. K. Ren, C. Wang, Q. Wang, Security challenges for the public cloud. IEEE Internet Comput. **16**(1), 69–73 (2012)
3. J. Idziorek, M.F. Tannian, D. Jacobson, The insecurity of cloud utility models. IT Prof. **15**(2), 22–27 (2013)
4. J. Hong et al., TAFC: Time and attribute factors combined access control for time-sensitive data in public cloud. IEEE Trans. Serv. Comput. 2017. [Online] https://doi.org/10.1109/TSC.2017.2682090.
5. K. Xue, J. Hong, Y. Xue, D.S.L. Wei, N. Yu, P. Hong, CABE: A new comparable attribute-based encryption construction with 0-encoding and 1-encoding. IEEE Trans. Comput. **66**(9), 1491–1503 (2017)
6. J. Bethencourt, A. Sahai, B. Waters, Ciphertext-policy attribute- based encryption, in *Proceedings of IEEE Symposium Security Privacy (SP)* (2007), pp. 321–334
7. T.V.X. Phuong, G. Yang, W. Susilo, Hidden ciphertext policy attribute-based encryption under standard assumptions. IEEE Trans. Inf. Forensics Security **11**(1), 35–45 (2016)
8. L. Zhou, Y. Zhu, A. Castiglione, Efficient *k*-NN query over encrypted data in cloud with limited key-disclosure and offline data owner. Comput. Secur. **69**, 84–96 (2017)
9. S. Yu, Y. Tian, S. Guo, D.O. Wu, Can we beat DDoS attacks in clouds? IEEE Trans. Parallel Distrib. Syst. **25**(9), 2245–2254 (2014)
10. C. Hoff. Cloud Computing Security: From DDoS (Distributed Denial of Service) to EDoS (Economic Denial of Sustainability). Accessed 6 March 2018. [Online]. Available: https://www.rationalsurvivability.com/blog/?p=66.
11. S. Yu, K. Ren, W. Lou, Attribute-based content distribution with hidden policy, in *Proceedings 4th Workshop Secure Networks Protocols (NPSec)* (2008), pp. 39–44
12. K.Xue etal., RAAC:Robustandauditable accesscontrolwithmultiple attribute authorities for public cloud storage. IEEE Trans. Inf. Forensics Secur. **12**(4), 953–967 (2017)
13. S. Hu, Q. Wang, J. Wang, Z. Qin, K. Ren, Securing SIFT: Privacy-preserving outsourcing computation of feature extractions over encrypted image data. IEEE Trans. Image Process. **25**(7), 3411–3425 (2016)
14. B. Waters, Ciphertext-policy attribute-based encryption: an expressive, efficient, andprovablysecurerealization, in *PublicKeyCryptography— PKC* (Springer, Berlin, Germany, 2011), pp. 53–70
15. Y. Qiao, T. Li, S. Chen, Fast Bloom filters and their generalization. IEEE Trans. Parallel Distrib. Syst. **25**(1), 93–103 (2014)
16. Q. Zhang, L. Cheng, R. Boutaba, Cloud computing: State-of-theart and research challenges. J. Internet Services Appl. **1**(1), 7–18 (2010)
17. L. Harn, J. Ren, Generalized digital certificate for user authentication and key establishment for secure communications. IEEE Trans. Wirel. Commun. **10**(7), 2372–2379 (2011)

18. J. Idziorek, M. Tannian, D. Jacobson, Attribution of fraudulent resource consumption in the cloud, in *Proceedings 5th IEEE International Conference Cloud Computer (CLOUD)* (2012), pp. 99–106
19. K. Xue, W. Chen, W. Li, J. Hong, P. Hong, Combining data owner-side and cloud-side access control for encrypted cloud storage, in IEEE Trans. Inf. Forensics Secur. **13**(8) (2018)
20. S. Hohenberger, B. Waters, Online/Offline Attribute-Based Encryption, in *Public-Key Cryptography—PKC* (Springer, Berlin, Germany, 2014), pp. 293–310
21. S. Tarkoma, C.E. Rothenberg, E. Lagerspetz, Theory and practice of bloom filters for distributed systems. IEEE Commun. Surveys Tuts. **14**(1), 131–155, 1st Quart. (2012)
22. A. Broder, M. Mitzenmacher, Network applications of bloom filters: A survey. Internet Math. **1**(4), 485–509 (2004)
23. F. Wang, J. Mickens, N. Zeldovich, V. Vaikuntanathan, Sieve: Cryptographically enforced access control for user data in untrusted clouds, in *Proceedings 13th USENIX Symposium on Networked Systems Design and Implementation* (2016), pp. 611–626

Providing Security for Cloud Computing Platform Using IP Address Randomization

K. Durgaprasadvarma

Abstract Distributed Denial of Service (DDoS) assaults represent an important warning toward basic communications and same to internet benefits. Here paper suggests IP Address Randomization, a moving objective defense mechanism with the aim of verifies authentic customers besides flood DDoS attacks. This project utilizes gathering of active packet indirection proxies toward passing information between real customers along with the protected servers. Our structure is able to successfully inhibit outer attackers' endeavors to legitimately bombard the network base. Subsequently, attacker's determination requires on the way to conspire through malicious insiders within discovering secrecy proxies after that initiating attack. However, moving objective defense mechanism can segregate insider assaults as of innocent clients through ceaselessly "moving" secrecy proxies toward latest network area whereas recognizing client to intermediary assignments. We build up a greedy shuffling computation to limit the quantity of proxy reassign (shuffles) while amplifying assaults detachment.

Keywords IP address randomization · Moving objective defense · Secrecy proxies · Shuffling

1 Introduction

Arbor Networks [1] have revealed critical increment during the commonness about enormous scope Distributed Denial of Service (DDoS) assaults as of late. In 2010, the biggest detailed data transfer capacity accomplished through a flood-based DDoS assault came to 100 Gbps. In interim, expense about playing out a DDoS assault has ended up being shockingly low down. A tendency Micro's document has uncovered to cost intended for 1-week DDoS administration might be present because small as $150 on Russian black marketplace. Various systems have been proposed previously to forestall or alleviate DDoS assaults [2]. Sifting-based methodologies

K. Durgaprasadvarma (✉)
VNR VignanaJyothi Institute of Engineering and Technology, Hyderabad, India
e-mail: durgaprasadvarma1994@gmail.com

C. Kiran Mai et al. (eds.), *Proceedings of International Conference on Advances in Computer Engineering and Communication Systems*, Learning and Analytics in Intelligent Systems 20,
https://doi.org/10.1007/978-981-15-9293-5_33

utilize pervasively conveyed channels to square undesirable traffic sent to the ensured hubs. Ability-based barrier components try to compel the asset utilization through the sender inside limit allowed via beneficiaries. Protected overlie arrangements mediate an overlay system to roundabout bundles among customers and the ensured hubs, intending to ingest and sift through assault traffic. In any case, these static resistance frameworks either depend on worldwide sending of extra functionalities resting on internet switches or else need huge, vigorous virtual arrange toward endure every compounding assaults. Also, a number be yet powerless against refined assaults, for example, clearing and versatile flooding assaults. Right now, suggests IP Address Randomization [3], an active DDoS guard system to embraces moving objective guard technique to ensure unified online administrations. Specifically, IP Address Randomization presents DDoS versatility intended for approved as well as confirmed customers about protection-sensitive administrations, for example, web-based banking and e-finance. IP Address Randomization [3] utilizes a layer of mystery moving intermediaries to intervene all correspondences among customers along with the secured relevance servers. The system-level channels encompassing the application servers just permit traffic from the legitimate intermediary hubs to arrive at the ensured servers. Intermediary hubs in IP Address Randomization have two significant qualities. To start with, all intermediary hubs are "mystery" within their IP address be disguised as of overall population with only known through real customers later than fruitful verification. Every authentic customer is given the IP address about working intermediary at some random time toward maintain a strategic distance from pointless data spillage.

We concern obtainable evidence of-work plans to secure the customer verification channel. Second, intermediary hubs are "moving." When a functioning intermediary hub is assaulted, it is supplanted by another hub at an alternate area, and the related customers are moved to elective intermediaries. We show that these attributes not just empower us to relieve savage power DDoS assaults, yet in addition enable us to find and seclude vindictive insiders that disclose the area of mystery intermediaries to outside assailants [4–10].

We do so by means of rearranging (repositioning) customers' task to new intermediary hubs when their unique intermediaries are enduring an onslaught. We create calculations to precisely gauge the quantity of insiders and modify customer to-intermediary task in like manner to salvage most blameless customers after each mix. Our answer does not depend on worldwide selection on internet switches or cooperation across various ISPs to work. Neither do we rely upon asset copious overlay arrange to outmuscle high data transfer capacity assaults and to give adaptation to internal failure. Rather, we exploit our intermediaries' mystery and portability properties to fight off ground-breaking assailants. This involves lower arrangement costs while offering significant protective readiness, bringing about a powerful DDoS insurance [11–20].

1.1 Background

This information is expected to empower increasingly educated choices about the utilization regarding system security innovation to ensure crucial internet and other IP-based foundation. The study yield fills in as a general asset for the internet tasks and building network, recording data on the work of different foundation security methods and different patterns. It likewise gives the immediate perceptions, bits of knowledge and recounted encounters of respondents that might be of an incentive to other people.

Operational system security issues the everyday parts of security in business systems are the essential focal point of study respondents. In that capacity, the outcomes gave right now planned to speak to certifiable concerns more precisely as opposed to the hypothetical and rising assault vectors tended to and estimated somewhere else. Denial of Service (DoS) assaults keep on undermining the dependability of systems administration frameworks [21–25].

The design is developed utilizing a mix of secure overlay burrowing, directing by means of reliable hashing, and sifting. We decrease the likelihood of effective assaults by (I) performing serious separating close to secured arrange edges, pushing the assault point edge into the center of the system, where rapid switches can deal with the volume of assault traffic, and (ii) bringing haphazardness and namelessness into the design, making it hard for an assailant to target hubs along the way to a particular SOS-ensured goal. Utilizing straightforward logical models, we assess the probability that an assailant can effectively dispatch a DoS assault against a SOS-secured organize. Our investigation shows that such engineering decreases the probability of an effective assault to infinitesimal levels [26–30].

Denial of service by server asset weariness has become a significant security danger in open interchanges systems. Open key validation does not totally ensure against the assaults in light of the fact that the verification conventions frequently leave ways for an unauthenticated customer to expend a server's memory space and computational assets by starting an enormous number of conventions runs and inciting the server to perform costly cryptographic calculations. We show how stateless validation conventions and the customer riddles of Jules and Brainard can be utilized to forestall such assaults [31–34].

We present Tor, a circuit-based low-dormancy unknown correspondence administration. This second-age Onion Routing framework tends to impediments in the first structure by including flaw-less forward mystery, clog control, registry servers, respectability checking, configurable leave arrangements, and a down to earth plan for area concealed administrations by means of meeting focuses. Tor takes a shot at this present reality internet, requires no extraordinary benefits or bit changes, requires little synchronization or coordination among hubs, and gives a sensible tradeoff between secrecy, ease of use, and productivity. We quickly portray our encounters with a global system of more than 30 hubs. We close with a rundown of open issues in unknown correspondence [35–38].

2 Methodology

The general design about IP Address Randomization, we describing a concept to save application server from DDOS attacks where malicious users can inject or place insider attacks and then this insider attack identify IP address and Port No of Application server and then send to malicious users and by using that IP and port no malicious users send continuous or huge size request to application server and application server will busy in processing those huge requests and raise DDOS error for genuine clients.

To avoid this problem, we are using four different types of entities which will not allow insider attacks from identifying correct or working IP address and port no of application server or proxies.

Authentication Server: Client request will be first taken by Authentication Server and then by random shuffling greedy algorithm, it will choose one proxy server from group of proxies and then assign that proxy for that client. Here proxy is choosing using random shuffling process so if malicious user place any insider attack then that attack will not able to identify which proxy server is using application server IP and port so it will never steal IP and port number of application server. Insider attack get confuse because of random shuffling process and cannot able to attack on proxy or application server.

Proxy Server: Proxy server accept request from client and then analyses request, if size of request data is in the processing range of application server then only it will allow client request to reach secure application server. If request size goes beyond application server limit then it will figure out DDOS attack occurs to JAM application server, so proxy server drops request and report to client or authentication server as it is in under attack.

Application Server: This entity accepts request from client and then process request and send response back to client. In this application, we are allowing client to upload txt files to application server and then application server will receive txt file and store. If size txt file goes beyond limit or client send huge size of file, then proxy will not be sent to application server. All received data will saved inside "Receive Data" folder.

Proxies will keep on moving from one proxy to other proxy, so it is called as Randomization.

Client: This is a client application that can be normal or malicious user who upload request to server for processing. Malicious users attack can be detected and drop by proxies.

3 Proposed Architecture

An outline of our proposed architecture IP Address randomization is depicted in Fig. 1. The proposed framework has a sequential procedure partitioned in forwarding the packets from client to application server.

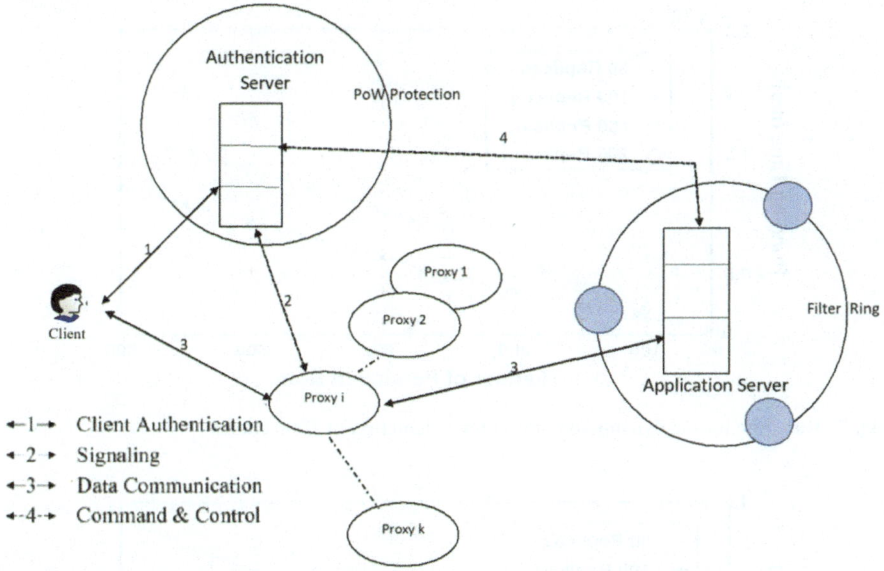

Fig. 1 IP address randomization architecture

To implement this project, we design following modules:

Application Server: we design one application server to process request.

Authentication Server: we design one authentication server to accept request from client and assign random proxy to client.

Proxies: We design proxy servers from which authentication server choose one random and assigned to client for processing.

Client: design one client to send request to application server (Figs. 2 and 3).

3.1 Greedy Algorithm

A greedy algorithm is an algorithm that follows critical thinking heuristic that depends on locally ideal decision at every stage. In tremendous issues, a greedy methodology creates that estimates a comprehensively ideal arrangement in a sensible measure of time. The benefit of utilizing the greedy calculation is observed by differentiating Figs. 2 and 4 Displays the attack rate Fig. 3. Even though small deflections in packets shuffling will take place, the greedy calculation required just a couple of milliseconds to register the customer to server assignments. Consequently, the greedy calculation is appropriate as the runtime calculation to control rearranging activities against on-going DDoS assaults.

Greedy algorithm is used for processing the client-to-proxy assignment. The primary function is called Greedy Assign.

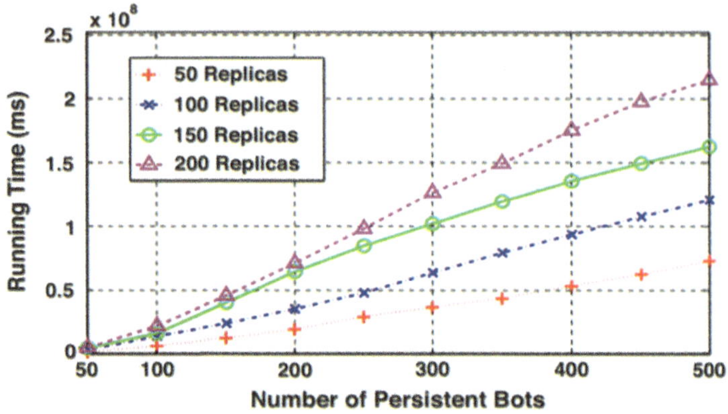

Fig. 2 Run time for the dynamic programming calculation of 1000 clients

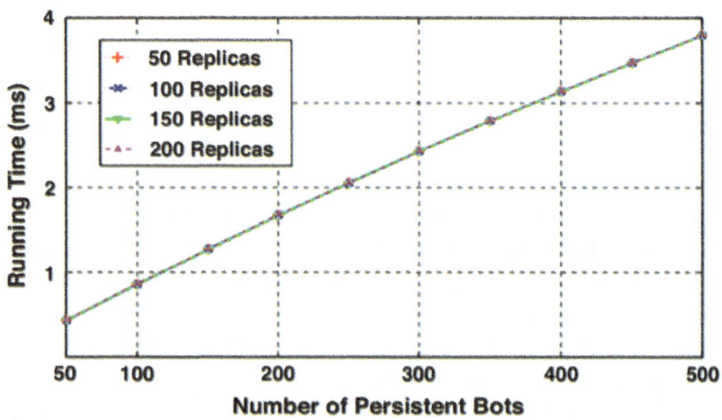

Fig. 3 Run time for the greedy calculation of 1000 clients

$$E(Nca) = Nc - E(Ncu).$$

Since in Equation E(Ncu) the aggregate of pieces (i.e. pjAj) for all shuffling intermediaries processed similarly, we mainly perform optimal investigation for each segment. To make arbitrary proxy j, Aj may be any value in between [0, N − 1]. Aj not be N, if else, each one will be assaulted when there an insider is locally available from the estimation of Ni may influence the ideal decision of Aj, for a specific Ni, we identify every conceivable estimation of Aj and select the one (ω) that expands pjAj. This subroutine is depicted in technique MaxProxy of calculation. Under our greedy methodology, we relegate ω clients to whatever number intermediaries as could be allowed. Function Greedy Assign is called recursively to allot the rest of the customers to the remaining intermediaries. This calculation will end based on three conditions. Initially, if there were a great number of intermediary hubs left

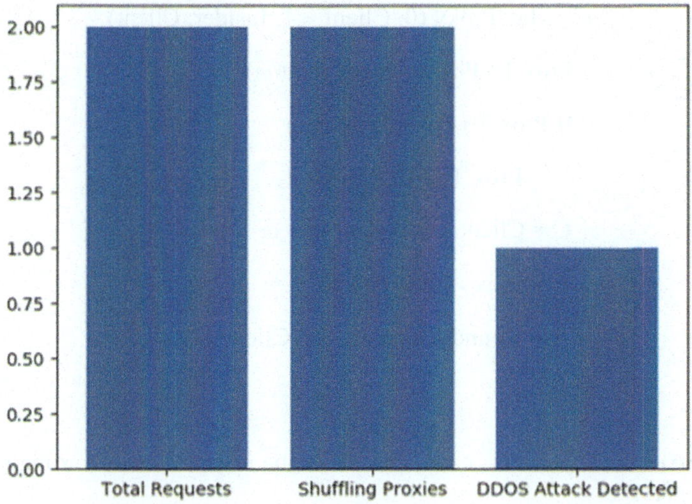

Fig. 4 Displays the attack rate

than clients, every client will be allocated a selective proxy node. Second condition will take place, if there only single proxy is available, all residual clients will be delegated to it. The third condition takes place if there were, when normal no of residual insiders is adjusted to 0, every single outstanding client will be equitably allocated for load adjusting.

Algorithm: Computing Client to Proxy Assignment

```
function Greedy Assign (Client, Insider, Prox)
        if Client ≤ Prox
            Assign 1 elite proxy to every client
        else if Prox = 1
            allocate all clients to the proxy
        else if Insider = 0
            Equally circulate Client over Prox
        else
```

$$\omega = \text{Max Proxy } (0, \text{Client} - 1, \text{Insider}, \text{Client})$$

$$\text{Prox To Fill} = \text{floor } (\text{Client}/\omega)$$

$$\text{If Prox To Fill} \geq \text{Prox then}$$

$$\text{Prox To Fill} = \text{Prox} - 1$$

$$C = \text{Client} - \text{Prox To Fill} * \omega$$

$$P = \text{Prox} - \text{Prox To Fill}$$

$$A = \text{Round } ((\text{Insider} * C)/\text{Client})$$

Fill Proxies with ω clients each. Distribute the rest proxies according to Greedy Assign (C, A, P).

4 Results

The result obtained in this paper shows that how we save application server from DDOS attacks where malicious users can inject or place insider attacks. Here we make a system that will not allow insider attackers from identifying correct or working IP address and port no of application server or proxies, and make Application server free from fake requests.

In the above Fig. 2, graph x-axis represents the number of total numbers of request received and total two proxies and number of detected attacks.

5 Conclusion

We suggest IP Address Randomization, a structure with the intention of utilizes active secret proxies while moving target to alleviate network flooding DDoS assaults. Toward arrive at ensured administration, verified customers are allotted to intermediary nods that will carry out packet transmitting and session policing. At the point while DDoS assault, the verified customers associated with the assaulted proxies be reassigned out toward elective proxy at real time, empowering them toward avoid continuous assault keep up get to the secured administration. With IP Address Randomization, we can successfully conceal the shielded basic administrations from outside assailants. Sophisticated attackers can just utilize insiders toward find our proxy hubs in addition to assault them. IP Address Randomization makes an effective rearranging system toward isolate insider assisted assaults. Our simulation demonstrates that IP Address Randomization be able to ensure larger part about innocent customers as of DDoS assaults helped through several insiders with few quantities of shuffle. Our safeguard system is successful, and simple-to-send, and versatile against solid and advanced DDoS assaults, what's more, it is financially savvy than existing static DDoS safeguard arrangements. Accordingly, our methodology offers a progressively engaging DDoS safeguard arrangement that can be offered as a help by existing cloud providers and ventures with a gigantic network footprint.

References

1. D. Anstee, D. Bussiere, Worldwide infrastructure security report viii (2012) [Online]. Available: https://www.arbornetworks.com/report
2. Layered intelligent ddos mitigation systems (2011) [Online]. Available: https://www.arbornetworks.com/ddos/Layered%20Intelligent%20DDoS%20Mitigation%20Systems.pdf
3. J. Sun, K. Sun, DESIR: Decoy-enhanced seamless IP randomization, in *IEEE INFOCOM 2016 - The 35th Annual IEEE International Conference on Computer Communications* (2016), pp. 1–9
4. J. Ullrich, K. Krombholz, H. Hobel, A. Dabrowski, E. Weippl, IPv6 security: Attacks and countermeasures in a nutshell, in *8th USENIX Workshop on Offensive Technologies (WOOT 14)* (San Diego, CA: USENIX Association, 2014). [Online]. Available: https://www.usenix.org/conference/woot14/workshop-program/presentation/Ullrich
5. M. Wright, S. Venkatesan, M. Albanese, M.P. Wellman, Moving target defense against DDoS attacks: An empirical game-theoretic analysis, in *ACM Workshop on Moving Target Defense* (2016)
6. E. Miehling, M. Rasouli, D. Teneketzis, Optimal defense policies for partially observable spreading processes on bayesian attack graphs, in *Second ACM Workshop on Moving Target Defense* (2015)
7. R. Zhuang, A.G. Bardas, S.A. DeLoach, X. Ou, A theory of cyber attacks: A step towards analyzing MTD systems, in *Second ACM Workshop on Moving Target Defense* (2015)
8. R. Zhuang, S.A. DeLoach, X. Ou, Towards a theory of moving target defense, in *First ACM Workshop on Moving Target Defense* (2014)
9. H. Maleki, S. Valizadeh, W. Koch, A. Bestavros, M. van Dijk, Markov modeling of moving target defense games, in *ACM Workshop on Moving Target Defense* (2016)
10. D. Evans, A. Nguyen-Tuong, J. Knight, *Effectiveness of Moving Target Defenses* (Springer, Berlin, 2011)
11. J.H. Jafarian, E. Al-Shaer, Q. Duan, Adversary-aware IP address randomization for proactive agility against sophisticated attackers, in *2015 IEEE Conference on Computer Communications (INFOCOM)* (2015), pp. 738–746
12. S. Achleitner, T.F.L. Porta, P. McDaniel, S. Sugrim, S.V. Krishnamurthy, R. Chadha, Deceiving network reconnaissance using SDNbased virtual topologies. IEEE Trans. Netw. Serv. Manage. **14**(4), 1098–1112 (2017)
13. S. Basu, G. Kannayaram, S. Ramasubbareddy, C. Venkatasubbaiah, Improved genetic algorithm for monitoring of virtual machines in cloud environment, in *Smart Intelligent Computing and Applications* (Springer, Singapore, 2019), pp. 319–326
14. R. Somula, R. Sasikala, Round robin with load degree: An algorithm for optimal cloudlet discovery in mobile cloud computing. Scalable Computing: Practice and Experience **19**(1), 39–52 (2018)
15. R. Somula, C. Anilkumar, B. Venkatesh, A. Karrothu, C.P. Kumar, R. Sasikala, Cloudlet services for healthcare applications in mobile cloud computing, in *Proceedings of the 2nd International Conference on Data Engineering and Communication Technology* (Springer, Singapore, 2019), pp. 535–543

16. R.S. Somula, R. Sasikala, A survey on mobile cloud computing: mobile computing+ cloud computing (MCC= MC+ CC). Scalable Comput.: Practice Exp. **19**(4), 309–337 (2018)
17. R. Somula, R. Sasikala, A load and distance aware cloudlet selection strategy in multi-cloudlet environment. Int. J. Grid High-Perform. Comput. (IJGHPC) **11**(2), 85–102 (2019)
18. R. Somula, R. Sasikala, A honey bee inspired cloudlet selection for resource allocation, in *Smart Intelligent Computing and Applications* (Springer, Singapore, 2019), pp. 335–343
19. S. Nalluri, S. Ramasubbareddy, G. Kannayaram, Weather Prediction Using Clustering Strategies in Machine Learning. J. Comput. Theor. Nanosci. **16**(5–6), 1977–1981 (2019)
20. K.S. Sahoo, M. Tiwary, P. Mishra, S.R.S. Reddy, B. Balusamy, A.H. Gandomi, Improving end-users utility in software-defined wide area network systems. IEEE Trans. Netw. Serv. Manag. (2019)
21. K.S. Sahoo, M. Tiwary, B. Sahoo, B.K. Mishra, S. RamaSubbaReddy, A.K. Luhach, RTSM: response time optimisation during switch migration in software-defined wide area network. IET Wirel Sens Syst (2019)
22. R. Somula, K.D. Kumar, S. Aravindharamanan, K. Govinda, Twitter sentiment analysis based on US presidential election 2016, in *In Smart Intelligent Computing and Applications* (Springer, Singapore, 2020), pp. 363–373
23. K.B.K. Sai, S.R. Subbareddy, A.K. Luhach, IOT based air quality monitoring system using MQ135 and MQ7 with machine learning analysis. Scalable Comput.: Practice Exp. **20**(4), 599–606 (2019)
24. R. Somula, Y. Narayana, S. Nalluri, A. Chunduru, K.V. Sree, POUPR: properly utilizing user-provided recourses for energy saving in mobile cloud computing, in *Proceedings of the 2nd International Conference on Data Engineering and Communication Technology* (Springer, Singapore, 2019), pp. 585–595
25. R. Vaishali, R. Sasikala, S. Ramasubbareddy, S. Remya, S. Nalluri, Genetic algorithm-based feature selection and MOE Fuzzy classification algorithm on Pima Indians Diabetes dataset, in *2017 International Conference on Computing Networking and Informatics (ICCNI)* (IEEE, 2017), pp. 1–5
26. R. Somula, R. Sasikala, A research review on energy consumption of different frameworks in mobile cloud computing, in *Innovations in Computer Science and Engineering* (Springer, Singapore, 2019); I.P. Kumar, S. Sambangi, R. Somukoa, S. Nalluri, K. Govinda, Server security in cloud computing using block-chaining technique, in *Data Engineering and Communication Technology* (Springer, Singapore, 2020), pp. 913–920
27. I.P. Kumar, V.H. Gopal, S. Ramasubbareddy, S. Nalluri, K. Govinda, Dominant color palette extraction by K-means clustering algorithm and reconstruction of image, in *Data Engineering and Communication Technology* (Springer, Singapore, 2020), pp. 921–929
28. S. Nalluri, R.V. Saraswathi, S. Ramasubbareddy, K. Govinda, E. Swetha, Chronic heart disease prediction using data mining techniques, in *Data Engineering and Communication Technology* (Springer, Singapore, 2020), pp. 903–912
29. R.V. Saraswathi, S. Nalluri, S. Ramasubbareddy, K. Govinda, E. Swetha, Brilliant corp yield prediction utilizing internet of things, in *Data Engineering and Communication Technology* (Springer, Singapore, 2020), pp. 893–902
30. S. Maddila, S. Ramasubbareddy, K. Govinda, Crime and Fraud detection using clustering techniques, in *Innovations in Computer Science and Engineering* (Springer, Singapore, 2020), pp. 135–143
31. K. Rakshitha, A.S. Rao, Y. Sagar, S. Ramasubbareddy, Demonstrating broadcast aggregate keys for data sharing in cloud, in *Innovations in Computer Science and Engineering* (Springer, Singapore, 2020), pp. 185–193
32. S. Ramasubbareddy, T.A.S. Srinivas, K. Govinda, S.S. Manivannan, Comparative study of clustering techniques in market segmentation, in *Innovations in Computer Science and Engineering* (pp. 117–125). (Springer, Singapore, 2020)
33. S. Ramasubbareddy, T.A.S. Srinivas, K. Govinda, S.S. Manivannan, Crime prediction system, in *Innovations in Computer Science and Engineering* (Springer, Singapore, 2020), pp. 127–134

34. C. Pradeepthi, V.V. Geetha, S. Ramasubbareddy, K. Govinda, Prediction of real estate price using clustering techniques, in *Emerging Research in Data Engineering Systems and Computer Communications* (Springer, Singapore, 2020), pp. 281–289
35. K. Sathish, S. Ramasubbareddy, K. Govinda, Detection and localization of multiple objects using VGGNet and single shot detection, in *Emerging Research in Data Engineering Systems and Computer Communications* (Springer, Singapore, 2020), pp. 427–439
36. B.L. Devi, V.V. Bai, S. Ramasubbareddy, K. Govinda, Sentiment analysis on movie reviews, in *Emerging Research in Data Engineering Systems and Computer Communications* (Springer, Singapore, 2020), pp. 321–328
37. I.P. Kumar, S. Sambangi, R. Somukoa, S. Nalluri, K. Govinda, Server security in cloud computing using block-chaining technique, in *Data Engineering and Communication Technology* (Springer, Singapore, 2020), pp. 913–920
38. S. Nalluri, S. Ramasubbareddy, G. Kannayaram, Cloud application security based on enhanced MD5 algorithm. J. Comput. Theor. Nanosci. **16**(5–6), 2022–2027 (2019)

Smart Wireless Black Box with Intelligent Facial Recognition System for Prevention of Accidents and Theft of Vehicles Using Raspberry Pi Along with Sensors Based on IoT

V. Brijendra Varma, B. V. Kiranmayee, L. Ashwik Reddy, S. Sravan Kumar, and P. Sandeep Varma

Abstract With the increased number of vehicles on road, the number of road accidents is increasing day by day. Every year, approximately 1.35 million people die in those accidents. In most of those accidents, people can be saved if taken to the hospital at the earliest, but due to lack of information regarding the time and place of the accident it may not be possible. This paper aims at building a smart black box system, which can help in reducing the road accidents and increasing the chances of saving injured people's lives through accident detection using MEMS accelerometer and thereby alerting the nearest hospital service and the police control room with the exact location of the accident with the help of GPS Sensor and Twilio SMS service. This smart wireless black box comes with an intelligent facial recognition software, which prevents vehicle theft by alerting the owner of the vehicle with the image of the unauthorized person captured through camera. There is also a built-in pre-accident detection mechanism in this black box, which prevents drunk and thereby stopping the vehicle's motor and alerting the nearest police control room and the driver's relatives. This black box system captures various environmental parameters through sensors from the surroundings such as temperature, humidity, obstacle distance, pulse of the driver, location coordinates of the vehicle and uploads all the data to the cloud and also provides the visualization of the data gathered through the

V. B. Varma · B. V. Kiranmayee (✉) · L. A. Reddy · S. S. Kumar · P. S. Varma
VNR VJIET, Department of CSE, Hyderabad, India
e-mail: kiranmayee_bv@vnrvjiet.in

V. B. Varma
e-mail: brijendra.varma@gmail.com

L. A. Reddy
e-mail: luckyashwik@gmail.com

S. S. Kumar
e-mail: shaganti.sravan456@gmail.com

P. S. Varma
e-mail: pathapati.sssv@gmail.com

C. Kiran Mai et al. (eds.), *Proceedings of International Conference on Advances in Computer Engineering and Communication Systems*, Learning and Analytics in Intelligent Systems 20,
https://doi.org/10.1007/978-981-15-9293-5_34

sensors. If any parameter value exceeds the limit then this black box system alerts the corresponding authority via SMS.

Keywords Vehicle black box · Accident detection · Theft detection · Facial recognition · Raspberry pi · Sensors · Thingspeak cloud

1 Introduction

1.1 Internet of Things

IoT (Internet of things) is a future that stands between humans and devices. In future, we will use IoT technology to automate our daily usage tasks with the help of devices connected to one another. The term IoT refers to a situation where networking and computing power extend to devices, sensors and objects, which we use in our daily lives. Devices can communicate to each other with minimum intervention from humans. They generate, transfer and understand the data with other devices by using sensor inputs and react with sensor outputs. By leveraging IoT technology, the world can pretty much become a connected place.

1.2 Motivation

According to the World Health Organization's (WHO) list of top 10 death causes, Road Accidents take the ninth place [1]. Every year almost 1.2 million people die and 20–50 million people get injured due to road accidents. Coming to India, it has 1% total vehicles in the world but accounts for 6% of accidents in the globe. Nearly, 73% of all the deaths caused from road accidents from the south and south-east Asia region are from India. Road accidents in India stand 12th place for premature deaths and 10th place for person disability. In 2018 alone, India had 467,044 reported road accidents, an increase of 0.5% from 464,910 in 2017. Among them, a 1.51 lakh death toll has registered alone in 2018, according to the road ministry's data as shown in Fig. 1 [2].

Based on the analysis of the accident records, major causes for road accidents in India are over-speeding and rash driving by drivers consuming excess alcohol. In spite of many campaigns against over-speeding, this problem continues to exist in several counties. With the advancement of IoT technology, we can leverage data collecting devices called sensors and then build a system which can monitor several parameters that are related to the cause of accidents and thereby prevent them [3]. Collectively that monitoring system is called a Black Box, which generally exists in Air Planes for post-crash analysis.

In this paper, we propose a black box system to detect collisions with the help of a MEMS Sensor installed and then immediately alert the nearest ambulance system

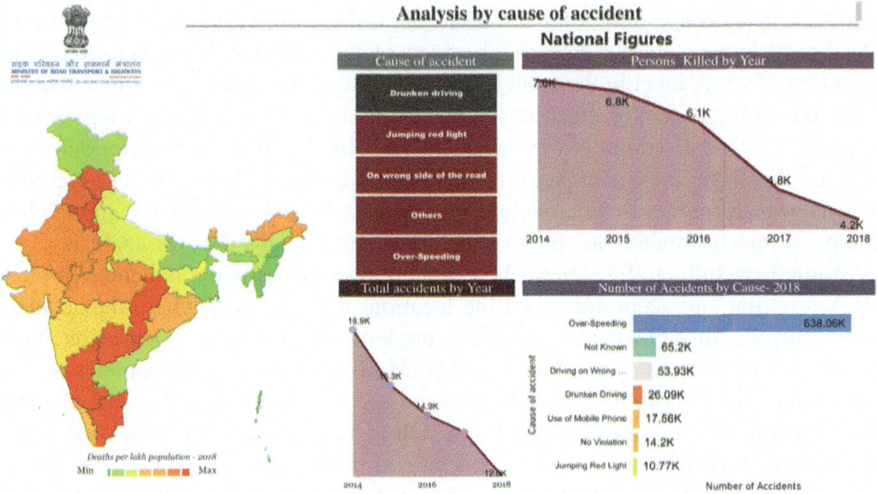

Fig. 1 Road ministry's data for accident from 2014 to 2018. *Source* https://morth.nic.in

with the exact location of the vehicle with GPS sensor and also capture the incident using a camera installed for detailed analysis of the accident that has occurred. We would like to reduce the accidents caused by drunken drivers by installing an alcohol sensor, which detects the excess alcohol consumption of the driver and then the motor of the vehicle gets turned off thereby not permitting the drunk driver to drive the vehicle. Simultaneously, an SMS alert will be sent to the Police by indicating the location of the vehicle. Not only accidents but also vehicle theft is a major concern for the public. Theft of vehicles is an opportunistic crime committed in almost every country in the world and causes major financial loss to people. It is very essential to build a system, which can alert the owner of the vehicle while a thief is trying to steal the vehicle. We propose a facial recognition system to recognize the owner of the vehicle with the help of a camera and permit only the authorized people to enter into the car. If the person is not authorized, then the captured image will be sent to the owner indicating the theft of the vehicle and also alerts the owner with a buzzer sound.

2 Literature Review

In order to know the existing techniques and their significance in building a black box for the main purpose of accident detection, a literature survey is conducted.

Kassem et al. [4] described a black box system with sensors and a microcontroller, which acts as the brain of this black box. Here mainly two approaches are depicted. One is with the detection and storing of data gathered using sensors attached to the microcontroller. The second one deals with the visualization of the recorded data. For detection of environment parameters, there are various sensors attached like Speed Sensor, which detects the speed of the vehicle, Water Sensor for the detection of water level in the vehicle's surroundings, Accident Sensor to detect the collision, Belt Sensor for ensuring the seat-belt safety of the driver, Light Sensor for analyzing the direction of the vehicle and finally the Brake Sensor for detecting whether the driver pushes brake or not during the collision. Now all these recorded data are transferred to the microcontroller's EEPROM and then it is transmitted to a computer interface where a VB.Net program can utilize this data for analysis.

Watthanawisuth et al. [5] described an alert system that is implemented through black box that can accurately send the location of the vehicle where the accident has occurred. This black box system is implemented using Accelerometer, GPS device, Microcontroller unit (MCU) and GSM Module. Here accelerometer is used for detection of fall of the vehicle by monitoring the three axis of the vehicle and it is attached to Microcontroller (MCU), which is used to store and process the data signal generated from the accelerometer in real time. GPS device is used to locate the vehicle where the accident has occurred and GSM Module enables that location of the vehicle to be sent as SMS alert. This black box system has a specialty of nonlinear fall of the vehicle in which the ground speed of the vehicle is calculated in order to prevent from sending false SMS alerts.

Das et al. [6] described a robust and sophisticated black box system. This black box system comes with Eye blink Monitoring, Alcohol detection, Automatic speed control mechanism and Accident alert system. Microcontroller (MCU) acts as the brain of this black box and all the sensors are attached to it. This system has Eye blink sensor for detecting the abnormal eye blinking rate of the driver, alcohol sensor for detection of excess alcohol level of the driver and an Ultrasonic proximity sensor which calculates the distance between the vehicle and the obstacle. This black box system also comes with a GSM module for locating the vehicle and thereby can send SMS alerts if any of the sensors attached to it detect any abnormalities.

Anil Kumar et al. [7] described an Intelligent black box system along with its implementation. The main advantage of this black box comes with the inclusion of Cloud storage where the location coordinates are sent continuously for every 5 s. This system operates with a microcontroller and multiple sensors attached to it. Sensors include Alcohol Sensor for detecting excessive alcohol level of the driver, Accelerometer Sensor for monitoring the 3 axes of the vehicle, Ultrasonic sensor for calculating the distance between the vehicle and the obstacle, Panic buttons for the driver emergency and a GPS and GSM modules for locating the vehicle and sending SMS alert to the programmed entity respectively whenever accident occurs.

Motivation for Proposed System:

The proposed system in this paper is the advancement of the above four papers with the actual implementation process of a smart black box system. In this paper, we provide special features of the black box that are not present in any of the above four papers such as facial recognition system using a camera in the black box which is very much essential for preventing vehicle theft. Also, this proposed black box system comes with various other sensors like pulse sensors, temperature sensors, etc., which collect data and it can be uploaded to cloud for post-accident analysis. This black box monitors the heart beat of the driver using pulse sensor and send SMS alert if the driver gets a cardiac arrest. Instead of GSM module for sending SMS, we are using Twilio SMS API Service, which is much more sophisticated with many built-in libraries and debugging tools that come in handy for developers.

3 Proposed System

In this paper, we provide a smart wireless black box implemented with the help of Raspberry-pi which acts as the brain of this black box system by storing and processing the data that is gathered through various sensors like temperature sensor, pulse sensor, MEMS accelerometer, Alcohol sensor, Ultrasonic sensor, IR sensor, GPS, Camera and Motor as shown in Fig. 2.

The proposed black box system consists of two main modules as follows:

1. Face Detection and Recognition system for **theft prevention** using Haar Cascade Classifier depicted in Fig. 3.
2. Smart Wireless Black Box system for **Accident prevention** using various sensors.

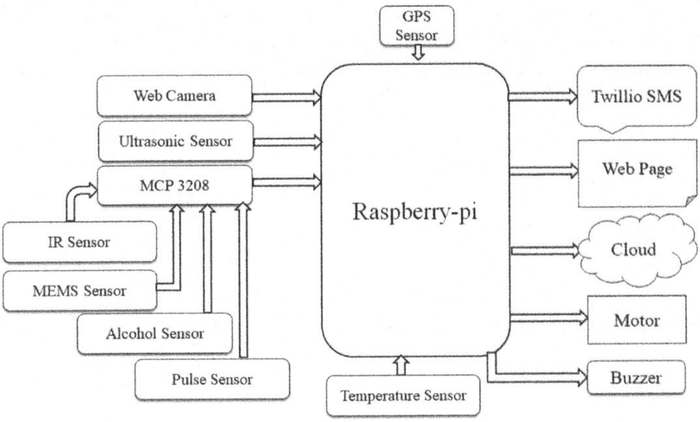

Fig. 2 Hardware architecture of the proposed black box system

4 Implementation

4.1 Face Detection and Recognition System to Prevent Theft of Vehicle

In order to prevent the theft of the vehicle, we have come up with a Face Authentication application which we authorize the driver, when trying to start the vehicle. We are using Haar Cascade machine learning algorithm for Face detection which is quick and robust. It was proposed by Paul Viola and Michael Jones in their paper [8]. In order to authorize the driver, first we need to create a face detection system. After detecting the face and highlighting it with a rectangle, we need to recognize the face. For performing the recognition, there are three algorithms available—Eigenface, Fisherface and Local Binary Pattern Histogram. Here in our project, we are using Fisherface recognition algorithm, since it is fast in recognition and also occupies less space in the memory [9].

Step 1: Import necessary modules

The modules include cv2, numpy, os, sys and image module. The module OS is used for the extraction of names of the image from the directory and a number that is extracted from the image is used as a label for the face present in that image.

Step 2: Create a Facial Detection System

In order to create a face detection system, we need to create a camera object **cv2.VideoCapture()** to capture the face of the driver and load the cascade

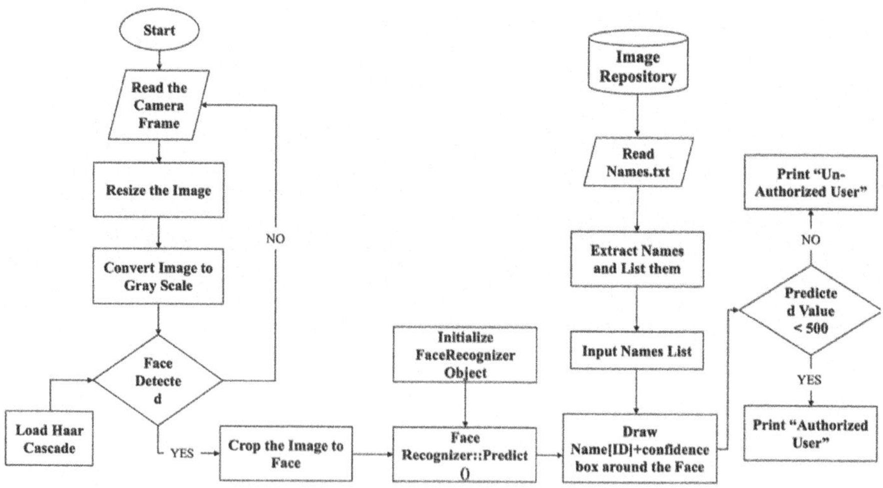

Fig. 3 Flowchart of the facial recognition module for theft prevention

classifier using **cv2.CascadeClassifier()** which is present in the file **haarcascade_frontalface_default.xml**. Now for identification of the face, we need to convert the captured face image to **grayscale** using **cv2.cvtColor()** and detect the features of the face using **CascadeClassifier.detectMultiScale().** After gathering the features of the face image, we can draw the rectangle around the face using **cv2.rectangle()** and label the face within the rectangle using **cv2.putText()**.

Step 3: Create Face Recognizer Object

After detecting the face, we need to recognize that face by indicating the name of the person as a label in the rectangle formed. For that, we need to create a Face Recognizer object. We have used **FisherFace Recognizer** which consumes less memory and is very fast. Using this FaceRecognizer object, we can leverage functions like **FaceRecognizer.train(), FaceRecognizer.predict()**, etc.,

Step 4: Prepare a training dataset and train the images

We can create a dataset of images with their corresponding labels using a function that takes in the absolute path of the image directory and takes two input parameters—image, label. After that we can use a predefined function called **FaceRecognizer.train()** for training the images.

Step 5: Testing

We can perform testing of the FaceRecognizer by comparing the predicted label with the actual label. We can get the actual label from the OS module.

4.2 Smart Wireless Black Box System for Accident Detection of Vehicles

Step 1: First, the Legitimate Owners of the Vehicle Must Register Their Face in the System. If Already Registered, then Go to *Step 2*.

Step 2: After the System (Raspberry Pi) is Turned "ON," the Sensors—MEMS Accelerometer, Ultrasonic Sensor, Pulse Sensor, Temperature Sensor, Alcohol Sensor, GPS Get Initialized. As the Steps Follow as in Fig. 4.

Step 3: All the sensors collect the data continuously for every 5 s and are displayed on the monitor as depicted in Fig. 6. These data include position of vehicle (MEMS), alcohol level of driver (alcohol sensor), temperature and humidity level inside the vehicle (temperature sensor), pulse of driver (pulse sensor) and vehicle to obstacle distance (ultrasonic sensor).

Step 4: The Sensors Data Including Location Coordinates of the Vehicle Are Uploaded to Cloud in the Respective Channels Created, Where the Data Are Visualized in Graphs as Shown in Figs. 9, 10 and 11.

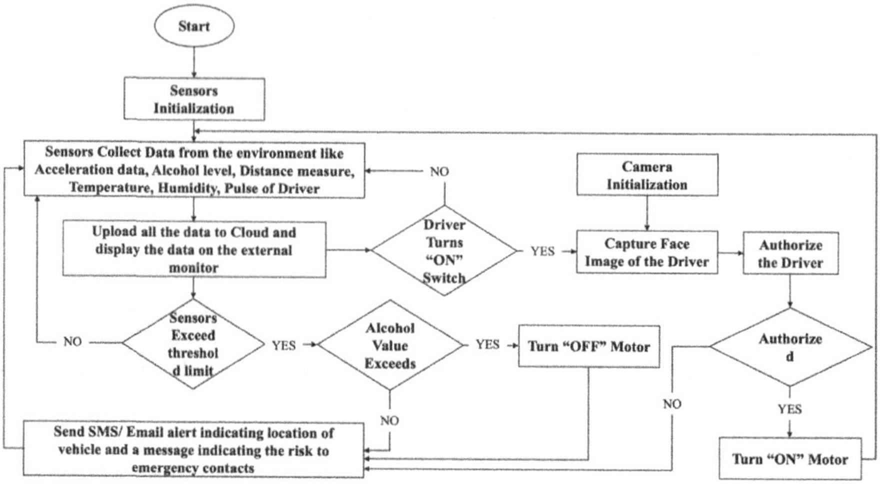

Fig. 4 Operational flow chart of smart wireless black box system

Step 5: If Driver Turns "ON" Ignition Switch of the Vehicle, then His Face is Captured and Authorized as Shown in Fig. 7.

- If Driver is Authorized then Motor is turned "ON."
- Else Alert the Owner of the vehicle via Email with the captured face image.

Go to *Step 3*.

Step 6: If any abnormal value of sensor data is detected, i.e., if any sensor data exceed the threshold limit, then immediately SMS alert is sent to the concerned authority with the risk associated to the vehicle/driver along with the location coordinates of the vehicle using Twilio SMS API service as shown in Fig. 8. If NOT, go back to *Step 3*.

Step 7: If Collision is Detected Through MEMS Sensor, then Capture the Incident Using Camera and Email, the Incident Image to Concerned Authority for Future Analysis and Go to *Step 6*.

Step 8: If the Alcohol Level Exceeds the Threshold Limit, then Immediately Stop the Vehicle Motor and Go to *Step 6*.

5 Experimental Results

The hardware setup is shown in Fig. 5 and experimental results of our Black Box system are depicted below in figures.

Fig. 5 Hardware setup of black box system

Fig. 6 Output of the black box system

6 Conclusion

This paper has proposed and demonstrated a smart and efficient wireless black box system by overcoming the limitations of existing black boxes as presented in our detailed survey. Our Black Box system can prevent accidents and save the lives of the injured people in the vehicle who met with an accident using IoT technology.

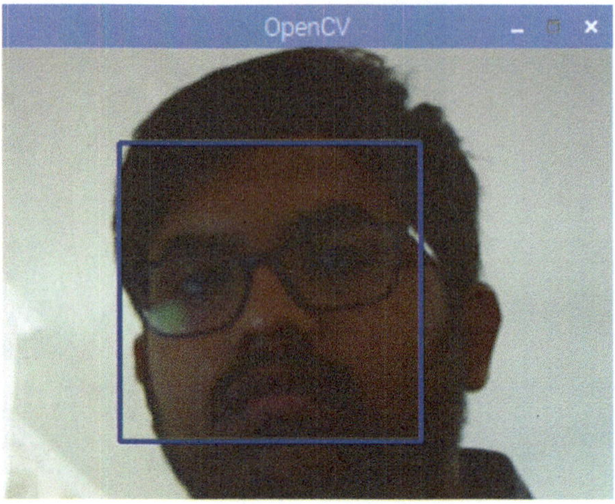

Fig. 7 Training of the face for face recognition

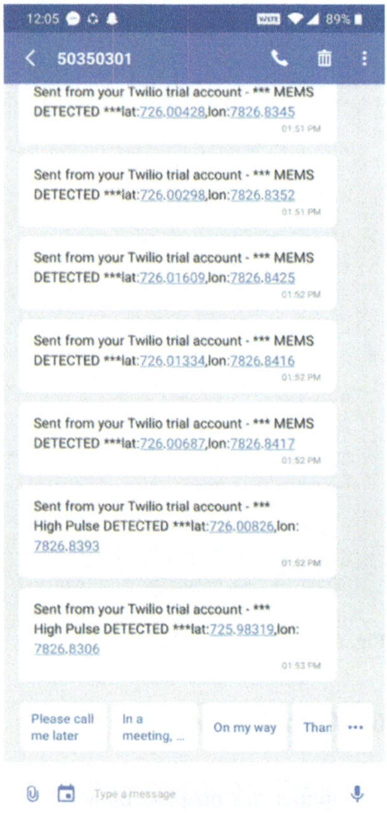

Fig. 8 SMS alert indicating the vehicle location and risk associated

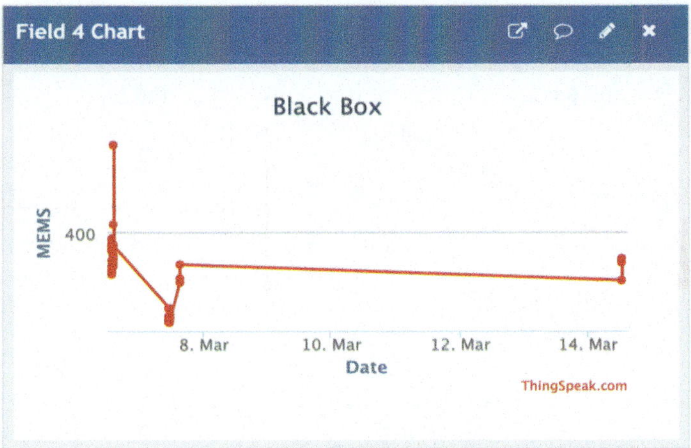

Fig. 9 SMS alert indicating the vehicle location and risk associated

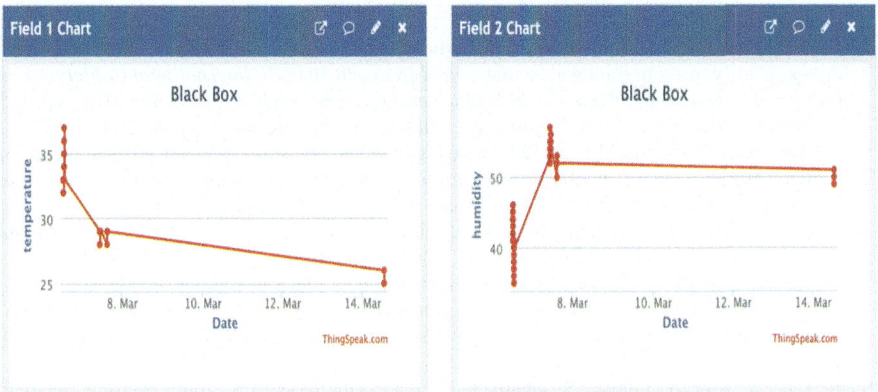

Fig. 10 Temperature and humidity data sent to cloud for further analysis

This black box not only applies for accident prevention but also extends to theft prevention application. Our Black Box works as pre-accident detection system by preventing the drunk driver from driving the vehicle with the help of Alcohol sensor. We have also made an effort to monitor the health of the driver by monitoring his/her pulse with Pulse sensor by which we can save the driver and others lives by informing the hospital services at an early stage. We also provided with the detailed analysis of vehicle surroundings with sensors data by visualizing them in separate channels. Using our black box system, there can be significant reduction in the occurrences of accidents.

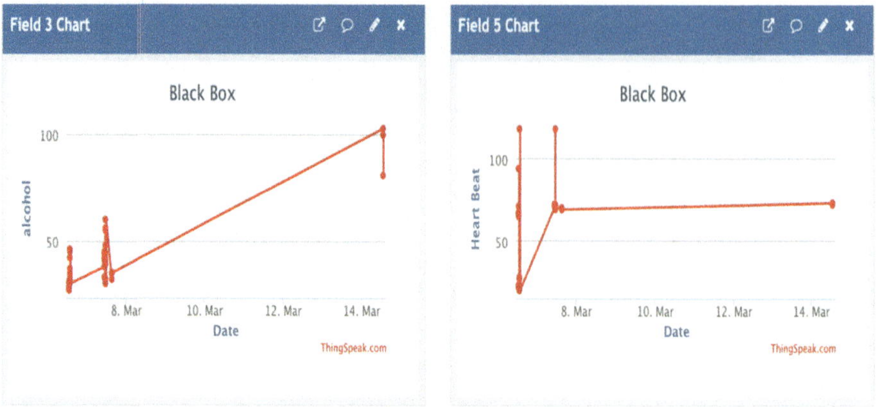

Fig. 11 Alcohol and pulse data sent to cloud for further analysis

References

1. J.C.F. Marquez, D.J.B. Ronquillo, N.B. Fernandez, V.C. Quevedo, Analysis of vehicle crash injury-severity in a superhighway: a markovian approach, in *IEEE International Conference on Industrial Engineering, Management Science and Application (ICIMSA)* (Jeju, 2016), pp. 1–5
2. Ministry of Road Transport and Highways. https://morth.nic.in/state-wise-data
3. L. Mingsong, Study on traffic accident investigation and site survey evidence based on motor vehicle insurance, in *IEEE Eighth International Conference on Measuring Technology and Mechatronics Automation (ICMTMA)* (Macau, 2016), pp. 660–662
4. A. Kassem, R. Jabr, G. Salamouni, Z.K. Maalouf, Vehicle black box system, in *2nd Annual IEEE Systems Conference* (Montreal, Que., 2008), pp. 1–6
5. N. Watthanawisuth, T. Lomas, A. Tuantranont, Wireless black box using MEMS accelerometer and GPS tracking for accidental monitoring of vehicles, in *Proceedings of IEEE-EMBS International Conference on Biomedical and Health Informatics* (Hong Kong, 2012), pp. 847–850
6. A. Das, A. Ray, A. Ghosh, S. Bhattacharyya, D. Mukherjee T.K. Rana, Vehicle accident prevent cum location monitoring system, in *IEEE 8th Annual Industrial Automation and Electromechanical Engineering Conference (IEMECON)* (Bangkok, 2017), pp. 101–105
7. M. Anil Kumar, M. Venkata Suman, Y. Misra, M. GeethaPratyusha, Intelligent vehicle black box using IoT. Int. J. Eng. Technol. **7**(2.7), 215–218 (2018). https://doi.org/10.14419/ijet.v7i2.7.10296
8. P. Viola, M. Jones, Rapid object detection using a boosted cascade of simple features, in *Proceedings of the 2001 IEEE Computer Society Conference on Computer Vision and Pattern Recognition. CVPR 2001* (Kauai, HI, USA, 2001), pp. I-I
9. V. Garg, K. Garg, Face recognition using haar cascade classifier. Int. J. Emerging Technol. Innovat. Res. (JETIR) **3**(12), 140–142 (2016). ISSN: 2349–5162
10. J.A. Chandrashekhar, K.P. Venkatesh, K.A. Birendra, M.S. Kasar, Black box for vehicles. Int. J. Res. Eng. Appl. Manag. (IJREAM) 47–50 (2017). ISSN :2454–9150
11. I. Kadri, B. Kadri, M. Beladgham, O. Dahmane, A robust vehicle accident detection algorithm using acceleration, in *Second International Conference on Life and Engineering Sciences (ICOLES)* (Istanbul, 2019), pp. 139–145

Indian Sign Language Recognition Using Convolutional Neural Networks

Keerthi Reddy Velmula, Indira Linginani, Kalali Bhargav Reddy, Palsa Meghana, and Aounti Aruna

Abstract The Communication plays an essential role in our daily life. People who are hearing and/or speech impaired find it difficult to communicate with others. To aid the deaf and dumb in communicating with others sign languages are used. There are many sign languages. In this paper we worked with Indian sign language. A convolutional neural network is used to identify the images and classify them. Convolutional neural network comes under deep learning algorithms. Various actions are performed in each layer of the neural network to classify the image correctly. Employing convolutional neural networks increases the accuracy of the system. The data set consists of hand gestures which are already processed and the model is trained with the dataset. The model takes hand gestures as input and converts those signs into text. Real time video feed is used as input. The model can be further trained to detect more signs depending on the efficiency of the device.

Keywords Sign language recognition · Indian sign language · Neural networks · Convolutional neural networks

K. R. Velmula · I. Linginani (✉) · K. B. Reddy · P. Meghana · A. Aruna
Department of CSE, VNRVJIET, Secunderabad, India
e-mail: Indira_l@vnrvjiet.in

K. R. Velmula
e-mail: Keerthireddy.velmula@gmail.com

K. B. Reddy
e-mail: chintubargav@gmail.com

P. Meghana
e-mail: meghana2839@gmail.com

A. Aruna
e-mail: arunandg@gmail.com

C. Kiran Mai et al. (eds.), *Proceedings of International Conference on Advances in Computer Engineering and Communication Systems*, Learning and Analytics in Intelligent Systems 20,
https://doi.org/10.1007/978-981-15-9293-5_35

1 Introduction

Even before the Agricultural revolution which took place 3,000 years ago, back when humans were still foragers, sign language was the essential mode of communication for exchanging information about the surroundings, animals roaming around the place, any game happening in the territory. As time passes, humans have discovered another language i.e., the language of the sounds. Similar to the Sign language, which uses distinct signs mapped for distinct information; sound language too mapped distinct information but enabled humans to convey more information. But this doesn't apply to every human being. Humans, who are unable to produce sound or hear, are complete aliens to these languages. Hence, sign language was utilized by such individuals as the medium for exchanging information. Any language spoken by an individual shall be shared among the members of the community. It is self-evident that all the members of the community may not understand the language spoken. It is continuously passed on to the next generation of speakers and becoming subjects to unavoidable changes over time. Language is a medium of communication, formal or informal, by mapping bits of actions, places, things, ideas and emotions to various alterations of physical elements such as sound, human hands and nowadays, writing. Language of sounds, i.e., the spoken language enables us to create various sounds of distinct frequency, pitch and volume. Since, humans developed a system for organizing, understanding and altering that language, it is easy to share thoughts irrespective of the timeline. Since, alterations made in the sound can also be communicated by the same means. In contrast to the former, sign language doesn't have any rigid system for organizing it. Also, a person who used sign language which existed 60 years ago may not understand the sign language used by the present generation. This is so because alterations made in the due course of time cannot be mediated to other members. Also, everyone in the community need not to learn the sign language due to two reasons namely, it is hectic to use and secondly, the language of sounds is much more user-friendly. If the person who uses sign language only, needs to convey essential information and others could not understand, then the situation itself becomes clumsy. It is essential that the hand-signs used among the community shall be understood by almost all of the individuals.

In India, it is estimated that about 1million people are deaf and about a10 million people are hard-hearing people. No formal census data are available. Based on these statistics, one in every five people who are deaf lives in India. One of the uses of sign language is in helping communication among members of the Deaf and the hearing community with the common people. The hearing impaired people find it very difficult to express feelings to the normal people, as the normal people are unaware about the sign language used by the deaf and dumb people. In deaf clique, there exists a transmission barrier amongst the folks and also with the outside world.

To break free of the transmission barrier between the folks using the sign language and the common people, we need to create a common platform or a common interface as a means of communication. This means of communication shall be built only the physical elements familiar to humans so that one could learn, interpret, create and alter the channel of thought. Human senses are the only elements that exist in the entire universe through which one can interpret the universe. Hence, the interface between the deaf and dumb community and the spoken language community must also be based on these human senses only. The only sense common to these distinguished communities is vision. Hence, the interface is based on vision and the system needs to map the human hand-signs and movements to the language of sounds so that the exchange of the information, formal or informal, could be possible. The simplest approach is to map hand-signs to words used in the spoken language. This may look simpler and efficient at first, but it has its own drawbacks. Firstly, no sign could be mapped to a specific symbol. In sign language, a sentence is formed by performing a sequence of a specific set of hand-signs. This sequence may be different in different situations like the sequence of hand-signs for representing an urgent issue and for a normal issue would be different even though both mean the same event. The simplest approach which has been explained earlier was the interpretation of static features. Real-time is always dynamic in the sense that symbols change with respect to color, shape, orientation, visibility and time. Keeping all these in mind, we opted for a novel approach of deciphering sign-language and converting them to spoken language terms.

Dynamicity is a characteristic of the human brain. It is only possible because of the extreme complexity of the human neural networks. In the process of this approach, we made use of the letters in the Indian sign language words with variant features of size, shape, color and style so that dynamic features are being recognized by the system. The system is fed with images of the Indian sign language words so that a model is created which in turn helps in deciphering the symbols. It takes frames of the image. These frames are organized into vectors called descriptors. The descriptors are evaluated by making use of the model. A classifier is present in the system which classifies the descriptors based on the dynamic features. The use of CNN enables the dynamicity of these classifiers. After retrieval, classification of the input, the data is processed and a valid interpretation of the data is provided as the output in a real-time scenario. This would enable us to decipher the sign language and convert them to sound language thus acting as an interface with the two separated communities.

2 Literature Review

Pankajakshan et al. [1] contrast in images is low and cannot easily find the edge detection and its borders. So, they proposed an algorithm which assists to discover a vast range of edges in an image which is known as multistage detection algorithm. They use canny algorithm to adapt to various environments. They also perform various stages of image processing steps. The system has to identify the captured image to get the results. Here they use YUP_640 × 480 resolution input frame.

Ashish et al. [2] defines hull algorithm which is used to execute for number recognition and finger point detection. RGB images are captured by the webcam, but they are too sensitive for different light conditions so the RGB information is converted into YCbCr.

Salian et al. [3] the execution process will be first preprocessing the dataset but the dataset has random width and height and also it does not possess any aspect ratio of 1:1. According to implementation they first resize the images to a size of 256 × 256 in the dataset to obtain an aspect ratio of 1:1 using hull detection algorithm.

Dutta et al. [4]. The Turbo ASR approach is used for digital communication that lies in the exchange of reliability. Adopting the turbo principle to the domain of automatic speech recognition.

Sridevi et al. [5]. The system takes out attributes from the database images for recognition and also for features to train the category classifiers. whereas the database alphabets name is labeled by the training images. In this dataset each of the images is transformed into a binary feature. it does not assign with dynamic gestures which require high accurateness. Processing is slow does not take quick actions.

Bantupalli et al. [6]. The system create a dataset of various signs from the American sign language data set but the model endured from loss of accuracy with the addition of faces of signers differ, the model turns out training inaccurate attributes from the videos. and it also performed poorly when there was a contrast in clothing.

Dimitrios et al. [7]. Designed to overcome the work on sign language detection depended on data gloves or inaccurate processing of video sequences and skeletal data which is directly measured from obtrusive sensors. It deals with the SLR system which helps to overcome the processing of hand and body skeletal data.

Muthu et al. [8]. Defines the system to contain a camera to capture images and designed a transferable unit for more comfortable to the user. Used FCM algorithm to group similar data items. The fuzzy c-means prediction algorithm is used to classify the new data items in the process.

Murtaza et al. [9]. They used a leap motion sensor to economical Universal series and mini portative where as the Bus device used for recording gestures and tracing. In this dataset each static hand gesture was recognized with the help of ANN, cross-correlation and also Geometric Template matching.

3 Proposed System

To overcome the problems of existing systems this paper provides a solution for recognizing Indian sign language using convolutional neural networks which takes real time video as input.

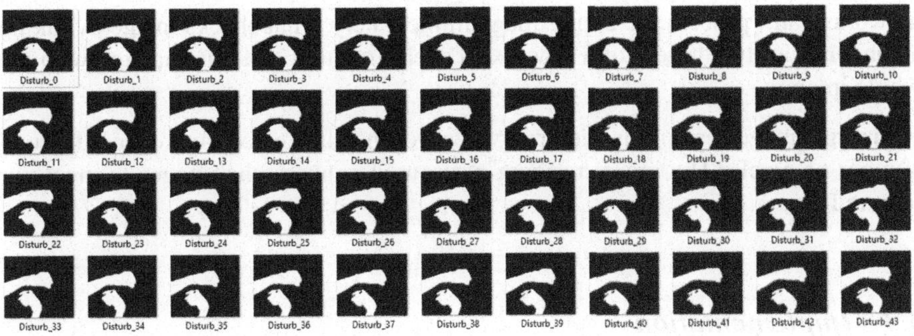

Fig. 1 Training data for disturb gesture

3.1 Data

The dataset used for the project is created from our own. The code for creating the dataset is given which on executing creates 1000 images for each gesture. The obtained dataset is divided as training data and testing data. We are considering 25 gestures of Indian sign language which requires 25,000 images. The gestures are recorded using a webcam using a plain background (Fig. 1).

3.2 Preprocessing

Gaussian Blur: Images taken may contain different noises. Image smoothing methods help in lessening the noise. Gaussian filters hold the characteristics of having no overshoot to a step function input data while lessening the inflation and fall time. In terms of image processing, any fine edges in images are smoothed while reducing too much blurring.

Segmentation: Image Segmentation is performed in order to extract features from a particular gesture. Image segmentation includes separating a visual contribution to segments to improve picture examination. Each segment symbolizes parts of objects and contains sets of pixels. Image segmentation is also performed to classify pixels into bigger components. Wiping out the need to analyze about individual components as units of perception. Segmentation distinguishing portions of image and get awareness about what object they have a place with. Segmentation places the purpose behind implementing classification and Object detection. Semantic segmentation tabulates all the components of an image into worthwhile classes of objects. These classes are interpretable and relate to genuine classification. This is otherwise called dense prediction since it forecast the meaning of each pixel.

Segmentation along with convolutional neural networks is incorporated by taking input as segments of an Image to a convolutional neural system. Convolutional neural networks can't process the entire image immediately. That checks image by taking a gander at a little "filter" of a few pixels each time that gets summarized with the whole image.

Resizing: Image resizing is required when you need to extend or reduce the total number of pixels. The image resizing performed converts the images to meet our required size.

3.3 Implementation

CNN: In order to reach the human neural networks and to break through the static features of the computer, a variant of the neural networks called the convolution neural networks (CNN) is being employed. Convolutional Neural Networks has been proved as a very potent class of models for the image identification problems [10]. Convolutional networks are solely neural networks which in the place of general matrix multiplication employs convolution [11]. Convolutional layers are present in the hidden layers of CNN. They convolve with the multiplication or the dot product.

System Architecture: The proposed system uses CNN, it consists of three layers. Each input is passed through various convolution and pooling layers succeeded by fully connected layers. There are a total of 7 convolutional layers, 7 pooling layers and 2 fully connected layers. For the pooling process, we use max-pooling as only the greatest value in a confined neighborhood of the feature map resides. To contain video data, the max-pooling is conducted in two dimensions. The pooling operation lessens the size of the images while conserving their important features (Fig. 2).

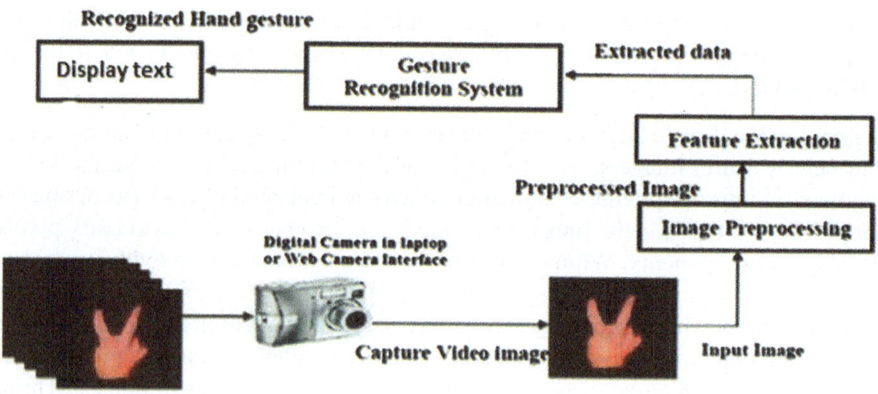

Fig. 2 System architecture

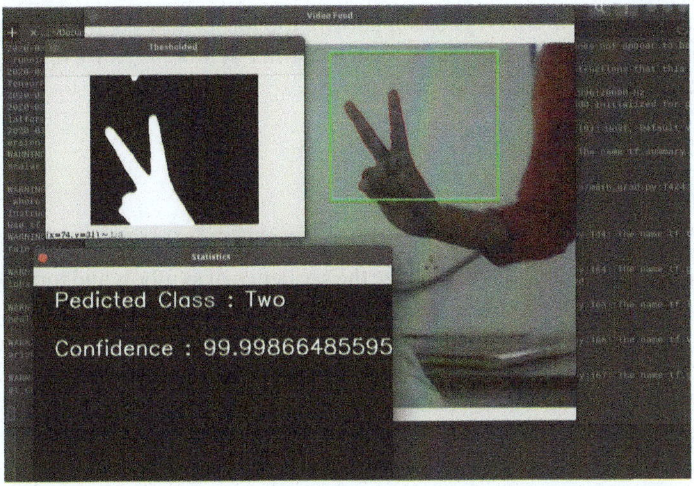

Fig. 3 Image with threshold extraction and result of recognizing symbol

Training and Testing:

Training: From the total collected data 90 percent of data (approx 1000 images) is used for training. The key points which are extracted from each image are sent for training the model where an algorithm is used specified by the user. Here we use Convolutional Neural Network to train the data.

Testing: From the dataset, the remaining 10 percent of data (approx 100 images) is used for testing. Here we follow the same procedure; we extract the key points from images and test the accuracy of the recognition of the system.

4 Results

The model was tested for 25 Indian sign language gestures using real time video input. Confidence was calculated using test data. The testing was performed using a wide range of hands and environments. The observation made is that the environment should be plain to correctly classify the gesture. The accuracy was based on the number of images used to train the model for a gesture (Fig. 3).

5 Conclusion

The Project shows that convolutional neural networks can be applied to correctly classify hand gestures. This project works in real time with around 1000 images to train a symbol and around 100 images to test the symbol. As long as there is computational compatible GPU to supporting the tensorflow deep learning libraries the system can distinguish between many distinct symbols. However sequence of symbols cannot be recognized as sentences.

6 Future Scope

The project can be extended to recognize the sequence of symbols and form sentences. That outcomes of the system can be converted in speech and in addition to it we can add a virtual assistant to help user. Currently this project works for Indian sign language we can extend it into other sign languages so once that is achieved the system can use the virtual assistant to translate the sentences into any language possible so that can be used by any required person all around globe.

References

1. P.C. Pankajakshan, B. Thilagavathi, Sign language recognition system using artificial networks, Matlab 2014 version, in *IEEE* (2015)
2. A.S. Nikam, A.G. Ambekar, Sign language recognition using image based hand gesture recognition techniques using image processing and IDE, in *IEEE* (2016)
3. S. Salian, I. Dokare, D. Serai, A. Suresh, P. Ganorkar, Sign Language recognition Using machine learning and image processing, in *IEEE* (2017)
4. K.K. Dutta, S.A.S. Bellary, Indian sign language recognition Using machine learning and matlab, in *IEEE* (2017)
5. P. Sridevi, T. Islam, U. Debnath, N.A. Nazia, R. Chakraborty, C. Shahnaz, Sign Language recognition for Speech and Hearing Impaired by Image processing in matlab, in *IEEE* (2018)
6. K. Bantupalli, Xie , American sign language recognition using deep learning and computer vision, in *IEEE* (2018)
7. D. Konstantinidis, K. Dimitropoulos, P. Daras, Sign language recognition based on hand and skeletal data Using Deep learning and linear dynamic system, in *IEEE* (2018)
8. H. Muthu Mariappan, V. Gomathi, Real time recognition of Indian sign language Using HCL and Fuzzy c-means clustering, in *IEEE* (2019)
9. Z. Murtaza, H. Akmal, W. Afzal, H.E. Gelani, A.Z. Ul, M.H. Gulzar, Human computer interaction based on gestural cues recognition Using leap motion controller and SLTC techniques, in *IEEE* (2019)
10. A. Karpathy et al., Large-scale video classification with convolutional neural networks, in *IEEE Conference on Computer Vision and Pattern Recognition (CVPR)* (2014)
11. A. Goodfellow, Y. Bengio, A. Courville, *Deep Learning* (MIT Press, 2016), p. 326

Text Recognition in Complex Images Using Deep Learning Models: A Survey

Thuraka Gnana Prakash and Vemparala Sravani

Abstract Text recognition in images is a very challenging task. It is used for guiding visually impaired persons and also in semantic meaning. In this survey, we are going to explore different techniques implemented in the previous years on complex background images. It gives the analysis of related works and also emphasizes the performance of the image recognition. Image complexity cannot be defined in a precise manner but we can describe its parameters such as backgrounds, noise, illuminating conditions, textures, fonts, etc. This survey also highlights few benchmark datasets that are used in this review. Through this review, we can analyze various problems present in this field by contrasting their features.

Keywords Complex images · Image processing · Text recognition

1 Introduction

These days multimedia play a vital role in one's life. Lot of information is shared through images, videos, etc. While these images differ from one another, in an image, if the text color is in contrast to the background color then the task is very easy. Generally, an image with a plain background is called simple image. In an image, the complexity lies in the text, background, light conditions, reflection, Hazy, natural surroundings and other disturbance factors. Image complexity can be broadly classified into two types, scene complexity and text complexity.

T. G. Prakash · V. Sravani (✉)
Department of CSE, VNRVJIET, Secunderabad, India
e-mail: vemparalasravani@gmail.com

T. G. Prakash
e-mail: gnanaprakash_t@vnrvjiet.in

C. Kiran Mai et al. (eds.), *Proceedings of International Conference on Advances in Computer Engineering and Communication Systems*, Learning and Analytics in Intelligent Systems 20,
https://doi.org/10.1007/978-981-15-9293-5_36

Fig. 1 Examples of scene text complexity

1.1 Scene Complexity

- Background—The backdrop of the image plays a vital role. Background may be of different colors, effects due to which the variation of foreground and the background of the image are very less, for example, foreground text may have same color of the background.
- Hazy [1]—Due to the fog the text is not clear and cannot be read.
- Reflection—When the flash of the camera reflects on the image. It makes the text invisible.
- Natural surroundings—When an image of a restaurant name is captured, along with the text the environment is also involved in the frame.
- Light conditions—If an image contains text with low contrast or the image is captured with low intensity of light the text may not be visible clearly.

If we want to read or convert it into text, due to the dazzling effect of the surroundings and objects present in the scene the text cannot be recognized. Through simple OCR (Optical character recognition) [2], we cannot recognize text in images containing complexity but rather can be used after removing the complex background (Fig. 1).

1.2 Text Complexity

Text in an image can differ in orientation, font size, style, language, textures, etc.

- Orientation [3]—Texts are present in any direction.
- Size—Can be of any number.
- Style—Can be of any printed format. Example: Times new Roman, Arial, Calibri, etc.
- Color—Text can have same or different colors.
- Textures—Comprises design in the text.
- Language—English, Chinese, French, etc.

Many methods are implemented for text detection, recognition and extraction. This is useful but tough task. A familiar and popular method is OCR [2]. This method

Fig. 2 Examples of text complexity

acquires text from simple images like documents. This cannot be used for complex images rather can be implemented after converting complex to simple image. This paper aims to provide the methodology of the text recognition systems for complex background images (Fig. 2).

2 Different Approaches for Recognizing Text in Complex Images

There are different methods for detecting the text in images. Each model has its own strategy for recognition on various images. Here are few methods implemented on Hazy images, complex background images, etc. in detail.

Liu [4] discussed about text detection-based methods namely character, word, text-line based methods and its uses. According to their framework, the process is performed in two steps. The input image is divided into two sets, i.e., Word stroke which predicts the text region, these text regions may contain small false regions. False regions are removed using Text center of the image. Gaussian map is used to define the block. Then the combination of these two results in the bounding box of the text. Through this model, word and line-level results are acquired. They have gained better results in Multi-oriented and inconsistent text using standard datasets.

Luoa [3] came up with MORAN, which stands for Multiobject rectified attention network. It is classified into MORN (multiobject rectification network)—used for removing the complexity of the image by adjusting the characters and making it simple for the next steps. Through this, the text is aligned uniformly. Now the text region is ready for recognition. ASRN (Attention-based sequence recognition network) is used to read the text and predict the outputword. CNN-LSTM network is implemented for ASRN. This model has achieved good results for irregular text. Authors have focused on Text complexity rather than scene complexity (Fig. 3).

Ansari [1] has proposed the recognition framework for haze images. As haziness is a disturbance factor in an image due to which the text cannot be read clearly. In this method, image processing techniques are implemented as the text layer must be separated from the haziness and should be converted to recognition format. To remove the haziness, bilateral filter is applied for the image. Once the disturbances are

Fig. 3 Rectification of curved text

removed from the image the text can be localized easily. This is one of the methods for scene complexity.

Zhong [5] processed on the multioriented text by changing the position of the text into horizontal line. Authors have divided this process into two steps. First, implementing region proposal network for detecting the text in the image, forwarding the regions to the extraction using recurrent network. Here the work is done in two ways: text and non-text classification, then region extraction. They have worked on scene complexity, text orientation and implemented Region proposal network through which the text regions are detected and then forwarded to the Recurrent network. This helps in extracting the text from the detected region and achieved good performance for benchmark datasets.

Liu [6] divided into text detection, localization and recognition. In text detection, different methods based on texture, connected component [7] and deep learning models like CNN (convolution neural network), VGG, LSTM (Long short term-based memory), etc. Recognition methods based on word and characters. An end to end recognition system is a fusion method of text detection and recognition. This method is used to execute an input image, process the image using above models and obtain final output. They also discussed text, scene complexity and its removal techniques, [8] Binarization of images, Preprocessing actions for background removal. Authors have also compared the performance of the models with the help of bench mark datasets.

Venkatesan [9] focused on color inversion techniques that simplify the complex backdrop images by background removal such that the text can be easily recognized. Model has obtained output for blurred images, differently oriented text. Connected components implemented for filtering text and nontext regions. As text do not vary much and lie in the same line they can be easily filtered. From those regions, the text is localized and processed for extraction. For categorizing text regions, K-means clustering is implemented.

Vyankatesh [10] evaluated the horizontal text. Thresholding is applied to the input image. Image processing techniques are implemented for background removal then the connected component analysis [11] is implemented for text extraction. After the rectangles are formed on the image, text is extracted one by one by checking its neighbors (Fig. 4).

Sun [11] analyzed the position of the text based on features and machine learning. These are performed based on texture, Features, edge information. Through these parameters, the text is recognized. There are many models starting from 2012. In

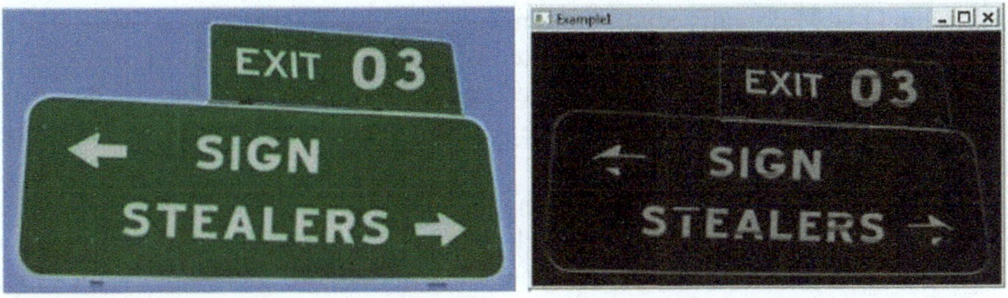

Fig. 4 Applying threshold for a color image

ALEX net overfitting is avoided and uses dropout to ignore selected neurons which are individual. VGG net is denser than the above model. It is an enhancement for the above model. Google net implements average pooling. RCNN performs text detection to extraction but takes long time for computation, which is resolved in Faster RCNN by simplifying pooling layers of region of interest. YOLO (You look only once) is more advanced model than the before as it does not implement any intermediate regions. Single shot Multi box detector (SSD) covers even small regions. It uses small convolution filter. This model is more precise than YOLO but not suitable for detecting text in natural scene directly. This is rectified using Textboxes which changes the size of the filter. EAST [12] removes unnecessary intermediate steps. In CRNN (Convolutional recurrent neural network), features of images are extracted using CNN and combine with BLSTM (Bi directional long short-term memory) which gives probability distribution of column features.

Ghoshal [8] implemented complete conversion of color image into simple image, which consists of only black and white. The concept of the authors in this method is, for each region, there is a variation in the data, for the region containing text is different from the nontext. Using edge detection techniques, the text boundary is found. Noise removal operations are performed as they may cause deviation in the result [10]. Morphological operations are performed as it works better on binary images. Then the image is forwarded to the trained model for output (Fig. 5).

Wei [13] worked on poor quality images. Work is done on OCR using deep learning model. Eliminating the disturbances in the image and performing the extraction process. CNN model is implemented for text recognition. They have achieved better results than other noise removal operations. Authors have analyzed different noise patterns and their removal techniques.

Fig. 5 Binarization of image

In Unar [14], Maximally Stable extremal region (MSER) [15] is used for detecting regions of text. This is based on connected component analysis [16]. Its working is in such a way that the whole text is colored uniformly therefore the text can be easily recognized. Few advancements [17] are performed to this method to improve the result.

Tang [18] implemented deep learning model in a cascaded form. Images are binarized [8] and fed to the first model called D-net (detection network), which extracts the features then fed to the segmentation for text and nontext classification, processed further for classification of text regions for extraction. The resultant images from the above step are cropped, which makes the image simpler.

Zhou [12] worked on Rectangle box detection. After thresholding, the detected regions are divided into four channels for extracting the text and then the rotation operations are applied to form horizontal text. PVA-Net, VGG (visual geometry group) models [11] are used for training the data. PVA-net is a light network, it is used as a Substitute for the extraction of features in the Faster R-CNN. The performance obtained between these two nets is compared by training with dataset. The model focuses on irregular text and also on natural scene images. Those images are very challenging. They also compared the performance of their models with the other existing models.

Su [19] discussed about feature extraction and deep learning networks. Different features need to be extracted and need to apply RNN [19, 20] as it cannot directly extract the text due to the complexity. The word image is converted into features using HOG [17] and SIFT (Scale Invariant Feature transform). HOG performance is better than SIFT therefore column features are extracted using HOG. One column of image is considered for processing. Now the model is trained RNN (Recurrent Neural Network) and LSTM (Long Short-term Memory) is implemented. As RNN takes long time for training to reduce this problem LSTM is implemented. The score for each lexicon word is calculated separately using RNN model. At last, the combination of these networks is used for final output.

He [17] proposed model comprises of Text attentional CNN and Contrast-enhanced MSER (Maximally Stable extremal regions) [15]. Text Attentional network is used for filtering text and nontext regions. It is trained with challenging datasets [21, 22]. Due to the complications in the background, the text regions that are obtained using MSER are deformed leading to incorrect result. As there is contrast variation in the regions, the contrast is increased for dominant regions but small regions may not be considered. For these small size regions, contrast is improved. All these regions are computed. The model acquired good results for arbitrarily shaped text and natural scene images.

Ye [23] focused on incidental scene text and its methods, features extraction and connected components. In these methods, they generally detect dense characters, for this, neural network is implemented by filtering the text and nontext regions. Same range of characters should be maintained for text localization as the less intense characters may not be recognized. Binarized images [6, 8, 16] are also created for reducing scene complexity. Converting the single line into words, namely, text line segmentation is performed for recognizing the text easily, then training the neural

Fig. 6 Results of recognized text

networks, recognizing the text [24] character wise and obtaining the output. Multi-oriented text is processed to change inclined text into the straight line. Multilingual text is detected using the above techniques as they focus only on the text but not the language, while recognizing tesseract OCR is used for language detection. The training is given for each language to process the content (Fig. 6).

3 Applications

1. It is mainly helpful for visually impaired persons [23] who can listen to the text which they cannot view. Using the above recognition models, the text is extracted and converted into audio.
2. It is also used in checking passport details, and extracting information.
3. Recognizing license plates.
4. Converting the printed text into electronic format.
5. In hospitals, the administration needs to store a large amount of patient details including insurance and health forms. Instead of manual entry by using the above techniques, information can be stored and accessed easily.

Table 1 Datasets and their complexity

Datasets	Horizontal text	Multioriented text	Scene complexity	Noisy
ICDAR 2003	✓	✗	✓	✓
ICDAR 2005	✓	✗	✗	✗
ICDAR 2011	✓	✗	✓	✗
ICDAR 2013	✓	✓	✓	✓
ICDAR 2015	✓	✓	✓	✓
ICDAR 2017	✓	✗	✓	✗
MSRA TD500	✓	✓	✓	✗
SVT	✓	✗	✓	✗
COCO-Text	✓	✓	✓	✗
IIIT5K	✓	✓	✗	✗

4 Datasets

ICDAR 2003 [25], ICDAR 2005 [26], ICDAR 2011 [22], ICDAR 2013 [21], ICDAR 2015, MSRA-TD500, COCO Text, SVT, IIIT5k, are the datasets used by the above methods (Table 1).

5 Comparison of Results

The following are evaluation metrics such as Precision, recall, F-Score, Accuracy of various datasets (Tables 2 and 3).

Table 2 Results of COCO-Text, SVT, ICDAR 2003

Model	P	R	F	P	R	F	P	R	F	A
Lin [24]	0.64	0.57	0.61	0.68	0.68	0.66	–	–	–	0.88
Liu [6]	0.50	0.32	0.39	0.68	0.53	0.60	–	–	–	0.96
Ghulam [27]	–	–	–	0.85	0.77	0.81	0.87	0.79	0.83	0.97

Table 3 Evaluation of ICDAR 2005, 2011

Model	P	R	F	Model	P	R	F
Tong [17]	0.87	0.73	0.89	Tong [17]	0.91	0.74	0.82
Lucas [26]	0.62	0.67	0.62	Lin [24]	0.88	0.82	0.85
Simon [26]	0.60	0.60	0.58	Zhong [5]	0.89	0.898	0.894

Table 4 Performance evaluation of ICDAR 2013, 2015, MSRA TD500

Model	P	R	F	P	R	F	P	R	F
Zhong [5]	0.93	0.87	0.80	0.87	0.80	0.83	0.81	0.78	0.79
Tong [17]	0.93	0.73	0.82	–	–	–	0.76	0.61	0.69
RRPN [11]	0.95	0.88	0.91	0.84	0.77	0.80	0.82	0.69	0.75
Seg link [11]	0.87	0.83	0.85	0.73	0.76	0.75	0.86	0.70	0.77
Pixel link [11]	0.88	0.87	0.88	0.85	0.82	0.83	0.83	0.73	0.778

6 Conclusion

A brief view of text recognition systems under complexity and other disturbances—recognizing images from complicated background and separating the foreground from background using image preprocessing techniques. For a complex background image, the basic priority is background and noise removal techniques explained in [6, 8, 10]. The deep learning model is used as a second step for most of the methods. Different models [11] are used for recognition and their performance on the benchmark datasets, which are discussed clearly in Tables 3 and 4, respectively. A quick analysis of complexities of different datasets is discussed in this paper. Though there is a lot of advancement in this area, there is no complete successful model for users. This area requires more and more work in future.

References

1. G.J. Ansari, J.H. Shah, M. Sharif, S. ur Rehman, A novel approach for scene textextraction from synthesized hazy natural images (Springer, 2019)
2. T.K. Hazra, D. Pratapsingh, Nikunjdaga, Optical character recognition using KNN on custom image dataset, in *IEEE* (2017)
3. C. Luoa, L. Jina,b, Z. Suna, MORAN: a multi-object rectified attention network for scenetext recognition (Elsevier, 2019)
4. Z. Liu, W. Zhou, H. Li, Scenetext detection with fully convolutional neural networks (Springer, 2019)
5. Z. Zhong, L. Sun, Q. Huo, Improved localization accuracy by LocNet for faster R-CNN based textdetection in natural scene images (Elsevier, 2019)
6. X. Liu, G. Meng, C. Pan, Scenetextdetection and recognition with advances in deep learning: a survey (Springer, 2019)
7. H. Zhang, K. Zhao, J. Guo, Text extraction from natural scene image: a Survey (Elsevier 2013)
8. R.G. Anandarup Roy, A. Banerjee, B.C. Dhara , Swapan K. Parui, A novel method for binarization of scenetext images and its application in text identification (Springer, 2018)
9. A.V. Karpagam, M. Manikandan, Text extraction from natural scene images using Renyi entropy, in *IET* (2019)
10. V.V. Rampurkar, S.K. Shah, G.J. Chhajed, S.K. Biswash, An approach towards text detection from complex images using morphological techniques, in *IEEE Conference* (2018)
11. Y. Sun, A. Dawut, A. Hamdulla, A review: text detection in natural scene image (2018)
12. X. Zhou, C. Yao, H. Wen, Y. Wang, S. Zhou, W. He, J. Liang, EAST: an efficient and accurate scene text detector, in *IEEE* (2017)

13. T. Chiang Wei, U.U. Sheikh, Improved optical character recognition with deep neural network, in *IEEE* (2018)
14. S. Unar, X. Wang, C. Zhang, C. Wang, Detected text-based image retrieval approach for textual images. J. Eng. IET (2018)
15. D. Joshy, M. Anishin Raj, A survey on MSER based scenetextdetection, in *IRJET* (2018)
16. Chaitanya, R. Kulkarni, Ashwini B. Barbadekar, TextDetection and recognition: a review, in *IRJET* (2017)
17. T. He, W. Huang, Member, IEEE, Y. Qiao, J. Yao, Text-attentional convolutional neural network for scene textdetection, in *IEEE Transactions on Image Processing* (2016)
18. Y. Tang, X. Wu, Scene textdetection and segmentation based on cascaded convolution neural networks, in *IEEE transactions on Image Processing* (2017)
19. B. Su, S. Lu, Accurate recognition of words in scenes without character segmentation using recurrent neural network (Elsevier, 2017)
20. H. Liang, X. Sun, Y. Sun, Text feature extraction based on deep learning: a review (Springer, 2017)
21. D. Karatzas, F. Shafait, S. Uchid, M. Iwamura, ICDAR 2013 robust reading competition (2013)
22. A. Shahab, F. Shafait, A. Dengel, ICDAR2011 robust reading competition challenge2: Reading text in scene images (2011)
23. Q. Ye, D. Doermann, TextDetection and recognition in imagery: a survey, in *IEEE Transaction* (2015)
24. H. Lin, P. Yang, F. Zhang, Review of scene text detection and recognition (Springer, 2019)
25. Lucas SM, ICDAR 2003 robust reading competitions (2003)
26. Lucas SM, ICDAR 2005 Text locating competition results, in IEEE (2005)
27. G. Jillani Ansari, J. Hussain Shah, M. Yasmin, M. Sharif, A novel machine learning approach for scenetext extraction (Elsevier, 2018)
28. C. Yi, Y. Tian, Portable camera-based assistive text and product label reading from hand-held objects for blind persons, in IEEE (2014)

Improving the Traffic Control System by Modifying the Geometry of Roads and Squares by Removing the Discontinuity of Space

Rakesh Kumar Tripathi, Yash Ratnaparkhi, and Deepak Sukheja

Abstract This paper presents a review of the existing geometry of the roads and squares and proposes a continuous topological model to solve the existing and upcoming traffic problems. As it is seen that due to discontinuity in traffic flow, public suffers a lot. After a deep study, it was found that the Discontinuity of spaces exists in this existing model and also it is topologically incorrect. The Discontinuity of spaces exists due to the reason that all the traffic signals situated on one road are not equidistant and cross roads create a chaotic situation. Therefore, we introduce a topologically proven model in which we tried to remove discontinuity between spaces and suggested some minor constraints which can be enforced as a part of discipline.

Keywords Topology · Continuity of spaces · Transport · Traffic · Design of roads

1 Introduction

In the last decade, the population of India has elevated itself from 1,234,281,170 in 2010 to 1,380,004,385 in 2020 and so is the increase in the number of vehicles registered. As a result of this, there is also rapid increase in the number of accidents, transportation problems in the form of increase in traffic congestion, air pollution, delays and emission of greenhouse gases. The long queues of vehicles result in idling condition, also cause extra fuel consumption. For this, different traffic control measures are taken into action by the Traffic Control Board of India. The existing

R. K. Tripathi
Dr. APJ Abdul Kalam University Indore, Indore, India
e-mail: drrakeshkumartripathi@gmail.com

Y. Ratnaparkhi (✉)
Department of Computer Science and Engineering, IPS Academy, Indore, India
e-mail: yashratnaparkhi2@gmail.com

D. Sukheja
Department of Computer Science and Engineering, VNR VJIET Hyderabad, Secunderabad, India
e-mail: deepak_s@vnrvjiet.in

C. Kiran Mai et al. (eds.), *Proceedings of International Conference on Advances in Computer Engineering and Communication Systems*, Learning and Analytics in Intelligent Systems 20,
https://doi.org/10.1007/978-981-15-9293-5_37

traffic management technology has been utilized by many people all over the India in several areas. With the development of transport technology, security of transport system has become an important issue, as accidents have been increased day by day over the past few years. Hence, it is very essential to find an effective way to manage the traffic system. In the present world, we are having traditional traffic management system. They are good within them. Still they are lacking few parameters in some aspects. The main motive of research paper is updating the existing traffic technology and improving the efficiency of traffic controlling. The existing technology includes Enforcement of traffic police, Visual stimuli (traffic signals and its rules) and some audible sources for seeking the concentration of the public. The theory proposed in this paper deals with some alterations in the construction of roads such that it could remove traffic signals thereby reducing the work load of traffic police.

Here, the case is related to that situation where there are two roads facing each other and having high traffic and the other two roads that have comparatively very low traffic. A normal traffic scenario consists of each signal with some waiting time. Now the road that has low count of vehicles gets free earlier and the busy road with a large count of vehicle passing through gets blocked for that particular time interval, resulting in traffic congestion and loss of fuel. On an average, in such condition, a vehicle consumes more than 20% of the fuel around that time.

A suggested model based on principle of continuity of topological spaces is proposed in this paper. Few researchers contributed and used topology in designing various models.

Norbert [1, 2] says **Topology denotes the connectivity between**, for example, the rooms of a building and their links like doors, or walls.

Vella [3] proposes topology as a generalization of graph theory. Similarly, it also connects two more different packets, which divide a long road in many parts. This connectivity may be of different levels. A square connects two packets of straight road but it may also be considered traffic connected to each packet by its boundary common surface.

Boltcheva et al. [4] suggested a model for smooth and easy traffic. But to design these packets, only a few actually used the topological methods. Raghothama et al. [5, 6] discussed about Constructive topological representations.

This paper presents a modified version of it that can remove the discontinuity of space. The discontinuity in space can be defined as it involves a large number of research areas and therefore, this paper focuses on those facts that we conclude to be the most relevant. The main motive of this paper is to study the achievements attained in the last years and to give an overview of possible directions toward future research in the field of traffic management, thereby reducing the number of accidents per year. It also deals with reducing the waiting time of the vehicle owners waiting over the signals.

2 Preliminaries

2.1 Limit Point

A point x is a limit point of a set A if every open set containing x meets A.

2.2 Definition of Topological Space

Let X be a set. A set Tx of subsets of X is called a topology for X if it satisfies the following three properties:

1. X and Ø are Elements of Tx.
2. The unions S of every subsets S of Tx is an element of Tx.
3. The intersection $A \cap B$ of every two elements A, B of Tx is an element of Tx.

2.3 Properties of Topological Space

A subset A of a topological space X is called closed if X - A is open in X.

Then closed sets satisfy the following properties

1. ϕ and X are closed
2. A, B closed $\Rightarrow A \cup B$ is closed
3. $\{Ai \mid i \in I\}$ closed $\Rightarrow \bigcap_{i \in I} A_i$ is closed.

2.4 Continuity for Topological Spaces

Definition A Function f: X $\rightarrow$ Y is said to be continuous if the inverse image of every open subset of Y is open in X. In other words, if V € Ty, then its inverse image $f^{-1}(V)$ € Tx.

A map f: X $\rightarrow$ Y between topological spaces is a homeomorphism or topological isomorphism if f is a continuous bijection whose inverse map f^{-1} is also continuous.

Remark By the remark above, such a homeomorphism induces a one-one correspondence between T_X and T_Y.

Theorem 1 Let f be the identity map from (R2, d2) to (R2, d ∞). Then f is a homeomorphism.

Proof Since every open set is a union of open neighborhoods, it is enough to prove that the inverse image of an ε neighborhood is open. This ε neighborhood is an open square in R2 which is open in the usual metric.

A similar proof shows that the image of an ε neighborhood in the usual metric (an open disc) is open in d ∞.

In general, if X is a set with two topologies T_1 and T_2 then the identity map $(X, T_1) \rightarrow (X, T_2)$ is continuous if T_1 is stronger (contains more open sets) than T_2.

3 Survey

A systematic survey and study of data related to number of vehicles waiting for their signal at high court square, Indore (Madhya Pradesh), was performed for a week and the data obtained are noted in Table 1. Figure 1 shows the map of the road and the square, which was surveyed. Figures 2 and 3 are the photographs taken of that particular square. The whole study was performed in three different time slots where waiting time of each signal was noted and accordingly the number of vehicles waiting was calculated. The readings were noted in a gap of 40 min for 2 h and an average of the reading was tabulated. Likewise, there were three time slots chosen from the 24 h' schedule. The time slots were:

1. 10.30 to 12.30 (office time)
2. 14.30 to 16.30 (noon time)
3. 18.30 to 20.30 (evening time)

The percentage of the total count of different vehicles, i.e., for car, bus, auto and all other two-wheeler was calculated in each time interval for morning, afternoon and evening hours (Fig. 4). A proper pie chart of volume of the vehicles at the signalized intersection is shown in Figs. 5, 6 and 7. From the pie chart, the main information extracted was that in any of the three time slots, the two-wheelers were in majority with around 65% of the total traffic and the four-wheelers count was 16%. Figure 4 shows different count of vehicles emerging from different ends of the road at that particular square.

4 Analysis of Data

Table 1 gives the information of the amount of vehicles passing and the volume of the traffic present in particular time interval, it shows that the volume of the traffic approaching from regal square and Treasure Island road in a particular time interval was much higher and comprised of 115 vehicles (80 two-wheelers and 35 four-wheelers) from each road as compared with the other two roads. The delay time for the three signals more than 65 s and from the high court, a very little amount

Table 1 Data of the volume of vehicle in different time intervals

Day		Monday			Tuesday			Wednesday			Thursday			Friday		
Signal from		Regal square	Yeshwant Niwas Rd.	Treasure Island Mall	Regal square	Yeshwant Niwas Rd.	Treasure Island Mall	Regal square	Yeshwant Niwas Rd.	Treasure Island Mall	Regal square	Yeshwant Niwas Rd.	Treasure Island Mall	Regal square	Yeshwant Niwas Rd.	Treasure Island Mall
Stay time		70 Sec	65 Sec	70 Sec	70 Sec	65 Sec	70 Sec	70 Sec	65 Sec	70 Sec	70 Sec	65 Sec	70 Sec	70 Sec	65 Sec	70 Sec
10.30 to 12.30	Two wheeler	78	61	79	82	66	84	84	59	82	80	68	79	85	64	90
	Four wheeler	35	25	32	38	22	36	34	26	31	34	25	35	32	23	42
14.30 to 16.30	Two wheeler	73	51	69	71	60	72	75	55	60	78	52	65	77	53	66
	Four wheeler	39	22	36	36	25	32	34	29	43	33	24	39	35	26	48
18.30 to 20.30	Two wheeler	83	63	81	80	58	85	86	67	88	79	60	84	84	62	82
	Four wheeler	42	31	42	38	34	39	46	39	32	38	32	37	34	36	36

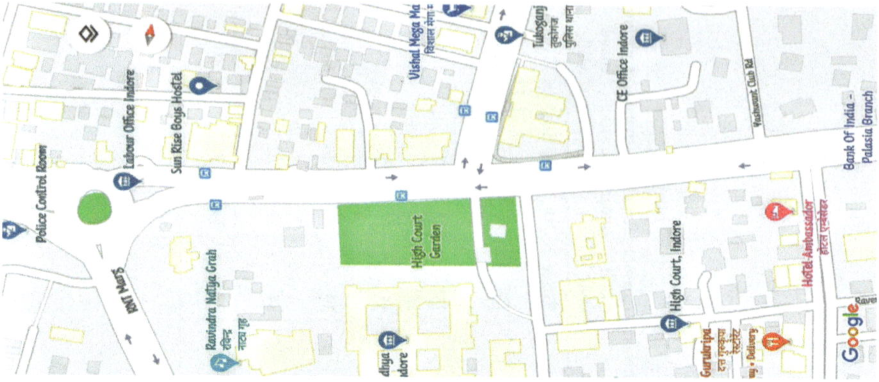

Fig. 1 Map of the road surveyed [7]

Fig. 2 Volume of vehicles from Yeshwant road

Fig. 3 Volume of vehicles from Treasure Island mall road

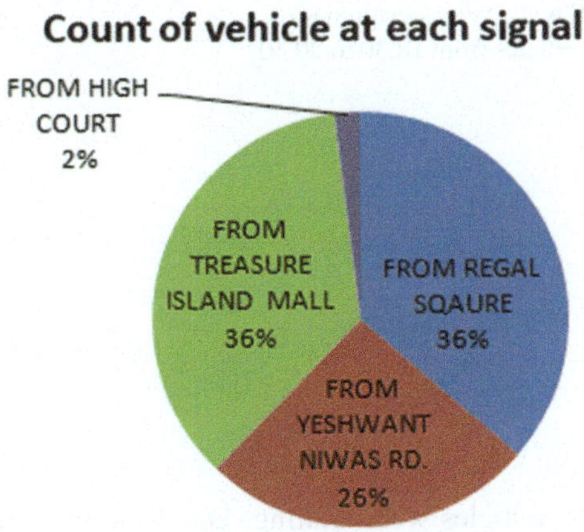

Fig. 4 Total counts of different vehicles at signals

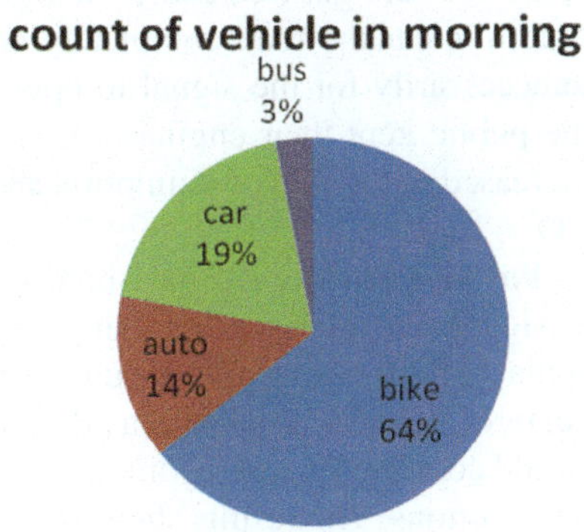

Fig. 5 Average volumes of vehicles from 10.30 to 12.30

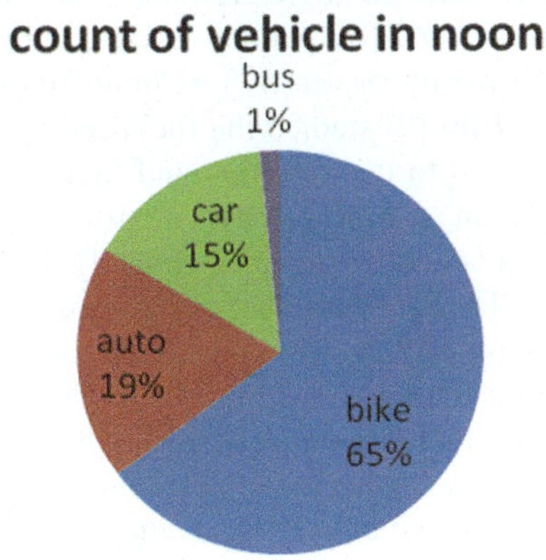

Fig. 6 Average volumes of vehicles from 14.30 to 16.30

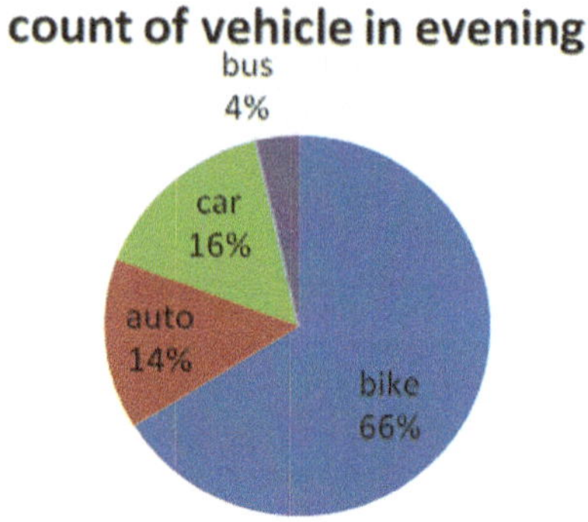

Fig. 7 Average volumes of vehicles from 18.30 to 20.30

of vehicles were waiting. Due to long waiting time, the road with less volume of vehicle would get free earlier and the roads with high traffic would get insufficient time to clear the road completely and thus the others had to wait for a long time unnecessarily for the signal to open. Also it was observed that more than 70% of the public kept their engines on while waiting for the signal to turn green, which increased in the fuel consumption and the release of harmful gases such as NOx and CO.

Pal and Sarkar [8] measured the total fuel consumption of different vehicles at signalized intersections using high accuracy instruments. Five red light traffic signals of different traffic volume were surveyed and the fuel loss during idling and delay of vehicles was estimated. On the basis of results obtained, an average fuel would cost Rs. 19, and an amount of Rs. 175.00 was exhausted per day at signalized intersections. The results showed that the delay of vehicles was more than 60 s that led to loss of 389.68 L of diesel and 810.38 L of petrol. Total loss output work found was Rs. 61,072 per day and Rs. 2, 22, 91,198 per annum. The total fuel consumption for petrol car was 573 ml/hr and diesel car was 705 ml/hr.

Lim [9] studied the fuel used up during idling diesel trucks. The data generated related to the emission and fuel consumption from the study give the information that in an average case, a bus could emit 8224 g/hour of CO_2, 144 g/hour of NOx and could use up approximately 0.82 gallon/hour diesel.

The data were analyzed regarding the average fuel consumption [10]. Only petrol-driven vehicle was considered for the test case. The count of vehicles was measured with their idling time. Since almost all the vehicles had their engines on, so the total fuel consumption during the idling time was calculated and tabulated. Now an ideal two-wheeler vehicle consumes about 0.0389 ml and a four-wheeler consumes an average of 0.21 ml, if they stand ideally for one second, that means 140 ml and 756 ml of fuel is exhausted by two- and four-wheeler in 1 h, respectively. The total fuel consumed in one cycle of the traffic was found to be 0.7467 ml. If we take the average cost of this fuel calculated per hour during idling, it would cost around Rs. 4477. Hence the amount would increase to Rs. 53,724, Rs. 76, 068, Rs. 1,611,720 and Rs. 19,340,640 if we calculate it per day, per week, per month and per year,

Table 2 Total fuel consumption and cost during idling

Signal	Type of vehicle	Count of vehicle	Time (in Sec)	Consumption of fuel (ml/sec	Price of fuel (Rs/ml)	Cost (in rupees)	Total cost (in an hour)
1	Two wheeler	80	70	0.0389	0.078	16.99152	509.7456
	Four wheeler	35	70	0.21	0.078	40.131	1203.93
2	Two wheeler	61	65	0.0389	0.078	12.030603	360.91809
	Four wheeler	25	65	0.21	0.078	26.6175	798.525
3	Two wheeler	79	70	0.0389	0.078	16.779126	503.37378
	Four wheeler	32	70	0.21	0.078	36.6912	1100.736

respectively. While surveying the total cost of fuel consumed during idling, time was calculated and was properly tabulated in Table 2.

5 Solution of the Problem

Existing geometry of the square is shown in Fig. 8. According to the figure in an existing geometry, a normal square consists of vehicles moving in all four directions and following the set of rules related to traffic signals.

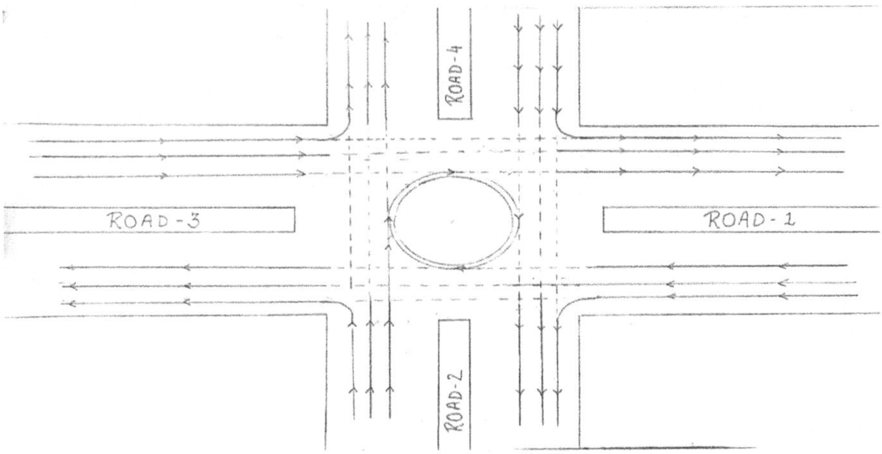

Fig. 8 Existing geometry of signalized intersection

5.1 Suggested Model

Let us consider three signal points P, Q, R in a sequence on a straight road and we say a packet between traffic signals P and Q as a space X and packet between traffic signals Q and R as a space Y. Now on going from X to Y a discontinuity arises because of many constraints like spaces do not have equal sizes and if Y is of smaller size than X, then all the elements (vehicles) of X can not move easily into Y which is almost similar as the discontinuity in two topological spaces. One more thing that is responsible for the discontinuity is intersection between roads and therefore to keep a basic principle of continuity of space in mind we proposed some geometrical changes and shifting the positions of squares so that a discontinuity between spaces can be removed.

Figure 9 is the suggested model that removes discontinuity under certain conditions of traffic rules and with the help of specially designed speed breakers, the speed limits can be controlled as per the need of the traffic.

Figure 9 represents the modified geometry of the signalized intersection. It consists of different lanes named from A, B, C, D, E, F, G, H, I, J. Also some additional modifications that are made in the diagram regarding this intersection such as speed breakers are named as 1, 2, 3, 4, 5, 6, 7, 8 and the U turns are labeled as X, X′, X″, X‴. Each labeling is explained below.

(i) LABELS A, B, C, D, E, F, G, H, I, J are the alphabets assigned to the different roads.

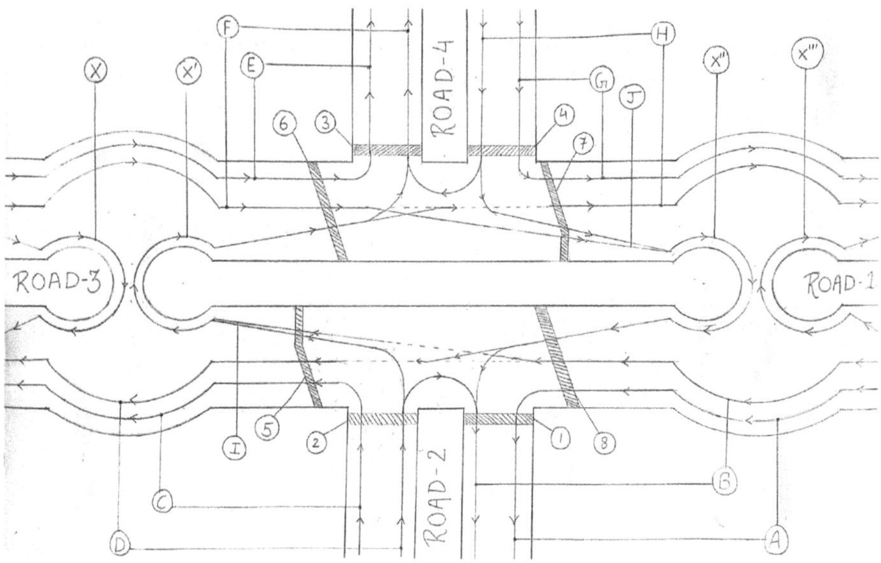

Fig. 9 Modified geometry of signalized intersection

(ii) LABELS 1, 2, 3, 4, 5, 6, 7, 8 are the numbers assigned to the speed breakers [11]. Weber [12] in his paper discussed the standard for the geometry design of the speed humps. Speed breakers are used to control the speed of the vehicles and to avoid accident and mishaps. The whole concept based on speed breakers is discussed by Patel et al. [13], Kadiyali [14]. The speed breakers are divided as:

- 1, 2, 3, 4 are straight speed breakers
- 6, 8 are tilted speed breakers
- 5, 7 are half-straight and half-tilted speed breakers.

(iii) LABELS X, X′, X″, X‴ are the U turns.

5.2 *Explanation of the Diagram*

Case I

Let us consider a case where the vehicles are commencing from road 1.

(a) Now, the vehicles that have to go to their respective left side will follow the lane "A", encounter breaker "8" and breaker "1" while turning left and keep following the lane "A".
(b) The vehicles that have to go straight will follow the lane "B", slow down their speed at the breaker "8" and after crossing the breaker 5, will join the lane "D" and go straight.
(c) The vehicles that have to go toward their right side will follow the lane "B," slowing down their speed at the breaker "8" and will switch to the lane "I," and encounter the breaker "5". Then the vehicles will take a U turn at the point marked as X‘ and then while switching to the lane "F" will encounter a breaker "6" and go right.
(d) The vehicles that have to take a U turn will follow the "step c" but rather than turning toward left, will continue to go straight and join the lane "H".

Case II

Considering the case where the vehicles are commencing from road 2.

(a) Now, the vehicles that have to go to their respective left side will follow the lane "C", encounter breaker "2" and breaker "5" while turning left and keep following the lane "C".
(b) The vehicles that have to go straight will follow the lane "D" and will slow down their speed at the breaker "2", take a turn to join the lane "I" and encounter the breaker "5". Then the vehicles will take a U turn at the point marked as X′ and then while switching to the lane "F" will encounter a breaker "6" and go right.
(c) The vehicles that have to go toward their right side will follow the "step b" but rather than turning toward left, will continue to go straight and join the lane "H".

(d) The vehicles that have to take a U turn will follow the lane "D", encounter breaker "2," take a U turn and after crossing the breaker "1" will switch and follow the lane "B".

Case III

Considering the case where the vehicles are commencing from road 3.

(a) Now, the vehicles that have to go to their respective left side will follow the lane "E", encounter breaker "6" and breaker "3" while turning left and keep following the lane "E".
(b) The vehicles that have to go straight will follow the lane "F", slow down their speed at the breaker "6" and after crossing the breaker 7, will join the lane "H" and go straight.
(c) The vehicles that have to go toward their right side will follow the lane "F", slowing down their speed at the breaker "6" and will switch to the lane "J", and encounter the breaker "7". Then the vehicles will take a U turn at the point marked as X'' and then while switching to the lane "B" will encounter a breaker "8" and go right.
(d) The vehicles that have to take a U turn will follow the "step c" but rather than turning toward left, will continue to go straight and join the lane "D".

Case IV

Considering the case where the vehicles are commencing from road 4.

(a) Now, the vehicles that have to go to their respective left side will follow the lane "G", encounter breaker "4" and breaker "7" while turning left and keep following the lane "G".
(b) The vehicles that have to go straight will follow the lane "H" and will slow down their speed at the breaker "4", take a turn to join the lane "J" and encounter the breaker "7". Then the vehicles will take a U turn at the point marked as X'' and then while switching to the lane "B" will encounter a breaker "8" and go right.
(c) The vehicles that have to go toward their right side will follow the "step b" but rather than turning towards left, will continue to go straight and join the lane "D".
(d) The vehicles that have to take a U turn will follow the lane "H," encounter breaker "4", take a U turn and after crossing the breaker "3" will switch and follow the lane "F".

6 Conclusion

Using the concept of continuity of topological spaces, we have made a traffic controlling model in which the spaces are made continuous. By modifying the geometry of the traffic controlling, the following advantages can be achieved as:

- No traffic signals required
- No fuel exhaustion when idling
- Continuity of the vehicles is maintained
- Less requirement of police enforcement
- Less chance of accident due to proper speed breakers
- Less expenses regarding fuel consumption, traffic installment cost, etc.

References

1. N. Paul, Applications of continuous functions in topological CAD data, Geodetic Institute of Karlsruhe (GIK), KIT, Englerstraße 7, DE 76131 Karlsruhe. E-mail: norbert.paul@kit.edu
2. P.E. Bradley, N. Paul, Using the relational model to capture topological information of spaces. Comput. J. **53**(1), 69–89 (2010). https://doi.org/10.1093/comjnl/bxn054. https://comjnl.oxfordjournals.org/content/53/1/69.abst
3. A. Vella, A fundamentally topological perspective on graph theory, Ph.D. thesis, University of Waterloo (2005). https://hdl.handle.net/10012/1033
4. D. Boltcheva, D. Canino, S.M. Aceituno, J.-C. Léon, L.D. Floriani, F. Hétroy, An iterative algorithm for homology computation on simplicial shapes, Comput. Aided Des. **43** (11), 1457–1467 (2011). https://doi.org/10.1016/j.cad.2011.08.015. https://www.sciencedirect.com/science/article/pii/S0010448511002144
5. S. Raghothama, Constructive topological representations, in Proceedings of the 2006 ACM symposium on Solid and physical modeling, SPM '06, ACM, New York, NY, USA, 2006, pp. 39–51. https://doi.org/10.1145/1128888.1128894
6. S. Raghothama, V. Shapiro, Topological framework for part families. ASME J. Comput. Inf. Sci. **32**, 8–9 (Mar.) (2003)
7. "Directions for driving from Regal square to Hukumchand square via Gandhi road, Indore, India" Google Maps, accessed March 12, 2020. https://goo.gl/maps/XHrmRibaFj8Aco9UA
8. P. Manish, S. Dipankar, Delay fuel loss and noise pollution during idling of Vehicles at signalized intersection in Agartala city, India. Int. Inst. Sci. Technol. Inst. **2**(6), 2222–1719 (2012)
9. H. Lim, Study of exhaust emissions from idling heavyduty diesel trucks and commercially available idle-reducing devices submitted at U.S. Environmental Protection Agency (2012)
10. CRRI Report (2009) Study on Losses of Petroleum Products at Traffic Intersections due to idling of vehicles at Delhi
11. IRC: 99 – 1988, Tentative guidelines on the provision of speed breakers for control of vehicular speeds on minor roads. The Indian Road Congress (1996)
12. P.A. Weber, Towards a Canadian Standard for the Geometric Design of Speed Humps, Dept. of Civil and Environmental Engg., Carleton University, Ottawa, Canada, April, 1998
13. M. Patel, Prof. (Dr.) P. J. Gundaliya, A study on speed breakers submitted at International Journal of Advance Engineering and Research Development 4(3), March (2017)
14. L.R. Kadiyali, Traffic engineering and transport planning, 8th Edition (Khanna Publishers, 2015), pp 456

Smart Ambulance System Using Fire-Based Cloud Technology and Laser Beam Mechanism

Y. Deepthi, K. Radhika, B. Tejaswi, and C. S. L. Vijaya Durga

Abstract The ambulance services in India are struggling with a dense traffic and the vehicles that are using for ambulance services are not technically build to save lives and guide themselves to reach destiny. We are coming with a amicable solution to get rid of traffic jams issue at the time of ambulance services, though ambulance vehicle has equipment, which supports for the injured person to do first aid, the main motto of the ambulance is to reach hospital in minimum time to save lives of people. We are proposing path-guided mechanism to install laser beams in front of the ambulance vehicle to guide the path of the vehicle and to guide people in front of the vehicle will move away from the laser beam. Most of the time we find the people in front of the ambulance confused to give way to ambulance by installing laser beams will guide people in front of the vehicles. The laser beams that guide only people in front of the vehicle by considering importance of the ambulance to reach on time. We are proposing this system by considering the percentage of people using smart android handsets. Our systems work on android capability reaching Google notifications alerts. With this alert, we are guiding public 500 mts around path of the ambulance vehicle will be guided with a fixed notification of free left to the ambulance by using Fire-based Cloud Messaging Service. We are in a system where implementing any major changes is not so easy but the above-proposed systems can be implemented without any changes in the existing system (Amer et al. in Coalition game for emergency vehicles re-routing in smart cities, 2018).

Y. Deepthi (✉) · K. Radhika · B. Tejaswi · C. S. L. Vijaya Durga
Department of CSE, Institute of Aeronautical Engineering, Dundigal, Hyderabad, Telangana, India
e-mail: deepthisagar7@gmail.com

K. Radhika
e-mail: radhi.rk2007@gmail.com

B. Tejaswi
e-mail: tej.519@gmail.com

C. S. L. Vijaya Durga
e-mail: vijayadurga.csl@gmail.com

C. Kiran Mai et al. (eds.), *Proceedings of International Conference on Advances in Computer Engineering and Communication Systems*, Learning and Analytics in Intelligent Systems 20,
https://doi.org/10.1007/978-981-15-9293-5_38

Keywords Laser beam path-guided mechanism · Ambulance services · Fire-based cloud technology

1 Introduction

Our proposed system is based on effective communication, which is based on "How a receiver feels at time of receipts of a message will influence how the message is interpreted." There are so many examples in daily life that communication plays an important role. When we see person met with accident, there will be a group of people surrounded to see the incident but only few will react for action, only few because of communication and knowing what next to be done on situation, we are trying to communicate people how are intended to act, but fails by lack of communication.

As we proposed these systems by considering so many situations and mentalities of people, everyone will have good intention to give way to ambulances; only thing we see is the communication gap between sender and receiver. To overcome and guide people in front of the vehicle, these systems will help to move people away from guide path of ambulances. Laser beam guide system is very easy to install in vehicle with low cost. It will be effective on traffic jams and to guide people with laser beam who are in front of vehicle, as laser beam falls on other vehicle, there will be effective communication will reach to person that he is in the restricted path of ambulance and will move away from path vehicle which result driver can move in one direction and get better view of path. Every system has some limitations as the effective path of laser guide will only influence people in front of the vehicle. Any way siren of ambulances will be effective to get attention of people but this cannot give path to guide for people. We are upgrading the system in addition to siren in ambulances. When considering speed of vehicle, the system is not so effective, but the main motto is to reach vehicle at the shortest time to destiny. We have concentrated to reduce actual time, which is lost on path during traffic and fixing it. The interesting point here is we see how communication helps to give way to ambulance. We have got inspired by one of real-time situation, where ambulance reached 418 km (Mangalore to Kochi) in 5 h to save life of 5-year-old boy. They were success to reach only by the help of people who left free path to ambulance by providing proper way and communication. This incident proves that effective communication definitely helps to reduce time taken to reach destination.

Another system is using Google feedback system that gives communication to ambulance. This system works on android application that allows Google notification to generate on other vehicles. All android subscribers will get notification, Google send the notification as text message can also possible to send voice note on the mobile. By keeping in mind of iconic siren of ambulance were not going with voice notification from Google. We are coming with simple and easy solution to implement in the existing system.

2 Literature Survey

The authors in [1] have referred CGA-PSO mechanism to route emergency vehicles in smart cities. They have used PSO algorithm to reach destination at high speed. They have used VANET communication systems and particle swarm algorithm. The authors in [2] have used only one camera at 1 km marking. They have implemented this in only one side of the junction. They have designed a system of traffic clearance for emergency vehicle using image processing in MATLAB especially by using a new mode called blue mode. The authors in [3] describe two mechanisms, viz. patient health monitoring system and intelligent traffic control system. In health monitoring system, the patient's vital health issues such as ECG, heart rate and body temperature are supervised. These data will be forwarded to the hospital for analysis. In traffic control part, RF transmitter on the ambulance will interconnect with the RF receiver mounted on the signal post. The authors in [4] have used RFID reader, NSK EDK-125-TTL and PIC16F877A system-on-chip to get the RFID tags fixed to the vehicle. It determines the number of vehicles that passes on a specified direction during a specified time. It also observes the network congestion, and hence the green light time for that path. The authors in [5] have determined traffic signal preemption using wireless mobile-phone transmission. The system is a low cost. The authors in [6] have used a multiobjective maximal covering location model, which solves the issue of observing the best base position for a limited number of vehicles so that the service level objectives are minimized. The authors in [7] have used alarm system that can make the ambulance position open. They have used smart phone IOT on ambulance dashboard. The alarm goes to people who have downloaded the application will get information of ambulance approach. There are no efforts in reducing the traveling time of ambulance. The authors in [8] have used IOT and android application for traffic light controlling and make arrangements at hospital for emergency patient by assigning priorities. The authors in [9] implemented application with different sensors to identify the status of health condition of the patient. The different health variants will be sent to hospital database by using cloud technology. They studied different types of MAC protocols for data delivery. The smart ambulance using IOT consists of several modifications concentrated on ambulance equipment's and controlling traffic signals. As the number of traffic signals is high to control adopting receivers on all signals and maintenance is too generic. The cost of implementation is very high. The authors in [10] used heuristic method [11] and Ant colony optimization methods and compared them with variant data instances using a guided local search by distributing ambulances with minimal destination reaching time.

3 Techniques

3.1 Android Systems by GPRS (Satellite Interface)

The Firebase Cloud Messaging (FCM) is a mechanism to send messages up to 4 KB at no charge. It is a mobile notification application provided by Google that enables third-party server applications to deliver messages to receivers using different mobile platforms. It notifies the receivers with message to move left when the ambulance is approaching. It is implemented in two steps. In the first step, we have to set up firebase on server side; the FCM sends messages to target people using Firebase Admin SDK. In the second step, develop message handling features for the client application. The receivers who are near to ambulance receive the messages. FCM provides efficient communication channel to deliver messages from ambulance server application to user mobile notifications. The Push notifications are text messages that pop on the mobile.

The proposed system mainly concentrates on android users. In complete urban population, 80% of people are using android-based mobile phones, there are some interesting technologies were used, which simplify bike riding. There are new helmets coming with Bluetooth enabled and which converts all text messages to voice that rides no need to see screen for message. Other manufactures using odometer of vehicle connected with mobile and message can be seen on odometer were rider and can read all messages on odometer. India is developing country from so many years, because development and inventions are never ending. We are simplifying our way of living. We see all development in making people comforts not them to save lives and protect them.

In this, we are using android system by satellite interface, India using Catrosate2 satellite camera, which is able to see 3-feet object on land in precise way. There are many communication satellites that give all comforts on mobile usage, we have our own navigation system in India (Mapma India) Google notification, which works and depends on communication satellite and allowing GPRS to identify people and notify them. People who are using android smartphone will be under surveillance of third EYE of Catrosate2. This is the communication system we proposed to save lives of people who are in ambulance. If this system can save 1 min also then the doctor can treat the patient in early time. So that it saves the life of the patient. Nowadays, generally in fraction of seconds, lives may lose in daily life because of delay time of reaching ambulance to hospital due to traffic.

We are calculating the time saved by normal ambulance and ambulance fitted with the proposed system. The time saved in real-time operation with the proposed system gives effectiveness of the proposed systems. The people will be identified by GPRS who are there in 500 m around the path of ambulance. Google supporting system will be connected to ambulance, which allows sending messages to people in front of ambulances as notification, we will give graphical representation on time saved by the proposed system.

3.2 Laser Beam for Path Guide to Ambulance

We are using laser beam for guiding path to ambulance, in front of the vehicle laser path-guided mechanism will be fitted. As vehicle is moving, laser beam on front will guide vehicle path and allow people to know exactly where ambulance is moving. Laser beam falls on vehicle, which restricts path of ambulance, which will allow vehicles to move away from path and the free path is given to ambulance. The laser beam orifices will be made good enough to see on day light in effective way. By laser beam, we can also make images, but the purpose of getting free path will not be achieved, only by using single laser orifices will be possible to get path.

Laser beam mechanism needs 4.13 V, will be powered by battery 12 V of the vehicle, the system is used only when vehicle is in running condition so this system will not affect battery of the vehicle because battery in vehicle will charge by alternator while running. Laser beam orifices depend on the lens, we are using high and precious concave lens will allow beam quality and effective in day light also. Color of laser beam is kept as RED because these colors give more visibility then other colors, length of beam is kept as 500 m that allows people around it to see the exact path allowed for ambulance. Laser beam is used because visual communication will be more effective than voice and reading communication, visual image will get more clear communication in faster way to any group of people. We see lot of miscommunication in traffic jams in India, everyone wants to give way to ambulance but everyone expects other people move away. These all confusions will able to solve by implementing laser beam mechanism.

Laser beam that is fitted to vehicle is so compatible when ambulance is taking a turn beam also moves along with steering. Whenever driver moves steering, wheel beam moves in same direction. It gives more comfort to driver and people in front of the ambulance. Cantilever arms are fitted to steering rack of the vehicle to laser beam mechanism, which guides laser beam whenever driver moves steering wheel, it is directly attached to rack will move beam in same direction. This system is very easy to install and not so expensive and easy to operate, where driver should not make any effort to operate because it is fitted and operated with steering system. Steering angle sensor will operate beam, so there is zero effort of driver that allows him to concentrate only on driving vehicle which is the primary task of driver. Laser beam mechanism is shown in Fig. 1.

As shown in Fig. 1, steering wheel and laser beam coordinate with steering angle sensor fitted in vehicle, which allow steering and beam in same direction, beam direction changes along with steering direction by steering angle sensor which is operated with switch provided in steering column. At signals we see traffic light will be operated by person will not be able to identify the path of vehicle, they always make green on straight line and ambulance move right or left by passing traffic every time. By seeing laser beam direction, traffic person comes to know about the ambulance exact path and makes signal light green (Fig. 2).

Fig. 1 Path-guided laser beam mechanism

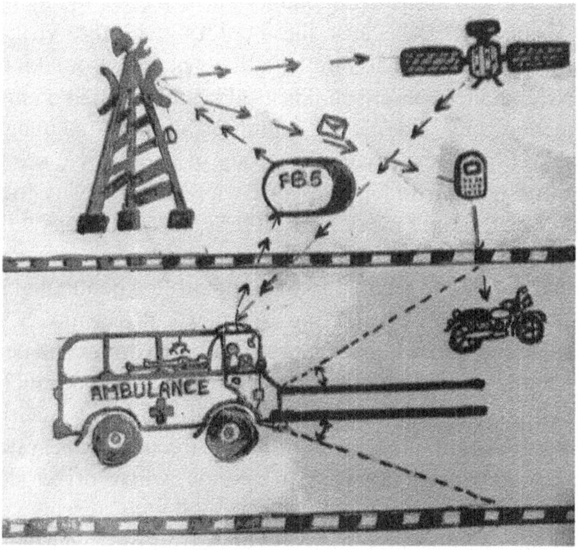

Fig. 2 SMS-based android alert system

4 Flow of Activity

1. The proposed system consists of Google trackers in the ambulance, which allows sending SMS to the people in front of the ambulance. The criterion of getting SMS or notification to android mobile phones from Google through GPRS is called Fire-based Cloud Messaging.
2. The person with android-enabled mobile phones gets a notification of message moves to free left. The people move toward left ambulance get free path to move in heavy traffic condition.

3. The proposed system works on the principle of Google feedback system already in existence. The ambulance is tracked by GPRS (Catrosat2), Satellite name, which gives the exact position of the vehicle and traffic condition ahead.
4. We are using another system because above system cannot reach all, we have put laser beam mechanism, which is mounted in front of the vehicle beam will fall on vehicle in front of ambulance, people will move away from laser beam when it falls on other car, its create impression of the path of ambulance free path is created by vehicle in front of the ambulance.
5. These systems are powered by battery of vehicle and laser beam orifice is kept good enough to see in day light, color of beam is kept as red, which is more visible. These beams are connected with steering rack, which allows move in same direction where driver is moving vehicle by using steering wheel.

Operation of Ambulance

Accuracy is less that is up to 65% only. Ambulance service is started with 15 vehicles in Telangana state now its increased to 315 vehicles, which saved most of lives in city, were people meet with accident every one call 108 vehicle and its easy and its take location of the caller and comes to exact local in no time. Vehicle is increased in number because it has saved lives and reduces the time to attend the person meet with accident. Ambulance service operated by GVK EMRI (Emergency Management and research Institute) was established in 2005. The proposed systems will be updated and will take as social responsibility to make it in operation by getting funds under charitable trust and interested people to involve in system.

5 Results

In Fig. 3, we have calculated distance by time to get efficiency of the proposed system by taking the real-time average taken under consideration. The proposed system is reaching 4 km distance in 20 min, whereas the existing system is taking 25 min for 4 km.

6 Conclusion

We are proposing this system by taking considerations of problems in real life by coming across ambulance operators. These two systems are proposed to overcome ambulance operators (number of interactions—20 operators) issues. The main motto is to reduce traveling time of ambulance by applying easily adoptable technologies. The cost of laser beam mechanism is very low and no additional driver efforts are required to operate. The SMS-based android asset system works with a basic principle of Google feedbacks, which is reliable. The proposed system having more

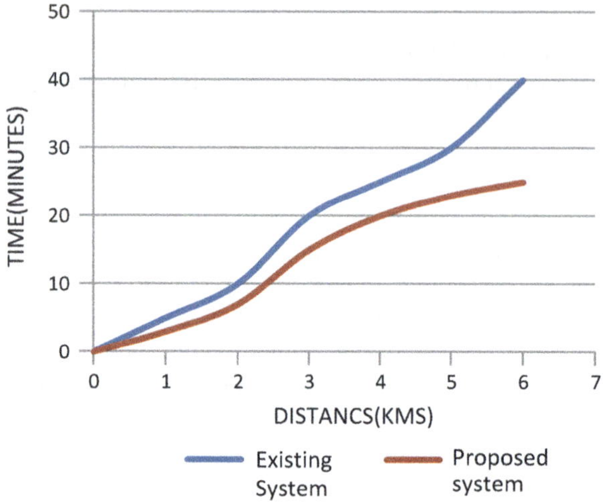

Fig. 3 Distance by time graph

effective tools than the existing systems, which reduces the traveling time of ambulance to destination. In the future, we are planning to install mechanism in the existing ambulance.

References

1. H.M. Amer, A. Hayder, A. Kashoash, A. Kemp, L. Mihaylova, Coalition game for emergency vehicles re-routing in smart cities. IEEE 10thSAM - ieeexplore.ieee.org (2018)
2. Nithyakumar, P. Aswin, Traffic clearance for emergency vehicles using priority mode. Int. Res. J. Eng. Technol. **4**(3) (2017)
3. G. Beri, P. Ganjare, A. Gate, A. Channawar, V. Gaikwad, Intelligent ambulance with traffic control. Int. J. Elect. Electron. Comp Syst. **4**, 43-46 (2016)
4. R. Sundar, S. Hebbar, V. Golla, Implementing intelligent traffic control syst for congestion control ambulance clearance, and stolen vehicle detectioning. IEEE Sens. J. **15**, 1109–1113 (2015)
5. Z. Sahar, Shatnawi, Balqees A Mohammad, Alaa I Kalaf, H.R. Al-zoubi, A wireless mobile-phone approach to traffic signal preemption for faster service of emergency vehicles. Int. J. Comput. Appl. **46**(3), 35–41 (2012). Published by Foundations of Computer Science, New York, USA
6. C. Ara, H. Selim, I. Ozkarahan, A fuzzy multi-objective covering-based vehicle location model for emergency services. Comput. Oper. Res. **34**(3), 705–726 (2007)
7. T. Koba Yashi, F. Kimura, K. Arai, Smart ambulance approach alarm system using smartphone, in *IEEE International Conference on Consumer Electronics* (2019)
8. S. Deshmukh, S.B. Vanjale, IOT based traffic signal control for reducing time delay of an emergency vehicle using GPS, in *4th International Conference on Computing, Communication Control and Automation, ICCUBEA* (2018)
9. O. Udawant, N. Thombare, D. Chauhan, A. Hadke, D. Waghole, Smart ambulance system using IOT, in *International Conference on Big Data, IOT and Data Science* (2017)

10. A.S.V. Praneel, T. Srinivasa Rao, M. Ramakrishna Murty, A survey on Accelerating the classifier Training Using various boosting schemes within casecades of bossted ensembles, in *International Conference with Springer SIST series*, Vol. 169, pp. 809–825 (2019)
11. M. Benabdouallah, C. Bojji, O. EI Yaakoubi, Deployment and redeployment of ambulances using a heuristic method and an ant colony optimization—Case study, in *Third International Conference on Systems of Collaboration (SysCo)*, IEEE (2016)

Hand Gesture Recognition to Implement Virtual Mouse Using Open Source Computer Vision Library: Python

Gummadi Sai Mahitha, Banala Revanth, Gaddam Geetha, and Ramavath Sirisha

Abstract Hand Movement Identification plays a vital part in a human–machine interconnection and to interact with a computer in a most effortless way. As many modern improvements occurring in today's world, such as Natural language processing, Bio-metric Authentication, Face detection, etc., which can be frequently seen in our Tablets, iPads, Computers and smart phones. In the same way, Hand Movement Identification was a contemporary method of Human–Machine Interconnection, that is, the mouse cursor of the system can be controlled just on appearing our figure's before the computer's web camera. These finger gestures are captured and controlled through a Colour Detection technique of webcam. This system allows us to direct the system pointer by using our finger bearing colour caps or tapes and the operations like dragging of files and the left click would be performed by using distinct finger gestures. It also performs the transfer of files among two PC's in a single similar network. This developed system makes use of only a less resolution webcam which acts as a sensor for tracking the user's hands in two dimensions. This system would be developed by using a Library named Open Source Computer Vision (OpenCV) and Python Server Programming. Based upon the idea of virtual mouse implementation, the paper was introduced. In the paper, we will provide a complete description of all the methodologies along with the libraries and packages that are used for the implementation of a Virtual Mouse.

G. Sai Mahitha (✉) · B. Revanth · G. Geetha · R. Sirisha
Department of Computer Science and Engineering, Institute of Aeronautical Engineering, JNTUH, Hyderabad, India
e-mail: saimahitha.29@gmail.com

B. Revanth
e-mail: revanthreddy1677@gmail.com

G. Geetha
e-mail: geethareddy0412@gmail.com

R. Sirisha
e-mail: sirisirisha998@gmail.com

C. Kiran Mai et al. (eds.), *Proceedings of International Conference on Advances in Computer Engineering and Communication Systems*, Learning and Analytics in Intelligent Systems 20,
https://doi.org/10.1007/978-981-15-9293-5_39

Keywords Hand movement identification · Human–machine interconnection · Colour detection · OpenCV · Virtual mouse

1 Introduction

Although there are different types of Mouse's available on the market, the most effective and impressive type of personal communication was through finger gesture, which was a language accepted worldwide. This was so impressive in such a way that people with disabilities would easily understand it. The real-time finger gesture system was developed in this work. Exploratory structure of the pc utilises only a pc's webcam, which has a high-quality recording factor, this captures the hand gestures of the person and performs the actions accordingly like left click, dragging and opening the files, etc. In this proposed system, there is no use of any hardware device, the only thing needed is the webcam and the two fingers of our hand. Two green caps or green taps are placed to the thumb and the index figures of our hand and the hand movements are made infront of the webcam. So that, the webcam captures the movements of the fingers with green caps and works similarly to mouse.

2 Literature Survey

The below survey was done by a few researchers, which mainly focuses on the Implementation of Virtual Mouse using different aspects. It also indicates the programming language used and the key feature and the functionality of Virtual Mouse. The following subsections describe each of them using the references.

2.1 Virtual Mouse Control by Webcam for the Disabled

As technology is growing very rapid, the techniques like image processing, eye tracking, human–machine interaction have a wide range of usage. The Image Processing is mainly used for processing of the images and it can also perform operations on the image in order to get an enhanced image or it can also be helpful in extracting the important and useful data from the image. In the following paper, the authors mainly focused on developing the virtual mouse, which may help all the people, including the disabled one's [1]. The main requirement for this development is the webcam of the system [2, 3]. This model is developed with the help of eye-tracking system. The user performs the eye movements, those movements are captured by the webcam and the processing of the input given by the user can be done [1]. It considers all the captured data as software. This system is mainly for the people who could not utilise the mouse with the help of their hands. This model is actually obtained with the help of the MATLAB code, colours and the eye-tracking system for the control of mouse by using a webcam. The virtual menu was also developed in order to make the usage of this mouse easy.

2.2 Virtual Mouse Control with Hand Gesture Information Extraction and Tracking

The virtual mouse proposed in this system is based on the hand gestures. To implement this mouse, we should mainly have the knowledge in the HCI, that is, human–computer interaction [3]. The information gets extracted from the hand gestures and the tracking of those hand movements can be done properly in order to get the accurate movement of the mouse [4]. This implementation mainly follows the three important steps, which includes.

1. Tracking of Hand Gestures
   The Study of tracking the hand gestures is generated by using a Continuously Adaptive Mean Shift Algorithm, which is simply known as Camshift Algorithm [4].
2. Extracting the features of Hand regions
   This can be done using a bag of visual words.
3. Classifying those extracted features
   All these features are classified with the help of SVM, that is, Support Vector Machines.

The tests are performed in order to check the working of the system and this is successful in the execution.

2.3 Simple and Robust Implementation of a Relative Virtual Mouse Controlled by Head Movements

This following paper mainly introduces a virtual mouse based on the movements of our head. The head movements are captured by a less resolution and a very low-cost web camera. This device is mainly developed for the purpose of non-contact alternative [5]. This is mainly used by the people who are having mobility impairments. This device is implemented and the observations are done in order to check the performance [6]. The results show that the performance of this device is relatively closer to the digital joystick.

2.4 Adaptable Virtual Keyboard and Mouse for People with Special Needs

The authors of this paper mainly concentrate on helping people with the other alternative devices or methods like functional writing or using the device without speech, etc. The Area which covers all these advancements is known as alternative communication [7]. The Major goal of this project is to implement a new solution which allows all the people with the special needs to utilise all the functions of the mouse and the keyboard using the sensors. The sensors are connected to the user's body which is used for the motor control [7]. The developed application consists of a mouse and a keyboard emulator, which is adaptable and configurable. The software catches all the signals that are connected to the Universal Serial Bus, that is, USB using the joystick buttons. The command processor is used to send the mouse and keyboard commands to the windows application. The tests are conducted using a cerebral palsy in order to check the accuracy and the results are satisfactory.

3 Problem Definition

3.1 Process Flow

Figure 1 depicts the workflow of the Virtual Mouse Implementation. It starts from the customer who opens a system to the installing the python software and executing the commands needed for the implementation.

3.2 Modules Implementation

1. Collect useful information.
2. Recheck devices (like webcam) for their proper working, as it is mandatory to use a webcam for the implementation.
3. Collection tape or finger ribbon, which will fit to your fingers.
4. Import all the required packages like NumPy, OpenCV, tkinter, pynput.mouse, Matplotlib, Scikit-learn and pandas.
5. Implement the Open Gesture Operation.
6. Fine Tuning.

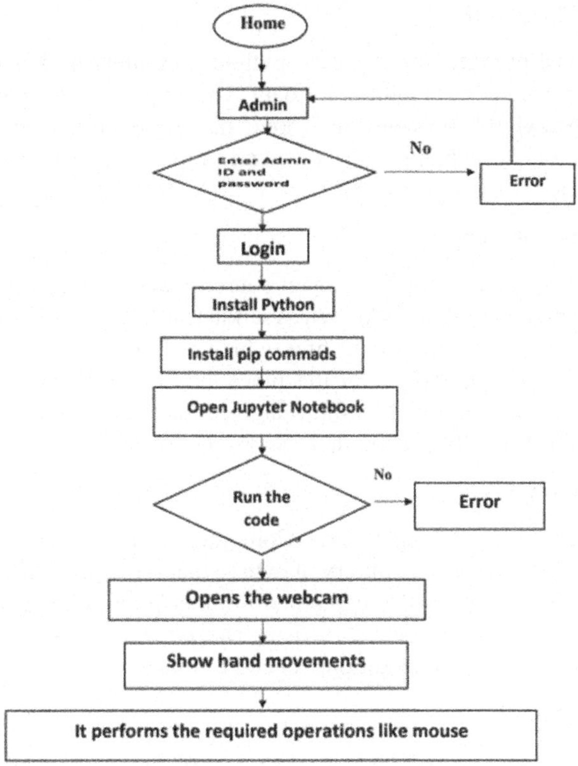

Fig. 1 Process flow of the virtual mouse

3.3 Libraries Description

NumPy:

NumPy is an open-source library in the python programming language. It is generally used for processing large arrays and matrices with multiple dimensions. It has a wide collection of many tools and functions, which are used for operating these arrays.

Pandas:

An open-source library which is used to perform analysis and data manipulation operations using its high-performance data structures is known as pandas. Pandas is built along with the NumPy library as pandas would not be operated without NumPy. Pandas provides the easiest and the useful way for creating, wrangling and manipulating the data. It is also one of the most relevant time series data solution. Python with pandas will be used in a vast range of fields, which includes academics, analytics, statistics, commercial, finance, etc.

Matplotlib:

The plotting library in the python programming language is known as matplotlib. It provides 2D plotting of data. This matplotlib is an extension to the NumPy. This Matplotlib provides an Application programming interface for imbedding all the plots into an application using a graphical user interface like wxpython or tkinter, etc.

Scikit—learn:

The software and the open-source machine learning library in the programming language python is known as scikit-learn. It contains many clustering, regression and the classification algorithms, which includes k—means clustering algorithm, random forests, support vector machines, etc. It also provides various range of supervised and unsupervised learning algorithms. This library operates along with the other scientific libraries, namely SciPy and NumPy.

OpenCV:

Open CV (Open Source Computer Vision Library) is an open-source software machine learning library of python that is mainly aimed at real-time computer vision. This library is also contained in other programming languages like C++ and Java. This library is created by using 'import cv2' in the python code. It is primarily used to perform all operations related to Images.

Pynput:

The package which is used to control and monitor all the movements of the mouse is known a pynput.

Tkinter:

The python languages provide multiple sources to develop the graphical user interface. Among all the sources provided by python, the tkinter is the most common and the most useful method. It creates the GUI very quickly and in a most evasive way.

Jupyter notebook:

This Jupyter Notebook is a free application which allows users to share and create the documents which contains live code, expressions, equations, narrative text and the visualisations. You need to install this Jupyter notebook. Then, open this Jupyter notebook and you need to select the code from the location where code is present. Then compile and execute the code in the kernel.

3.4 Installing Libraries

To install the required libraries, you need the type the following commands in the command prompt.

- pip install numpy
- pip install pandas
- pip install matplotlib
- pip install scikit-learn
- pip install pynput
- pip install tkinter
- open jupyter notebook.

After opening jupyter notebook, you need to select the code and then execute and run the code in the kernel.

3.5 System Implementation

This project is done with a zero cost, that is, the entire project executes only with the help of this code. This Virtual Mouse works for a Lifetime. There is nothing like Lifespan and durability as it is not a hardware device.

We are executing this code in the jupyter Notebook, which is a free application. This is used to create and share the documents. As we need a webcam to complete this task, it enables the webcam and starts executing.

We can put the green caps or green tapes to our hands and we can make the necessary movements in front of our webcam. So that the webcam identifies the green colour and works according to the hand movements (Figs. 2, 3, 4, 5 and 6).

The code used for the implementation of this Virtual Mouse is as follows:

Fig. 2 Code for importing libraries and setting upper and lower bounds

```
In [ ]:  # # IMPORTING LIBRARIES
         import cv2
         import numpy as np

         # Setting lower bound and upper bound

         #LowerBound=np.array([33,80,40])
         #upperBound=np.array([102,255,255])

         '''for green
         lowerBound=np.array([40,80,40])
         upperBound=np.array([65,255,255])

         skin color
         lowerBound=np.array([0,80,40])
         upperBound=np.array([20,255,255])

         yelloish
         lowerBound=np.array([20, 80, 100])
         upperBound=np.array([40,200,300])

         for red
         lowerBound=np.array([170,120,150])
         upperBound=np.array([190,255,255])

         for blue
         lowerBound=np.array([110,150,100])
         upperBound=np.array([120,200,200])
```

Fig. 3 Code for video capturing

```
In [ ]:  # FOR VIDEO CAPTURE
         cam= cv2.VideoCapture(0)
         kernelOpen=np.ones((5,5))
         kernelClose=np.ones((20,20))

         #font=cv2.InitFont(cv2.FONT_HERSHEY_SIMPLEX,2,0.5,0,3,1)
         font = cv2.FONT_HERSHEY_SIMPLEX, 1,(255,255,255)
```

Fig. 4 Code for Importing Libraries for virtual mouse controller, GUI and setting up Global variables

```
In [ ]:  # IMPORTING LIBRARIES for virtual mouse controller and GUI
         import cv2
         import numpy as np
         from pynput.mouse import Button, Controller
         import tkinter as tk
         #import wx
         mouse=Controller()
         #Global variables Setup
         #now that we already have all the libraries lets setup all the variables and objects

         #app=wx.App(False)
         #(sx,sy)=wx.GetDisplaySize()

         root = tk.Tk()
         sx = root.winfo_screenwidth()
         sy = root.winfo_screenheight()
         (camx,camy)=(320,240)


         lowerBound=np.array([33,80,40])
         upperBound=np.array([102,255,255])
         # FOR VIDEO CAPTURE
         cam= cv2.VideoCapture(0)

In [ ]:  kernelOpen=np.ones((5,5))
         kernelClose=np.ones((20,20))
         pinchFlag=0
```

Fig. 5 Code for converting BGR to HSV Color conversion, creating the mask and morphology

```
In [ ]:  while True:
             ret, img=cam.read()
             img=cv2.resize(img,(340,220))

             #convert BGR to HSV
             imgHSV= cv2.cvtColor(img,cv2.COLOR_BGR2HSV)
             # create the Mask
             mask=cv2.inRange(imgHSV,lowerBound,upperBound)
             #morphology
             maskOpen=cv2.morphologyEx(mask,cv2.MORPH_OPEN,kernelOpen)
             maskClose=cv2.morphologyEx(maskOpen,cv2.MORPH_CLOSE,kernelClose)

             maskFinal=maskClose
             conts,h=cv2.findContours(maskFinal.copy(),cv2.RETR_EXTERNAL,cv2.CHAIN_APPROX_NONE)

             if(len(conts)==2):
                 if(pinchFlag==1):
                     pinchFlag=0
                     mouse.release(Button.left)
                 x1,y1,w1,h1=cv2.boundingRect(conts[0])
                 x2,y2,w2,h2=cv2.boundingRect(conts[1])
                 cv2.rectangle(img,(x1,y1),(x1+w1,y1+h1),(255,0,0),2)
                 cv2.rectangle(img,(x2,y2),(x2+w2,y2+h2),(255,0,0),2)
                 cx1=x1+w1//2
                 cy1=y1+h1//2
                 cx2=x2+w2//2
                 cy2=y2+h2//2
```

Fig. 6 Code for morphology

```
        cx=(cx1+cx2)//2
        cy=(cy1+cy2)//2
        cv2.line(img, (cx1,cy1),(cx2,cy2),(255,0,0),2)
        cv2.circle(img, (cx,cy),2,(0,0,255),2)
        mouseLoc=int(sx-(cx*sx/camx)), int(cy*sy/camy)
        mouse.position=mouseLoc
        while mouse.position!=mouseLoc:
            pass
    elif(len(conts)==1):
        x,y,w,h=cv2.boundingRect(conts[0])
        if(pinchFlag==0):
            pinchFlag=1
            mouse.press(Button.left)
        cv2.rectangle(img,(x,y),(x+w,y+h),(255,0,0),2)
        cx=x+w//2
        cy=y+h//2
        cv2.circle(img,(cx,cy),(w+h)//4,(0,0,255),2)
        mouseLoc=int(sx-(cx*sx/camx)), int(cy*sy/camy)
        mouse.position=mouseLoc
        while mouse.position!=mouseLoc:
            pass
    cv2.imshow("cam",img)
    cv2.waitKey(2)
```

4 Results

4.1 *Input*

- There is no specific input given to the system. The only input is our fingers containing tapes. We need to show our fingers in front of webcam and the webcam recognises the colour tapes and performs the actions according to the gestures.
- To get this done, you need to install packages like Open-CV, Matplotlib, NumPy, Pandas, scikit—learn, pynput, tkinter and some of the useful packages and algorithms in machine learning.

4.2 *Output*

- The output and the functionality of the virtual mouse is given in the form of figures below.

Figure 7 shows the greens tapes which are put to our fingers. These fingers are shown in front of the webcam and then, the webcam recognises the green colour and performs all the operations accordingly as our normal physical mouse.

In Fig. 8, the webcam identifies the green colour tapes to our fingers. After identifying, it measures the borders, the distance between two fingers and also identifies the diagonal distance in order to make the necessary movements based on our hand gestures.

In Fig. 9, we have selected the file in the background and based on our hand gestures, the operations on this file are performed.

We can observe that the dragging of file from one location to another location is performed as shown in Fig. 10.

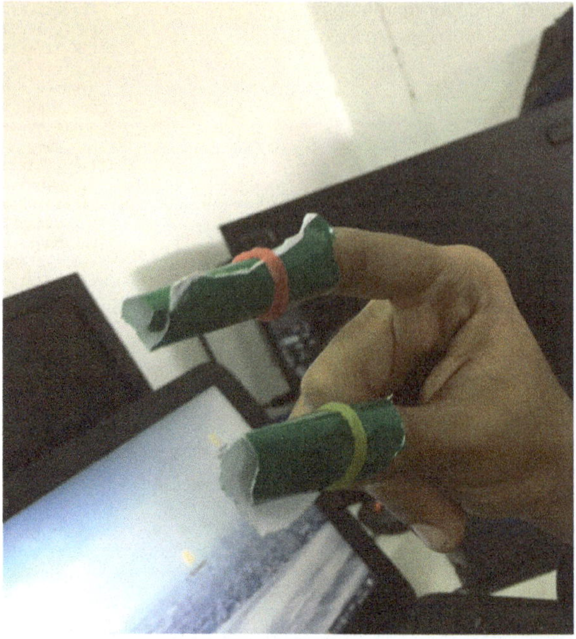

Fig. 7 Fingers bearing green colour caps

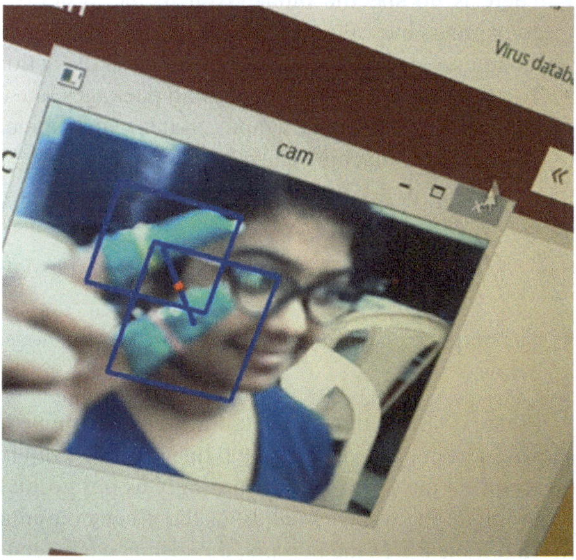

Fig. 8 Calculating the distance between the fingers

Fig. 9 File selection

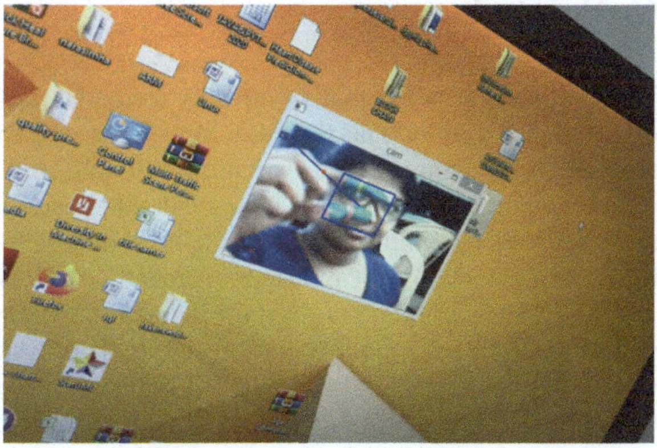

Fig. 10 File movement

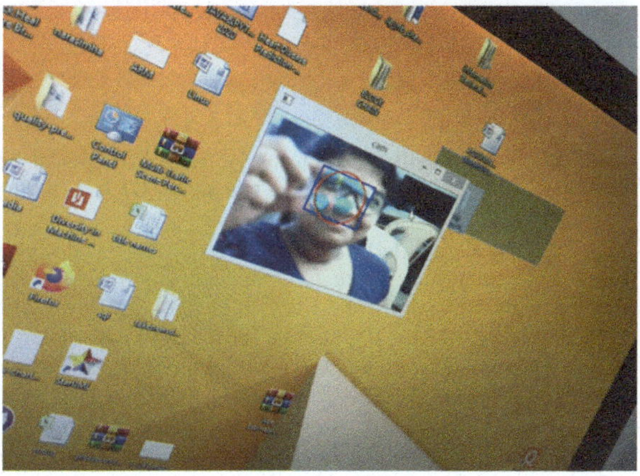

5 Future Scope

The development of techniques like this is growing rapidly. In the future, this virtual mouse is also enhanced and the whole new version of the virtual mouse is developed. The virtual mouse works only by the simple gestures, without performing some of the packages and the code in python and other high-level languages, it is also possible to develop a virtual keyboard and also to perform a virtual monitor operations which works similar to touch screen. In the future enhancements, the virtual keyboard and the virtual mouse are developed in such a way that, there is no need to use or touch our real mouse or keyboard. In the future, the size of the pc also reduces a lot in such a way that the entire pc fits in our hand.

6 Conclusion

This proposed system mainly works by using the packages of computer vision like OpenCV and NumPy. This OpenCV Module forms the masks which will differentiate the colours by using a colour variation technique and identifies a green colour caps which are put to our hands. By recognising the green caps and the movements of the fingers, the functionality of the mouse accordingly. This proposed model is helpful in the implementation of mouse movement by utilising some of the packages like 'mouse' which would be utilised for the mouse movement by using the coordinates which are linked for identifying the colour. This system not only provides ease of use for the systems, but also provides for many different applications. And also, the OpenCV was serving all the customers with distinct accessed forms and a whole lot of different models that would make the life much easier.

References

1. R.S. Baştuğ, B. Yeşilkaya, M. Unay, A. Akan, Virtual mouse control by webcam for the disabled, in *Medical Technologies National Congress (TIPTEKNO), Magusa*, pp. 1–4 (2018). https://doi.org/10.1109/TIPTEKNO.2018.8597147
2. X. Xue, W. Zhong, L. Ye, Q. Zhang, The simulated mouse method based on dynamic hand gesture recognition, in *8th International Congress on Image and Signal Processing (CISP)*, Shenyang (2015), pp. 1494–1498. https://doi.org/10.1109/CISP.2015.740812
3. S. Thakur, R. Mehra, B. Prakash, Vision based computer mouse control using hand gestures, in *International Conference on Soft Computing Techniques and Implementations (ICSCTI)*, Faridabad (2015), pp. 85–89. https://doi.org/10.1109/ICSCTI.2015.7489570
4. R.Ö. Doğan, H. Doğan, C. Köse, Virtual mouse control with hand gesture information extraction and tracking, in *23nd Signal Processing and Communications Applications Conference (SIU)*, Malatya (2015), pp. 1893–1896. https://doi.org/10.1109/SIU.2015.7130228
5. T. Palleja et al., Simple and robust implementation of a relative virtual mouse controlled by head movements, in *Conference on Human System Interactions* Krakow (2008), pp. 221–224. https://doi.org/10.1109/HSI.2008.4581437
6. A. Mhetar, B.K. Sriroop, A.G.S. Kavya, R. Nayak, R. Javali, K.V. Suma, Virtual mouse, in *International Conference on Circuits, Communication, Control and Computing*, Bangalore (2014), pp. 69–72. https://doi.org/:10.1109/CIMCA.2014.7057759
7. A. Henzen, P. Nohama, Adaptable virtual keyboard and mouse for people with special needs, in *Future Technologies Conference (FTC)*, San Francisco, CA (2016), pp. 1357–1360. https://doi.org/10.1109/FTC.2016.7821782

Image Fusion Using LUKR in Multi-Modal Authentication

P. E. S. N. Krishna Prasad, Pavan Kumar Kalluri, Suresh Yadlapati, and Sagar Yeruva

Abstract Now-a-days data are available in different forms and obtainable from multiple sources. Even though more number of algorithms are available but it is vulnerable to the hackers. Hence it is required to improve the security levels of the algorithm so that data security can be increased. This kind of security is expected more in banking, railway, hospital and like more number of areas. Most of the cases, security is available at a single level in the existing algorithms and at the same time, it is important how to extract the features. In this paper, we have presented how security can be implemented in multi-level and features can be extracted in Lower-Upper (LU) decomposition and Singular value decomposition (SVD). For security purposes, extracted features can be decrypted by using Khatri-Rao Product in both the cases. When comparing the results LU decomposition with Khatri-Rao (LUKR) gives better results than Singular value decomposition with Khatri-Rao (SVDKR).

Keywords Lower-upper decomposition · Singular value decomposition · Khatri-rao product · Kronecker product

P. E. S. N. Krishna Prasad (✉)
SV College of Engineering, Tirupati, India
e-mail: surya125@gmail.com

P. K. Kalluri · S. Yadlapati
PVP Siddhartha Institute of Technology, Kanuru, Vijayawada, India
e-mail: pavankalluri@pvpsiddhartha.ac.in

S. Yadlapati
e-mail: sureshyadlapati@gmail.com

S. Yeruva
VNR Vignana Jyothi Institute of Engineering and Technology, Hyderabad, India
e-mail: sagar_y@vnrvjiet.in

C. Kiran Mai et al. (eds.), *Proceedings of International Conference on Advances in Computer Engineering and Communication Systems*, Learning and Analytics in Intelligent Systems 20,
https://doi.org/10.1007/978-981-15-9293-5_40

1 Introduction

Biometric [1–3] plays the main role to identify or verify an individual. For these identification human traits like eyes, face, fingers like others play a major role. Most of the time research can be done by taking a single trait. In the case of unknown person identification, traits are compared with the existing database traits. In this process of identification, the following steps are followed.

In the *first step*, perform the pre-process. The goal of pre-processing is to improve the given input image for extracting the main traits from the image. In the *second step,* features can be extracted from the pre-processing image.

In this paper, Lower-Upper (LU) decomposition and Singular value decomposition (SVD) methods used for feature extraction. Once features are extracted, decoding can be obtained for security purposes. For this decoding Khatri-Rao product, which is an application of can be used. After performing feature extraction matching can be done. For this matching Mean Square Error (MSE) was used. At the end, decision can be taken based on the distance between training and testing by comparing with the threshold value.

1.1 Lower–Upper (LU) Decomposition

In Mathematics, any given matrix can be factorised into Lower triangular matrix (L) and Upper triangular matrix (U). In the Lower Matrix (L) above the diagonal elements are zeros. Similarly in the Upper triangular matrix (U) below the diagonal elements are zeros. Let us consider the matrix Z, its LU decomposition [4–6] is

$$\mathrm{Z} = \mathrm{LU} \tag{1}$$

i.e.

$$\begin{bmatrix} z_{11} & z_{12} & z_{13} \\ z_{21} & z_{22} & z_{23} \\ z_{31} & z_{32} & z_{33} \end{bmatrix} = \begin{bmatrix} l_{11} & 0 & 0 \\ l_{21} & l_{22} & 0 \\ l_{31} & l_{32} & l_{33} \end{bmatrix} \begin{bmatrix} u_{11} & u_{12} & u_{13} \\ 0 & u_{22} & u_{23} \\ 0 & 0 & u_{33} \end{bmatrix} \tag{2}$$

1.2 Singular Value Decomposition (SVD)

The singular value decomposition [7–9] of a matrix Z is the product of three matrices

$$Z = UDV^T \tag{3}$$

where U and V are orthonormal matrices and D is the Diagonal matrix with positive values. Calculating the SVD of a matrix is finding the Eigen values and eigenvectors of ZZ^T and Z^TZ. The columns of matrix V are formed by the eigenvectors of Z^TZ and the columns of matrix U are formed by ZZ^T. These Eigen values represent the features of the image.

1.3 Khatri-Rao Product

Khatri-Rao product [10–13] is a variant of the kronecker product. It is a matching column wise kronecker product. Let $K \in R^{A \times B} andL \in R^{C \times B}$ are two matrices with the same size than their Khatri-Rao product is defined as $K \nu L$. The size of the resultant matrix is (AC) × B is defined by

$$K \nu L = \left[k_1 \otimes l_1 k_2 \otimes l_2 \ldots k_{b\otimes} l_b\right] \quad (4)$$

1.4 Mean Square Error

Let us consider a vector Ai which contains n number of prediction values and also Ai contains the n number of original values. Then the Mean Square Error (MSE) [14, 15] is defined as

$$\text{MSE} = \frac{1}{n} \sum_{n}^{i=1} \left(A_{obs,i} - A_{\text{model},i}\right)^2 \quad (5)$$

2 Methodology

In this paper, mainly two techniques were compared regarding feature extraction. One is Lower-Upper (LU) decomposition and the other is Singular value decomposition. Once features were extracted from the training set these features were fused and form a single image which has all the features of different images. Later fusion was decrypted using the Khatri-Rao product.

The same thing was done on the testing set also. Once values are obtained from training and testing, the threshold is set. Based on the threshold, categorise the image whether it is a known image or an unknown image.

In the first method, LU factorisation technique was used for feature extraction and Khatri-Rao which is an application of the kronecker product was used for decoding and encoding. Here the features are nothing but eyes, nose, mouth and dimension of the face. This decoding is used for security. Mean square error (MSE) technique was used for distance calculation between training and testing images.

In the second method, SVD was used for feature extraction from the given images. The same Khatri-Rao was used for encoding and decoding. Mean square error (MSE) technique was used for distance calculation between training and testing images.

The results are obtained by running the MATLAB code in DELL i3 system with 2 GB RAM and 500 GB Hard disk. Here key sizes would be considered as 8 × 8, 16 × 16 … 64 × 64. In this paper, Yale face dataset and CASIA fingerprint databases are used for identification and recognition.

3 Results

Table 1 shows the results of Feature extraction by using LU factorisation and decoding can be done by using the Khatri-Rao product on the face and finger images.

By observing the above table threshold can be set as d = 0.19. If the values are greater than 0.19 testing images are treated as unknown images. Otherwise testing images are treated as known images. Figure 1 shows the graphical representation of Table 1 results.

Table 2 provides the values of feature extraction by using Singular value decomposition (SVD) [7, 8]. Here for similar and dissimilar images most of the key sizes got the same values. So that cannot conclude anything from the given Table 2. This method cannot be suitable for identification. Figure 2 shows the graphical representation of Table 1 results.

Table 1 Feature extraction using LU factorisation

S.No	Key size	Similar	Dissimilar
1	8 × 8	0.439359	0.180398
2	16 × 16	0.252205	0.122428
3	24 × 24	0.413670	0.182049
4	32 × 32	0.271018	0.135972
5	40 × 40	0.220332	0.121331
6	48 × 48	0.249772	0.129889
7	56 × 56	0.243794	0.129512
8	64 × 64	0.259174	0.137399

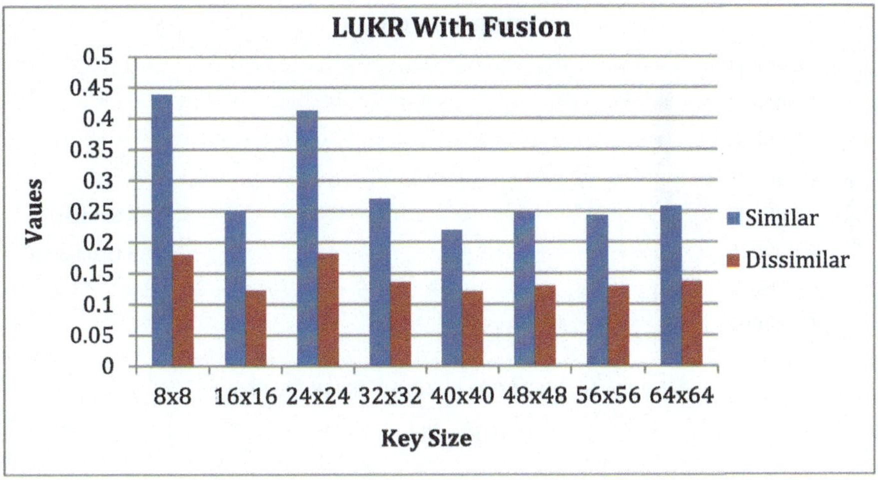

Fig. 1 Graphical representation of LUKR

Table 2 Feature extraction using SVD

S.No	Key size	Similar	Dissimilar
1	8 × 8	0.000001	0.000037
2	16 × 16	0.000001	0.000001
3	24 × 24	0.000000	0.000000
4	32 × 32	0.000000	0.000000
5	40 × 40	0.000000	0.000000
6	48 × 48	0.000000	0.000000
7	56 × 56	0.000000	0.000001
8	64 × 64	0.000000	0.000001

4 Conclusion

Now-a-days security is a more important aspect in many fields. Most of the time for accessing the security system single level authentication is used everywhere like a pin number, pattern, Irish, voice and face. But most of the time a single level is more vulnerable to the hackers. For that reason, this paper presents multi-level authentication using face and finger. In this paper, LU factorisation and SVD techniques can be used for feature extraction and Khatri-Rao products can be used for decoding the features. But by observing the results LU factorisation is more appropriate when comparing the results with SVD for feature extraction.

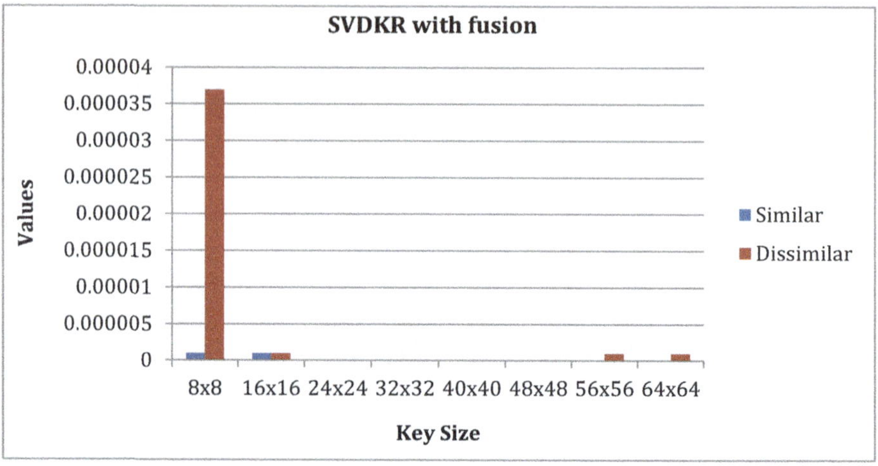

Fig. 2 Graphical representation of SVDKR

References

1. A.K. Jain, A. Ross, S. Pankanti, Biometrics: a tool for information security. IEEE Trans. Inf. Forens. Secur. **1**(2), 125–144 (2006)
2. K. Jain, A. Ross, S. Prabhakar, An introduction to biometric recognition. IEEE Trans. Circuits Syst. Video Technol. Special Issue Image Video-Based Biomet. **14**(1), Jan, 4–20 (2004)
3. Z. Korotkaya, Biometric person authentication: Odor, 1–6
4. P. Bientinesi, J.A. Gunnels, M.E. Myers, E.S. Quintana-Ort´ı, R.A. van de Geijn, The science of deriving dense linear algebra algorithms. ACM Trans. Math. Soft. **31**(1) (March), 1–26 (2005)
5. J.A. Gunnels, G.M. Henry, R.A. van de Geijn, A family of high-performance matrix multiplication algorithms, in *Computational Science - ICCS 2001, Part I*, eds. by V.N. Alexandrov, J.J. Dongarra, B.A. Juliano, R.S. Renner, C.K. Tan. Lecture Notes in Computer Science 2073. (Springer-Verlag, 2001), pp. 51–60
6. Enrique s. Quintana-ort´i and robert a. Van de geijn, Updating an LU factorization with pivoting. ACM Trans. Math. Softw. **35**(2), July, 1–16 (2008)
7. R. Sadek, SVD Based image processing applications: state of the art, contributions and research challenges. Int. J. Adv. Comput. Sci. Appl. IJACSA **3** (2012). https://doi.org/10.14569/IJACSA.2012.030703
8. S. Brunton, J. Kutz, Chapter 1: singular value decomposition (2019)
9. R.A. Sadek, SVD Based image processing applications: state of the art, contributions and research challenges. Int. J. Adv. Comput. Sci. Appl. **3**(7) (2012)
10. S. Liu, O.T.Z Trenkler, K.-R. Hadamard, Kronecker and other matrix products. Int. J. Inf. Syst. Sci. (2008)
11. A. Ploymukda, P. Chansangiam, Khatri-Rao products for operator matrices acting on the direct sum of hilbert spaces. J. Math., Article ID 8301709, 7 pages (2016)
12. Zeyad al zhour and adem kilicman, Matrix equalities and inequalities involving khatri-rao and tracy-singh sums. J. Inequal. Pure Appl. Math. **7**(1), Article 34 (2006)
13. H. Yang, G. He, Some properties of matrix product and its applications in nonnegative tensor decomposition. J. Inf. Comput. Sci. **3**(4), 269–280 (2008)
14. J. Yu, J. Amores, N. Sebe, Q. Tian, A new study on distance metrics as similarity measurement. Proc. ICME. 533–536 (2006). https://doi.org/10.1109/ICME.2006.262443
15. X. Gu, P. Angelov, D. Kangin, J. Principe, A new type of distance metric and its use for clustering. Evolving Syst. **8** (2017). https://doi.org/10.1007/s12530-017-9195-7

A Review on Comparative Analysis of Add-Shift Multiplier and Array Multiplier Performance Parameters

P. Kishore, P. Sravanthi, G. Deepthi, and A. Rashmitha

Abstract As in today's era, the speed of the digital systems can be estimated by the performance of the multiplier. This paper aimed to came up with a few novel techniques to estimate the speed and performance of the multipliers that are most effectively implemented to design our real-time digital filter applications. In most of the multipliers, the Add and Shift multiplication technique is used, because of its simplest architecture. Also, another commonly used multiplier is the Array multiplier. Though the FPGA implementation is considered as a most efficient tool, there are limitations for the components to be implemented on the FPGA which considers as a major issue. Thereby, both the multipliers are being designed and the simulation results such as timing analysis and also the FPGA components (i.e. LUTs, Delays, Flip-flops, etc.) that are been needed for each kind of multiplier are estimated using the Verilog HDL (simulation and synthesis). Thus, the effective approach to design best multiplier for a digital system has been concluded upon the comparison of performance parameters.

Keywords MAC · Shift-add technique · Array multipliers · Digital signal processor

P. Kishore · P. Sravanthi (✉) · G. Deepthi · A. Rashmitha
Department of ECE, VNR VJIET, Hyderabad, Telangana, India
e-mail: sravanthiparepalli@gmail.com

P. Kishore
e-mail: kishore_p@vnrvjiet.in

G. Deepthi
e-mail: gopudeepthi79@gmail.com

A. Rashmitha
e-mail: rashmithasweety35@gmail.com

C. Kiran Mai et al. (eds.), *Proceedings of International Conference on Advances in Computer Engineering and Communication Systems*, Learning and Analytics in Intelligent Systems 20,
https://doi.org/10.1007/978-981-15-9293-5_41

1 Introduction

Now-a-days, multipliers are playing a key role in the modern digital signal processing, and in applications like image and video processing. The multipliers are the most fundamental blocks in many complex and high-performance systems [1]. Almost 70% instructions in micro-processors and many DSP algorithms perform addition and multiplication. In the conventional method of multiplication, the partial products are formed and these partial products are summated to form a final product. This leads to large power consumption and also the delay will be a major constraint. As these operations constitute most of the operation time of the system. Hence there is a thrust for acquiring high-speed multipliers as the demand for high-speed and lower power applications are being increased. The main aim is to acquire a high-performance multiplier with low power by replacing the existing shift and add method of multiplication [2, 3]. The major high-speed multipliers used in recent applications are add-shift multiplier technique and array multiplier technique.

1.1 Add-Shift Multiplier

Add-Shift is a basic conventional technique that is most commonly used in many of the applications due to its simplest architecture and high speed. The algorithm involves simple steps of binary addition and shifting operation which yields the final product in the specified clock cycles as shown in Fig. 1. The n-bit multiplication needs n clock cycles in order to acquire the final product [4]. For example, a 4-bit Add-Shift multiplier requires 8-bit product register, 4-bit multiplicand and multiplier registers, and a 4-bit adder. To store the sum of partial products, the product register will be used. Thereby for each shift of the multiplicand to the left, an 8-bit register is needed thereby the contents in the register are shifted to the right.

1.2 Array Multiplier

The array multiplier is used for simultaneous additions of various product terms obtained. To acquire the product terms an array of AND gates is connected before

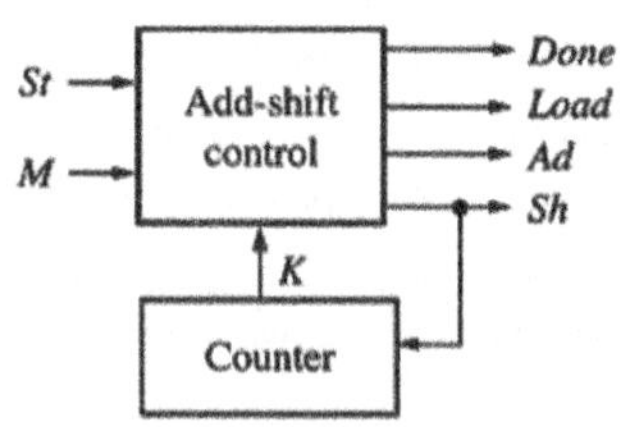

Fig. 1 Add-Shift multiplication technique

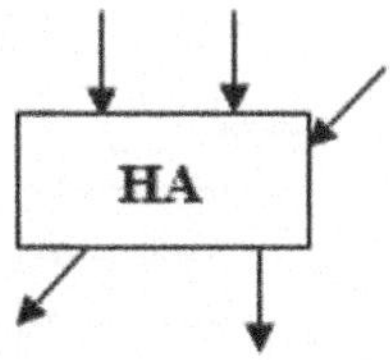

Fig. 2 Basic block of array multiplier

the adder array. Array multiplier is a fast method to multiply two binary numbers as the multiplications can be done at one micro-operation. Figure 2 depicts the half adder which is the basic building block of array multiplier.

To carry out addition operation, a carry propagate adder is used. In this process, N-1 adders are required for N-bit multiplier. The addition may be serial or parallel based on the type of adder used. In high-performance multipliers, the carry propagate adders can be replaced with carry save adders or carry skip adders. Equations 1–4 gives the multiplication procedure of 2-bit array multiplier.

$$P(0) = A(0)B(0) \tag{1}$$

$$P(1) = A(1)B(0) + B(1)A(0) \tag{2}$$

$$P(2) = A(1)B(1) + C(1) \tag{3}$$

$$P(3) = C(2) \tag{4}$$

Here, $P(0)$, $P(1)$, $P(2)$, $P(3)$ are the partial products, $C(1)$ and $C(2)$ are the carry's generated during the addition of the $P(1)$ and $P(2)$ terms.

2 Proposed Algorithms

2.1 Add-Shift Algorithm

The binary multiplication significantly requires adding and shifting technique. In shift and add, the partial product is added immediately when it is obtained. It will eliminate the addition of more than two numbers at a time. In 4-bit multiplication, two 4-bit registers are required for multiplicand and multipliers and an 8-bit register for storing the final product. The product register is used to store the partial products which can also be called as the accumulator. Unlike a conventional multiplication, the product register is shifted to right instead of shifting left after every addition function. Figure 3 shows the block diagram of 4-bit add-shift multiplier.

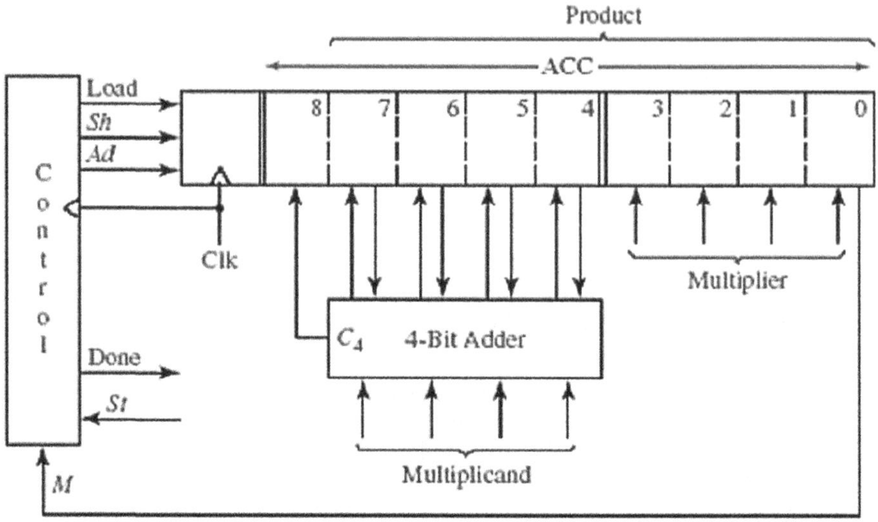

Fig. 3 Block diagram of Add-Shift Multiplier

As represented in the block diagram shown in Fig. 3, the addition process initiated by accepting multiplicand and multiplier bits from the accumulator and multiplicand register. The sum and carry bits of the addition will be given to accumulator and if the add signal (Ad) is active then the adder starts adding by accepting the inputs and the result is stored in the accumulator when the next rising edge of the clock. Then, the multiplicand will be added to the result of adder which is stored in the accumulator. When the multiplicand is added to the accumulator, a carry (C4) will be generated and generally, this is the left-most bit of the product register.

If the control circuit after the signal (St = 1), the proper sequence of add and shift is carried out. When the current multiplier bit is M = 1, then the multiplicand is added to the product register and then the shifting is carried to the right and if M = 0, the addition operation is skipped and only the right shift is performed by the circuit. The shift signal (Sh) will shift the contents in the product register to shift a bit right for every rising edge of clock. The operation of the Add-Shift Multiplier is briefly represented by a flow chart shown in Fig. 4. The signals St, M, Sh, Load and K are clearly determined by the operating conditions that depend on either binary 1 or binary 0. Thereby performing the specific operation at every time step of the rising edge of the clock signal. Also, the control circuit is to be designed properly in order to perform the correct order of operation of adding and shifting of the multiplicand and product, respectively.

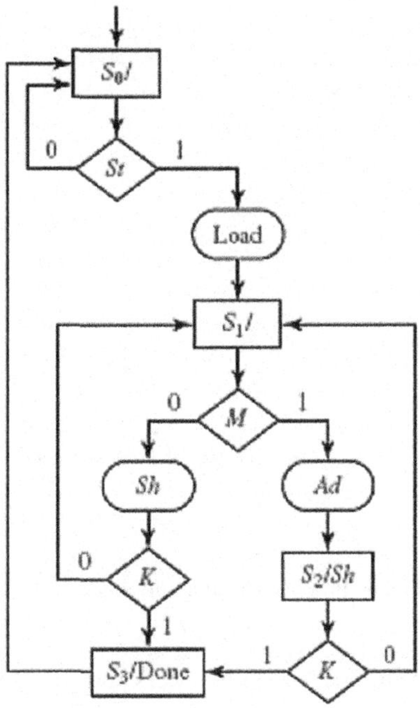

Fig. 4 Flow chart of add-shift multiplier

2.2 *Array Multiplier Algorithm*

The Array multipliers are the fastest and familiar multipliers used over a wide range of applications [5, 6]. The array multiplier performs the multiplication process in three major steps which involve: (1) Partial products generation, (2) Accumulation of those partial products to finally get them into two rows which are added by using carry save adder. (3) Finally, the last two rows are added effectively by the adder structure. The block diagram of an array multiplier is shown in Figs. 5 and 6.

As the partial products are obtained in a parallel manner which improves the performance of the output. For a mXn array multiplier, it needs mXn AND gates,

Fig. 5. 4X4 array multiplication

				X_3	X_2	X_1	X_0
				Y_3	Y_2	Y_1	Y_0
				X_0Y_3	X_0Y_2	X_0Y_1	X_0Y_0
			X_1Y_3	X_1Y_2	X_1Y_1	X_1Y_0	
		X_2Y_3	X_2Y_2	X_2Y_1	X_2Y_0		
	X_3Y_3	X_3Y_2	X_3Y_1	X_3Y_0			
P7	P6	P5	P4	P3	P2	P1	P0

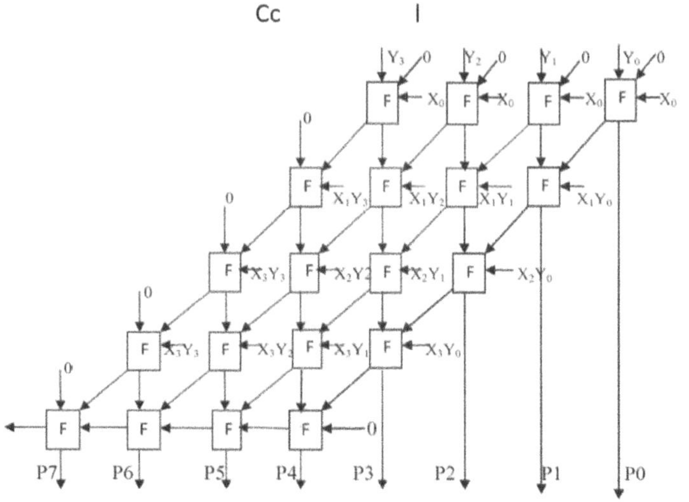

Fig. 6 Proposed array multiplier

n-number of Half Adders and (m-2)Xn full adders and a total of (m-1)Xn adders. For example, a 4 × 4 array multiplier needs 16-AND gates, 4-Half Adders and 8-Full Adders (total of 12 Adders).

3 Results and Discussion

The simulation results of the proposed shift-add multiplier and array multiplier are obtained by using Xilinx ISE Design Suite 14.7 as shown in Figs. 7 and 8.

The area, speed and power of both the multipliers are compared in this section by estimating their device utilisation, time delay and power consumption summaries,

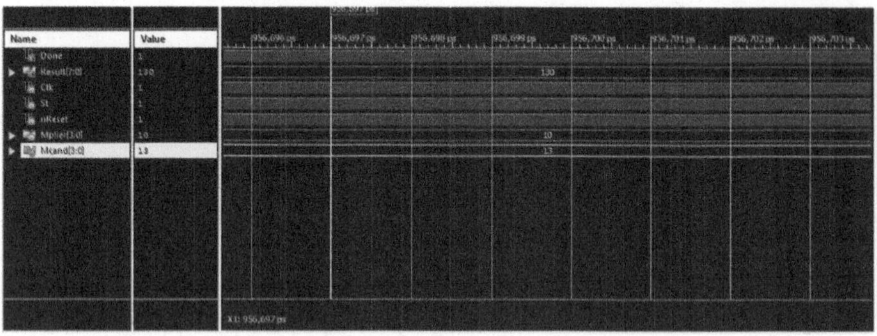

Fig. 7 Simulation results of shift-add multiplier

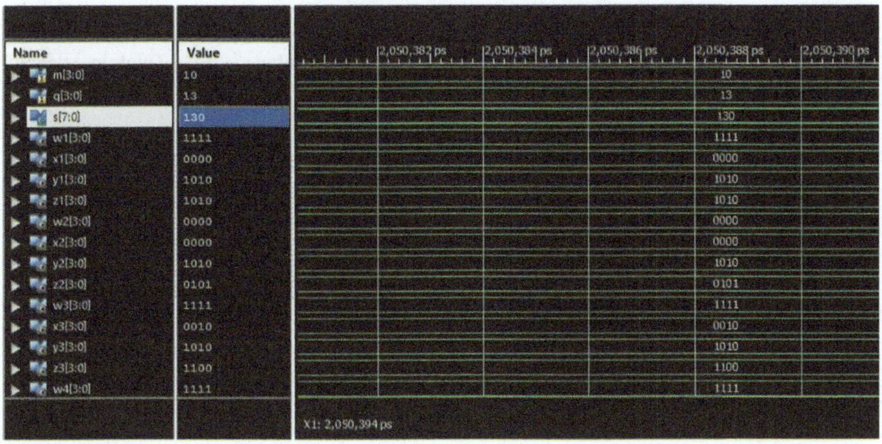

Fig. 8 Simulation results of array multiplier

respectively [7]. The device utilisation and design summary of Add-Shift Multiplier and proposed array multiplier are shown in Tables 1 and 2, respectively.

Tables 3 and 4 depict the timing and power analysis of the designed adders. From the results tabulated in tables it is observed that the proposed Array Multiplier is having better performance characteristics with respect to delay, power and area optimisation.

Table 1 Design summary of add-shift multiplier

Logic distribution	Used	Available	Utilization (%)
Number of occupied LUT slices	22	63,400	0
Number of LUT slices containing related logic	22	22	100
Number of LUT slices containing unrelated logic	0	22	0
Number of bonded IOBs	16	170	9

Table 2 Design summary of array multiplier

Logic distribution	Used	Available	Utilization (%)
Number of occupied LUT slices	15	63,400	0
Number of LUT slices containing related logic	18	22	81
Number of LUT slices containing unrelated logic	0	22	0
Number of bonded IOBs	12	170	14

Table 3 Time delay analysis of multipliers

S. No	Multiplier	Logic delay	Route delay	Total time delay
1	Add-Shift multiplier	0.955 ns (12.9%)	6.413 ns (87.03%)	7.368 ns
2	Array multiplier	0.486 ns (13.6%)	3.099 ns (86.4%)	3.585 ns

Table 4 Power estimation of multipliers

S. No	Multiplier	Power consumption
1	Add-Shift multiplier	54 mW
2	Array multiplier	42 mW

4 Conclusion

This paper depicts the clear concept of two typical multipliers. It can be justified that the parallel multiplier, that is array multiplier is better than a serial multiplier, that is add-shift multiplier in terms of power, area and delay parameters. In array multiplier, the total area utilisation is less than the add-shift multiplier. Hence the system performance will be faster which reduces the time delay [8, 8]. As the multipliers are the most important components of many systems and thereby it is necessary to find a better solution in designing a perfect multiplier in order to maintain a better system performance. Hence always a multiplier with less area and low power should be designed for optimal and better performances of the real-time system applications.

References

1. S.S. Nichat, S.J. Honade, P.V. Ingole, Design of digit serial FIR filter using shift add architecture, in *2014 International Conference on Smart Structures and Systems (ICSSS-2014)* (2014), pp. 90–93
2. L. Guckert, E.E. Swartzlander, Optimized memristor-based multipliers, in *IEEE Transactions on Circuits and Systems–I: Regular Papers* (2016), pp. 1–13
3. A. Mittal, A. Nandi, D. Yadav, Comparative study of 16-order FIR filter design using different multiplication techniques. IET Circuits Devices Syst. **11**(3), 196–200 (2017)
4. A. Mohapatra, A. Bisoyiv, Design of novel multipliers-vedic and shift-add for IEEE 754–2008 single precision floating-point unit in high speed applications, in *2019 IEEE International Symposium on Smart Electronic Systems (iSES) (Formerly iNiS)* (2010), pp. 160–163
5. M. Mottaghi-Dastjerdi, A. Afzali-Kusha, M. Pedram, A low-power low-area multiplier based on shift-and-add architecture. IEEE Trans. Very Large Scale Integration (VLSI) Syst. **17**(2), 302–306 (2009)
6. L. Han, H. Zhang, S.-B. Ko, Decimal floating-point fused multiply-add with redundant internal encodings. IET Comput. Digital Tech. **10**(4), 147–156 (2016)
7. A.A. Wahba, H.A.H. Fahmy, Area efficient and fast combined binary/decimal floating point fused multiply add unit. IEEE Trans. Comput. **2**, 1–14 (2016)

8. J. Choi, B. Yong Kong, I.-C. Park, Retrain-less weight quantization for multiplier less convolutional neural networks, in *IEEE Transactions On Circuits And Systems–I: Regular Papers* (2019), pp. 1–11
9. M. Garrido, Member, IEEE, Fahad Qureshi, Member, IEEE, Oscar Gustafsson, Senior Member, IEEE, Low-complexity multiplierless constant rotators based on combined coefficient selection and shift-and-add implementation (CCSSI). IEEE Trans. Circuits Syst.—I: Regular Papers **61**(7), 2002–2012 (2014)

Efficient Online-Task Scheduling in Distributed Computing Environments Using Least Average Load Variance Algorithm

Maniza Hijab and Avula Damodaram

Abstract For the case of online tasks, one of the most important parameters that need to be optimal is the task response time. To achieve high resource utilisation and achieve a good load balance across grid resources, algorithms should not compromise in terms of response time. Load balancing in itself is a challenge due to the heterogeneity of grid resources. Randomised algorithms effectively improve resource utilisation but can create a load imbalance naturally due to the randomness of its input space. When incorporating a specific level of load across all the grid resources, the response time may become worsened. In the proposed work, to optimise the response time a least average load variance-based algorithm is used. This algorithm is successful in providing a lower bound on the load across grid resources while ensuring a better response time compared to other methods proposed in the literature. The simulation results using standard workload format datasets reveal that the proposed algorithm outperforms existing solutions in terms of improvement of minimum resource utilization by 10–20% with an optimal response time.

Keywords Grid computing · Load balancing · Resource utilisation · Task response time

1 Introduction

Grid computing resources deliver to users across a range of applications. Most of the user tasks in real-time demand an optimal response time. The grid resources with better feedback values for response time attract more user requests, but then also create a load imbalance at some resources. The resource providers use load

M. Hijab (✉)
CSE Department, Muffakham Jah College of Engineering and Technology, Hyderabad, Telangana, India
e-mail: hijabmaniza@gmail.com

A. Damodaram
School of Information Technology, JNTUH, Hyderabad, Telangana, India
e-mail: damodarama@gmail.com

C. Kiran Mai et al. (eds.), *Proceedings of International Conference on Advances in Computer Engineering and Communication Systems*, Learning and Analytics in Intelligent Systems 20,
https://doi.org/10.1007/978-981-15-9293-5_42

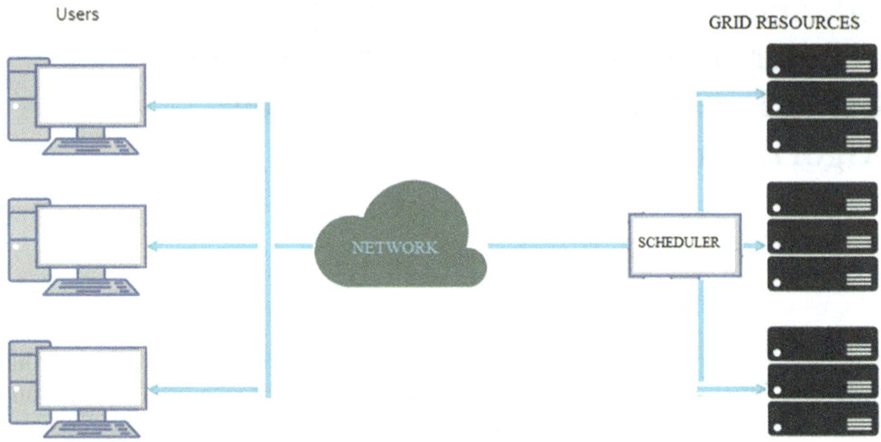

Fig. 1 Example of a grid with users and grid resources

balancing techniques to address this issue by classifying the user requests into varied categories. These user request categories are based on the application response time demanded. Figure 1 depicts the grid scenario with users and resources.

2 Related Work

The issue of unbalanced load between grid resources is addressed for a long time in the literature [1]. It is established that randomised numerical algorithms can often be organised to better exploit modern computer architectures [2]. The more commonly studied algorithms for centralized scheduling are Least Connection, Round Robin, Weighted Round Robin for Flat topology [3]. The researchers came up with many other approaches for balancing load in distributed systems which can be found in the literature and are given in [1–9].

In [1], one such technique proposed is a new methodology, which uses a hierarchical structure. The proposed technique used a task-grained load matching, tried migration of local tasks than global ones which brings a reduction in the communication costs. Load matching is achieved at two levels, initially at Intra-cluster level and subsequently at Intra-grid level. The task processing is carried by modules or nodes inside a grid-cluster, while the role of a cluster administrator is assigned to an imaginary processing unit called 'virtual-node'. The grid administrator executes tasks at all the clusters at the elevated level. Various Task selection criteria are used to bring a change in the output, for example FIFO, Longest_Remaining_Time, shortest_remaining_time, Random and LIFO. The Load transfer decision plays a prominent role in bringing diversified heuristics in terms of the variation between request and response of load and that of communication cost.

In [4], strategy based on clusters load prediction status is used to dynamically balance the load when a new job/task arrives.

In [5], the proposed methodology used cross load matching policy called HLBP, for grid-based resource settings and is a hybrid of static and dynamic load balancing. It propels the load matching with the help of a mediator that has the exclusive responsibility to assemble all the significant information with all resources, such as CPU usage and memory utilisation. For the purpose of load harmonising, the mediator forwards all the task grained data to the scheduler. These tactics withheld the use of a scheduler to manage actions interrelated between generated jobs and resources.

In [7], the load balancing technique displayed is an expanded form of user requirement conscious goal-driven load-matching technique. The key characteristics of the system are its assessment of the grid system parameters, such as scattered fault-tolerant arrangements and practical task relocation. To improve the system in terms of both elasticity and dependability, a reactive duplication design is used in the proposed methodology. The authors found it relevant to bring together the benefits of fault-tolerant scheduling and grid load scheduling.

In [8], a newly introduced, prediction for load balancing method, considers user precedence and tries to predict the future state of a grid resource cluster in terms gridlets held for computation. This enables the movement of the gridlet selection for computation in the right direction and does not need gridlet relocation. The algorithm is highly complex and requires a lot of processing from the backend to monitor the cluster usage.

In [9], the latest load matching technique is to create a list of thresholds of gridlet occupancy. When searching for a feasible resource, apart from being a feasible resource, the selected grid resource may be checked for the current utilisation of task processing. These thresholds do not consider the weights of the grid resources; hence provide a balance in spreading of the gridlets over the grid resources up to a lower level only.

The present work is aimed at using least average load variance-based approach towards implementing a load balance algorithm that connects across heterogeneous grid resources while aiming at maintaining a prescribed level of load across all grid resources. In summary, our algorithm (i) has optimal response time compared to existing algorithms, (ii) does load balancing to a specified level of load across the grid resources and (iii) delivers a good delay performance (tardiness), better than the delay performance of the prior known policies, that is when compared with [9]. The details of the system are given in the next section.

3 Proposed Work

3.1 System Description

The important elements of the proposed system when implementing a joint optimisation of resource utilisation and load-balanced task scheduling in grid systems are as follows.

- Grid Resource: Each grid resource needs to uphold a FIFO queuing mechanism, which provisions the jobs (gridlets) for processing. The queuing span is the amount of gridlets at a grid resource at time t.
- Tasks: The task arrivals are read from a log file in the SWF format which specifies the tasks parameters like the submit time, waiting time, number of CPUs requested and deadline among many other parameters.
- Task Scheduler: Basically upon task arrival, the scheduler samples grid resources and sends the gridlets to the least-occupied resource (i.e. grid resource with the smallest queue).

3.2 Least Average Load Variance Algorithm

Randomized algorithm spread the resource selection uniformly across all grid resources. At the same time, these algorithms are the worst performers when a specific kind of resource needs to be picked in one go. These may fail to balance load across a grid system. The proposed Least Average Load Variance based Load Balancing (LLVBLB) Algorithm (Table 1) is introduced to provide a lower limit to load across grid resources while ensuring a better response. LLVBLB works in two steps, the

Table 1 Algorithm listing

```
Algorithm 1: Least Average Load Variance Based Load Balancing
Input: A queue of Tasks and a set of Resources
Output: Task assignment with enhanced Load Balancing
Data: Resource set R, Task queue G,
for each resource in the grid do
    Find the total Fraction of busy-PEs in the grid
Find the average load across grid
for each task in G do
    for each resource in R do
        if required number of PEs is available then
            Calculate the load difference with average load across grid
    Select the resource with least average load variance
    Update the total Fraction of busy-PEs for the selected resource
    Update the average load across the grid
```

first step—computing the average load across the grid and the second step—finding the resource with the least average load variance, is depicted in next section.

Computing the Average Load across the Grid. Each grid resource when processing user requests dedicates a set of machines composed of a number of Processing Elements (PEs). These Processing Elements are termed as Busy PEs. The fraction of Busy PEs to the total number of PEs available at a resource signifies the load at a resource R. The average load across the grid is computed through the loads at each resource.

Finding the Resource with the Least Average Load Variance. Whenever a grid resource has enough number of free PEs to process a user request (i.e. gridlet), it is scrutinized for a further check. It is done by comparing the load at a given resource to the average load across the grid among all grid resources. The resource with least average load variance is selected for scheduling a task.

To study the tradeoffs between minimum average resource utilisation and balancing the load across the grid resources, a set of algorithms were implemented. The schemes are as follows.

Weighted Randomized Resource Allocation (WRRA): It is the basic random resource selection scheme with a helper to spread load across all resources with different utilisation levels. The output of this scheme is the highest resource utilisation and a varying level of load across the grid.

Weighted Randomized Resource Allocation with one target load level (WRRA1): Random resource selection + helper with only one utilisation level (say 50% load). The output of this scheme is better resource utilisation (but lesser than basic scheme) and a tight level of load across all grid resources.

Improved Load Balanced using Weighted Randomized Round Robin (ILBWRRR) resource selection scheme: This scheme used randomized variant of weighted round robin algorithm. A proportionate weight is assigned to the resource which has more vacated (free) PEs.

Improved Load Balanced using Weighted Round Robin (ILBWRR) resource selection scheme: This scheme used a variant weighted round robin algorithm. A proportionate weight is assigned to the resource which has more vacated (free) PEs.

4 Simulation Setup

The ALEA [10] simulation framework provides basic features to implement scheduling algorithms both queue and schedule based. The parent for ALEA framework is the GridSim [11], a famous gridlet arrangement simulator. The rich-featured framework selected for this study deals and models varied common parameters considered essential in voluntary computing.

The proposed system consists of a grid with heterogeneous resources (R). Each resource has a set of machines (M) which in turn has a set of processing elements (PEs). The number of machines associated with each resource is a variable. Each machine is characterized by the presence of processing elements that vary in their architecture and speeds. Testing has been done using the real-time dataset from Gaia Cluster log whose details are in [12]. For example, it has a resource named Gaia [1–60], with 60 machines, each with 12 processing elements of Intel Xeon L5640 architecture and speeds of 2.26 GHz. Gridlets (tasks) can be scheduled on to these resources as and when launched by user applications. Here implemented the schemes discussed in the previous section using real-time datasets of tasks available in standard workload format that were run on the Gaia grid [12]. The selected real-time dataset spans a period of 12 weeks of user-submitted gridlet information. The dataset has a collection of 51,987 gridlets [9]. The parameters and gathered the results for analysis in our experimental simulations are presented in the next section.

5 Results and Discussion

5.1 *The Tradeoffs Between Resource Utilisation and Load Balance*

Average utilisation is used to measure the improvement in utilisation levels of grid resources. Table 2 shows the performance of LLVBLB algorithm in comparison with P2SDLB [9]. The average utilisation of LLVBLB scheme is 51.73% compared to 42.86% of P2SDLB. The highest utilisation level reached is at 70.44% while the lowest is at 41.80% compared to 30.09% by P2SDLB scheme. LLVBLB manages to get a better lower bound on the lowest utilisation level when compared to others. Hence the minimum and maximum load-balanced scheme is achieved with LLVBLB. The other schemes strike a balance for the joint optimisation of the utilisation and load balance.

Table 2 Comparison of resource utilisation for different schemes

Res/Alg	P2SDLB (%)	WRRA1 (%)	WRRA (%)	ILBWRRR (%)	ILBWRR (%)	LLVBLB (%)
gaia-[120	30.09	41.54	35.49	40.49	39.47	41.80
gaia-[1-	41.84	42.48	42.26	43.49	42.12	42.92
gaia-[80-	42.82	44.70	48.91	44.50	45.50	45.03
gaia-73	44.26	49.49	54.96	49.88	52.17	47.54
gaia-[63-	44.59	53.09	60.65	56.01	56.00	49.84
gaia-[75-	45.31	54.45	63.43	56.08	59.10	51.97
gaia-74	45.93	58.16	64.89	56.28	60.34	64.28
gaia-[61-	48.03	61.81	67.16	57.56	66.43	70.44
Avg. Util	42.86	50.72	54.72	50.54	52.64	51.73

The spread of the load across the grid resources is depicted using other variants of weighted round robin scheme, that is ILBWRRR and ILBWRR algorithms. P2SDLB [9] scheme that falls flat by losing resources in most cases The highest utilisation level reached by ILBWRRR is at 57.56% while the lowest is at 40.49%. The highest utilisation level reached by ILBWRR is at 66.43% while the lowest is at 39.47%.

5.2 The Improvement in Response and Other Important Parameters of Grid Task Scheduling

Figure 2 (left) shows the performance of LLVBLB algorithm for minimum utilisation level achieved. In terms of response time Fig. 2 (right), the LLVBLB algorithm performs better than the P2SDLB [9]. The WRRA1 has the highest response time (right) value as it tries to search for the resource that balances the load as near as possible to the specified value. The proposed LLVBLB algorithm efficiently delivers better value for response time as it uses the least variance method for comparison.

Figure 3 shows the performance of LLVBLB algorithm compared to other variants. In terms of average utilisation, Fig. 3 (left) the LLVBLB algorithm performs better than the P2SDLB [9]. It is because it successfully identifies the potential resources for task scheduling. The WRRA1 has the highest wait time value Fig. 3 (right), as

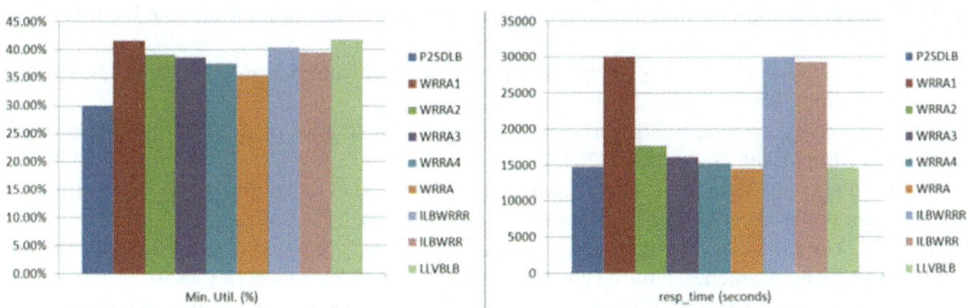

Fig. 2 Comparison of minimum utilisation achieved and response time across the grid

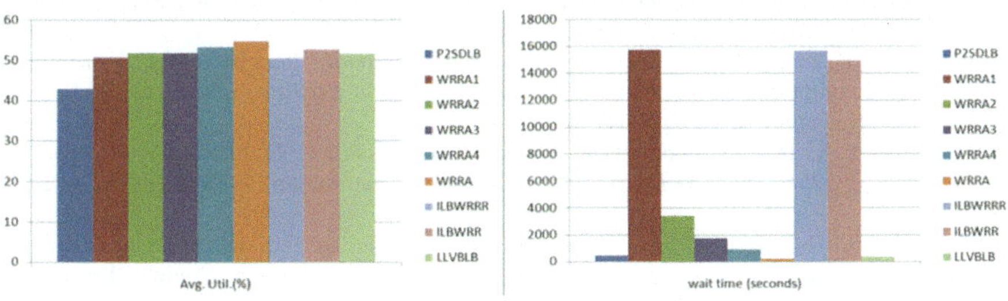

Fig. 3 Comparisons of average utilisation achieved and wait time across the grid

it tries to search for the resource that balances the load as near as possible to the specified value. The proposed LLVBLB algorithm efficiently delivers better value for wait time as it uses the least variance method for comparison.

6 Conclusion

Using randomized numerical algorithms results in better spread the load across the grid resources. This algorithm was successful in providing a lower bound on the load across grid resources while ensuring a better response time compared to other methods proposed in the literature. The simulation results using standard workload format datasets supported the claims and outperformed existing solutions in terms of improvement of minimum resource utilisation by 10–20% with an optimal response time. It is intended in the future work to model user estimation of task runtime and its effect on task wait time.

References

1. B. Yagoubi, Y. Slimani, Dynamic load balancing strategy for grid computing. Trans. Eng. Comput. Technol. **13**, 260–265 (2006)
2. R.M. Karp, An introduction to randomized algorithms. Discr. Appl. Math. **34**(1–3), 165–201 (1991)
3. S. Khan et al., Load balancing in grid computing: taxonomy, trends and opportunities. J. Netw. Comput. Appl. **88**, 99–111 (2017)
4. Y.C. Lee, A.Y. Zomaya, Energy efficient utilization of resources in cloud computing systems. J. Supercomput. **60**(2), 268–280 (2012)
5. K.-Q. Yan, S.-C. Wang, C.-P. Chang, J.S. Lin, A hybrid load balancing policy underlying grid computing environment. Comput. Standards Interf. **29**(2), 161–173 (2007)
6. Y. Li, Y. Yang, M. Ma, L. Zhou, A hybrid load balancing strategy of sequential tasks for grid computing environments. Future Gener. Comput. Syst. **25**(8), 819–828 (2009)
7. J. Balasangameshwara, N. Raju, Performance-driven load balancing with a primary-backup approach for computational grids with low communication cost and replication cost. IEEE Trans. Comput. **62**(5), 990–1003 (2012)
8. D. Thakor, B. Patel, PDLB: an effective prediction-based dynamic load balancing algorithm for clustered heterogeneous computational environment, in P. K. Sa et al. (eds.), *Recent Findings in Intelligent Computing Techniques* (2019), pp. 593–603
9. D. Thakor, B. Patel, P2s_dlb: Pluggable to scheduler dynamic load balancing algorithm for distributed computing environment, in *Emerging Trends in Expert Applications and Security* (Springer, Singapore, 2019), pp. 347–355
10. D. Klusáček, Š. Tóth, G. Podolníková, Complex job scheduling simulations with Alea 4, in *Proceedings of the 9th EAI International Conference on Simulation Tools and Techniques* (2016)
11. R. Buyya, M. Murshed, Gridsim: a toolkit for the modeling and simulation of distribute-dresource management and scheduling for grid computing. Concurr. Comput. Pract. Exp. **14**(13–15), 1175–1220 (2002)

12. ULHPC, The gaia cluster hpc at university of luxemburg. Accessed 26 March 2020. https://hpc.uni.lu/systems/gaia/

A Survey on Real-Time Automated Attendance System

K. Jhansi Lakshmi Bai, K. Sreemae, K. Sairam, B. Praveen Kumar, and K. Saketh

Abstract Regular attendance marking is a standard and important practice for student performance checks in schools and colleges. Traditionally, the faculty work manually to ensure attendance in the classroom, which is time consuming. Many attendance management systems are introduced in the market to track the attendance of the students. Earlier attendance marking methods focused on RFID systems, Raspberry pi punch card systems, swipe card systems, fingerprint recognition. Old practice attendance systems are fairly inefficient today to keep track of student attendance. Overcoming this can be made easier by using face recognition and face detection methodology, it can be rendered in a smarter way. With the implementation of this attendance scheme, it will become impossible for the students to skip classes without the knowledge of the staff. Problems of students unintentionally marked present though being absent and proxies can be resolved.

Keywords Face detection · Face recognition · Viola-jones algorithm · PCA (principal component analysis) · SVD (single value decomposition) based algorithm · Fisher face and eigen face methods

K. J. L. Bai (✉) · K. Sreemae · K. Sairam · B. Praveen Kumar · K. Saketh
Department of CSE, VNR VJIET, Secunderabad, India
e-mail: jhansi_cse@vnrvjiet.in

K. Sreemae
e-mail: sreemae_k16@vnrvjiet.in

K. Sairam
e-mail: sairam_k16@vnrvjiet.in

B. Praveen Kumar
e-mail: praveenkumar98279@gmail.com

K. Saketh
e-mail: sakethkunchakarra@gmail.com

C. Kiran Mai et al. (eds.), *Proceedings of International Conference on Advances in Computer Engineering and Communication Systems*, Learning and Analytics in Intelligent Systems 20,
https://doi.org/10.1007/978-981-15-9293-5_43

1 Introduction

One of the important aspects of the classroom is attendance. For validation of the student attendance record, staff should need a suitable system for consistently approve and maintain the attendance record. In general, there are two types of framework for tracking student attendance, that is Manual Attendance System (MAS) and AAS (Automated Attendance System).

The teachers can encounter difficulties in authorising and holding the record of each student in the classroom all the time, in case of manual attendance system. Although teachers note down the present students in their register, it is an exhaustive task involving a waste of time and effort. It may even appear that a student present be marked absent or proxies are called out. Specifically, it becomes a cumbersome and challenging task to mark not only the attendance physically but the cumulative attendance of each student in a class of high teacher–student ratio. Bio-metric methods provide an authorised person to gain access based on physiological aspects, such as iris, fingerprint. This is a means to heighten the security and prevent unapproved individual access.

Similarly, a system proposed on facial detection appears to garner much technological development. In recent years, improving facial recognition has drawn a great many interests. It is a crucial application of image analysis. Therefore a viable framework that will automatically mark student attendance through face recognition can be executed. Smart attendance system assists faculty in saving time by marking attendance automatically and informing the attendance of academic performance.

There are two phases of automatic attendance system. The first phase is student registration and the second phase is marking attendance in the respective classes. Students enrol with their respective names, roll numbers and facial pictures. The acquired facial images support this model as training data. The model extracts the facial features from the images during training and stores the information used for facial recognition in the student database.

The sequence of steps involved are acquiring student facial images, preprocessing, face detection, face recognition and updating the attendance database or generating attendance report. Preprocessing involves scale factor in which image is scaled since some faces may be larger than some faces in an image as some people sit closer to the camera than others. Facial detection is identifying faces captured in an image, a very popular algorithm involved in face detection is Viola-Jones algorithm which is a box boundary detection. Facial recognition is validating the detected face against the stored facial images. Widely known facial recognition techniques are PCA (Principal component analysis), LBPH (Local Binary Pattern Histogram), SVD (Single value decomposition)-based algorithm, Fisher face, Eigen face methods.

Face recognition-based automatic attendance management system offers several benefits over earlier attendance management systems (Fig. 1).

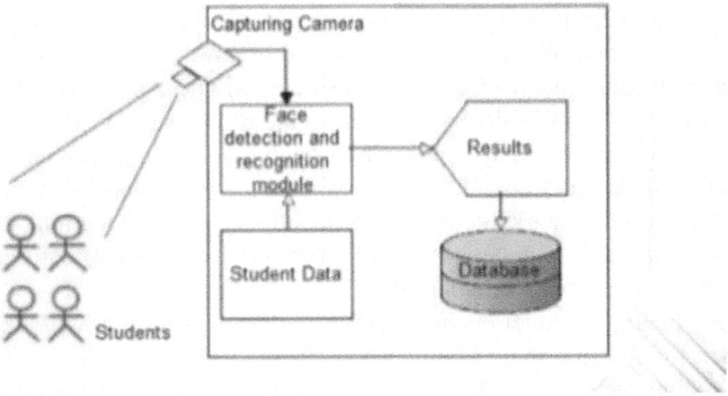

Fig. 1 Schematic diagram of AAS

- Saves both time and effort.
- Eliminate proxies.
- Prevent human error.
- Reduce paperwork.
- Auto-generation of reports.

2 Related Work

This section involves work related to the automatic attendance management system proposed by several authors.

In [1], a high-end camera is installed outside the classroom. Its purpose is to grant the students access inside the classroom. Firstly, the students enrol with their roll no., name, face Id which is stored in the student database. Face detection is done by referring to the 68 landmarks of the face, the Viola-Jones algorithm is applied for box-based face detection. After successful detection of the student's face, the next step is face recognition. For this phase, we implement PCA. It reduces the number of measured parameters for recognition by considering the principal parts of the learning pictures. Every image is represented as a linear weighted eigenvector called Eigen face which encapsulates the essential features of the face. Upon verification, the students are allowed to enter the classroom. Otherwise, he will go back to the enrolment step. The 2nd camera Installed inside the classroom has a clear and visible sight of all the present students. This solves the problem of proxies. After the lecture is completed, the total number of students present in the classroom are awarded attendance.

In [2] RFID reader is installed outside the classroom and a Raspberry Pi camera is installed in the classroom, faculty when entering the class have to swipe the given RFID card and the camera starts recording a video of the class. From the video, images are captured and sent to cloud (AWS), Faces are detected and recognised using AWS

Rekognition. To recognise the faces, students must register earlier with their pictures along with the roll numbers which will be stored in AWS cloud in form of buckets. After recognising, the students who are in the image will be marked as present, others will be marked as absent. SMS will be sent to the mobiles of HOD, parents of the absentees. Faculty and Students can login through the web interface and can check their weekly and monthly attendance and can see the percentage RFID card given to the faculty is used to mark attendance for faculty. For Face detection, Viola Jones algorithm consisting of four steps is used. They are selecting Haar features, Integral Images, Ada-boost and Cascade classifier.

In [3], Raspberry Pi camera is installed inside the classroom. It will take pictures for a certain time-period that is being set by faculty. The pictures are then sent to the Image enhancement module. There Face Count Algorithm is used where the image is converted into grey-scale and scaling is done as some person faces would be larger and some person faces would be smaller depending upon their distance from the camera that is being installed in front of the classroom. After that faces are detected using a detection algorithm which uses a moving window to detect objects. After detection, the facial space is separated from the background image and is compared with the facial spaces that are already being stored in the database, depending upon the matching person is identified and marked as present. The system is designed in such a way that the list of absentees is being generated for any given day. Template matching concept is used for face recognition and detection. OpenCV has the available libraries and SQLite database is used.

In [4], Camera which is connected to Raspberry Pi is installed in the classroom, Images are captured and faces are detected using Haar Cascade classifiers and then the detected faces are recognised by comparing the facial features of the detected image and the features stored in the database. Depending on the match, the attendance is provided to the students who are in the images.

In [5], the necessary drivers for the web camera from MATLAB are installed. The model requires somewhere between 500 and 1000 captures of a person for high accuracy. It can be performed in two ways. The images from the video stream can be captured or each image manually by the web camera. The former approach yields better results in a shorter duration. Viola Jones algorithm is used. Training and testing dataset is split in the ratio of 0.2:0.8. MATLAD offers features to design and create the user interface. Components like buttons and plots are used to get the image frame of the web camera. Invoking callback functions to configure the web camera for capturing & saving the image is performed. Edit boxes and static boxes allow for user inputs and display results to the end user. A registration form needs to be filled out by students which includes name, roll no, dept, password and student facial image. After signing in the portal, the student is provided with several options such as take attendance, view attendance, reset attendance and logout.

Table 1 Results of four different face recognition techniques

S.No	Recognition method	Recognition (%)
1	PCA + SVD + Fisherface	99.5
2	SVD	96.6
3	Fisherface	98.4
4	*PCA*	80.7

In [6] The algorithm proposed is used on training two dataset models for the purpose of recognition. The first model undergoes fisherface for the feature extraction followed by the k-nearest neighbour for classification purposes. The second model undergoes the SVD technique. A subset is generated from the above resulting sets of both models. In the event of obtaining more than one distinct label in the generated subset, then ranking all the labels on the basis of minimum distance is carried out and the lowest-ranked label is chosen. The recognition rate results of four different face recognition techniques are tabulated. It is clearly observed that the combination of PCA, fisherface and SVD yields the best among the four aforementioned techniques (Table 1).

In [7], initially student must register. In the registration process when students 'click images' button, the camera is opened and continuously 60 sample images are taken and then the camera is closed. After taking 60 images, the model would be trained, features extracted from 60 sample images are stored in a face database with roll numbers. When students enter the classroom, the camera will take a video of the class for few minutes, from the video frame images are captured and has been sent to image enhancement module, the output of the image enhancement module is been sent for face detection algorithm, there the faces are detected and for face recognition the features extracted from image is compared against the stored features in the face database one by one, attendance is provided to the roll number for which the features match most. For feature extraction, local binary pattern histogram is used. Features are stored in a face database in form of histogram. It is implemented as a GUI based project.

In [8] After acquiring the photos directly from the camera, the RGB image is converted to HSV image by executed by rgb2hsv command in MATLAB. Each of the HSV features are extracted from the respective image to provide information about the colour of skin. A pixel whose components of S and H must satisfy the following constraints: $0.15 < = S < = 0.9, 0 < = H < = 0.25$ is deemed to be a pixel of the individual. Following the HSV step, the image is again converted back to RGB and the values of B,G,R, respectively must satisfy the conditions: $B > 20$, $G > 40$ and $R > 95$, Max (B,G,R)-Min (B,G,R) and $|R-G|>15$ and $R > G$ and $R > B$. If the pixel meets the conditions, then it is said to be white else it is of black colour. Black spots can create an issue in processing and the command 'imfill' of MATLAB can fill these black spots.

In [9], CCTV cameras are deployed outside the entrance of the classroom for capturing images of the incoming students. CCTV camera offers captured images of high quality as well as different view angles. The images are automatically sent to an android smartphone. The input images are matched with the existing images in the student database by 3D face recognition software technology. In this 3D technique, the received image is cropped, and canonical face matching is applied to obtain the various facial reaction of the particular image. The new canonical face image obtained is matched to award attendance and if the match appears to come out as false, then the image is added as a new student to the existing student database. Attendance is marked by sending the obtained information to the server which updates the attendance database. The authors of the paper have reported the attendance of four students. SMS will be sent to the parent's phone when the student is absent on a specific working day.

In [10], The methodology described in this paper follows the below steps-

- Creation of the face database.
- Live Video Activation
- Face detection
- Face recognition
- Attendance registering.

The enrolled student images in the database are the training dataset of the system. To obtain the region of interest, images are cropped to procure the face of the students. Students when registering have to submit five pictures of themselves, which will be used for training the model. A camera is installed near the blackboard of the classroom which captures the images of the attendee's real time. It is connected to PC and the necessary drivers are properly installed. The live video mode captures the pictures by frames. Once the reading is completed, the frames are sent for face detection. The object detection algorithm Viola-Jones is applied for face detection. It includes Haar features, which is similar to the convolution kernel. Individual feature outputs a value calculated by the difference of the sum of pixels under white rectangle to sum of the pixels under black rectangle. Value of the pixel (x, y) from the integral image is the cumulative sum of the pixels from the (0, 0) position to the (x, y) position (Fig. 2).

Ada Boost is applied for removing redundant features as all the Haar features are not relevant for detection. Thus, Ada boost determines the relevant features Cascading enables all the features to be grouped into categories of each 10 or more features. This is arranged in a hierarchical fashion as identifying the non-facial region is relatively easier than the facial region. The registration module allots the faculty username and password to access the student attendance reports (Fig. 3).

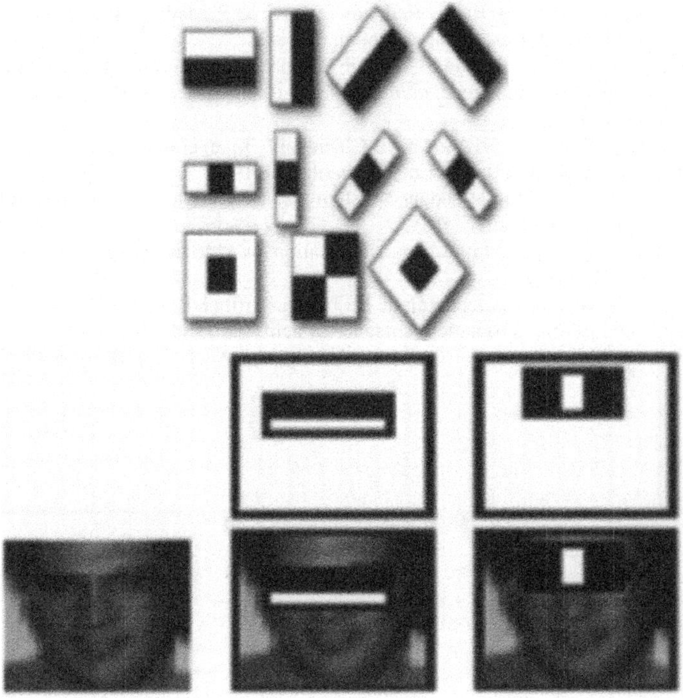

Fig. 2 Haar features

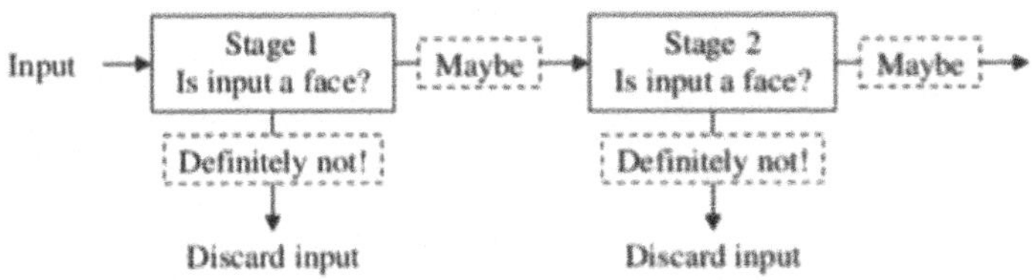

Fig. 3 Cascade of classifiers

References

1. S. Sawhney, K. Kacker, S. Jain, S. Narayan Singh, R. Garg, Real-time smart attendance system using face recognition techniques, in *9th International Conference on Cloud Computing, Data Science & Engineering*, IEEE (2019)
2. D. Narendar Singh, M. Kusuma Sri, K. Mounika, IOT based automated attendance with face recognition system. Int. J. Innov. Technol. Explor. Eng. (2019)
3. S. Amulya, Nisarga, P.S.G.R., S.A. Hariprasad, M.P. Sunil, Face detection and recognition for automatic attendance system. Int. Educ. Res. J. (IERJ) **4** (2018)
4. D. Vishal, Bikkad, S.B. Sonkamble, S.A. Mane, S.S. Sarada, Smart attendance system using face detection on Raspberry pi. Int. J. Innov. Res. Comput. Commun. Eng. (2018)

5. V. Kalyan Polamarasetty, M. Reddy Reddem, D. Ravi, M.S. Madala, Attendance system based on face recognition. Int. Res. J. Eng. Technol. (2018)
6. J. Dhamija, T. Choudhury, P. Kumar, Y. Singh Rathore, An advancement towards efficient Face Recognition using Live video feed, in *International Conference on Computational Intelligence and Networks* (2017)
7. A. Varshney, S. Singh, S. Srivastava, S. Chaudhary, Tanuja, Automated attendance system using face recognition. Int. Res. J. Eng. Technol. **6** (2019)
8. S. Kewalramani, Automatic attendance system by face recognition using machine learning. Int. J. Eng. Sci. Res. Technol. (2018)
9. K. MuthuKalyani, A. VeeruMuthu, Smart application for AMS using face recognition. Comput. Sci. Eng. Int. J. **3** (2013)
10. M. Deeksha, S. Seham, S. Geethanjali, K. Tanuja, Artificial intelligence based face detection and recognition for automatic attendance system. Int. J. Sci. Res. Comput. Sci. Eng. Inf. Technol. **4** (2018)

Vocal Analysis to Predict Suicide Tendency

Police Preethi Reddy, Chalumuru Suresh, Venkata Krishna Rao, K. Sai Chandana, S. Sowkya, and R. Akhila

Abstract Depression and suicide are one of the public health concerns. It has been recognized as a prominent cause of disability and burden worldwide, which is hard to identify. The major indications in an individual regarding suicide and depression are homebound conditions, which include lower life quality, relationships and multiple responsibilities. The growth of suicidal rate is found more among the adolescent, To find this we are using the properties of acoustic speech that has probed into depression in adolescents and finds the suicidal tendency and it also outperforms the state of art techniques by both subjective and objective way where the review starts by building the case for speech for different situations including both the current treatment and methods leading to depression and suicidality. It generally focuses on how common paralinguistic speech characteristics are affected by depression which leads to suicide tendency where we manually extract the emotions of individual based on speech using Support Vector Machine (SVM), and Mel Frequency Cepstral Coefficients (MFCC) classifiers.

Keywords Depression · Acoustic speech · Suicide · SVM · MFCC

1 Introduction

Depression and suicide are one of the prevailing cause of death overall in the world. There is no single way which shows the outcome of suicide tendency so that we can follow that particular analysis to detect suicidal tendency. So to know regarding the suicides and what actually lets a person take such an extreme step has come to know with the recent studies regarding suicides which show that there is an unprecedented increase in the rate of suicides is because of depression, which cannot be identified easily. According to the reports about 15% of the disorder is due to mental or physical issues leading to depression which then may lead to are suicide. So how to identify this suicide tendency? There are many methodologies and the methodology we are

Police Preethi Reddy (✉) · C. Suresh · V. K. Rao · K. S. Chandana · S. Sowkya · R. Akhila
Department of Computer Science, VNRVJIET, Secunderabad, India
e-mail: preethireddypolice11@gmail.com

C. Kiran Mai et al. (eds.), *Proceedings of International Conference on Advances in Computer Engineering and Communication Systems*, Learning and Analytics in Intelligent Systems 20,
https://doi.org/10.1007/978-981-15-9293-5_44

following is identifying the emotion using speech. Human being voice is very flexible and carries a various types of emotions [1]. Emotion in speech consists much about human behaviour. We can understand the intentions of human thoughts whether they are happy, sad or pleasant with the help of human speech. People can easily detect the emotion of another person, but the area of emotion detection using speech through machine learning is a quiet large open area for research [1]. We started our research on emotion in speech by identifying one emotion. Further, we systematically carry out the classification of emotions in speech notes. In analysis of emotion, we start by the data of speech and use the best methodology for analysis and classification of emotions where we select the features which are necessary to differentiate the emotion. We also consider the training data and testing data to construct a model with which we classify the emotions using speech. There are various risk factors which lead to depression and then may be to suicide some of the risk factors are listed below [2].

Risk factors in depression analysis

Basically, risk factor analysis is nothing but the study of causes of depression in human being and the way to solve it [3]. There are three types of risk factors in depression analysis.

Biological and Setting

Biological factors may include genetics, age, gender and culture.

Psychological Factors

Multiple Psychological factors, such as assumption, belief, problem solving, hopelessness, optimism, rumination.

Social Factors

Social factors may include early trauma and loss, marriage and relationship issues, social rejection, avoidance and negative feedback.

2 Literature Review

Speech type and emotions a new computational methodology for detecting depression [4] (STDD), High Accuracy with desirable sensitivity ratio, Acoustic correlates of depression and examined the emotions (positive, negative, neutral) using traditional machine learning classifiers these all are performed to identify the different emotions which are used in identifying the emotional behaviour change in a person.

Not only the traditional classifiers are used to identify the emotions now identification and classification of emotion can be performed by Artificial Intelligence [1] thereby making it a bit easy in identifying the emotions for this the AI used the Natural Processing Language (NLP). This NLP is applied on the twitter data to get a complete analysis of different emotions leading to depression.

To get more accurate data on emotions the same AI is applied on various social media sources like twitter posts [5], twitter feeds from various data streams where the different warning indications of suicide are identified from the posts [6]. The main aim is to identify sudden changes in the online behaviour of an individual user. So that it would be easy to rectify the changed behaviour.

AI is been used to a great extent to study the suicidal behaviour of an individual but with growing technology and advances many methods have been identified to prevent and detect suicide. A study also shows how actually the impact of emotions on intonation patterns (more specifically F0/pitch contours) is addressed and a very traditional method to identify this suicidal tendency is the feasibility study of Hospital-based data of the patient and their family, this study approximately identified around 114 individuals.

This speech emotion is a key for classification addressing different frequencies, energy contour, duration of silence and quality of the speech notes so that they can be classified into a specified category.

3 System Architecture

In the existing system, various methodologies are used to find the suicidal tendency like studying the historical data of a person, hospital data of a person, one to one interaction with the person suffering from the loneliness which may sometimes lead to depression. These methods though did not help to a great extent are useful in finding certain areas which might lead a person to suicide (Fig. 1).

In our proposed system, we are going to use various speech notes consisting of various emotions. At a high level, each speech file from the database is processed and used to build vectors consisting of labels that are either normal or abnormal [7]. Once done with the extraction of features we use the feature selecting algorithms to find the most appropriate results and apply these results to various models [8].

The architecture used is a top-down approach where the user gives some sample speech notes to the system collected from a particular database. The data collected will be optimized to remove any sort of disturbance and noise so that further formulation of data is accurate and easy.

Once optimisation and formulation of data are done it is given to the classifiers like MFCC and SVM which are used in the classification of emotions based on input. Training and testing model is constructed from the output provided and then the mapping of emotions is performed and if the emotion of anxiety goes beyond the threshold value then the output is displayed as depressed and we can conclude that the person might commit suicide.

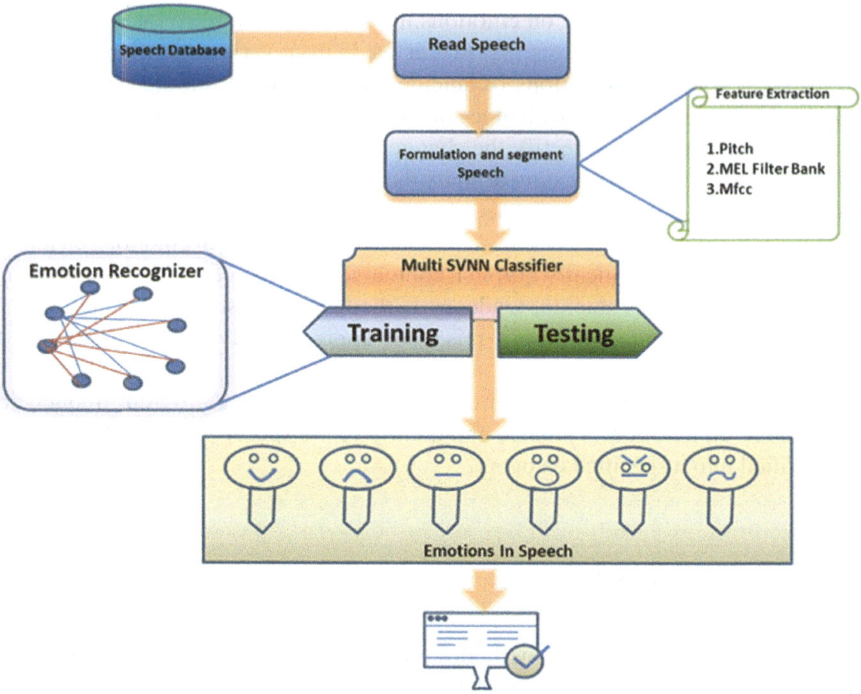

Fig. 1 Suicidal ideation

4 MFCC

Mel Frequency Cepstral Coefficients is a feature extraction technique it is accomplished by changing the speech waveform to a form of parametric representation [9]. The requirement for feature extraction is a sliding window of 40 ms wide to extract audio features.

- **A/D Conversion**: A/D conversion samples the audio clips and digitizes the content, i.e. converting the analogue signal into discrete space.
- **Pre-emphasis**: Pre-emphasis boosts the amount of energy in the high frequencies.
- **Windowing**: Windowing involves the slicing of the audio waveform into sliding frames.
- **Discrete Fourier Transform (DFT)**: We apply DFT to extract information in the frequency domain.
- **Log**: Logarithm is applied to Mel filter bank output. DCT is applied to obtain MFCC.

5 SVM Classifier

It is a supervised machine learning classifier which has two sessions; one is a training session in which we will train to the SVM classifier for a different type of depression stages and normal stage. As there is training session it is called as supervised machine learning algorithm. Second one is the testing session in which we will test multiple samples to validate our proposed work. The kernel function of SVM to use to get a more accurate result (Figs. 2 and 3).

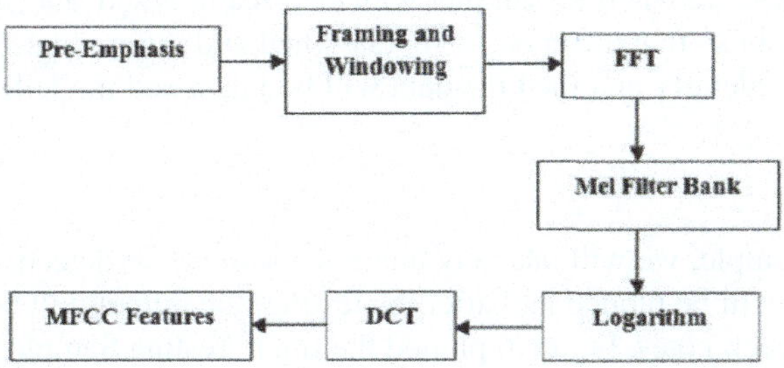

Fig. 2 Genetic Flow of MFCC

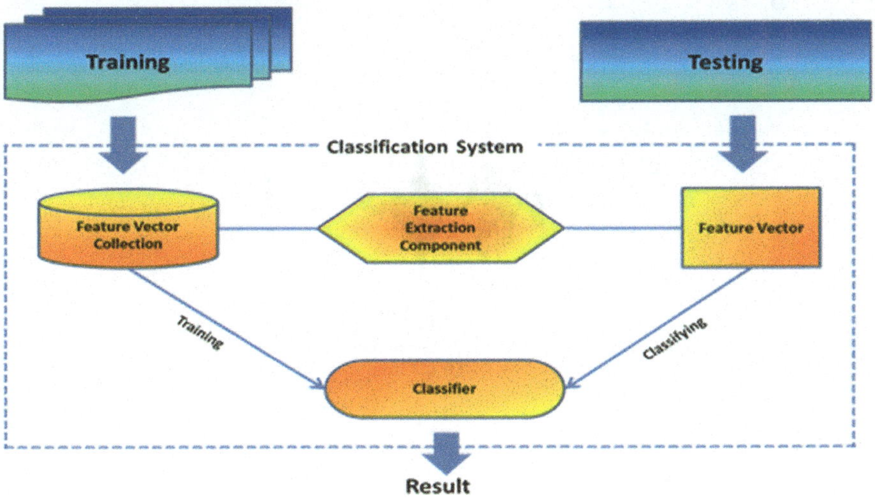

Fig. 3 Classification system using SVM

6 Experiment Analysis

From the figure, we can identify that the speech notes are taken as input and given for processing to the system once the processing is done we apply the extraction function to the data. Once the data is taken as input which is identified in the figure the feature extraction is performed where the speech notes are classified using the classifiers like MFCC and SVM and a training model is developed (Fig. 4).

Training Session

In training session, we will train samples with their features and label them according to the requirement so that it can be easily classified and differentiated which makes the data easy to identify and then the data will be taken and modelled into training data (Fig. 5).

Testing session:

In the testing sample, we will take any unknown sample for detection the category and the sample will be plotted and after the testing, the output will be shown based on the input speech notes. Figure 6 plotted the Input Testing Sample.

Figure 7 shows the final results in Message Box.

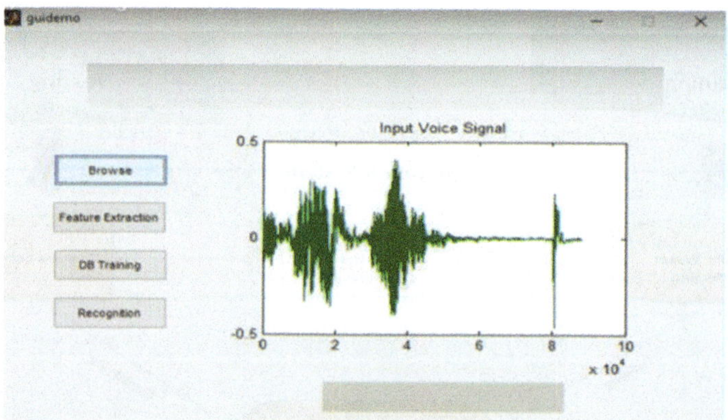

Fig. 4 Input voice signal

Fig. 5 Training data

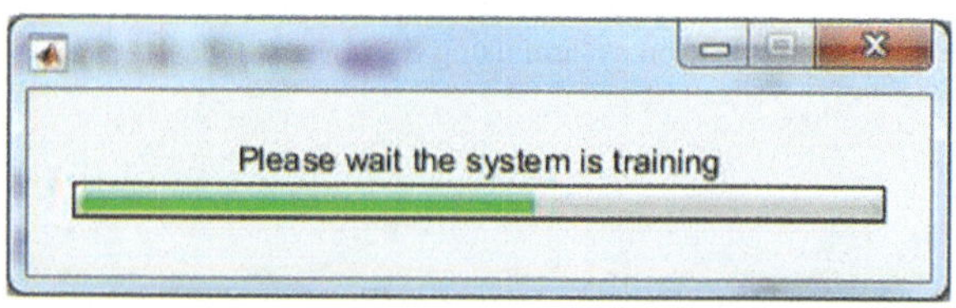

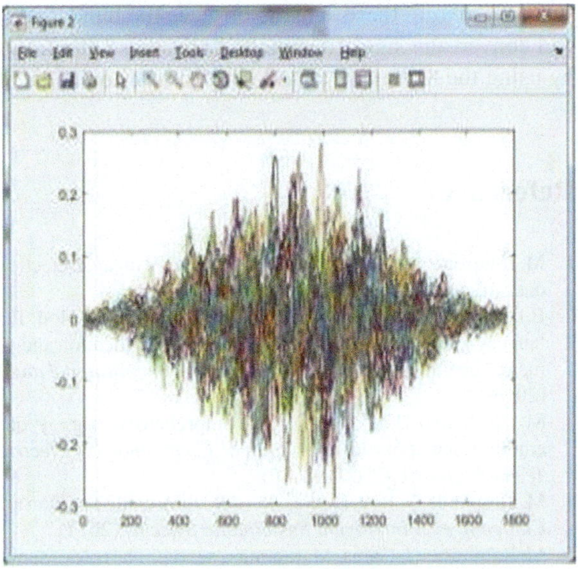

Fig. 6 Testing data plot

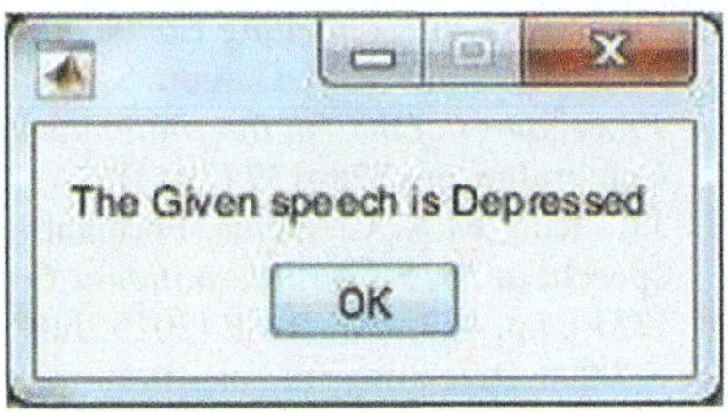

Fig. 7 Test result

7 Conclusion

We designed a technique which classifies the input speech data into normal person speech or depressed person speech (abnormal) by performing data preprocessing step, feature extraction step and classification step. For the segmented sample, we will extract features such as MFCC, pitch, Cepstrum to get enough information about the input data. Detected features are classified to get the final results of the signal category.

In future, we would like to enhance the application with speech recognition technique, wherein there would be no need for manually sign in. The application should detect the voice of the user and understand the emotion through his/her voice. Also, we would like to include the Natural Language Processing system to increase the conversations and analyse the user response, through the bot's reply. In a further case of deaf and dumb, we provide them with a simple solution by providing then with little or complete speech development using a speech recognizer which can be used to identify the speech of deaf-mute. This speech recognition can be performed by sensors wherein these sensors observe the hand gestures and sense the environment and convert the signs to signals. Later these emotions can be converted into speech by using the technique called Hidden Markov Model (HMM).

References

1. M. JohnsonVioules, B. Moulahi, J. Aze, S. Bringay, Detection of suicide—related posts in twitter data streams. IBM J. Res. Dev. (2018)
2. B.L. Cook, A.M. Progovac, P. Chen, B. Mullin, S. Hou, E. Baca-Garcia, Novel use of natural language processing (NLP) to predict suicidal ideation and psychiatric symptoms in a text-based mental health intervention in Madrid, in *Computational and Mathematical Methods in Medicine* (2016)
3. M. Nachamai, J.A. Paulose, A comprehensive survey on features and methods for speech emotion detection, in *International Conference on Electrical, Computer and communication Technologies (ICECCT)* (2015)
4. M. Despande, V. Roa, Depression detection using emotion artificial intelligence, in *International Conference on Intelligent Sustainable Systems* (2017)
5. M. Rambocas, J. Gama, Marketing research. The role of sentiment analysis, in *The 5th SNA-KDD Workshop11.* University of Porto (2013)
6. R.S. Sudhakar, M.C. Anil, Analysis of speech features for emotion detection, in *International Conference on Computing Communication Control and Automation* (2015)
7. M. De Choudhury, M. Gamon, S. Counts, E. Horvitz, Predicting depression via social media, in *Proceedings of the 7th International Conference on Weblogs and Social Media,* ICWSM 2013, Cambridge, pp. 128–137 (2013)
8. J.C. Kim, M.A. Clements, Formant-based feature extraction for emotion classification from speech, in *2015 38th International Conference on Telecommunications and Signal Processing (TSP),* pp. 477–481. IEEE (2015, July)
9. S. Johar, Psychology of voice in emotion, in *Affect and Personality in Speech*, pp. 9–15. Springers, Cham (2016)

Intelligent Traffic Light Management System

Somavarapu Jahnavi, G. Prasanth, D. Priyanka, A. Sneheth, and M. Navya

Abstract The present traffic light management system causes a lot of problems such as long delays and congestion. The existing traffic signal system is in such a way that the signal is given for a lane using a timer circuit irrespective of the density of traffic in that lane, due to this the waiting period of a lane with more number of vehicles will be increased. In order to increase the efficiency, we need to consider the traffic information at the junctions as an input to decide the signal. In this paper, we propose a model to improve efficiency of traffic management by taking real-time traffic information. The model takes the video input from each lane at the intersection and decides the signal considering the following three factors:

- number of vehicles
- type of vehicle
- emergency vehicle.

The vehicles are classified using TensorFlow, and density across the lanes is calculated based on the count and weightage of the vehicles. Along with this, we are considering the emergency vehicles, and the waiting time is also set for the lanes which can be used for a lane with the least number of vehicles waiting for a longer time. Using these factors, signals are decided across the lanes.

S. Jahnavi (✉) · G. Prasanth · D. Priyanka · A. Sneheth · M. Navya
Department of CSE, VNR Vignana Jyothi Institute of Engineering & Technology, Hyderabad, India
e-mail: jahnavi_s@vnrvjiet.in

G. Prasanth
e-mail: prasanthvg3@gmail.com

D. Priyanka
e-mail: priyankac1411@gmail.com

A. Sneheth
e-mail: sneheth2500@gmail.com

M. Navya
e-mail: machanavya10@gmail.com

C. Kiran Mai et al. (eds.), *Proceedings of International Conference on Advances in Computer Engineering and Communication Systems*, Learning and Analytics in Intelligent Systems 20,
https://doi.org/10.1007/978-981-15-9293-5_45

Keywords Traffic management · Weightage · Density · Emergency vehicle · TensorFlow

1 Introduction

In modern life, we are facing many problems due to traffic congestion while going to offices, schools, and colleges. The main reason behind it is the pre-timed circuit which is used to give signals at a particular junction. The present system does not consider the number and type of vehicles on a lane while giving a signal. Hence, it leads to a higher waiting time for a lane which has more number of vehicles. Even the lane with the least number of vehicles will be given a green signal for the same amount of time which is usually fixed. Thus, leading to wastage of time. Considering all these factors, we have proposed a system which considers number and type of the vehicle for issuing the signal instead of a pre-timed circuit.

1.1 Image Processing

Image processing means processing an image by using a computer. Using image processing, we apply certain algorithms and enhance the image obtained and retrieve the relevant information. Image processing involves the following steps:

1. Image is obtained using the image acquisition tools.
2. Image is manipulated and altered.
3. Enhanced image with specific features is obtained.

An image is defined on a two-dimensional plane with a function "F(x,y)" where x and y are spatial directions, and the value of "F" at any value of (x,y) is known as the intensity of that image. An image contains several finite elements where each element has an intensity at a location known as a pixel.

2 Literature Review

Kawade et al. [1] Proposed a system using OpenCV as a software and the concept of image processing. Blob detection and color thresholding techniques are implemented. The algorithm used is Viola Jones algorithm. The main limitation in this system is it uses only a single camera for monitoring traffic at an intersection. Moreover, it does not categorize the vehicles.

Liang et al. [2] Proposed a Deep Reinforcement Learning Network for Traffic light cycle control. The model represents the traffic scenario as states, and the intersection is divided into blocks. Markov decision process is used to model the traffic light duration change. The main limitation in this system is the release of one lane is not dependent on traffic of other lanes.

Linganagouda et al. [3] Proposed a system based on sensors and transmitters. It controls the traffic density of cross-roads by increasing the duration of the green signal. IR transmitter and receiver are used to provide dynamic traffic control. Main limitations are it is fully based on sensors and transmitters and the sensor may not be able to detect each and every vehicle. Moreover, it does not categorize the vehicles.

Hejun et al. [4] Proposed an intelligent traffic control system based on traffic flow. They have used DSP technology and designed the system hardware with the corresponding software to control traffic light at intersections.

3 Objectives

1. To improve the existing Traffic management system.
2. To provide signal priority to emergency vehicles.

4 Implementation

4.1 Proposed System

In our approach, we detect the vehicles from a frame in the source video at an instance, and apply an existing identifier for each of the vehicles. Later, we count them individually. Along with this, we also store the waiting time of each lane so that we can avoid a lane with less vehicles to wait for a longer time. When an emergency vehicle like an ambulance, fire engine is on a lane, that lane will be given the highest priority regardless of waiting period or density of lanes. We have used python for coding as it has many libraries which can be used in raspberry and also for image capturing. We also used OpenCV along with a TensorFlow lite API in order to capture images, and also identify the vehicles and also classify them according to their type. In order to represent the traffic signals, we have used a traffic signal board which has four lanes with LED lights representing traffic lights. This board will be connected to the raspberry board and will receive the instructions from the algorithm (Fig. 1).

4.2 Implementation

To implement this proposed system, we used Raspbian buster OS, python 3.7.3 as a software and Raspberry pi 3 Model B board and traffic signals board which is used to imitate the traffic signals as a hardware. We have 4 modules in this system.

Module 1: Video is captured through camera and is divided into frames.

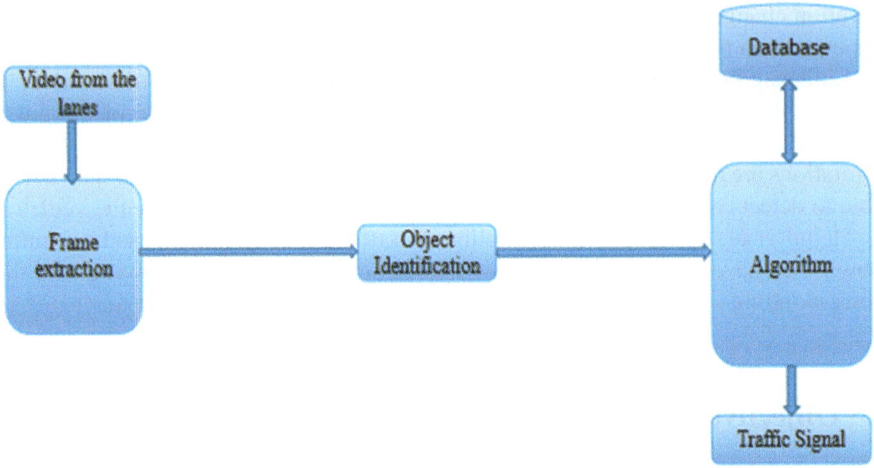

Fig. 1 System architecture of intelligent traffic light management system

Module 2: The obtained frames are processed and moving objects are detected through TensorFlow lite API. Using this API, the different vehicles are identified according to their type.

Module 3: In this module, we consider the type and weightage of a vehicle, the waiting time of a lane, and the max waiting time. Emergency vehicles like ambulances and fire extinguishers are given the highest priority. Else density for each lane is calculated based on vehicle weightage and count. Then the waiting time for each lane is checked against the maximum waiting time.

Module 4: Based on the output from the previous module the lane which should be released (green signal) is decided.

4.3 Methodology

The application designed by us is implemented with the following methodology.

1. The input video is read using the Video-Capture object of OpenCV. OpenCV is a library which is mainly used in computer vision. It supports different frameworks like TensorFlow, Torch, PyTorch.
2. The obtained video is divided into frames and the frames obtained are further processed (Fig. 2).

Fig. 2 Frame extraction from video input

3. The frames are processed to remove the salt and pepper noise using median filter. It takes the entire signal as input and replaces each entry with the median of neighboring entries (Fig. 3).
4. The obtained filter frame is in BGR format. In order to send it to the TensorFlow lite API, it is converted into RGB format (Fig. 4).
5. The above frame is sent to TensorFlow lite API for vehicle detection. TensorFlow is a python open-source library for numerical computation used in deep learning. TensorFlow is used to train, run neural networks for image recognition, and also to identify objects. Here, we have used the TensorFlow lite API in order to identify the objects and also categorize the vehicles. TensorFlow lite has mainly consists of two components.

i. TensorFlow lite interpreter which runs optimized models on embedded Linux systems.
ii. TensorFlow lite converter which is used to convert TensorFlow models into efficient forms (Fig. 5).

Fig. 3 Frame after applying Median Filter

Fig. 4 Frame after converting to RGB format

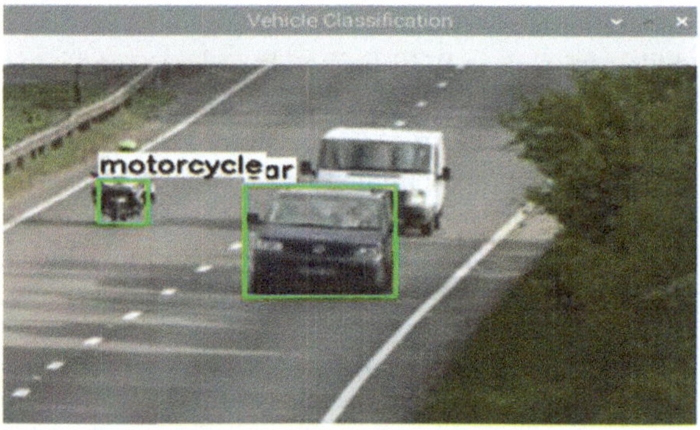

Fig. 5 Frame after identification of vehicles

6. After the vehicles are categorized, the density is calculated based on the count and type of the vehicles. In our system, we have given a weightage of 1 to motorcycles and 2 to four-wheelers and a weightage of 4 to heavy vehicles.

Algorithm 1: Density Calculation

```
Input: Frame from each lane
Output: Total density at each lane
Input parameters : input_details[ ][ ],Frame[ ]
Set density = [0,0,0,0]
For lane ← 0 to 3 do
    set den=0
    interpreter. set_tensor(input_details[0]['index'], Frame[lane])
    interpreter.invoke()
    set boxes = interpreter.get_tensor(output_details[0]['index'])[0]
    set classes = interpreter.get_tensor(output_details[1]['index'])[0]
    set len = range(len(classes))
    for i ← 0 to len-1 do
            set objectname=labels[int(classes[i])]
             if objectname == "cars"
            set den=den+2
            if objectname == "motorcycle"
            set den=den+1
            if (objectname =="bus") || (objectname =="truck")
            set den=den+4
   end for
      store the den value in density
      density[lane]=den
   end for
```

Thus, density is calculated across each lane using the algorithm.

5 Working

1. After the density is calculated, the waiting time of each lane will be updated in order to monitor the waiting time of each lane. Then the signal is decided based on the density and waiting time.
2. If an emergency vehicle across a lane is detected, signal is provided to that lane irrespective of the density (Fig. 6).

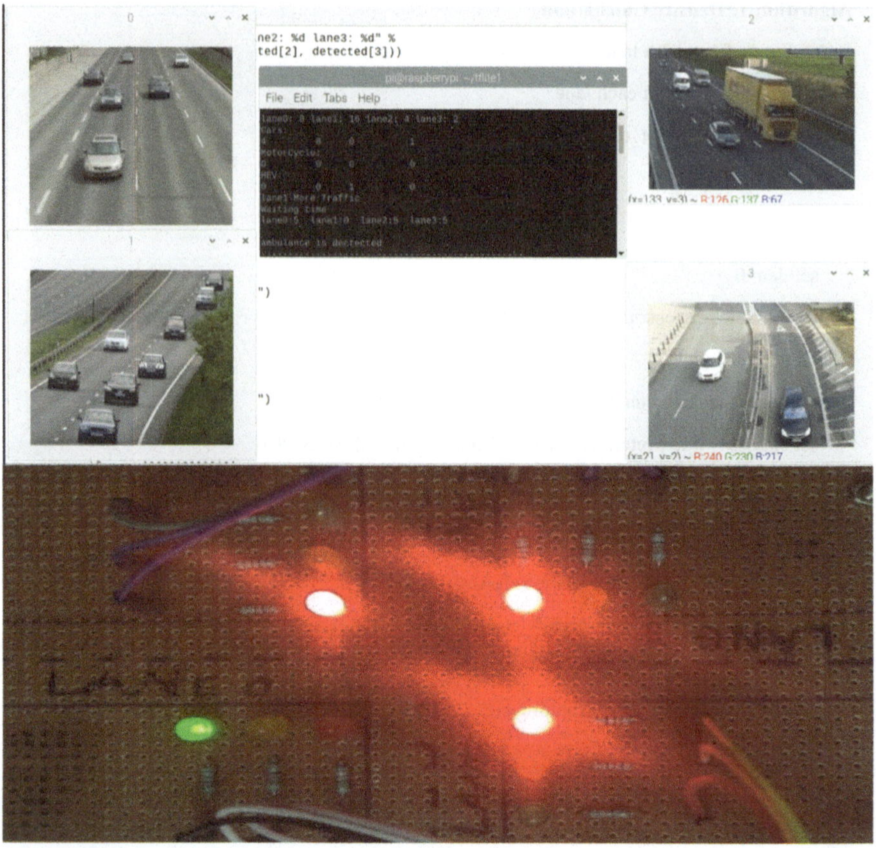

Fig. 6 Signal to emergency vehicle detected lane

3. If no emergency vehicle is detected, then the waiting time of all lanes is checked against the maximum waiting time. If the waiting time exceeds the maximum waiting time, then the signal is issued to that lane (Fig. 7).
4. If none of the above conditions meet, then the lane with highest density is given the green signal (Fig. 8).

6 Conclusion

In this paper, we propose to solve the traffic management problem by using OpenCV, TensorFlow API. The traffic information is gathered from cameras and used as the input to our algorithm. And we have also taken care that the emergency vehicles are given the utmost importance. This approach reduces the waiting time significantly and also provides a mechanism to give importance to the emergency vehicles.

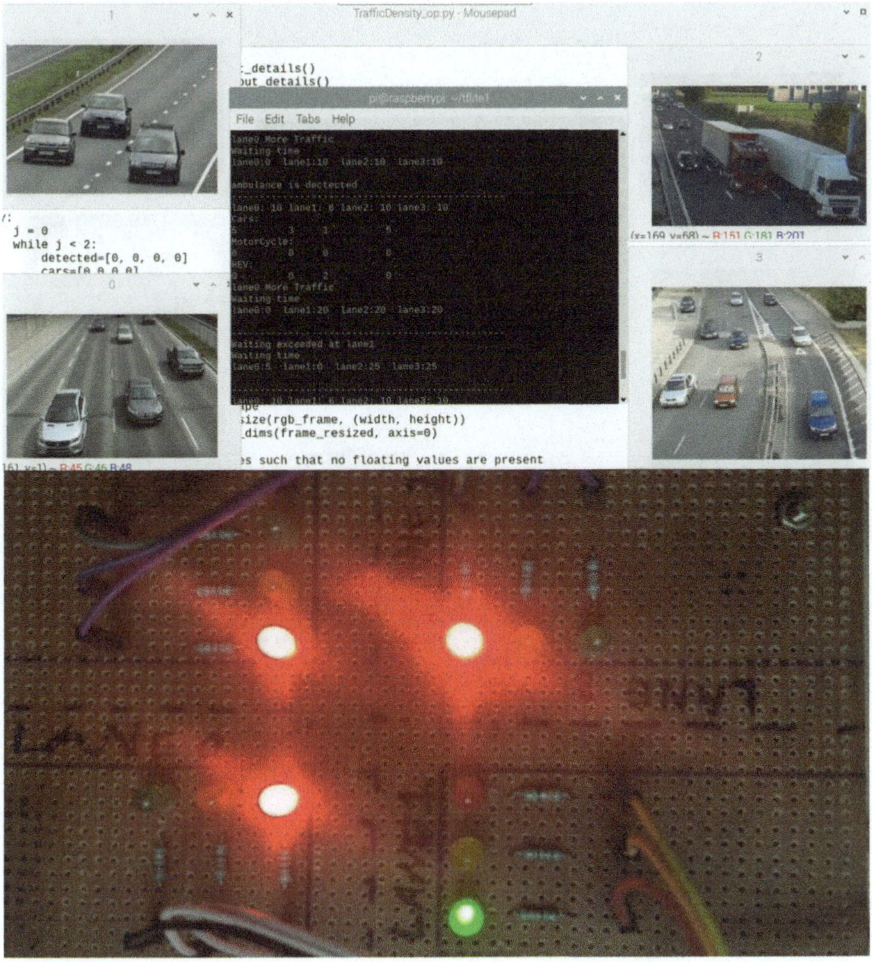

Fig. 7 Signal to lane with waiting time greater than maximum waiting time

7 Limitations and Future Scope

1. The quality of image can be improved with future technology and better cameras.
2. Also for each lane, images can be taken from all four directions in order to capture the vehicles which are behind larger vehicles for better management of the traffic.
3. The condition of the road can also be taken into consideration by the system while giving signals. In case of any violation of traffic rules by a vehicle, it can be immediately reported.

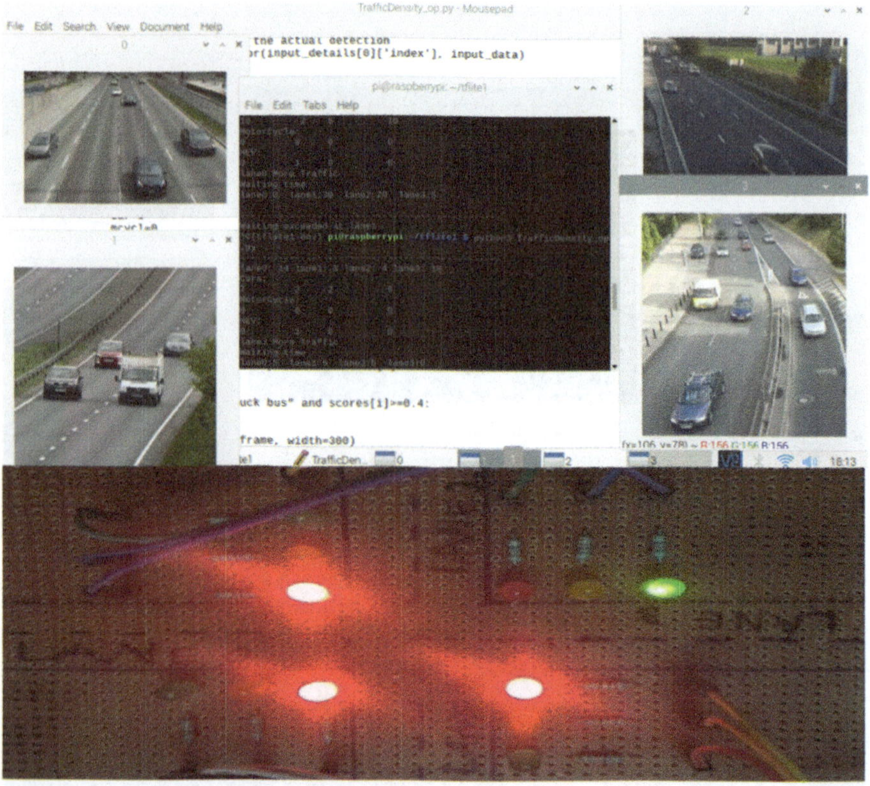

Fig. 8 Signal to lane with highest density

References

1. D. Kawade, S. Deshmukh, S. Gamare, A. Sankhe, Smart traffic control using OpenCV. IOSR J. Eng. (IOSRJEN) (2018)
2. X. Liang, X. Sheng Du, G. Wang, Z. Han, A deep reinforcement learning network for traffic light cycle control. IEEE Trans. Vehic. Technol. **68**(2) (2019)
3. R. Lingana Gouda, P. Raju, A. Patil, Automatic intelligent trafficcontrol system. Int. J. Adv. Res. Electr Electron Instrument Eng. **5**(7) July (2016)
4. W. Hejun, M. Changyun, Design of intelligent traffic light control system based on traffic flow. IEEE (2010)
5. https://www.TensorFlow.org/api_docs/python/tf/lite/Interpreter
6. https://docs.opencv.org/3.4/d8/dfe/classcv_1_1VideoCapture.html

An Effective Technique to Detect Dimensions of Pipe Under the Ground Using GPR Images

Gollapudi Ramesh Chandra, Kode Rajiv, and Bobba Basaveswara Rao

Abstract The detection of underground objects is the main component of a GPR system. For a given objects detection, the final interpretation of GPR image leads to the assessment of shape-based object detection and representation, so it could be accurate. The existing works were able to identify various object under the subsurface by analyzing the images obtained by GPR device. But very little work is done in identifying the various dimensions of the pipe such as radius, circumference, and length. This paper focuses on identifying radius of the pipe based on various features detected using different image processing techniques. The performance of the proposed algorithm is tested on 3-D synthetic pipe and raw data obtained by ground-penetrating radar images.

Keywords Underground object · Ground-Penetrating radar (GPR) · Pipe · Radius · Image processing

1 Introduction

A lot of work has carried out in detecting the various objects under the ground using subsurface imaging technologies. One of the popular subsurface imaging technology is Ground-Penetrating Radar (GPR), which is used to detect the various objects under the ground non-invasively. A growing number of GPR systems have been reported in recent years on the underground object processing, and detection has made deep dwell into the GPR research area. Underground object detection for the geographical surveys, landmine detection, oil exploration, mineral exploration, archeological

G. R. Chandra (✉)
Department of Computer Science and Engineering, VNRVJIET, Secunderabad, India
e-mail: rameshchandra_g@vnrvjiet.in

K. Rajiv
Department of IT, GRIET, Hyderabad, India

B. B. Rao
Computer Center, Acharya Nagarjuna University, Guntur, India

C. Kiran Mai et al. (eds.), *Proceedings of International Conference on Advances in Computer Engineering and Communication Systems*, Learning and Analytics in Intelligent Systems 20,
https://doi.org/10.1007/978-981-15-9293-5_46

surveys & extraction and civil aviation have recently become widespread applications in GPR processing using 3-D image processing techniques. In landmine detection, there is a requirement to get the outline shape of the object under the ground. Similarly in detecting the position, radius, shape, and dimensions of the pipe is a subject of interest for the civil people while laying the roads. In this regard, there is a necessity to generate a data having detailed information, which is required for analysis. Another requirement is representation of GPR data for detailed analysis and comparison. In this regard, 3-D models of pipes were designed, which is used for comparing the actual results with 3-D models with high accuracy and speed. Several papers reported usage of 3-D models [1–3] for cross-checking the performance of algorithm against actual GPR data. A little work is done for extracting features for detection of dimensions of the pipe buried in subsurface using the ground-penetrating radar technology. The major signature for detecting the dimensions of the pipe is "Hyperbolic Signatures". The hyperbolic signatures reflect the presence of circular objects in the subsurface imaging. Few literature is available for extracting hyperbolic features from GPR images [4–7]. The proposed method in this paper has discussed the detailed procedure for extraction of hyperbolic features, reconstructing a circular shape, and identifying the radius of the pipe under the subsurface from a B-scan images scanned by Ground-Penetrating Radar (GPR). For the processing of GPR data, scalar migration algorithms developed for three-dimensional seismic data are commonly used. However, the above algorithm does not relate the GPR transmitter and receiver antennas. An algorithm which takes the data of GPR and these data contains the characteristic of polarization and vectorial radiation. This algorithm is named as multi-component vector imaging algorithm [1, 8]. Streich and Kruk [9, 10] developed multi-component vector imaging scheme. Polarization of the electric field can be used to reduce unwanted reflections [11, 12]. Polarimetric stepped-frequency GPR and fully-polarimetric processing technique have also been developed considering the polarization [13] for GPR detection.

2 Previous Work

In the literature, a small number of published works still exist with concealed substance. In [14], the authors projected a procedure for digging out the hyperbolic signatures of concealed substance and therefore determining their location. In [15], determining the process is divided into three main processes: (1) pre-processing step which minimizes the noise and unexpected system defects; (2) image segmentation with artificial neural network classifier which will determine the part that contains the object deviations; and (3) Hough transform to determine the hyperbolic prototype. In [4], certain methods are determined to enrich the mark of buried objects. In [16], the authors used a fuzzy clustering approach. Molyneaux [5] demonstrated the capacity of the whole associated network in order to obtain the occurrence, shape, and depth of a pipe which utilize the neural network approach. Al-Nuaimy and Shihab [6, 17–19] have developed many automatic detection and interpretation method for

GPR such as neural network techniques. Shaw [20] used a neural network approach to mechanize as well as to make easy the preceding stage of the procedure and also established a fresh method for determining the diameter of reinforcing bars.

3 Proposed Work

Previous GPR object detections were performed using Machine Learning and Classification algorithms. From a careful study of the results of these surveys, it was determined that these methods were not satisfactory because they were not able to detect the radius of the object. Therefore, before implementing proposed work, a test field in which sample pipes and other subsurface objects were buried. Various methods of detections were analyzed in this environment, to determine which method was the mode effective at underground object detection. The MATLAB analysis tool is used to be convenient to implement and test the proposed work.

The procedure for features extraction is detailed below

a. Textures of GPR image pipe objects are matched with the images in the system parameters.
b. Scale the known simulated GPR image textures through identification, belonging to pipe dimensions.
c. Verify the matched pipe objects with known simulated pipe data for consistent unchanged features.
d. Localize the pipe objects at each GPR data location to measure the pipe structure textures parameter.
e. Pipe features gradients are measured at the GPR data location around each pipe textures and these are 2-D shaped into different local shapes at significant changes to the next level of GPR pipe extraction.

3.1 Features

There is a requirement to extract various features required for identification of dimensions of the pipe. Few of the important features are outlined here.

Pipe Penetration. The GPR sends the low-frequency signals which may penetrate roughly 5 meters–30 meters depending on the cost of the device and soil properties. For example, if the soil is having more moisture the conductivity is more, so signal loss is more. In this case, signal cannot penetrate into much depth. The performance of pipe penetration depends on the soil properties rather than pipe material and pipe depth. The 3-D pipe and its detection are shown in Figs. 1 and 2.

Speed of Detection: The proposed method can perform the pipe shape and dimension detection in less than a minute. The previously used GPR pipe detection methods cannot do within this time.

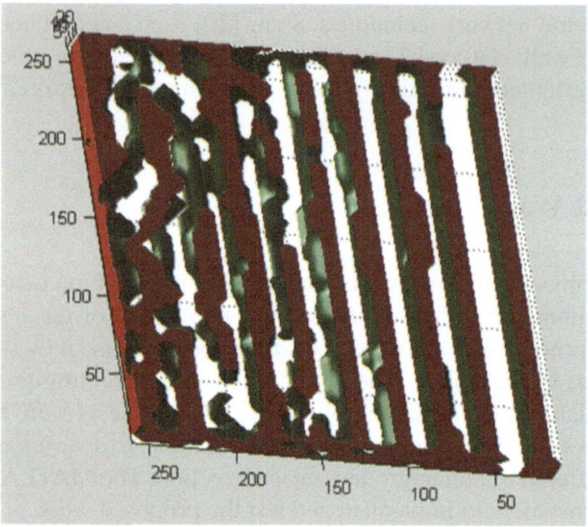

Fig. 1 Pipe horizontal view in GPR data

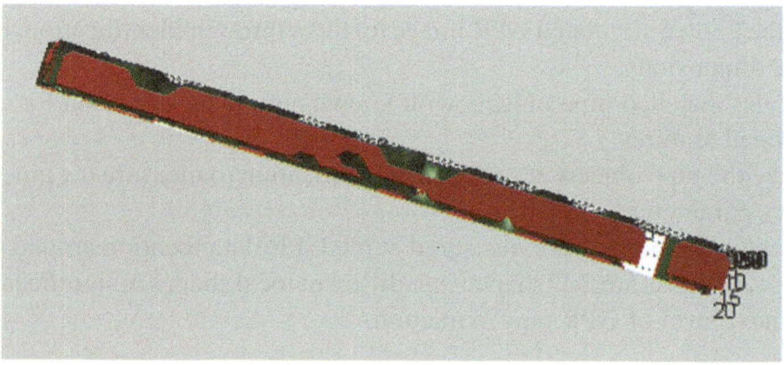

Fig. 2 Vertical view of pipe

Pipe detection capability: In the previously mentioned GPR field works and detection systems, several objects were identified, producing a success rate of 90%. This produces an importance to determine the pipe under deep-buried areas, and also in actual site areas in which the noise problems can be avoided. This process was proven tested in proposed method causing 100% success rate in pipe detection through synchronized GPR data with proper time-zero position, by tracing the 3-D data to a common time-zero position, to correctly align the GPR reflections correctly beneath the objects.

3.2 A Method for Detection of Radius of Pipe

The following steps outline a method for detecting the dimensions of the pipe:

a. The GPR layered data is processed using image filters, through which the pipe noise and unwanted damaged pixels are distinguished with the pipe location pixels. Wiener filter is used to perform this operation.
b. This step creates a Region of Interest (ROI) region to identify the pipe located regions. It is based on the technique that there is a small amount of pipe area and high area of ground area, and differentiates between them by identifying their image filter coefficient values, respectively.
c. As a function of pipe presence level, the 2-D voxel-level information of identified pipe areà is analyzed through 2-D spatial distribution, by computing the pipe features of the GPR date and ground regions, shows the pipe shape occurrence at each GPR layer located at a fixed geometric position relative to each other pixel-level area.
d. For the purpose of pipe analysis, the analysis of 2-D pipe pixels features between the pipe and ground regions is done to find out the pipe simulated matched features among them, if found they are recognized as pipe, else ground area.

The following steps outline the detection of radius of pipe:

(a) Scan the area of interest with ground-penetrating radar device which generates B-scan images.
(b) Scan all B-Scan GPR data into 3-D voxel representation
(c) Identify the regions of 3-D GPR data (as in Figs. 3 and 4).
(d) The regions of the subject of interest are identified based on the spatial accumulation of data. Whenever more white regions are accumulated spatially together,

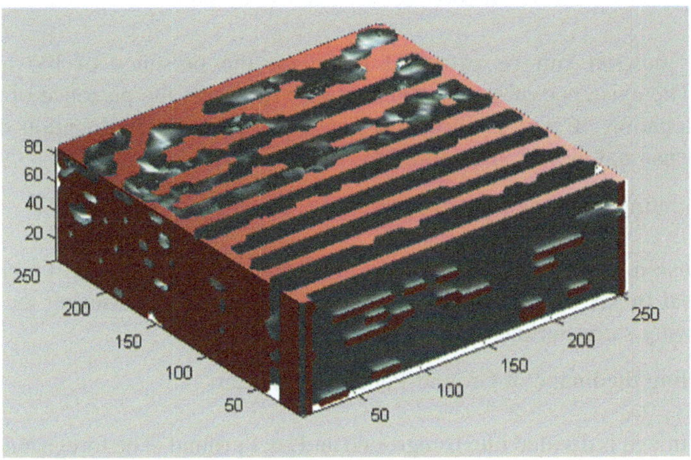

Fig. 3 Vertical region of 3-D GPR voxel

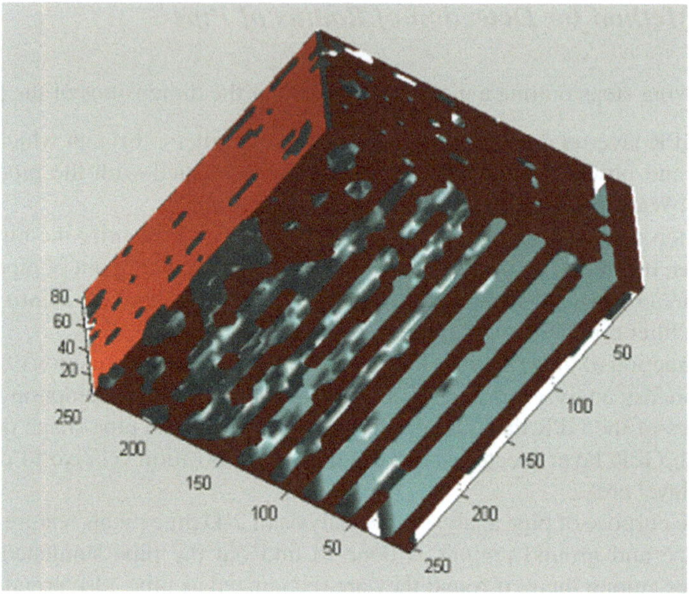

Fig. 4 Horizontal region of voxel

then these regions are interested in further processing. More technically this could be happening by repositioning the time zero in the vertical axis. By utilizing the said processing steps, the detected regions of three-dimensional ground-penetrating radar images are depicted in Fig. 5, and the location of the pipe is shown in Fig. 6.

(a) Extraction of Pipe Location:

The pipe location can be extracted based on the presence of the hyperbolic features. The existence of hyperbolic features represents the presence of pipe. The detected location of pipe is shown in Fig. 6. Identifying the location of pipe in three-dimensional view is shown in Figs. 7, 8, 9, 10 and 11.

(b) Calculating the dimensions of pipe:

The dimension of pipe could be identified by analyzing the location of pipe in three-dimensional image of the pipe location and subsequently measuring the distance. The following steps outline the Pipe Radius detection.

(1) Dividing the image of Ground-Penetrating Radar:

The GPR Image is divided into foreground and background. The foreground features considered here are presence of hyperbolic features and background image is considered as, where there is no presence of hyperbolic features. Based on the desired level,

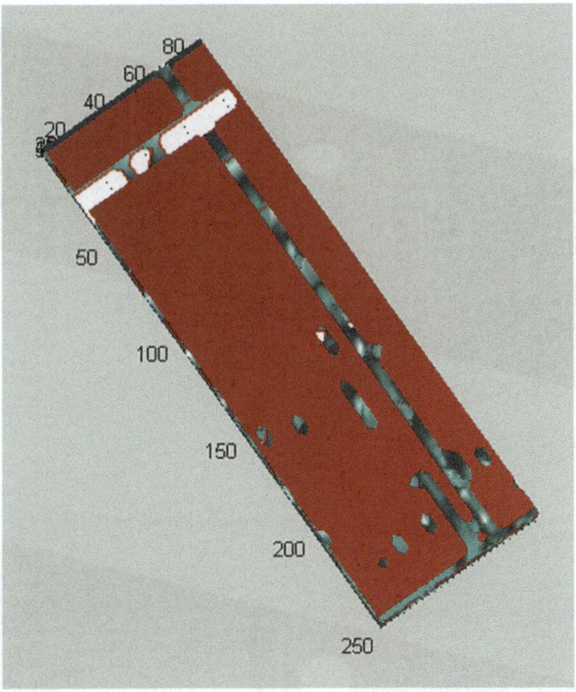

Fig. 5 View of pipe in voxel form

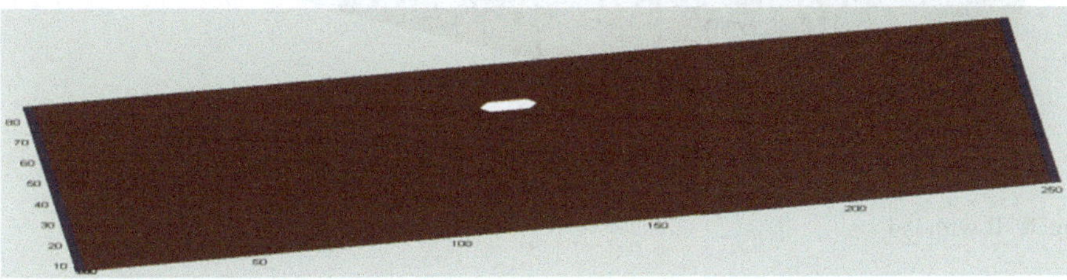

Fig. 6 Detecting the location of pipe

the threshold is selected due to the conversion of intensity image into a binary image and the obtained value among 0 and 1 concludes the pixels. Increment rate could be set to 0.01 and also set the threshold value to the finest value.

(2) Measuring pipe length:

Hole in the original image is demonstrated by an image segmentation and cleanup procedure which creates a distinctive cohesive blob. The actual image is of binary form which would be sufficient to build other functions into MATLAB. This could immediately evaluate the province and a host of diverse information.

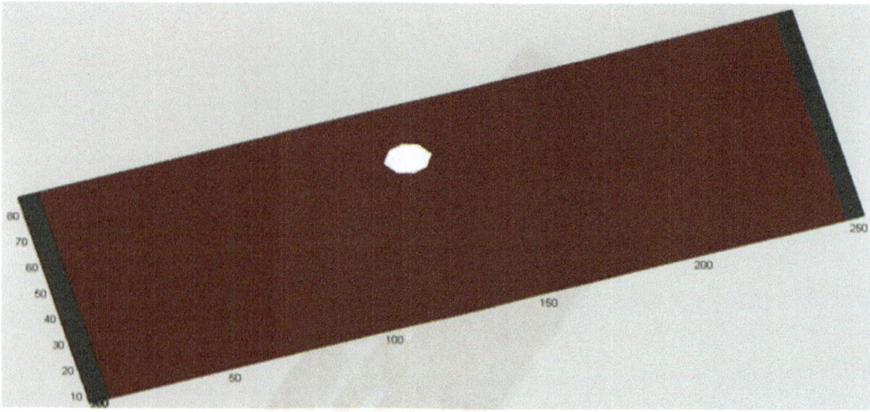

Fig. 7 II with 0.2

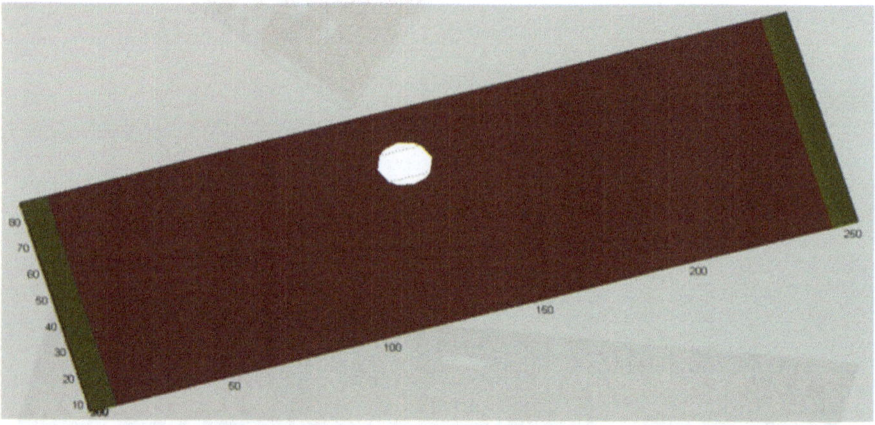

Fig. 8 II with 0.4

(3) Pipe Radius

Among the two figures (Figs. 12 and 13.), the estimated value was near to the physical measurement in synthetic data.

The calculation of pipe location is shown in Figs. 12 and 13.

4 Result

The dataset is acquired by MALA Easy Locator System at ground-penetrating radar and pipe cable locator test base.

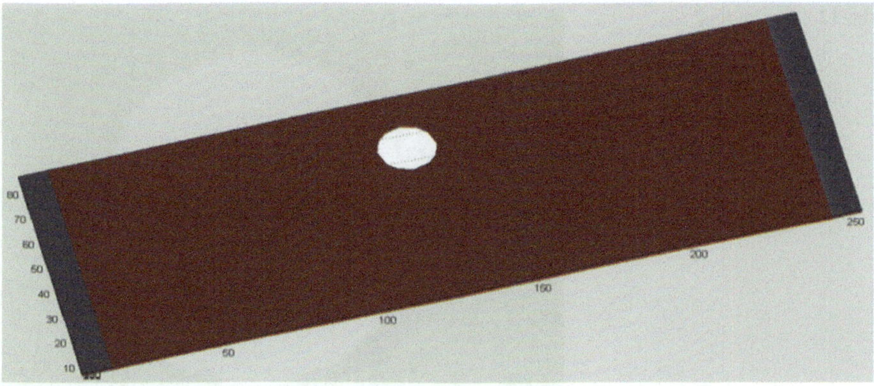

Fig. 9 II with 0.6

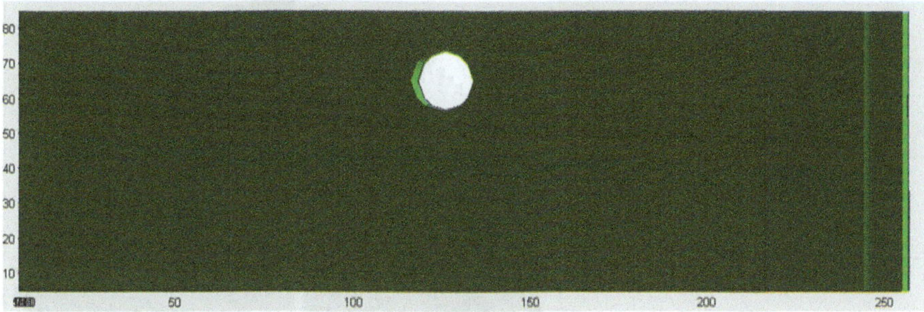

Fig. 10 Front view of pipe

Fig. 11 Pipe is identified with a red circle

To check the accuracy of the proposed algorithm, a synthetic three-dimensional pipe is created and then the algorithm is applied on the said data. Then the proposed algorithm discussed in this paper is applied to the actual data obtained with MALA GPR pipe cable locator dataset. As shown in Table 1, the radius of the buried pipe is 12.8 with an accuracy percentage of 99.22%. From this paper, it is determined that the accuracy is low for the smaller pipes for determining the radius when compared to the larger pipe size. Supplementary studies based on this approach have been proposed [2, 3, 7, 21]. However, since it is distinct in the construction industry, the correct

Fig. 12 Pipe radius

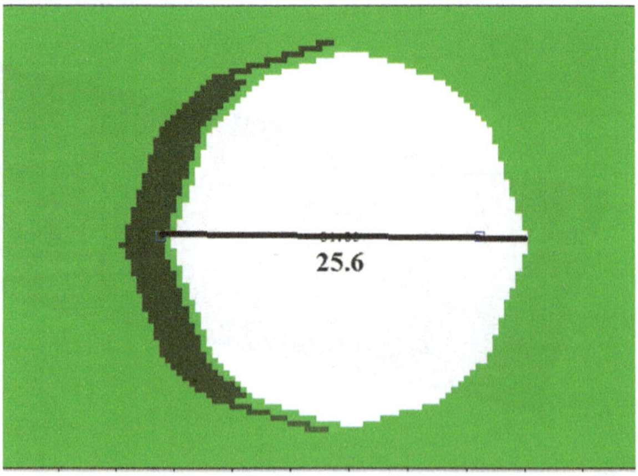

Fig. 13 Pipe width

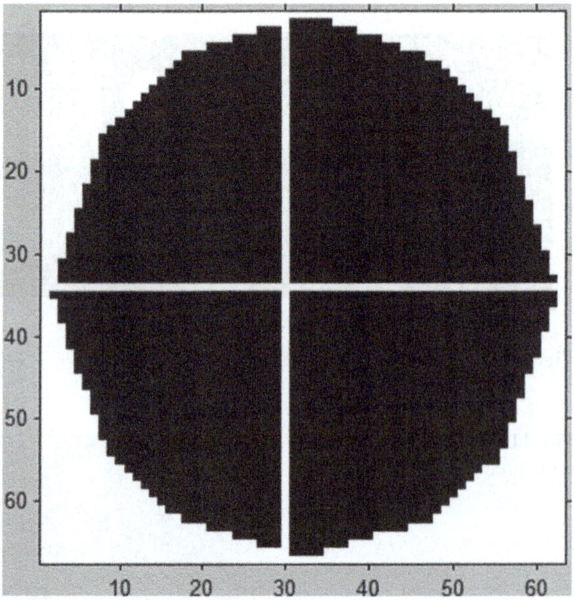

Table 1 The simulated radius distribution functions for a 25.4 mm radius pipe at a depth of 1 m

S. No	Pipe type	Radius obtained		Actual radius		Error	Accuracy (%)
		(mm)	(inch)	(mm)	(inch)		
1	Synthetic pipe	12.7	0.5	12.7	0.5	0.0	100
2	GPR	12.8	0.5	12.7	0.5	0.1	99.22

pipe sizes could be determined by referencing to the available database consisting of available pipe sizes.

5 Conclusion

There exists a technique to identify the presence of pipe in a certain location or not. But there are very few techniques to approximately measure the various dimensions of pipes such as radius and length. This paper discusses the methods for approximately measuring length and radius of the pipe under the subsurface non-invasively using ground-penetrating radar technology, image processing, and mathematical reconstruction techniques. For comparing the proposed method accuracy, it created a 3-D synthetic pipe data. The accuracy of the proposed method is compared on three-dimensional synthetic data and actual image sequences generated through ground-penetrating radar. The proposed work has to be further validated more ground truth data, which is not available abundantly. So there is a need to create a civil site with in situ data and respective validation of the data by concern experts.

Acknowledgements This work done in this paper is supported by Collaborative Research Fund (CRS) scheme of TEQIP-III under JNTU Hyderabad with Procs No. JNTUH/TEQIP-III/CRS/2019/CSE/02.

References

1. J. Van der Kruk, C.P.A. Wapenaar, J.T. Fokkema, P.M. Van Den Berg, Three-dimensional imaging of multicomponent ground-penetrating radar data. Geophysics **68**, 1241–1254 (2003)
2. K. Rajiv, G. Ramesh Chandra, B. Basaveswara Rao, 3D model based approach for data visualization. Int. J. Comput. Sci. Inf. Secur. (IJCSIS) **15**(3), March (2017)
3. K. Rajiv, G. Ramesh Chandra, B. Basaveswara Rao, Geometrical properties of 3-D images and its uses in 3-D image processing. Int. J. Adv. Res. Comput. Commun. Eng. **5**(7), July (2016). ISO 3297:2007 Certified
4. S. Delbμo, P. Gamba, D. Roccato, A fuzzy shell clustering approach to recognize hyperbolic signatures in subsurface radar images. IEEE Trans. Geosci. Remote Sens. **38**, 1447–1451 (2000)
5. S. Shihab, W. Al-Nuaimy, Y. Huang, A. Eriksen, Automatic region-based shape discrimination of ground penetrating radar signatures, in *Proceedings of the Symposium on the Application of Geophysics to Environmental and Engineering Problems SAGEEP 2003*, San Antonio, USA (2003)
6. W. Al-Nuaimy, Y. Huang, A. Eriksen, V.T. Nguyen, Automatic detection of hyperbolic signatures in ground-penetrating radar data, in *Proceedings of SPIE: Subsurface and Surface Sensing Technologies and Applications III*, San Diego, Vol. 4491, pp. 327–335, August (2001)
7. K. Rajiv, G. Ramesh Chandra, B. Basaveswara Rao, GPR objects hyperbola region feature extraction. Adv. Comput. Sci. Technol. **10**(5), 789–804. ISSN 0973–6107 (2017). http://www.ripublication.com
8. J. Van Der Kruk, Multi-component imaging for different heights of source and receiver antennas, in *Conference Proceedings, 2nd International Workshop on Advanced Ground Penetrating Radar* **6** (2003)

9. R. Streich, J. Van Der Kruk, Accurate imaging of multicomponent GPR data based on exact radiation patterns. IEEE Trans. Geosci. Remote Sens. **45**, 93–103 (2007)
10. R. Streich, J.V.D. Kruk, A.G. Green, Vector-migration of standard co-polarized 3D GPR data. Geophysics **72**(5), 65–75 (2007)
11. J. Van Der Kruk, E.C. Slob, The influence of the soil on reflections from above surface objects in GPR data, in Proceedings Eight International Conference on Ground-penetrating Radar, pp. 453–457, Queensland, Australia, May 23–26 (2000)
12. O.K. Jensen, O.G. Gregersen, Stepped frequency GPR for utility line detection using polarization dependent scattering, in GPR 2000, Proceedings of the Eight International Conference on Ground Penetrating Radar, Vol. 4084, pp. 727–730, Gold Coast, Australia, May 2000
13. S.J. Radzevicius, E.D. Guy, J.J. Daniels, Pitfalls in GPR data interpretation: Differentiating stratigraphy and buried objects from periodic antenna and target effects. Geophys. Res. Lett. **27**, 3393–3396 (2000)
14. L. Capineri, P. Grande, J.A.G. Temple, Advanced image-processing technique for realtime interpretation of ground-penetrating radar images. Int. J. Imag. Syst. Technol. **9**, 51–59 (1998)
15. W. Al-Nuaimy, Y. Huang, M. Nakhkash, M.T.C. Fang, V.T. Nguyen, A Eriksen, Automatic detection of buried utilities and solid objects with GPR using neural networks and pattern recognition. J. Appl. Geophys. **43**, 157–165 (2000)
16. T.C.K. Molyneaux, S.G. Millard, J.H. Bungey, J.Q. Zhou, Radar assessment of structural concrete using neural networks. NDT & E Int. **28**(5), 281–288 (1995)
17. W. Al-Nuaimy, H. Lu, S. Shihab, A. Eriksen, Automatic mapping of linear structures in 3-dimensional space from ground-penetrating radar data, in *Ninth International Conference on Ground Penetrating Radar, GPR 2002, Proceedings of SPIE*, pp. 121–124 (2002)
18. S. Shihab, W. Al-Nuaimy, A. Eriksen, Image processing and neural network techniques for automatic detection and interpretation of ground penetrating radar data, in *Proceedings of 6th WSEAS International Multi-conference on Circuits, Systems, Communications and Computers*, Crete (2002)
19. M.R. Shaw, S.G. Millard, T.C.K. Molyneaux, et al., Location of steel reinforcement crimination of ground penetrating radar signatures, in *Proceedings of the Symposium on the Application of Geophysics to Environmental and Engineering Problems SAGEEP 2003*, San Antonio, USA (2003)
20. M.R. Shaw, T.C.K. Molyneaux, S.G. Millard, M.J. Taylor, J.H. Bungey, Assessing bar size of steel reinforcement in concrete using ground penetrating radar and neural networks. Insight: Non-Destructive Testing and Condition Monitoring **45**(12), 813–816 (2003)
21. K. Rajiv, G. Ramesh Chandra, B. Basaveswara Rao, A distinctive similarity rendering approach to reconstitute hyperbola apices in GPR images, in *Region 10 Conference, TENCON 2017 – 2017*. IEEE. https://doi.org/10.1109/tencon.2017.8227888
22. http://www.malagpr.com.au/new-asylocator-hdr-locator.html

Chatbot for College Website

G. S. Ramesh, G. Nagaraju, Vemula Harish, and P. Kumaraswamy

Abstract A Chatbot is a program that reenacts a conversation between a user and a machine. The machine has been embedded knowledge to identify the queries asked by the student and other users and making a decision itself to respond to the queries of the student and other users. The user just needs to ask the queries that are answered by the chatbot. These bots will be totally founded on a book-based UI that permits the client to type orders and get message just as content to discourse response. It can be more made sure about when it is coordinated with well-known web services. The school request chatbots will be fabricated utilizing counterfeit calculations that can undoubtedly comprehend and investigate clients message. The client can pose the inquiries that are any school-related activities through chatbot without genuinely accessible to the school for inquiry. By using Artificial Intelligence, the queries are answered by the system asked by various users. The user simply needs to enlist and login to the system. Natural language handling (NLP) is utilized for tokenizing, stemming, and separating the substance of the objection given by the client.

Keywords Natural language processing · Sentiment analysis · Synsets · WordNet

G. S. Ramesh (✉) · G. Nagaraju · V. Harish
Department of CSE, VNR VJIET, Secunderabad, India
e-mail: ramesh_gs@vnrvjiet.in

G. Nagaraju
e-mail: nagaraju_g@vnrvjiet.in

V. Harish
e-mail: harish_v@vnrvjiet.in

P. Kumaraswamy
Department of CSE, SR University, Warangal, India
e-mail: palleboina.kumar@gmail.com

C. Kiran Mai et al. (eds.), *Proceedings of International Conference on Advances in Computer Engineering and Communication Systems*, Learning and Analytics in Intelligent Systems 20,
https://doi.org/10.1007/978-981-15-9293-5_47

1 Introduction

Chatbot is a PC program that emulates human discussions in its normal organization. Chatbot for college website executive framework will be created utilizing man-made reasoning calculations that investigates clients inquiries. Clients will simply need to choose the class for questions and ask the inquiries to the bot that will be utilized for noting it. Man-made reasoning will be utilized to answer the client's queries. The clients will find fitting solutions to their questions.

The usage of the task on school condition is especially valuable for students, parents, and other visitors. They can get different data about the school like scholastics, school yearly day, sports day, occasions, etc. It will support the understudies and different clients to be refreshed about the school exercises (Fig. 1).

1.1 Artificial Intelligence

Man-made brainpower is a wide part of PC science. The objective of Artificial Intelligence is to make frameworks that can work keenly and autonomously. Man-made consciousness makes it workable for chatbots to learn by finding designs in information. Without preparing, these chatbots would then be able to apply the example to comparative issues or marginally various inquiries. This capacity gives them the knowledge to perform assignments, tackle issues, and oversee data without human intercession (Fig. 2).

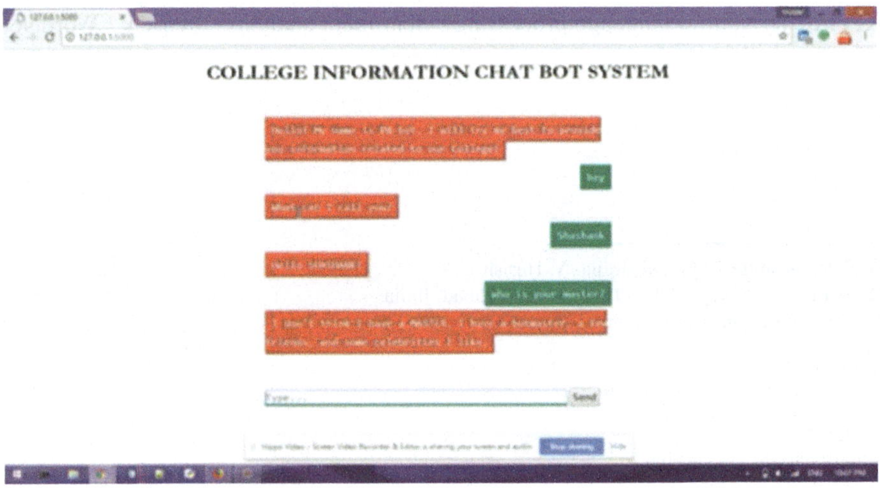

Fig. 1 College Chatbot

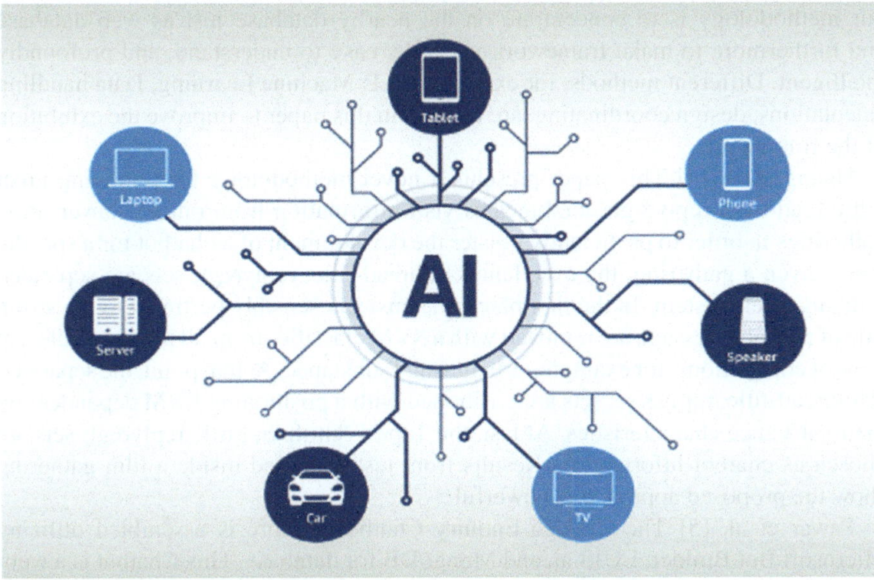

Fig. 2 Artificial intelligence

2 Literature Survey

Rane et al. [1] is utilized for computing the sentence similarity. The machine has been trained with the knowledge to find sentence/question and makes a conclusion itself to respond to the queries posed by the user. In this investigation of paper, the measurements of example coordinating on the content information and insights of packed example coordinating on compacted type of same content information are thought about. Here, the data is associated diagrams where the hubs contain the data and connection interrelates these data hubs. The structure semantics incorporates AIML determination language for writing data store with an end goal that talk bot configuration isolates the information vault from normal language interface part.

Bani et al. [2] In this paper, a proposition is done to clarify the plan of talk bot framework explicitly as an application with the goal that it encourages the new understudies to tackle all the issues they face and questions which raise in their brain when affirmation. Specifically, the proposition researches the usage of ALICE chatbot framework as an application named as school inquiry chatbot. A catchphrase-based human-PC discourse framework makes it conceivable that the client could talk with PC utilizing a characteristic language, for example, in English.

Hiremath et al. [3] The motivation behind this paper is to build up a robotized framework that answers the client inquiries for the benefit of a human for the instructive framework. It can reply to every single question posed by the end client. Existing chatbots, for example, Facebook chatbot, WeChat, Natasha from Hike, Operator, and so forth, were replying answer from their neighborhood database. In any case, our methodology is to concentrate on the nearby database just as web database and furthermore to make framework versatile, easy to understand, and profoundly intelligent. Different methods, for example, NLP, Machine Learning, Data handling calculations, design coordinating, are utilized in this paper to improve the exhibition of the framework.

Huang et al. [4] This paper presents a novel methodology for removing great <thread-title, reply> combines as visit information from online conversation gatherings in order to proficiently bolster the development of a chatbot for a specific area. Given a gathering, the excellent <thread-title, reply> sets are separated utilizing a fell system. In the first place, the answers sensibly pertinent to the string title of the root message are removed with a SVM classifier from all the re-handles, in view of connections, for example, structure and substance. At that point, the separated <thread-title, reply> sets are positioned with a positioning SVM dependent on their substance characteristics. At last, the Top-N <thread-title,reply> sets are chosen as chatbot information. Results from tests directed inside a film gathering show the proposed approach is powerful.

Pawar et al. [5] The College Enquiry Chatbot venture is assembled utilizing Microsoft Bot Builder, LUIS.ai, and MongoDB for database. This Chatbot is a web-based application that answers the understudy questions. Understudies simply need to inquiry to the bot and it will respond to their inquiries. Understudies can pose inquiries utilizing any English content organization. There is no particular arrangement the client needs to follow. The system makes use of inbuilt Artificial Intelligence to answer queries gave by LUIS.ai. The system responds using a compelling Graphical User Interface which infers as if a real person is talking with the client. To start with, bot will send question to the LUIS.ai. LUIS.ai will send its reaction to bot after that bot will get an incentive from database and afterward offer response to client question.

Jain [6] In this one, a suggestion is carried out to explain the arrangement of a chatbot especially custom-fitted as an application which will help new college understudies of Jaypee Institute Of Information innovation to understand all the issues they go up against and the requests which develop in their mind in the midst of and after the confirmation of the affirmation in schools. A stop words-based human-PC system makes it possible that the customer could chat with the PC using a trademark lingo, for example, in English. In this way, a school explicit chatbot is intended to inquiry an individual's questions during admission to various universities.

Gangrade [7] This paper presents the structure and improvement of a savvy voice acknowledgment talk bot. The paper presents an innovation demonstrator to confirm a proposed structure required to help such a bot (Android Application). Chabot can be depicted as a software that interacts with individuals using man-made brainpower. These software are used to perform assignments, such as quickly responding to users, educating the users and helping them explaining their queries. In this one, we present basic working principle and fundamental concepts of AI-based Chatbot to facilitate our College Information Retrieval System. Our Chatbot could successfully answer college-related questions with an additional bit of leeway that it likewise gives individual data like grades, etc.

Bhosale et al. [8] The motivation behind this paper is to build up a computerized framework which offers answer to client question in the interest of human for instruction framework. It can offer response to every single inquiry asked by end client. Existing Chabot, for example, Facebook chatbot, WeChat, Natasha from Hike, and so forth, were giving answer from their neighborhood database. Be that as it may, our methodology is to concentrate on neighborhood database just as web database and furthermore to make framework adaptable, easy to understand, profoundly intelligent. Different methods, for example neural system, NLP, design coordinating information mining calculations are utilized in this paper to improve the presentation of the framework.

Wahi et al. [9] This paper proposes a shrewd application for College Bus Tracking System, which runs on Android advanced cells and web. This empowers understudies to discover the area of the transport with the goal that they won't get late or won't show up at the stop too soon. The primary reason for this application is to give definite area of the understudy's separate transports in Google Maps other than giving data like transport subtleties, driver subtleties, stops, contact number, courses, and so on and to give area of understudy to their folks. This application might be broadly utilized by the understudies since Android PDAs have gotten normal and reasonable for all. It is an ongoing framework as the present area of the transport is refreshed each second as scope and longitude, which is gotten by the understudies through their application on Google maps.

Bala [10] suggested "Chatbot for school the board framework utilizing A.I". A chatbot means to make both user and a machine communicate with each other. The machine has been embedded knowledge to find queries and makes a conclusion on its own to respond to the queries of the user. Chatbot will be totally founded on a book-based user interface, permitting the client to type orders and get message just as content to discourse reaction. Chatbots are usually state full administrations, recollecting past orders so as to give usefulness. It very well may be used safely by a significantly bigger crowd when chatbots innovation is incorporated with mainstream web administrations. The school enquiry chatbots will be fabricated utilizing counterfeit calculations that investigate client's questions and comprehend the client's message. The response standard is coordinating the info sentence from a client. The client can pose the inquiry that is any school-related exercises through the chatbot without truly accessible to the school for enquiry.

3 Objectives

1. To actualize a bot framework that helps the understudies and different clients to get to the college-related data from anyplace with a web association.
2. To lessen the outstanding task at hand of school organization and other staff of noting all the questions of the understudies.

4 Implementation

4.1 Proposed System

- **NTLK (NLP) Processing and Sentiment Research for Complaint**:

At the point when client grumbling is given to the framework, NLP is enforced and feeling of the protest is recognized. The feeling of the words is discovered utilizing grammatical feature labeling and WordNet word reference. By using the slant investigation, invalidation level of a grievance is recognized. What's more, client objections are organized in like manner.

- **Search Questions in information database**:

When the invalidation level of the protest is recognized, moreover, the specific inquiry in the objection is identified utilizing WorldNet. As the grievance depiction can change from individual to individual. A similar inquiry might be posed uniquely in contrast to various clients. One client poses an inquiry so just and unmistakably while another client may pose a similar inquiry with all the more contrarily. So it is important to discover what is the specific specialized issue with the specific item to give a right arrangement.

- **Answer the Complaints**:

As depicted above at whatever point client presents a protest, the invalidation level and definite issue/question of the objection are recognized. At that point, it is watched that is there such inquiry enrolled in database. On the off chance that the appropriate response is discovered, at that point that answer is sent to that User. On the off chance that a specific inquiry isn't found in the database such inquiries are replied by administrator individual. When he addressed the inquiry the appropriate response is sent to that client. Also, that question alongside answer is put away in database so that at whatever point such inquiries will be posed with the goal that they get addressed straightforwardly from the database. Because of this administrator doesn't have to respond to same inquiry physically any longer Interface. WordNet is just a lexical and database that's semantic the English language. It is useful to bunch terms which are English the arrangement of equivalent terms called synsets, it offers quick definitions and utilization models, and documents different relations among these comparable sets or their individuals (Fig. 3).

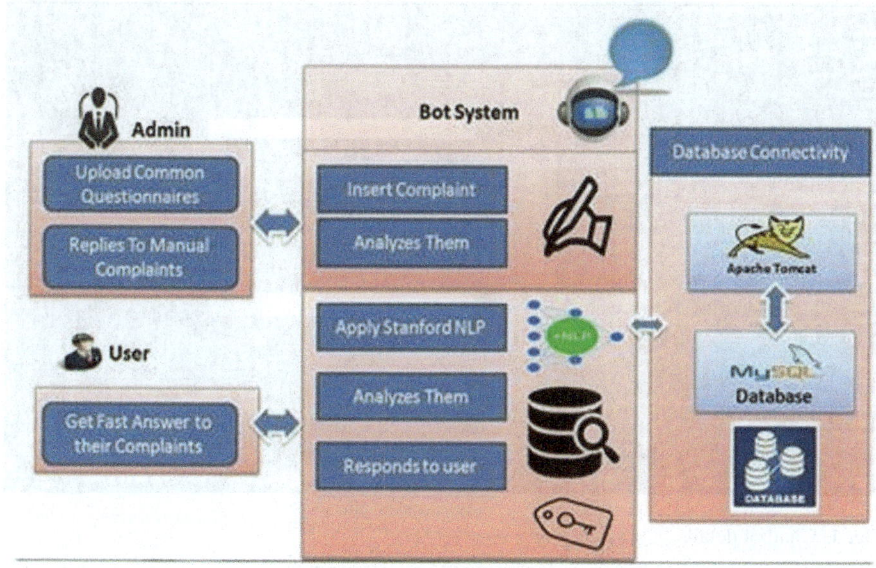

Fig. 3 Proposed system

4.2 Implementation

Porter Stemmer Algorithm:

Porter calculation that's stemming or ("Watchman stemmer") is just a means of expelling postfixes from terms in English. Expelling improvements consequently can be a task which is often particularly valuable in neuro-scientific data information data recovery (Figs. 4 and 5).

After would be the method of this calculation:-

- Gets free of plurals and -ed or -ing additions
- Turns terminal y to I if you find another vowel into the stem
- Maps twofold improvements to people being single -ization, -ational, and so forth.
- Handles improvements, -full, -ness, and so forth. Takes off -pest, -ence, and so forth. Evacuates a–e that's final.

1. Chatbot understands the personal queries asked by the users and responds itself using Artificial Intelligence.
2. Chatbot also responds to the user queries regarding technical knowledge also.
3. Chatbot gives college courses details such as syllabus, timetable, and grades of particular courses.
4. Chatbot also gives courses, departments, and professors Details.

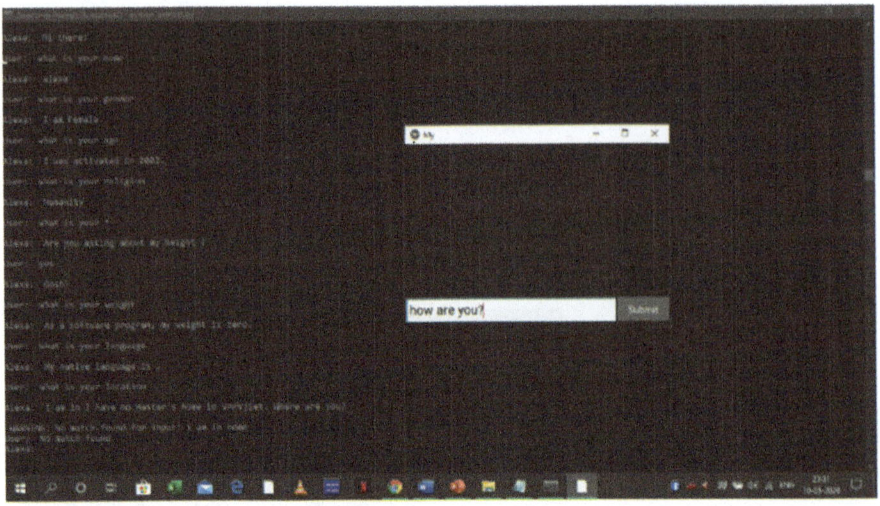

Fig. 4 Chatbot details

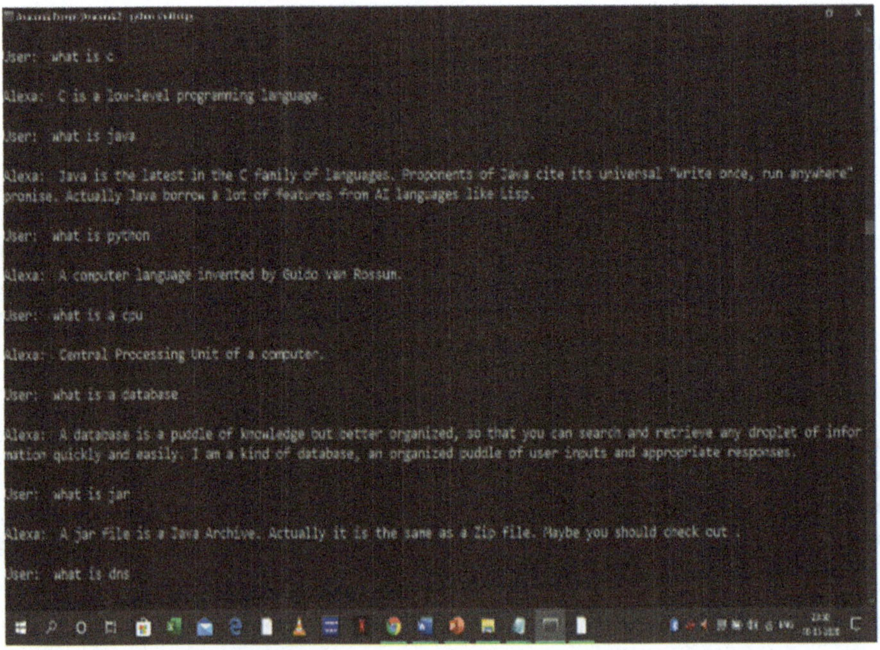

Fig. 5 Technical knowledge of Chatbot

4.3 Dataset

Collected 150 PYAIML datasets from Kaggle. These datasets are loaded while chatbot implementing. PyAIML is a markup language. The bot can able to remember the users and their previous chat which is functionally achieved through PyAIML. It is more scalable and intelligent that remembers and stores previous chats from the users.

5 Results

This project implements a chatbot which is useful for educational systems. This chatbot quickly gives the answers for an input query. The accuracy of answer and response time of this chatbot is more effiecient. We used an efficient algorithm for quick replies for an input query. Porter Stemmer Algorithm is used for pattern matching.

5.1 Term Order Similarity Between Sentences

We ought to look at a situation which specific delineate the significance of word request.

For instance, for two sentences:

T1: A canine hops over the apathetic fox.

T2: A fox hops over the languid pooch.

For estimating word request similitude of two sentences, the proposed measure is as per the following:

$$Sr = 1 - ||r1 - r2||/||r1 + r2|| \tag{1}$$

r1, r2 are word request vectors

r1 = {1,2,3,4,5,6,7}

r2 = {1,7,3,4,5,6,2}

The framework works fine and bot answers the inquiries of the users. Bot likewise answers in some amusing manner as appeared in execution to make less exhausting for the clients to collaborate with.

As shown in the implementation part, the chatbot responds to queries asked on technical knowledge, college details such as academics, annual day, sports day, time table, grade of courses, about faculty, departments. It also answers the queries related to the personal information, dates and emotions, universal truths. It can also answer some queries in a funny way and apologizes for the queries that are not answered (Fig. 6).

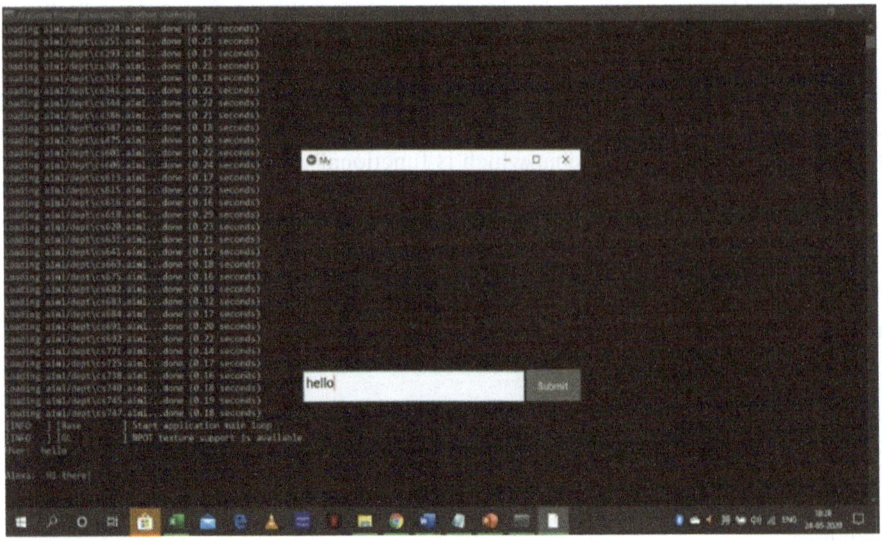

Fig. 6 Chatbot

6 Conclusion

We make a product instrument which will be utilized by any establishment to push the clients to openly transfer their questions. When the grumbling is enlisted in the database, programmed tokens are produced and passed on to the client through an instant message and email for additional following of the objection. Normal language preparing advancements are utilized for parsing, tokenizing, stemming, and sifting the substance of the grievance. The yield is taken care of to the calculation where the quality of the sentence is determined. The force of nullification is determined, which organizes the grumbling naturally for the specialist co-op to determine the grievance.

In thusly, the proposed framework will assist numerous establishments with ensuring quality help arrangement and client fulfillment with less human endeavors.

7 Future Scope

We might want to build the extent of the task in accordance aided by the e-governance (India digitilization) that's electronic associated with administration toward presenting chatbots over the different branches of the legislature. It can radically help diminish the red tapes, making administrations implied for the everyday citizens, progressively available to them.

Following enhancements are conceivable

1. Increase the extent of inquiries.
2. Ask client for clarifications on an inadequately design coordinated question.
3. Web creeping for chasing data mentioned in question progressively.

References

1. O. Rane, S. Pawar, P. Mehta, O. Wankhade, A web based college enquiry chatbot with results, in IEEE (2018)
2. B.S. Bani, A.P. Singh, College enquiry Chatbot using A.L.I.C.E, in IEEE (2017)
3. G. Hiremath, A. Hajare, P. Bhosale, R. Nanaware, Chatbot for education system, in IEEE (2018)
4. J. Huang, M. Zhou, D. Yang, Extracting chatbot knowledge from online discussion forums, in IJCAI (2007)
5. H. Pawar, P. Prabhu, A. Yadav, V. Mendonca, J. Lemos, College enquiry chatbot using knowledge in database, in IJRASET (2018)
6. P. Jain, College enquiry chatbot using iterative model, in IJSER (2019)
7. J. Gangrade, S.S. Surme, S. Somu, S. Raskonda, P. Gupta, A review on college enquiry chatbot, in IJESC (2019)
8. P. Bhosale, R. Nanaware, A. Hajare, G. Hiremath, Chatbot For education system, in IJETCS (2018)
9. S. Singh Wahi, K. Bhatia, D. Ukey, A. Barmate, Chat Bot for college management system, in ICEMESM (2018)
10. K. Bala, M. Kumar, S. Hulawale, S. Pandita, ChatBot for college management system using A.I (2018)

Comparative Analysis of Different Classifiers for Speech Emotion Recognition

A. Madhavi, Albert Priya Valentina, Karakavalasa Mounika, Boddeda Rohit, and Sheripally Nagma

Abstract Speech Emotion Recognition has been a vital topic of research in human–machine interface applications for many years. It attempts to make human–machine interactions more intelligent by understanding the emotional state of human beings. This paper presents an up-to-date survey of Speech Emotion Recognition discussing the important approaches regarding the use of different classification algorithms to recognize emotions. The focus is mainly on classifiers like Multilayer Perceptron (MLP), Support Vector Machine, Decision Tree, Random Forest, and Convolutional Neural Network (CNN). First, an acted emotional dataset, RAVDESS, will be discussed in detail. Second, the features that were extracted and selected will be addressed. Then, the focus is shifted to these classifier algorithms that categorize the input data into four classes of emotions: happy, angry, sad, and neutral. Each algorithm is implemented and its performance is compared with the others. Finally, conclusions about the best working model and limitations of each classifier used for Speech Emotion Recognition System are presented.

Keywords Speech emotion recognition · RAVDESS · MLP · CNN · Decision trees · Random forest · Support vector machine

A. Madhavi · A. Priya Valentina (✉) · K. Mounika · B. Rohit · S. Nagma
CSE, VNR Vignana Jyothi Institute of Engineering and Technology, Hyderabad, India
e-mail: priya.valentina21@gmail.com

A. Madhavi
e-mail: madhavi_a@vnrvjiet.in

K. Mounika
e-mail: karakavalasamounika@gmail.com

B. Rohit
e-mail: rohitboddeda@gmail.com

S. Nagma
e-mail: sheripallynagma46@gmail.com

C. Kiran Mai et al. (eds.), *Proceedings of International Conference on Advances in Computer Engineering and Communication Systems*, Learning and Analytics in Intelligent Systems 20,
https://doi.org/10.1007/978-981-15-9293-5_48

1 Introduction

Many systems have been designed over the years in an attempt to achieve human–machine interaction. To truly achieve interpersonal communication with a machine, emotional knowledge is key. There is enough evidence proving that emotion plays a huge role in properly understanding the other person. The scientific community believes that emotions are subjective and can be experienced by humans and other primates only. Human emotion doesn't have a tangible definition as it is an abstract concept that only human intuition can understand. These emotions are a result of the coordination of all the five senses and consciousness which enables human beings to empathize with each other. For a machine to understand such consciousness, is still a far-fetched goal [1].

By studying the current progress in the field of emotion recognition, various models were explored. These models were based on facial analysis [2], speech signals [3], and biological signals [4]. Biological signals like respiration rate, heart rate, and blood pressure are quite burdensome to obtain from numerous individuals. The convenience offered by speech signals in terms of accessibility and economic benefit makes them a viable source for emotion analysis. Machines that respond to humans through speech signals by speech processing techniques exist, but their incompetence to recognize the emotional state of the communicator remains to be a challenge [5]. Automatic Speech Recognition has had a lot of progress over the recent years with various computer-based systems deploying voice as a medium for communication. Amazon Alexa, Google Home, Apple Siri, etc., are some of the products that are widely known today in the market [6].

Speech Emotion Recognition has faced various issues over the years due to ambiguity. There is no clear idea as to what features should be selected to get the best results. This is attributed to the fact that to classify some emotions, certain extracted features tend to outstand and prove to be more useful than others [7]. Over the years, researchers have come up with various features of speech that contain emotional knowledge. They are energy, pitch, formant frequency, Linear Prediction Cepstrum Coefficients (LPCC), Mel-Frequency Cepstrum Coefficients (MFCC), and Modulation Spectral Features (MSFs) [8]. An inclination to combine features to form a feature set in an attempt to gain more emotional information has resulted in the redundancy of features and high dimensional complexity. Due to this, learning becomes complex and overfitting is likely to occur. Moreover, the uniqueness with which each person speaks poses a great challenge to overcome. People may express the same emotions in different sentences, tones, styles, and pace. Factors like gender also play a huge part as males and females have contradicting ways of expressing their emotions. Besides, the geographical and cultural aspects should be taken into consideration as people from different parts of the world have distinctive ways of conveying a message. This acoustic variability is a great hurdle to overcome due to its ability to influence the speech features to be extracted [9].

After many years of researchers pondering over the various aspects of human emotion, we have concluded that emotion has two dimensions. They are activation and valence [10]. Advancements in this field will pave the way for automatic translation systems, lie detection, intelligent toys, psychiatric diagnosis, and human–robot communication [6]. Emotion recognition will prove to be useful in web-based E-learning, banking, and call centers. Faculty in educational institutions can effectively make use of Speech Emotion Recognition Systems. By understanding the emotions of the students, they can plan their teaching approaches and lesson plans accordingly. The curriculum can be changed to best suit the interests of students and consequently improve the quality of education.

In this work, the four basic classes of emotions considered are: neutral, sad, angry, and happy. The speech emotion recognition system usually consists of five parts: Input speech signal, feature extraction, feature selection, classification, and recognition of emotion [11]. Classification involves taking pre-processed data samples and assigning them to a class of emotion. In addition, various classification algorithms are analyzed and the results which they yield for the speech emotion recognition system are compared. We have exclusively studied and implemented algorithms such as Decision Tree, Support Vector Machine, Random Forest, Multilayer Perceptron (MLP), and Convolutional Neural Networks (CNN).

The evaluation metrics that we have chosen to study for comparing different models are accuracy, logarithmic loss, epochs, and confusion matrix. An unbiased dataset that covers all the variations and incorporates a fair proportion of all the different classes is necessary to have an optimal model. The accuracy metric provides the ratio of the number of correct predictions to the total number of predictions. It gives an overall insight into the performance of each model but often fails to provide detailed information regarding errors. We also evaluate the log loss function for the best working model that gives us more subtle information like how the model deviates in its predictions from the correct values. The ideal number of epochs for each model is also determined. The generated confusion matrices for each model are studied to comprehend the number of accurate and inaccurate predictions for each class of emotion. All these metrics will be scrutinized to arrive at a suitable conclusion for the most appropriate working model.

2 Literature Review

Various Speech Emotion Recognition Systems are being developed making it a promising field of research in signal processing and pattern recognition [12]. Different classifiers such as Hidden Markov Model (HMM), Gaussian Mixture Model (GMM), Artificial Neural Network (ANN), K-Nearest Neighbors (KNN), and Support Vector Machine (SVM) for Speech Emotion Recognition were introduced in [9]. In this work, they compared the performance of these algorithms in both speaker-independent as well as speaker-dependent systems. It was found that HMM had an advantage over GMM due to its ability to work with temporal aspects of speech features. However, HMM had a drawback of making the selected features fit into the HMM framework. It was concluded from this work that speaker-independent systems varied significantly in detecting emotions and had less accuracy rate than speaker-dependent systems.

Speech Emotion Recognition using Deep Learning techniques was discussed in [6]. In this work, the IEMOCAP corpus dataset was used where speech signals with .wav format were converted into spectrograms for each emotion class. An Inception Net v3 model (consisting of three inception modules) which is a deep neural network model was used to achieve compaction as opposed to a numerical array. The proposed CNN model was implemented using TensorFlow. More than 400 spectrogram images from the data set were resized to 500×300. The model was trained with batch size set to 100, learning rate as 0.01, and delay of 0.1 for 10 epochs. It was observed that a low accuracy of 35.95% was achieved after 28 epochs. This is believed to be attributed to the small size of the dataset and the number of spectrograms used.

Research on Speech Emotion Recognition based on Deep Belief Network (DBN) and SVM is presented in [12]. This work integrates a deep belief neural network with SVM. The purpose was to automate the task of feature extraction. A DBN is formed by a series of Restricted Boltzman Machine (RBM) stacks which can reduce computational complexity. This work used it to recognize emotions from the Chinese language and achieved an accuracy of 86.5% which was much higher than that which was obtained from a conventional SVM model. Speech Emotion Recognition methods and cases were described in [13]. For this work, they have combined the Mel-frequency Cepstrum Coefficient (MFCC) and Modulation Spectral Features (MFS) [8] as a feature set. They implemented different models to classify emotions on Berlin and Spanish databases. First, they tried the Multivariate Linear Regression (MLR) with this combination feature set and obtained better results when compared with SVM. For the Spanish database, RNN had the best accuracy of 90.05% using the feature set. Nonetheless, it was concluded that combining the features does not yield good results as in this case, it had resulted in overfitting. A survey of Speech Emotion Recognition is given in [14, 11]. A Neural Network approach for language-independent emotion recognition is presented in [15].

3 Implementation

3.1 Database

Speech can contain certain parameters that are useful in gaining emotional information. Thus, the collected speech samples must preserve legitimacy [16]. It is a fundamental part of Speech Emotion Recognition which determines the quality of the model and the credibility of the results [11].

The sampling library of speech data must follow certain standards to be of any relevance:

1. The dialogue uttered must not contain any emotional bias that will affect the judgement.

2. The dialogues should be such that the same sentence can be expressed in various emotions for us to compare the emotional speech parameters [9].

Taking these constraints into consideration, we used the Ryerson Audio-Visual Database of Emotional Speech and Song (RAVDESS) Dataset for the Speech Emotion Recognition System. This is an example of an acted emotional dataset which is popular among many researchers and used for a majority of studies for emotion analysis [17]. This dataset has 7356 files rated by 247 individuals ten times on emotional validity [18]. Only audio files are considered for this comparative study of classifiers. It contains the speech samples collected from 24 professional actors (12 male and 12 female) who vocalized the emotional expressions at two levels of emotional intensity with an additional neutral expression. For our work, we focused only on classifying four classes of emotions (angry, happy, sad, and neutral) using this dataset. The dataset was sampled at 536 mb.

3.2 Pre-Requisites

Librosa Python library is used for analyzing music and audio. It has a flatter package layout, standardized interfaces, names, backwards compatibility, modular functions, and readable code. In addition, soundfile, sklearn, numpy glob, os, keras, tensorflow, and matplotlib libraries are imported.

3.3 Feature Extraction and Selection

Feature Extraction is one of the rudimentary steps in any automatic speech recognition system. It helps us to bin the background noise and focus on obtaining components that are essential for identifying semantic content that may help in determining the emotion [19]. There has been a lot of dialogue over the years, regarding the types of features that should be extracted [20]. In this work, the following are the features selected for extraction: Mel-frequency cepstrum coefficient, chroma, mel, contrast, and tonnetz.

Mel-frequency cepstrum coefficient (MFCC) represents the short-term power spectrum of a sound. They are a small set of features (10–20) which concisely describe the overall shape of the spectral cover. This cepstrum estimates the human ear's response more accurately because it has frequency bands that are equally spaced on the mel scale which allows for better speech (audio) recognition. To extract MFCCs, first the Fourier Transform for each frame of the speech signal is taken. This is followed by mapping the powers of the spectrum obtained onto the mel scale. Then, we calculate the logs of the powers at each of the mel frequencies and take the Discrete Cosine Transforms (DCTs) of them as a signal. The amplitudes obtained from the resulting spectrum are called the MFCCs. MFCCs are known to have good frequency resolution and robustness to noisy data. It has high recognition rate in low-frequency regions.

Chroma or Chromagram pertains to the 12 different pitch classes. If two pitches differ by an octave, then humans perceive them as similar. Tone height and chroma are the two components of pitch. One of the ways to extract chroma features from a speech signal is by using binning techniques combined with short-time Fourier transforms. For each frame of the signal, we collect all the chroma information and equate it to a single coefficient. A sequence of chroma features can be obtained within a time window.

Mel (Mel Spectrogram) is a Spectrogram with mel scale on its y-axis and time on its x-axis. To obtain this spectrogram, we first divide the speech signal into frames and compute the Fast Fourier Transform (FFT) for each frame. We also need to change the time domain to frequency domain. Next step is to generate a mel scale and to do this, we take the entire frequency spectrum and separate it into equally spaced frequencies as it is heard by the human ear. Finally, for each frame, we break up the signal to generate the frequencies on the mel scale.

Contrast determines the RMS difference (contrast) between the actual speech and background noise or between the spectral peak and the spectral depression for each frame of the speech signal.

Tonnetz is a pitch space defined by the network of relationships between audio pitches in the rise and fall of speech (audio) signals. The tonal centroid features for each frame of the speech signal will be computed.

For each feature, we will check if it exists and then call the corresponding function from librosa.feature.feature_name() (e.g.:librosa.feature.melspectrogram for mel). Next, we obtain the mean from the function. hstack() function from numpy is called which stacks arrays in a column fashion. Then, we store the obtained mean feature value in an empty numpy array.

3.4 Data Pre-Processing

Take the RAVDESS dataset and define a list to hold the emotions that are needed from available emotions.

Load the data.

If the emotion is present in the defined list then extract features like mfcc, chroma, mel.

Append the extracted features to list X and emotion to list Y.

Else continue to the next file.

Take this pre-processed dataset and split it into Training and Testing dataset as X_train,Y_train,X_test and Y_test, respectively.

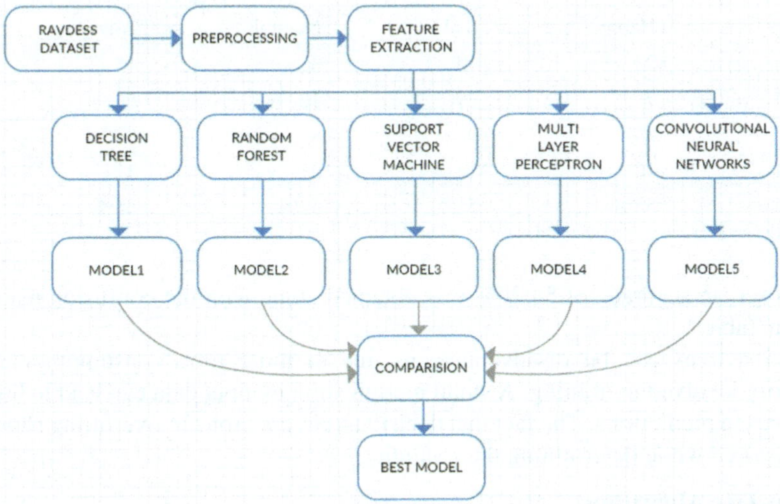

Fig. 1 System architecture

3.5 *System Architecture*

See Fig. 1.

3.6 *Techniques*

Various machine learning algorithms have been studied for emotion classification. We first train these models and use them to classify new data samples. Each algorithm has its advantages and disadvantages in providing the correct classification. Therefore, the working of the following five algorithms will be discussed and their performances will be compared.

Decision Tree Classifier

The decision tree algorithm is the simplest and one of the most popular algorithms that implement supervised learning. It is very convenient if you have noisy data as it does not affect the construction of the decision tree. A decision tree is basically a flow chart where every internal node represents a test and after each test, information is gained that results in data splitting into subsets. All the leaf nodes represent class labels or in this case, the emotional states [14].

A top-down greedy search approach with no backtracking is used to build decision trees. By importing the DecisionTreeClassifier from sklearn library, we trained the model with the training data set and max_depth value of 6. When validated with the

Table 1 Confusion matrix for decision tree model on RAVDESS dataset

Emotion	Happy	Sad	Neutral	Angry
Happy	23	13	3	15
Sad	7	29	12	2
Neutral	3	9	8	0
Angry	10	1	2	31
Precision	53%	56%	32%	65%

test data set, an accuracy of 54.16% was obtained along with the confusion matrix shown in Table 1.

Decision trees can be disadvantageous due to more time consumption and complexity involved in training. A small modification in input data can lead to huge differences in predictions. The fact that decision trees are prone to overfitting should not be ignored while determining their validity.

Decision Tree Algorithm

Train the DecisionTree model with the Training dataset(X).

DecisionTreeClassifier Function, i.e., DecisionTreeClassifier() by fixing the parameter max_depth = 6.

Store the arrival esteem into a variable.

Fit the training set information into DecisionTree using fit (X, Y) where X and Y are input and output labels, respectively.

Support Vector Machine

Support Vector Machine (SVM) is a machine learning algorithm that was originally intended for binary classification but has over the years evolved into a multi-class classification algorithm. Its prominence in applications like pattern recognition has increased significantly. Its main objective is to map the original feature set of low dimensions to a high dimensional feature space where optimum classification is achieved. This can be performed by kernel functions like linear, polynomial, Radial Basis Function (RBF) [21]. It works on the principle of classifying data by constructing N-Dimensional hyperplanes [22]. A proper understanding of how SVMs work is explained in [23]. Furthermore, detecting emotion from nonverbal features of speech and applying the method using Support Vector Machines is proposed in [24]. By selecting different features and consequently changing the model, SVM's performance can be improved [13].

For this work, the linear kernel function was used to train the model. The classes of emotions in the data were not well separated and overlapping occurred as a consequence. We chose the best regularization parameter 'C' for our data to reduce overfitting. The optimizer will choose a smaller margin hyperplane for large values of C and a larger margin hyperplane for small values of C. For C = 2, we obtained better

Table 2 Confusion matrix for SVM model on RAVDESS dataset

Emotion	Happy	Sad	Neutral	Angry
Happy	37	7	1	9
Sad	6	30	11	3
Neutral	4	5	11	0
Angry	9	5	0	30
Precision	66%	64%	32%	71%

classification results with a larger margin hyperplane between the classes of emotion when compared to other values of C. The model is validated against the test data set and an accuracy of 64.88% was obtained. The confusion matrix obtained is shown in Table 2.

SVM classifier can classify emotional states by a huge margin and is reputed to have high generalization capability [25]. Comparatively, it is more memory efficient. However, SVM is known to have poor performance in handling large data sets with noisy data.

Support Vector Machine Algorithm

Train the SVM model with the Training dataset(X).

SVM Function, i.e., SVC () by fixing the parameters kernel='linear' and C = 2

Store the arrival esteem into a variable.

Fit the training information into SVM using fit (X, Y) where X and Y input and output labels, respectively.

Random Forest

Random forest algorithm is an ensemble learning method that overcomes the problem of overfitting posed by the decision tree classification algorithm. This classification algorithm employs multiple decision trees generated to make a classification. It uses bagging techniques to extend its functionality. All the decision trees are constructed by taking only a part of the training data as input and their output is given accordingly. Random forest classifier predicts the class label by taking the majority voting of predictions made by the decision trees.

We implemented the Random Forest Classifier by giving our input and specifying the number of n_estimators as 60. An accuracy of 69% was obtained when validated against the test data set. The confusion matrix obtained is shown in Table 3.

Table 3 Confusion matrix for random forest model on RAVDESS dataset

Emotion	Happy	Sad	Neutral	Angry
Happy	35	9	1	9
Sad	6	39	4	1
Neutral	5	7	8	0
Angry	9	3	0	32
Precision	64%	67%	62%	76%

Random forest algorithm tends to cover all aspects obtained from the decision trees to give an unbiased generalized classification on the input data [26]. They give better accuracy than decision trees and can handle large sets of data efficiently. Methods for handling missing data and balancing errors in unbalanced data sets are provided by this classification algorithm. Nevertheless, random forests' predictions are quite difficult to understand and they tend to be more biased for those attributes containing more levels. Moreover, as we increase the number of n_estimators, the complexity of the model increases significantly and leads to inaccurate predictions.

Random Forest Algorithm

Train the Random Forest model with the Training dataset(X).

RandomForestClassifier Function, i.e., RandomForestClassifier() by fixing the parameters like n_estimators = 60 and random_state = 0.

Store the arrival esteem into a variable.

Fit the training information into Random Forest using fit (X, Y) where X and Y are input and output labels, respectively.

Multilayer Perceptron

Unlike other classification algorithms, such as Support Vector Machine, this algorithm fundamentally depends on a Neural Network for the classification. For our MLP model, we have considered one input layer, one hidden layer, and one output layer. The data from each hidden layer transforms linearly. However, having more hidden layers or linearity can cause overfitting. With caution, we have considered one hidden layer with 300 neurons. Our model was initialized with hyperparameters like batch size equal to 256 and alpha equal to 0.01 for 700 max_iterations. We have used the ReLu activation function and set the learning rate as "adaptive". After training the model, an accuracy of 68.45% was achieved when validated with the test data. The confusion matrix obtained is shown in Table 4.

MLP classifier is suitable when we are dealing with multiple features or a combination of features. It works by splitting input data into a layer of individual nodes. It offers great versatility for classification problems. Having said that, it is computationally expensive and time-consuming to train with CPU. Moreover, high dependence on training data can lead to overfitting.

Table 4 Confusion matrix for MLP model on RAVDESS dataset

Emotion	Happy	Sad	Neutral	Angry
Happy	41	3	3	7
Sad	5	25	15	5
Neutral	2	1	14	3
Angry	8	1	2	33
Precision	73%	83%	41%	69%

MLP Algorithm

Train the MLP model with the Training dataset(X).

MLPClassifier Function, i.e., MLPClassifier() with following

In this function, fix the following values for the parameters

alpha = '0.01',

batch_size = 256,

epsilon = 1e-08,

hidden_layer_sizes = (300,),

learning_rate='adaptive',

max_iter = 700.

Store the arrival esteem into a variable.

Fit the training information into MLPClassifier using fit (X, Y) where X and Y are input and output labels, respectively.

Convolutional Neural Network

Convolutional Neural Networks (CNNs) have a high capacity to extract features with the application of filters. For the implementation of the CNN algorithm, Keras API was used to develop the CNN model with input size as 180 × 1. The input is fed into the CNN architecture with a sequence of convolution and pooling layers. Our architecture consisted of 357,770 parameters. Librosa python library is used to convert the input data to the required format [27]. For our CNN model, we used four Conv1D layers, one MaxPooling1D layer, and one Dense layer. Softmax and ReLu were used as the activation functions and RMSprop as the optimizer. The learning rate (lr) was set to 0.0001 and decay as 1e-6. The compilation of the CNN model was done by specifying the loss function as "sparse_categorical_crossentropy" and evaluation metric as "accuracy". Training of the CNN model was done with batch size equal to 16 for 700 epochs. When validated with the test set, an accuracy of 77.38% was obtained.

Convolutional Neural Networks are computationally efficient in terms of memory and time because of parameter sharing. They tend to perform better than regular neural networks [29]. However, CNNs have high computational cost and training is slow if you don't have a good GPU. Also, they demand large training data to make accurate classifications.

CNN Algorithm

Initialize keras.models.Sequential() model.

Add a one-dimensional convolution layer with filters = 256, kernel size = 5. Also, provide the input_shape as an argument to the convolution layer.

Add an activation layer with activation function as ReLU next to the convolution layer.

Add another convolution layer with filters = 128 and kernel size = 5.

Add an activation layer with the same configuration as the former layer.

Apply Max Pooling to the convolution obtained from the previous layer with pool_size = 8.

Add two more convolution layers with filter = 128 and kernel size = 5 with activation function being ReLU to both the layers.

Flatten the convolution obtained. Feed the flattened convolution to dense layers followed by a final dense layer of 4 neurons. Use the softmax activation function at the end to classify the labels.

Use RMSprop optimizer with learning rate and decay set to 10^(−4) and 10^(−6), respectively.

Use sparse_categorical_crossentropy for computing the loss.

After configuring the CNN model as per the above steps, train the model with batch_size set to 16 and epochs = 700(recommended).

Use model.evaluate(x_test, y_test) method to obtain test accuracy.

4 Experimental Results

In this section, the experimental results are presented and discussed. The results show that the Decision Tree model has obtained an accuracy of 54.16% at max_depth = 6. Later, it decreased to 52.38% when max_depth was increased to 7 indicating that overfitting has occurred. Support Vector Machine model was trained with regularization parameter 'C' at C = 1, C = 2, and C = 3, and it was observed that the highest accuracy of 64.88% was obtained when C = 2. Random Forest model did relatively better in giving 69% accuracy after tweaking the parameters and setting

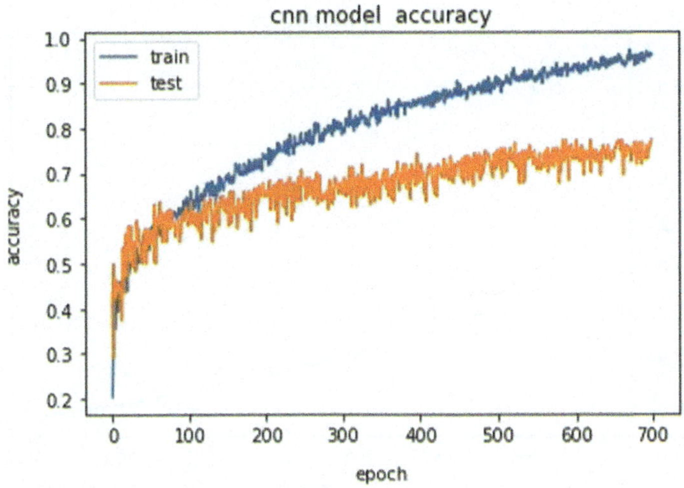

Fig. 2 Epoch versus accuracy

the n_estimators as 60. As for the MLP model, we initially had one hidden layer of 400 neurons and max_iterations set to 700 which gave an accuracy of 61.31%. As we decreased the number of neurons in the hidden layer to 300 and tuned other parameters, a high accuracy of 68.45% was obtained. Convolutional Neural Networks (CNN) model, by using 4 Conv1D layers, has obtained the best overall recognition accuracy of 77.38% in recognizing emotions through speech. The confusion matrices for Decision Tree, Random Forest, Support Vector Machine, and Multi-Layer Perceptron algorithms are shown in the tables above. Further examination of these tables shows us that precision was highest for the anger emotion indicating that it was the easiest to recognize. Neutral emotion, on the other hand, was the most difficult to detect for all the algorithms.

Since CNN had achieved the best accuracy in recognizing emotions when compared to others, we have decided to study the accuracy and the loss function. From Figs. 2 and 3, we can see that the accuracy increased and loss decreased drastically in the first 100 epochs. From Fig. 2, it is clear that the model approached the highest accuracy for 700 epochs. Similarly, from Fig. 3, we can conclude that loss decreased as epochs increased which is a good sign of learning. Hence, the CNN model is reasonably accurate in its predictions. The histogram comparing the accuracies of each algorithm is shown in Fig. 4.

5 Conclusion

In this paper, various state-of-the-art classification algorithms have been studied and implemented for Speech Emotion Recognition Systems. An understanding of the

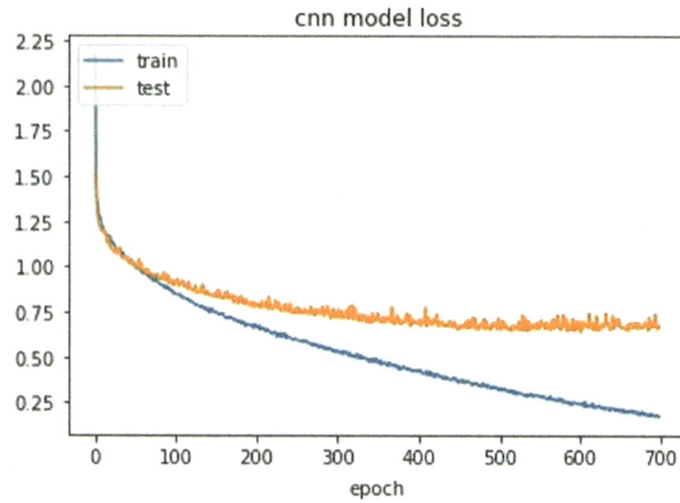

Fig. 3 Epoch versus Loss

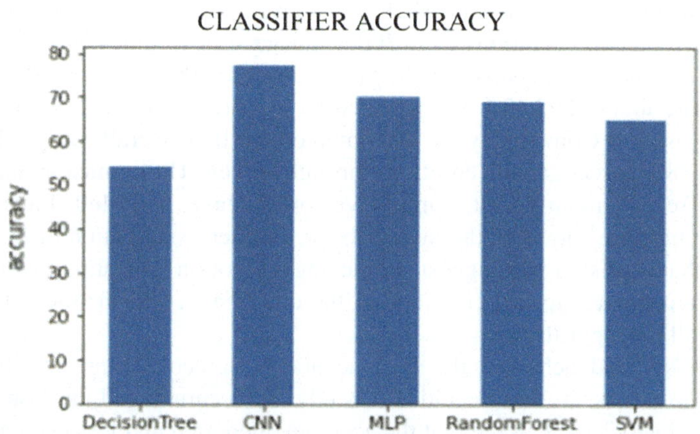

Fig. 4 Comparison of accuracies of different classifiers

obscurities and the drawbacks posed by the Speech Emotion Recognition Systems of today has been briefed. The features that were extracted and the RAVDESS dataset were also discussed. The diverse applications of emotion recognition ranging from efficient classroom teaching and lie detection to human–robot communication were informed. It is inferred from the results and the graph shown in Fig. 4 that Convolutional Neural Networks (CNN) model, which obtained an accuracy of 77.38%, is the most appropriate model for recognizing emotions through speech.

6 Future Scope

The types of features selected determine the accuracy of the model and its significance is described in [27]. Combination of features can yield higher accuracies in the future. However, combined features can lead to overfitting due to increased dimensionality and complexity [13]. Further standardization of the training data set and increasing the number of real-time samples can improve the validity of the dataset [28]. Speech Emotion Recognition Systems can be expanded to be made functional for multiple languages. Research into employing Deep Neural Networks for Speech Emotion Recognition Systems to detect more abstract emotions can be the next further step in this field. Emotion recognition through speech signals combined with facial analysis can bring more reliability to the system. In the future, these methods can be deployed for real-time Speech Emotion Recognition Systems.

References

1. H. Meng, T. Yan, F. Yuan, A. Wei, Speech emotion recognition from 3D Log-Mel spectrograms with deep learning networks. IEEE (2019)
2. H. Ali, M. Hariharan, S. Yaacob, A.H. Adom, Facial emotion recognition using empirical mode decomposition. Expert Syst. Appl. **42**(3), 1261–1277 (2015)
3. Z.T. Liu, M. Wu, W.H. Cao, J.W. Mao, J.P. Xu, G.Z. Tan, Speech emotion recognition based on feature selection and extreme learning machine decision tree. Neurocomputing **273**, 271–280 (2018)
4. M. Ragot, N. Martin, S. Em, N. Pallamin, J.M. Diverrez, Emotion recognition using physiological signals: Laboratory vs. wearable sensors, in *International Conference on Applied Human Factors and Ergonomics*. Springer, pp. 15–22 (2017)
5. C.M. Lee, S.S. Narayan, Towards detecting emotions in spoken dialogues. IEEE Trans. Speech Audio Process. (2005)
6. S.N. Roopa, M. Prabhakaran, P. Betty, Speech emotion recognition using deep learning. Int. J. Recent Technol. Eng. (2018)
7. Z. Li, A study on emotional feature analysis and recognition in speech signal. J. China Inst. Commun. (2000)
8. S. Wu, T.H. Falk, W.Y. Chan, Automatic speech emotion recognition using modulation spectral features. Speech Commun. **53**, 768–785 (2011)
9. A.B. Ingale, D.S.Chaudhari, Speech emotion recognition. Int. J. Soft Comput. Eng. (2012)
10. Z. Yongzhao, C. Peng, Research and implementation of emotional feature and recognition in speech signal. J. Jiangsu Univ. (2005)
11. M. El Ayadi, M.S. Kamel, F. Karray, Survey on Speech emotion recognition: features, classification schemes and databases. Pattern Recogn. (2011)
12. C. Huang, W. Gong, W. Fu, D. Feng, *A Research of Speech Emotion Recogntion Based on Deep Belief Network and SVM* (Hindawi, 2014)
13. L. Kerkeni, S. Youssef, M. Mbarki, K. Raoof, M. Ali Mahjoub, Speech emotion recognition: methods and cases study. ICAART (2018)
14. N. Ratna Kanth, S. Saraswathi, A suvery on speech emotion recognition. Adv. Comput. Sci. Inf. Technol. (2014)
15. M. Waqas Bhatti, Y. Wang, L. Guan, A neural network approach for human emotion recognition in speech, in *International Symposium on Circuits and Systems* (2014)

16. K. Scherer,, Vocal communication of emotion: a review of research paradigms. Speech Commun. (2003)
17. M. Swain, A. Routray, P. Kabisatpathy, Databases, features and classifiers for speech emotion recognition: a review. Int. J. Speech Technol. (2018)
18. https://www.kaggle.com/uwrfkaggler/ravdess-emotional-speech-audio
19. M. Swain, A. Routray, P. Kabisatpathy,Databases, features and classifiers for speech emotion recognition. Int. J. Speech Technol. (2018)
20. Q. Mao, M. Dong, Z. Huang, Y.Zhan, Learning salient features for speech emotion recognition using convolutional neural networks. IEEE (2014)
21. P. Shen, Z. Changjun, X. Chen, Automatic speech emotion recognition using support vector machine, in *International Conference in Electronic and Mechanical Engineering and Information Technology* (2011)
22. Y. Chavhan, M.L. Dhore, P. Yesaware, Speech emotion recognition using support vector machine. Int. J. Comput. Appl. (2010)
23. S.R. Gunn, *Support Vector Machines for Classification and Regression* [PhD thesis] (1998)
24. T.P. Robinson, Real-time recognition of affective states from nonverbal features of speech and its application in public speaking skill. IEEE (2011)
25. I.Chiriacescu, Automatic Emotion Analysis Based on Speech. M.Sc, THESIS Delft University of Technology (2009)
26. F. Noroozi, S. Tomasz, D. Kamińska, G. Anbarjafari, Vocal–based emotion recognition using random forests and decision tree. Int. J. Speech Technol. (2017)
27. Z. Ciota, Feature extraction of spoken dialogues for emotion recognition. ICSP (2006)
28. J. Zhu, X. Wu, Z.Lv, Speech emotionr recognition algorithm based on SVM. Comput. Sci. Appl. (2011)
29. O. Abdel-Hamid, A.R Mohamed, H. Jiang, L. Deng, G. Penn, D. Yu, Convolutional neural networks for speech recognition. IEEE/ACM Trans. Audio Speech Lang. Process (2014)

A Review of Contemporary and Future Renewable Energy Generation Technologies to Store Data in the Era of Cloud Computing and IoT

Keerthi Manikanthan and S. Pranavanand

Abstract Be it to perform work or to make machines perform work, store data, compute online, interact remotely, energy resources are crucial. With the growing industrialization, ease of agriculture, automation, and the development of transportation, the demand for energy resources is increasing day by day. Lots of data is generated, and it is essential to store both useful and junk data as separating the junk from useful data is a tedious process in automation. Already existent energy resources do not seem too promising to satisfy the growing needs as well as to support sustainability. Thus, a correlation exists between sustainability, development, and demand for energy. To meet the demand of today and future novel, green and smart energy resources are needed. This paper reviews various contemporary and futurist technologies that work on renewable energy to take a step forward toward sustainability.

Keywords Sustainability · Renewable energy · Energy generation

1 Introduction

Over the years, fossil fuels have been used for energy generation despite its hazardous effects on the environment. Alternatives are the need of the hour. But the important question is can alternative energy sources displace or replace fossil fuels?

Primarily, energy resources can be classified as renewable and non-renewable sources. It is well known that renewable sources are inexhaustive while on the contrary, non-renewable resources aren't [1]. Energy generation is essential to perform almost every task.

K. Manikanthan (✉) · S. Pranavanand
VNR Vignana Jyothi Institute of Engineering and Technology, Secunderabad, Telangana, India
e-mail: keerthimani01@gmail.com

S. Pranavanand
e-mail: pranavanand_s@vnrvjiet.in

C. Kiran Mai et al. (eds.), *Proceedings of International Conference on Advances in Computer Engineering and Communication Systems*, Learning and Analytics in Intelligent Systems 20,
https://doi.org/10.1007/978-981-15-9293-5_49

Energy can be generated in many ways—naturally or artificially. Based on the source used for generating energy, it can be classified as follows:-

- Energy generation using natural sources
- Energy generation using special materials
- Energy generation using strategic and smart technologies.

2 Energy Generation Using Natural Sources

Some energy generation techniques never go old. We can greatly reduce the amount of energy used by us by taking small measures that help us in the long run. Such measures include insulating buildings, following the 3Rs, i.e., reduce, reuse, and recycle, and using biomass, etc.

2.1 Biomass as an Energy Source

Biomass energy production is a well-known method to convert huge amounts of wastes into energy, along with a few environmental side effects [2]. Biomass energy is inexhaustible and maintainable, yet shares numerous qualities with non-renewable energy sources. While biomass can be legitimately consumed to acquire energy, it can likewise fill in as a feed-stock to be changed over to different fluid or gas fills (biopowers). These similitudes represent that the significant job biomass is relied upon to play in future vitality situations. Consequently, a late-rising technique is to create a bioprocessing plant and biochange advancements to change over biomass feedstock into clean energies [3].

2.2 Microorganism as an Energy Source

Similarly, microalgae are photosynthetic microorganisms. They contain organic compounds such as lipids, proteins, and carbohydrates. All of these are produced in very little time. There is a potential to convert them to useful biofuel and use them as a source of energy. Algae is easy to produce. It doesn't require a huge area or any other luxury. Moreover, processing the energy acquired has minimal requirements compared to producing energy. Research is on to convert algal oil usable in locomotives and automobiles as biofuels. The content of lipids and fatty acids depends on the conditions under which the culture of algae is grown. The emissions of CO_2 are utilized by the algae to carry out the process. Apart from its amazing ability to convert emissions such as CO_2 into biomass, algae liberate more oxygen for the atmosphere as compared to forests [4].

2.3 Microbial Fuel Cells Energy Source

Apart from all the artificial and non-biodegradable material energies, energy can be obtained from biodegradable, reduced compounds. Microbial Fuel Cells (MFC) are one such alternative. In this method, microorganisms are used as catalysts. An MFC converts energy, available in a bioconvertible substrate, directly into electricity. Bacteria use anode as an electron acceptor. The energy metabolism and bacteria activity are yet to be known clearly [5]. MFCs follow a similar concept to traditional fuel cells. However, the usage of metal catalysts is replaced by viable microorganisms that have biocatalytic capabilities, by using the energy stored in their chemical bonds [6].

2.4 Pressure-Retarded Osmosis (PRO) as an Energy Source

The salinity difference between freshwater and seawater has remained a hidden source of energy. Efficient energy-conversion technologies are required to put this into use. Several approaches to capture salinity-gradient energy are being developed, but the most promising are Pressure-Retarded Osmosis (PRO) and Reverse Electrodialysis (RED). Some of the methods that are under development, with greater potential than many existent energy harvesting techniques are PRO, RED, and MFCs, which are based on harvesting non-renewable materials from nature [7].

2.5 Wet Renewables as an Energy Source

When it comes to water resources, tidal energy is viable. The term 'wet renewables' is ordinarily used to allude to seaward wind energy advancements just as a tidal stream and wave energy plan. For nations with critical regions of beachfront waters, the usage of seaward and waterfront energy assets is alluring. The World Energy Board gauges that if under 0.1% of the sustainable power source inside the seas could be changed over into power it would fulfill the present world interest for energy over multiple times over (World Energy Board 2007). The innovation doesn't exist to use the majority of this energy asset, not least in light of issues related to the spatial confusion of the regions of interest with locales of the most noteworthy asset. All things considered, wet renewables are getting progressively monetary and it is normal that seaward energy assets will turn into a critical wellspring of sustainable power source sooner rather than later [8].

2.6 Robots for Energy Storage

The Solar-powered Exploration Rover (SOLERO) takes us toward space research. SOLERO's performance was tested on a model Mars-like environment. It is like a few solar-powered vehicles that run on Earth. However, due to the larger distance between the Sun and Mars, the amount of solar power available is much lower. Considering how beneficial robots maybe soon, robots can be made to run on renewable energy. Additionally, robots can be used to produce renewable energy [9]. In other words, a robot can become both source and a supply of energy. The Ecoppia E451 is a fully self-directed robot. The robot uses an on-board solar panel and a battery to store energy that allows cleaning at night [10].

Another such space exploration project is the Interplanetary Kite-craft Accelerated by Radiation of the Sun, IKAROS. IKAROS is the first-ever solar-powered sail craft, which was made by The Japan Aerospace Exploration Agency (JAXA). Solar sail missions are yet to be carried out. It has a membrane that extends over a span of 20 m in diameter. On the membrane, a-Si (amorphous silicon) solar cells are attached in a distributed manner throughout its area. The thin-film solar cells attached to the sail are lighter than solar cells used for spacecraft, and easier to handle, thereby making this mission work at lower costs in the solar cells market, whose growth is very significant in reducing global warming. They are undoubtedly the base of future solar power satellite systems [11].

3 Energy Generation Using Special Materials

3.1 Smart Memory Alloys for Energy Generation

A highly unique energy harvesting method is the usage of Smart Memory Alloys (SMA). Smart memory alloys are also called smart metals. They are materials that deform when heated to certain temperatures [12]. The deformed material reacts under heat and tends to return to its original shape. This reaction called the shape memory reaction generates mechanical work. Mechanical systems are built with SMA located at a certain position. The shaft attached to the mechanical structure with proper support generates electricity [12, 13].

3.2 Piezoelectric Transducers (PZT) as an Energy Source

Piezoelectric materials are materials of the present and are being studied because of their very specific and interesting properties. In fact, these materials have the flexibility to provide current from energy [14]. Piezoelectric Transducers (PZT) are employed in mechanisms to transfer vibrations and mechanical loads and stresses into current which will be stored and used to power other devices.

Piezoelectric transducers are accustomed to control sensors, actuators, MEMs devices, etc. The energy produced by the PZT is stored by employing a capacitor that permits immediate access to the stored energy. The energy generated by the vibration of the piezoelectric is shown to be a maximum of 2mW. These materials, therefore, are used as motion transferrers (mostly vibration to current) that will be stored and adapted to power other devices. By employing power harvesting devices, the portability of energy systems is greatly improved, making them advantageous over traditional energy technologies [15].

3.3 Thermal Energy Storage Using Phase Change Materials

Residential buildings are the interfaces between the indoors and the outdoors. A roof above the head is the most basic amenity for living. Hence, residential buildings require our attention when it comes to energy technologies. For the implementation of energy storage technologies, various factors such as energy density, efficiency, power output, toxicity, lifetime, reliability, self-discharge, duration of storage, and cost are to be considered. Thermal Energy Storage (TES) by using Phase Change Materials (PCM) has been gaining attention over the few decades [16].

PCMs are best placed in the walls, roof, and floor as they provide maximum area for the transfer of heat from and to the buildings [17]. With passive solar heating, heat is collected from the sun and stored in the PCM envelope. It is then released at night when it is cooler. PCMs are used to store the thermal energy for other utilizations [16].

3.4 Photovoltaic for Power Generation

Photovoltaic (PV) works on the principle of photovoltaic effect, where energy is captured by a solar cell and is thereby converted into electricity [18]. A combination of solar cells is in series and parallel. These combinations are called PV modules giving us our anticipated power output. Furthermore, these can be connected in series and parallel for mass energy production. The first (1st) generation solar cells are based on crystalline silicon technology and are widely used as of now. The second (2nd) generation solar cells, built on from thin-film materials, were industrialized as cheaper alternatives to crystalline silicon materials. They have a better light absorption capacity, for which very thin layers of materials are employed. Dye-sensitized solar cells (DSSCs), use organic dyes to absorb the solar energy that it receives from the sunlight. They are easier to manufacture and fabricate, lower in cost, and are very flexible. Different layers of the components are made, starting from a Fluorine doped Tin Oxide (FTO) coated glass substrate, $TiO2$ nanoparticles, dyes, electrolyte, and counter electrode [19].

4 Energy Generation Using Strategic and Smart Technologies

4.1 Smart Grid Technology

Out of many energy improvisations, the concept of smart grid technology is quite adaptable soon. In short, a smart grid is a network that supplies electricity by using digital communications technology. The digital communications technology is used to detect local changes and thereby make necessary moves accordingly [20]. SMART grid technology is widely viewed as a major development that is implementing renewable energy, wind energy in particular [21]. It is to be noted that the World Wide Fund (WWF) for nature has the vision to make the earth a completely renewable energy run planet by the middle of this century [2]. Demand Response (DR) is a strategy that involves the planned management of energy consumption corresponding to high- and low-energy requirements in a time of the day. Hence, it is highly possible that the effects of DR on electricity use can reduce dependency on fossil fuel-based electricity generation [21].

When it comes to the electricity system in a smart city, it can be subdivided into various levels [13], namely

- Generation level
- Grid level
- Customer level.

Distributed Generation (DG) is the concurrency of generation and consumption. When compared to a conventional centralized system, DG is less expensive, lowers complexity, dependency, and inefficiency related to power transmission and distribution systems. In the future, buildings can be DGs aided by renewable energy and storage systems, making Zero-Energy Building (ZEB) a probable outcome. DG-based renewable energy systems can provide clean energy to the buildings. Energy management software is designed to store and manage energy usage and modify them as per our requirements [22].

4.2 Energy Resources from Human Efforts

The world uses huge amounts of energy of kinds such as mechanical and electrical. Researchers have been trying to find out novel ways to generate clean and renewable sources of energy. Renewable energies are clean and replenishable, inexhaustive sources of energy. For generating this type of energy, sensor-based energy developments using the Internet of Things (IoT) have been devised. These sensor-based systems, such as piezoelectric sensors, body heat to an electric converter, and solar panel, are used to generate and are in turn connected to storage circuits for further utilization [20].

4.3 EVs as an Energy Storage Device and V2G Technology

Talking of smart grids brings us to Electric Vehicles (EVs). EVs are becoming very popular now due to the decrease in fuel consumption and emissions [2]. They provide valued services to power grids [18]. Smart grids will be the connecting link for EVs and Renewable Energy Resources (RES). EVs are taken as loads for which optimal charging can be achieved by planned technology and investment as means to arrange to charge time, to achieve peak-load shifting, improve the efficiency of the system, and decrease the impact on grid safety. EVs may be both distributed energy storage units as well as grid power supplies. This is done to expand the safe dependability of the power system. Two-way communication between EVs and the grid can be realized by the Vehicle to Grid (V2G) technology [2]. V2G is a system in which electric vehicles (of various types) connect with the power grid to provide peak power sinning reserves or regulation services. They can act as renewable energy storage and standbys in the future [18]. Compared with the traditional grid, a large quantity of distributed generations using Renewable Energy Sources (RES) as fuel is applied in the smart grid. Due to the alternating nature of distributed energy such as wind power and photovoltaic cell, it is difficult for these generations to balance the load without the modification of the smart grid [2].

4.4 Tapping the Exhaust Gas

The usage of locomotives and automobiles led to the production of exhaust gases that possess high kinetic energy. This kinetic energy of the wind gust is one form of energy that can be but is not out to use. The kinetic energy in wind gust produced by vehicles is an under-research complete energy utilization method [9].

4.5 Waste-to-Energy Technologies

Ever wondered how waste can be useful? Although it is oxymoronic, waste can be put to the best using the waste-to-energy technologies. Waste-to-Energy (WtE) technologies essentially use solid wastes and materials such as paper, biodegradable wastes, and plastics by carrying out thermochemical or biochemical processes. The thermochemical techniques consist of several processes which include combustion, gasification, thermal depolymerization, etc.

The biochemical processes consist of anaerobic digestion, hydrolysis, Mechanical Biological Treatment (MBT), or fermentation using enzymes that produce little heat in slow reaction times. Although some of the methods in this technology involve the burning of waste materials in an incinerator, some developed technologies are capable of extracting heat without direct burning. For maximum output and minimum emissions, modern plants use multiple thermochemical and biochemical processes [23].

Similarly, food processing wastes serve as a source of energy. Although biodegradable, when disposed of into water bodies or as such, it alters the Biochemical Oxygen Demand (BOD) and Chemical Oxygen Demand (COD). Food processing wastes have high potential energy which can be used to reduce the ill impacts. By using food processing wastes for energy generation, we are benefited both economically and environmentally. Thermal processing is the most common method for energy generation from food processing wastes. Incinerators are used for this purpose. There are many other methods, which include biological processing such as usage of anaerobic digestors, chemical processing such as biodiesel production [24].

In Indonesia, the utilization of rice husk for the generation of alternative clean fuel for power plants is under study. Indonesia is the third-largest rice producer in the world [25]. In Thailand, fruit processing residues are being studied for alternate fuel sources [26].

4.6 Nanotechnology Toward Energy Generation

Nanotechnology is the next big leap toward improvement. It has been used to design, synthesize, store, and process new energy materials. Nanomaterials with planned size, shape, and characteristics are to be used for sustainable energy technologies. As of now, Titanium Dioxide (TiO2) has been deeply researched for solar panel applications. Being a semiconductor, TiO2 makes it possible to achieve better performance in Dye-Sensitized Solar Cells (DSSCs) [27]. The main aim of the usage of nanomaterials is to enhance the performance of solar cells.

5 Conclusion

An exhaustive list of methods for energy generation using renewable technologies is discussed. Approximately, a minimum of 10 W power is required to power a smartphone. India alone has 65 crores smartphones and 40 crores feature phones. The energy required to power these devices will be approximately 8.5 GW per day. Almost all the smartphones use at least one form of Cloud or IoT service. The cost required to establish one MW solar power plant is very high. The above estimate throws light on the energy requirement and capital cost to establish. Along with energy source, energy harvesting and storage are also important. Implementing novel renewable energy techniques and making way toward betterment is essential for the growing pace of requirements. Switching to potential alternatives and novel energy generation and harvesting methods is indeed the hour and for better tomorrow.

References

1. A.A. Demirbas, M. Fatih Demirbas, Importance Of algae oil as a source of biodiesel. Proceed. IEEE **96**(9), 112–121 (2006)
2. J. Dai, M. Dong, R. Ye, A. Ma, W. Yang, A review on electric vehicles and renewable energy synergies in smart grid, in *2016 China International Conference on Electricity Distribution (CICED 2016)*, China, pp. 158–164
3. O. Ellabban, H. Abu-Rub, F. Blaabjerg.: Renewable energy resources: current status, future prospects and their enabling technology. Renew. Sustain. Energy Rev. **39**, 48–764 (2004)
4. M. Fatih Demirbas, Biofuels From algae for sustainable development. Appl. Energy **88**, 3473–3480 (2011)
5. K. Rabaey, W. Verstraete, Microbial fuel cells: novel biotechnology for energy generation. Trends Biotechnol. **23**(6), 291–298 (2005)
6. A.J. Slatea, K.A. Whiteheada, D.A.C. Brownsona, C.E. Banksa, Microbial fuel cells: an overview of current technology. Renew. Sustain. Energy Rev. **101**, 60–81 (2018)
7. B.E. Logan, M. Elimelech, Membrane-based processes for sustainable power generation using water. Nature **488**, 313–319 (2012)
8. A. Chris Frid et al., The environmental interactions of tidal and wave energy generation device. Environ. Impact Assess. Rev. **32**, 133–139 (2015)
9. O.P. Eze, R. Amali, Harnessing electric energy from vehicle induced wind gust, in *3rd International Symposium on Environmental Friendly Energies and Applications (EFEA)*, St. Ouen (2014), pp. 1–5
10. A. Hassan, M. El Habrouk, S. Deghedie, Renewable energy for robots and robots for renewable energy—A review. Robotica **2**(1), 1–29 (2019)
11. I.B.O. Mori et al., First solar power sail demonstration. Trans. JSASS Aerospace Tech. Japan **8**(27), 425–431 (2011)
12. H.Y. Ayirga, N. Kutucu, Ü.K. Terzi, Use of shape memory alloys as smart materials to harvest energy powered by renewable energy resources, in *5th International Istanbul Smart Grid and Cities Congress and Fair (ICSG)*, Istanbul (2015), pp. 175–179
13. D. Reddy, Shape memory alloys. **6**(7), 56–58 (2015)
14. R. Srivastava, N. Tiwari, A. Kumar, D. Sen, Power generation using piezoelectric material **2**(1), 12–19 (2015)
15. R. Chandra Garimella, V.R. Sastry, M. Shoeb Mohiuddin, Piezo-Gen- An approach to generate electricity from vibrations. Procedia Earth Planet. Sci. **11**, 445–456 (2005)
16. Z. (John) Zhai, M.L.L. Abarr, S.N.J. AL-Saadi, P. Yate, Energy storage for residential buildings: review and advances, in *APEC Conference on Low-carbon Towns and Physical Energy Storage*, China (2013), pp. 646–655
17. Y. Zhang, G. Zhou, K. Lin, Q. Zhang, H. Di, Application of latent heat thermal energy storage in buildings: state-of-the-art and outlook building and environment **42**(6) (2007)
18. T. Yiyun, L. Can, C. Lin, L. Lin, Research on vehicle-to-grid technology, in *2011 International Conference on Computer Distributed Control and Intelligent Environmental Monitoring*, pp. 1013–1016. Changsha
19. R.K. Koech, M. Kigozi, A. Bello, P.A. Onwualu, W.O. Soboyejo, Recent advances in solar energy harvesting materials with particular emphasis on photovoltaic materials, in *IEEE PES/IAS Power Africa*, pp. 627–632 2019. Nigeria (2019)
20. V. Puri et al., A hybrid artificial intelligence and internet of things model for generation of renewable resource of energy. IEEE Access **7**, 111181–111191 (2019)

21. T. Broeer, F.K. Tuffner, A. Franca, N. Djilali, A demand response system for wind power integration: greenhouse gas mitigation and reduction of generator cycling. CSEE J. Power Energy Syst. **4**(2), 121–126 (2018)
22. F.M. Bhutta, Application of smart energy technologies in building sector – future prospects, in *International Conference on Energy Conservation and Efficiency (ICECE)*, pp. 126–131. Pakistan (2017)
23. I. Kayes, A.H. Tehzeeb, Waste to energy: a lucrative alternative, in *1st International Conference on the Developments in Renewable Energy Technology (ICDRET)*, pp. 1–4. Dhaka (2009)
24. B. Digman, D.-S. Kim, Review: alternative energy from food processing wastes. Environ. Progr. **27**(4), 321–327 (2012)
25. M. Anshar, A.S. Kader, F. Nasir Anil, The utilization potential of rice husk as an alternative energy source for power plants in Indonesia. Adv. Mater. Res. **845**, 494-498 (2013)
26. A. Marcus Nagle et al., Fruit processing residues as an alternative fuel for drying in northern Thailand. Fuel **90**(2), 818–823 (2011)
27. C.-J. Liu et al., Preparation and characterization of nanomaterials for sustainable energy production. Fuels **4**(10), 5517–5526 (2010)

Author Index

C. Kiran Mai et al. (eds.), *Proceedings of International Conference on Advances in Computer Engineering and Communication Systems*, Learning and Analytics in Intelligent Systems 20,
https://doi.org/10.1007/978-981-15-9293-5

The manufacturer's authorised representative in the EU is Springer Nature Customer Service Centre GmbH, Europaplatz 3, 69115 Heidelberg, Germany. If you have any concerns regarding our products, please contact ProductSafety@springernature.com

Printed and bound by CPI Group (UK) Ltd, Croydon, CR0 4YY
30/09/2026
02225331-0006